Physics of Waves

William C. Elmore
Mark A. Heald

Department of Physics   Swarthmore College

Physics of Waves

McGraw-Hill Book Company

New York, St. Louis, San Francisco, London, Sydney, Toronto, Mexico, Panama

Physics of Waves

Library of Congress Catalog Card Number 68-58209

19260   73048

1234567890 MAMM 7654321069

Dedicated to the memory of

Leigh Page

*Professor of Mathematical Physics
Yale University*

Preface

Classical wave theory pervades much of classical and contemporary physics. Because of the increasing curricular demands of atomic, quantum, solid-state, and nuclear physics, the undergraduate curriculum can no longer afford time for separate courses in many of the older disciplines devoted to such classes of wave phenomena as optics, acoustics, and electromagnetic radiation. We have endeavored to select significant material pertaining to wave motion from all these areas of classical physics. Our aim has been to unify the study of waves by developing abstract and general features common to all wave motion. We have done this by examining a sequence of concrete and specific examples (emphasizing the *physics* of wave motion) increasing in complexity and sophistication as understanding progresses. Although we have assumed that the mathematical background of the student has included only a year's course in calculus, we have aimed at developing the student's facility with applied mathematics by gradually increasing the mathematical sophistication of analysis as the chapters progress.

At Swarthmore College approximately two-thirds of the present material is offered as a semester course for sophomores or juniors, following a semester of intermediate mechanics. Much of the text is an enlargement of a set of notes developed over a period of years to supplement lectures on various aspects of wave motion. The chapter on electromagnetic waves presents related material which our students encounter as part of a subsequent course. Both courses are accompanied by a laboratory.

A few topics in classical wave motion (for the most part omitted from our formal courses for lack of lecture time) have been included to round out the treatment of the subject. We hope that these additions, including much of Chapters 6, 7, and 12, will make the text more flexible for formulating courses to meet particular needs. We especially hope that the inclusion of additional material to be covered in a one-semester course will encourage the serious student of physics to investigate for himself topics not covered in lecture. Stars identify particular sections or whole chapters that may be omitted without loss of continuity. Generally this material is somewhat more demanding. Many of the problems which follow each section form an essential part of the text. In these problems the student is asked to supply mathematical details for calculations outlined in the section, or he is asked to develop the theory for related cases that extend the coverage of the text. A few problems (indicated by an asterisk) go significantly beyond the level of the text and are intended to challenge even the best student.

The fundamental ideas of wave motion are set forth in the first chapter, using the stretched string as a particular model. In Chapter 2 the two-dimensional membrane is used to introduce Bessel functions and the characteristic features of waveguides. In Chapters 3 and 4 elementary elasticity theory is developed and applied to find the various classes of waves that can be supported by a rigid rod. The impedance concept is also introduced at this point. In Chapter 5 acoustic waves in fluids are discussed, and, among other things, the number of modes in a box is counted. These first five chapters complete the basic treatment of waves in one, two, and three dimensions, with emphasis on the central idea of energy and momentum transport.

The next three chapters are options that may be used to give a particular emphasis to a course. Hydrodynamic waves at a liquid surface (e.g., water waves) are treated in Chapter 6. In Chapter 7 general waves in isotropic elastic solids are considered, after a development of the appropriate tensor algebra (with its future use in relativity theory kept in mind). Although electromagnetic waves are undeniably of paramount importance in the real world of waves, we have chosen to arrange the extensive treatment of Chapter 8 as optional material because of the physical subtlety and analytical complexity of electromagnetism. Thus Chapter 8 might either be ignored or be made a major part of the course, depending on the instructor's aims.

Chapter 9 is probably the most difficult and formal of the central core of the book. In it approximate methods are considered for dealing with inhomogeneous and obstructed media, in particular the Kirchhoff diffraction theory. The cases of Fraunhofer and Fresnel diffraction are worked out in Chapters 10 and 11, with some care to show that their relevance is not limited to visible light. Chapter 12 removes the idealizations of monochromatic waves and point sources by considering modulation, wave packets, and partial coherence.

Conspicuously absent from our catalog of waves is a discussion of the quantum-mechanical variety. Many of our choices of emphasis and examples have been made with wave mechanics in mind, but we have preferred to stay in the context of classical waves throughout. We hope, rather, that a student will approach his subsequent course in quantum mechanics well-armed with the physical insight and analytical skills needed to appreciate the abstractions of wave mechanics. We have also restricted the discussion to continuum models, leaving the treatment of discrete-mass and periodic systems to later courses.

We are grateful to Mrs. Ann DeRose for her patience and skill in typing the manuscript.

William C. Elmore
Mark A. Heald

Contents

Physics of Waves

one

Transverse Waves on a String

We start the study of wave phenomena by looking at a special case, the transverse motion of a flexible string under tension. Various methods for solving the resulting wave equation are developed, and the solutions found are then used to illustrate a number of important properties of waves. The emphasis in the present chapter is primarily on developing mathematical techniques that prove to be extremely useful in treating wave phenomena of a more complex nature. It will be found impressive to view in retrospect the rather formidable theoretical structure that can be based on a study of the motion of such a simple object as a flexible string under tension.

1.1 The Wave Equation for an Ideal Stretched String

We suppose the string to have a mass λ_0 per unit length and to be under a constant tension τ_0 maintained by equal and oppositely directed forces applied at its ends. In the absence of a wave, the string is straight, lying along the x axis of a right-handed cartesian coordinate system. We further suppose that the string is indefinitely long; later we shall consider the effect of end conditions.

Evidently if the string is locally displaced sideways a small amount and quickly released, i.e., if it is "plucked," the tension in the string will give rise to forces that tend to restore the string to the position of its initial state of rest. However, the inertia of the displaced portion of the string delays an immediate return to this position, and the momentum acquired by the displaced portion causes the string to overshoot its rest position. Moreover, because of the continuity of the string, the disturbance, which was originally a local one, must necessarily spread, or *propagate*, along the string as time progresses.

To become quantitative, let us apply Newton's second law to any element Δx of the displaced string to find the differential equation that describes its motion. To simplify the analysis, suppose that the motion occurs only in the xy plane. We use the symbol η for the displacement in the y direction (reserving the symbol y, along with x and z, for expressing position in a three-dimensional frame of reference). We assume that η, which is a function of position x and time t, is everywhere sufficiently small, so that:

(1) The magnitude of the tension τ_0 is a constant, independent of position.

(2) The angle of inclination of the displaced string with respect to the x axis at any point is small.

(3) An element Δx of the string can be considered to have moved only in the transverse direction as a result of the wave disturbance.

We also idealize the analysis by neglecting the effect of friction of the surrounding air in damping the motion, the effect of stiffness that a real string (or wire) may have, and the effect of gravity.

As a result of sideways displacement, a net force acts on an element Δx of the string, since the small angles α_1 and α_2 defined in Fig. 1.1.1 are, in general, not quite equal. We see from Fig. 1.1.1 that this unbalanced force has the y component $\tau_0(\sin\alpha_2 - \sin\alpha_1)$ and the x component $\tau_0(\cos\alpha_2 - \cos\alpha_1)$. Since α_1 and α_2 are assumed to be very small, we may neglect the x component entirely† and also replace $\sin\alpha$ by $\tan\alpha = \partial\eta/\partial x$ in the y component. Accord-

---

† In Sec. 1.11 it is found that the x component neglected here is responsible for the transport of linear momentum by a transverse wave traveling on the string.

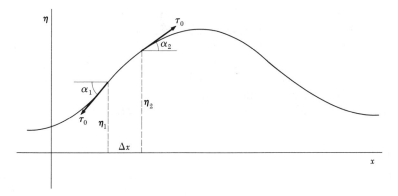

Fig. 1.1.1   Portion of displaced string. (The magnitude of the sideways displacement is greatly exaggerated.)

ing to Newton's second law, the latter force component must equal the mass of the element $\lambda_0 \, \Delta x$ times its acceleration in the y direction. Therefore at all times

$$\tau_0 \left(\frac{\partial \eta_2}{\partial x} - \frac{\partial \eta_1}{\partial x} \right) = \lambda_0 \, \Delta x \, \frac{\partial^2 \eta}{\partial t^2}, \tag{1.1.1}$$

where η, the mean displacement of the element, becomes the actual displacement at a point of the string when $\Delta x \to 0$. The partial-derivative notation is needed for both the time derivative and the space derivative since η is a function of the two independent variables x and t. The partial-derivative notation merely indicates that x is to be held constant in computing time derivatives of η and t is to be held constant in computing space derivatives of η.

Next we divide (1.1.1) through by Δx and pass to the limit $\Delta x \to 0$. By the definition of a second derivative,

$$\lim_{\Delta x \to 0} \frac{1}{\Delta x} \left(\frac{\partial \eta_2}{\partial x} - \frac{\partial \eta_1}{\partial x} \right) = \frac{\partial^2 \eta}{\partial x^2},$$

(1.1.1) becomes

$$\tau_0 \frac{\partial^2 \eta}{\partial x^2} = \lambda_0 \frac{\partial^2 \eta}{\partial t^2}. \tag{1.1.2}$$

We choose to write (1.1.2) in the form

$$\frac{\partial^2 \eta}{\partial x^2} = \frac{1}{c^2} \frac{\partial^2 \eta}{\partial t^2}, \tag{1.1.3}$$

where

$$c \equiv \left(\frac{\tau_0}{\lambda_0} \right)^{1/2} \tag{1.1.4}$$

will be shown to be the velocity of small-amplitude transverse waves on the string (*c* after the Latin *celeritas*, speed). We now turn our attention to developing methods for solving this *one-dimensional scalar wave equation* and to discussing a number of important properties of the solutions. Equation (1.1.3) is the simplest member of a large family of wave equations applying to one- two-, and three-dimensional media. Whatever we can learn about the solutions of (1.1.3) will be useful in discussing more complicated wave equations.

Problems

1.1.1   An elementary derivation of the velocity of transverse waves on a flexible string under tension is based on viewing a traveling wave from a reference frame moving in the *x* direction with a velocity equal to that of the wave. In this moving frame the string itself appears to move

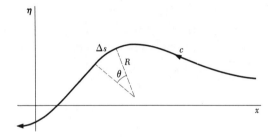

Prob. 1.1.1   String seen from moving frame.

backward past the observer with a speed *c*, as indicated in the figure. Find *c* by requiring that the uniform tension τ_0 give rise to a centripetal force on a curved element Δs of the string that just maintains the motion of the element in a circular path. Does this derivation imply that a traveling wave keeps its shape?

1.1.2   A circular loop of flexible rope is set spinning with a circumferential speed v_0. Find the tension if the linear density is λ_0. What relation does this case have to Prob. 1.1.1?

1.1.3   The damping effect of air on a transverse wave can be approximated by assuming that a transverse force $b \, \partial\eta/\partial t$ per unit length acts so as to oppose the transverse motion of the string. Find how Eq. (1.1.2) is modified by this viscous damping.

1.1.4   Extend the treatment in Prob. 1.1.3 to include the presence of an externally applied transverse driving force $F_y(x,t)$ per unit length acting on the string.

1.1.5   Use the equation developed in Prob. 1.1.4 to find the *equilibrium* shape under the action of gravity of a horizontal segment of string of linear mass density λ_0 stretched with a tension τ_0 between fixed supports separated a distance *l*. Assume that the sag is small.

1.1.6   The density of steel piano wire is about 8 g/cm³. If a safe working *stress* is 100,000 lb/in.², what is the maximum velocity that can be obtained for transverse waves? Does it depend on wire diameter?

1.2   A General Solution of the One-dimensional Wave Equation

A *partial* differential equation states a relationship among partial derivatives of a dependent variable that is a function of two or more independent variables. Such an equation, in general, has a much broader class of solutions than an *ordinary* differential equation relating a dependent variable to a single independent variable, such as the equation for simple harmonic motion. As with ordinary differential equations, it is often possible to guess a solution of a partial differential equation that meets the needs of some particular problem. For example, we might guess that there exists a sinusoidal solution of the wave equation (1.1.3) of the form

$$\eta = A \sin(\alpha x + \beta t + \gamma).$$

Indeed, substitution of this function in (1.1.3) shows that it satisfies the equation provided $(\beta/\alpha)^2 = c^2$. Although this solution represents a possible form that waves on a stretched string can take, it is far from representing the most general sort of wave, as the following analysis shows.

We rewrite (1.1.3) in the form

$$\frac{\partial^2 \eta}{\partial x^2} - \frac{1}{c^2}\frac{\partial^2 \eta}{\partial t^2} = \left(\frac{\partial}{\partial x} - \frac{\partial}{c\,\partial t}\right)\left(\frac{\partial}{\partial x} + \frac{\partial}{c\,\partial t}\right)\eta = 0, \tag{1.2.1}$$

where the differential operator operating on η has been split into two factors. This factorization is possible when c is not a function of x (or t). The form of these operators suggests changing to two new independent variables $u = x - ct$ and $v = x + ct$. It is easy to show that (Prob. 1.2.1)

$$\frac{\partial}{\partial x} - \frac{\partial}{c\,\partial t} = 2\frac{\partial}{\partial u} \qquad \frac{\partial}{\partial x} + \frac{\partial}{c\,\partial t} = 2\frac{\partial}{\partial v} \tag{1.2.2}$$

and therefore that (1.2.1) becomes

$$4\frac{\partial^2 \eta}{\partial u\,\partial v} = 0. \tag{1.2.3}$$

The wave equation in this form has the obvious solution

$$\eta(u,v) = f_1(u) + f_2(v), \tag{1.2.4}$$

where $f_1(u)$ and $f_2(v)$ are completely arbitrary functions, unrelated to each other and limited in form only by continuity requirements. We thus arrive at

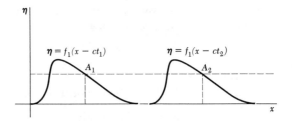

Fig. 1.2.1   Arbitrary wave at two successive instants of time.

d'Alembert's solution of the wave equation,

$$\eta(x,t) = f_1(x - ct) + f_2(x + ct). \tag{1.2.5}$$

Whereas we expect a second-order linear ordinary differential equation to have two independent solutions of *definite functional form*, which may be combined into a general solution containing *two arbitrary constants*, the wave equation (1.1.3) has *two arbitrary functions* of $x - ct$ and $x + ct$ as solutions. Because the wave equation is *linear*, each of these functions can in turn be considered to be the sum of many other functions of $x \pm ct$ if this point of view should prove useful. For example, it is often convenient to subdivide a complicated wave into many partial, simpler waves whose linear *superposition* constitutes the actual wave.

Let us now examine the properties of a solution consisting only of the first function

$$\eta(x,t) = f_1(x - ct). \tag{1.2.6}$$

Figure 1.2.1 shows the wave at two successive instants in time, t_1 and t_2. The wave keeps its shape, and with the passage of time it continually moves to the right. A particular point on the wave at time t_1, such as point A_1 at the position x_1, has moved to point A_2 at the position x_2 at the later time t_2. The two points have the same value of η; that is, $f_1(x_1 - ct_1) = f_1(x_2 - ct_2)$. This fact implies that $x_1 - ct_1 = x_2 - ct_2$. Hence

$$c = \frac{x_2 - x_1}{t_2 - t_1},$$

showing that the wave (1.2.6) is moving in the *positive* direction with the velocity c. By a similar argument, $f_2(x + ct)$ represents a second wave proceeding in the *negative* direction, independently of the first wave but with the same speed.

We have established, therefore, that the wave equation permits waves of arbitrary but permanent shape to progress in both directions on the string with the wave velocity $c = (\tau_0/\lambda_0)^{1/2}$. Although the wave equation (1.1.3) does not in itself restrict the amplitude and form of the wave functions f_1 and f_2 that satisfy it, the conditions under which the wave equation has been derived restrict the wave functions applying to the string to a class of rather well-behaved functions. They must necessarily be continuous (!) and have rather gentle spatial slopes; that is, $|\partial \eta / \partial x| \ll 1$. In contrast, (1.1.3) has mathematically acceptable solutions having discontinuities. The form of the wave (or waves) that occurs in a practical application of the present theory depends, of course, on the way in which the wave gets started in the first place, i.e., on the source of the wave. It is a characteristic of wave theory that many properties of waves can be discussed independently of the source of the waves.

Problems

1.2.1 Establish the operator formulas (1.2.2) by using the "chain rule" of differential calculus.

1.2.2 Verify that (1.2.5) is a solution of the wave equation by direct substitution into (1.1.3).

1.2.3 A long string, for which the transverse wave velocity is c, is given a displacement specified by some function $\eta = \eta_0(x)$ that is localized near the middle of the string. The string is released at $t = 0$ with zero initial velocity. Find the equations for the traveling waves that are produced and make a sketch showing the waves at several instants of time with $t \geq 0$. *Hint:* Find two waves traveling in opposite directions that together satisfy the initial conditions.

1.2.4 If, in Prob. 1.2.3, the string is given not only the initial displacement $\eta = \eta_0(x)$ but also an initial velocity $\partial \eta / \partial t = \dot{\eta}_0(x)$ when it is released, find the equations for the resulting waves.

1.2.5 Can you give physical significance to the answers found in Probs. 1.2.3 and 1.2.4 when t is negative?

1.2.6 A long string under tension is attached to a fixed support at $x = l$. The wave

$$\eta = A e^{-\alpha(x - ct)^2}$$

approaches the fixed end from the left and is reflected. Find an expression for the reflected wave. *Hint:* Find a second wave traveling in the negative direction such that at $x = l$ the combined amplitudes of the two waves vanish for all t.

1.3 Harmonic or Sinusoidal Waves

The analysis of the preceding section has shown that the wave equation is satisfied by *any* reasonable function of $x + ct$ or of $x - ct$. Of the infinite variety

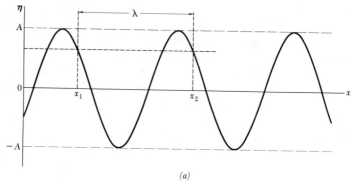

(a)

Fig. 1.3.1   Sinusoidal wave at some instant in time.

of functions permitted, we choose for closer study waves having a *sinusoidal* waveform. The reason the sine (or cosine) function occupies a key position in wave theory is fundamentally that *linear* mathematical operations (such as differentiation, integration, and addition) applied to sinusoidal functions of a definite period generate other sinusoidal functions of the *same period*, differing at most in amplitude and phase. Since many of the interesting applications of wave theory lead to these linear mathematical operations when formulated analytically, it is obvious that waves having a sinusoidal waveform lead to simple results. Later we shall discover that elastic waves in many media are not described by the simple wave equation (1.1.3). In such an event sinusoidal waves are found to have a wave velocity that depends on frequency. Nonsinusoidal waves are then found to change their shape as they progress, and it is only sinusoidal waves that preserve their functional form in passing through the medium.

Another important, but less fundamental, reason for giving emphasis to sinusoidal waves is based on the fact that the sources of many waves encountered in the real world vibrate periodically, thereby giving rise to periodic waves. Of the class of periodic functions, a sinusoidal function has the simplest mathematical properties. Furthermore, it can be shown that periodic functions of arbitrary form (and, as a matter of fact, aperiodic functions also) can be represented as closely as desired by the linear superposition of many sine functions whose periods, phase constants, and amplitudes are suitably chosen. A brief introduction to this branch of mathematics, known as *Fourier analysis*, is given in Secs. 1.6 and 1.7 and Chap. 12.

Let us therefore investigate various aspects of a sine wave of amplitude A traveling on a stretched string. We choose to express the wave initially by the equation

$$\eta = A \cos \frac{2\pi}{\lambda} (x - ct). \tag{1.3.1}$$

Mathematically such a wave has no beginning or end in either space or time. In practice the wave must have a source somewhere along the negative x axis, and be absorbed, without reflection, at a distant "sink" or *termination* along the positive x axis. In between source and sink at any time t_0, the wave, i.e., the shape of the string, has the form shown in Fig. 1.3.1. Because of the 2π periodicity of the cosine, the wave repeats itself in a distance such that

$$\frac{2\pi}{\lambda}(x_1 - ct_0) + 2\pi = \frac{2\pi}{\lambda}(x_2 - ct_0),$$

where x_1 and x_2 are the space coordinates of any successive points at which the wave has the same amplitude and slope. This condition reduces to $x_2 - x_1 = \lambda$, so that the constant λ introduced in (1.3.1) is the *wavelength* of the sinusoidal wave.

Similarly, at a given position x_0 the dependence of η on time has the periodic form shown in Fig. 1.3.2. Again the wave repeats itself, now in time; i.e., any point on the string is executing simple harmonic motion. As before, we may write (note the position of 2π)

$$\frac{2\pi}{\lambda}(x_0 - ct_1) = \frac{2\pi}{\lambda}(x_0 - ct_2) + 2\pi.$$

Defining $T \equiv t_2 - t_1$ to be the *period* of the sinusoidal wave, we find that the period and wavelength are related by

$$c = \frac{\lambda}{T}. \tag{1.3.2}$$

A sinusoidal wave evidently repeats itself in time at any position with the *frequency* $\nu \equiv 1/T$. To avoid the necessity of constantly writing the 2π that would normally occur in the argument of a sinusoidal vibration or wave, we

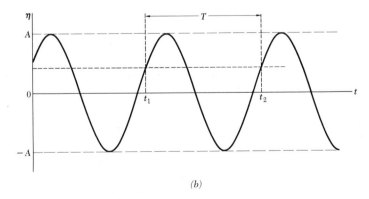

(b)

Fig. 1.3.2   The wave disturbance at a fixed position x_0.

"rationalize" the frequency by introducing the notation $\omega \equiv 2\pi\nu = 2\pi/T$. When it is necessary to distinguish which frequency is meant, we can use the adjective *angular* for ω and *ordinary* or *cyclic* for ν. It is also convenient to define a space counterpart of angular frequency,

$$\kappa \equiv \frac{2\pi}{\lambda},\tag{1.3.3}$$

which is termed the (*angular*) *wave number*, i.e., the number of waves in 2π units of length.† Then the sinusoidal wave (1.3.1) may be written in the equivalent but neater form

$$\eta = A\,\cos(\kappa x - \omega t).\tag{1.3.4}$$

The velocity of the sinusoidal wave may thus be written variously as

$$c = \frac{\lambda}{T} = \lambda\nu = \frac{\omega}{\kappa}.\tag{1.3.5}$$

We next introduce an extremely convenient representation of a sinusoidal wave based on the *Euler identity*

$$e^{i\theta} \equiv \cos\theta + i\,\sin\theta,\tag{1.3.6}$$

where $i \equiv \sqrt{-1}$. The sinusoidal wave (1.3.1) is evidently the *real* part of (consider A to be real)‡

$$\eta = A e^{i(\kappa x - \omega t)}.\tag{1.3.7}$$

Similarly the *imaginary* part of (1.3.7) could be used to represent the physical wave

$$\eta = A\,\sin(\kappa x - \omega t).\tag{1.3.8}$$

However, it is an unwritten rule (the *real-part convention*) that when a complex representation is used for a sinusoidal function, the *real* part of the complex quantity is the one that has physical significance. In electrical-engineering parlance, the term *phasor* is often used for this representation of an oscillatory physical quantity by a complex exponential.

The usefulness of the complex representation depends on a number of its properties.

---

† Spectroscopists often use the *ordinary* wave number, $1/\lambda = \kappa/2\pi$, in specifying spectral lines. We shall not make use of this alternative, however.

‡ For the most part we follow the physicists' convention of using $e^{-i\omega t}$, rather than $e^{+i\omega t}$, as the time factor in a sinusoidal wave such as (1.3.7). The sign in the spatial factor $e^{+i\kappa x}$ then agrees with the direction in which the wave is traveling. In electrical engineering it is customary to use $e^{+j\omega t}$ as the time factor (here the letter j stands for $\sqrt{-1}$ since the letter i is reserved for electric current).

(1) If the amplitude $\breve{A}$ of the complex wave is considered to be a complex number, signified by a cup over the symbol,†

$$\breve{A} \equiv A_r + iA_i \equiv A e^{i\alpha} \tag{1.3.9}$$

where $A = |\breve{A}| = (A_r{}^2 + A_i{}^2)^{1/2}$ and $\tan\alpha = A_i/A_r$, then the real part of the complex wave

$$\eta = \breve{A} e^{i(\kappa x - \omega t)} = A e^{i(\kappa x - \omega t + \alpha)} \tag{1.3.10}$$

represents the actual (physical) wave

$$\eta = A \cos(\kappa x - \omega t + \alpha). \tag{1.3.11}$$

We thus have a compact way of including a constant phase angle in the complex amplitude of the wave.

(2) Taking the real part (or the imaginary part) of a complex function is an operation that commutes with various linear operations on the function, such as differentiation, integration, and addition. Two simple examples are

$$\mathrm{Re}\left(\frac{de^{i\theta}}{d\theta}\right) = \frac{d}{d\theta}[\mathrm{Re}(e^{i\theta})] = -\sin\theta \tag{1.3.12}$$

$$\mathrm{Re}(e^{i\theta_1} + e^{i\theta_2}) = \mathrm{Re}(e^{i\theta_1}) + \mathrm{Re}(e^{i\theta_2}) = \cos\theta_1 + \cos\theta_2, \tag{1.3.13}$$

where Re denotes taking the real part of the function within the parentheses. Other examples can easily be supplied by the reader. Note, however, that $\mathrm{Re}(e^{i\theta_1}e^{i\theta_2}) \neq \mathrm{Re}(e^{i\theta_1})\mathrm{Re}(e^{i\theta_2})$, a problem to which we return in Sec. 1.8, when we become concerned with energy and power, which involve the squares and products of wave functions (see Prob. 1.3.3).

The real and imaginary parts of a complex quantity may be written more formally by introducing the *complex conjugate*

$$\breve{A}^* \equiv A_r - iA_i, \tag{1.3.14}$$

that is, the sign of the imaginary part is reversed. Then,

$$\begin{aligned}\mathrm{Re}(\breve{A}) &= \tfrac{1}{2}(\breve{A} + \breve{A}^*) \\ \mathrm{Im}(\breve{A}) &= \tfrac{1}{2}(\breve{A} - \breve{A}^*).\end{aligned} \tag{1.3.15}$$

---

† We choose the unconventional notation of placing a cup (˘) over a symbol to make explicit that it represents a complex quantity. Normally, a sophisticated reader is expected to deduce from the context when a quantity may be complex without benefit of a special notation, just as he assumes the real-part convention whereby (1.3.10) is in fact equivalent to (1.3.11). Other explicit notations for complex quantities may be found in the literature, e.g., the use of roman (nonitalic) type.

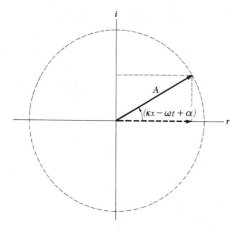

Fig. 1.3.3 The wave $\breve{A} = A e^{i(\kappa x - \omega t + \alpha)}$ represented in the complex plane. The actual physical wave is the projection on the real axis.

(**3**)   The differentiation or integration of $e^{i\theta}$

$$\frac{d}{d\theta} e^{i\theta} = i e^{i\theta} \qquad \int e^{i\theta}\, d\theta = -i e^{i\theta} \qquad\qquad (1.3.16)$$

simply multiplies $e^{i\theta}$ by i or by $-i$, respectively, whereas the corresponding operations on $\cos\theta$ and $\sin\theta$ change one function into the other, with asymmetrical changes of sign. This change of function is avoided when $e^{i\theta}$ is used. It follows that differentiation and integration of the complex wave (1.3.10) with respect to time reduce to simple multiplication and division, respectively, by $-i\omega$.

(**4**)   The complex wave (1.3.10) may be represented by the two-dimensional graph of Fig. 1.3.3. The radius vector, of length A, rotates clockwise in time and counterclockwise in space. The projection of the rotating vector on the

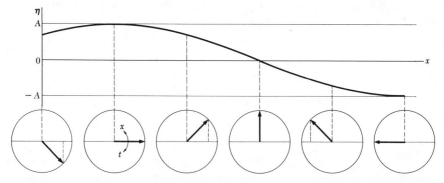

Fig. 1.3.4   A positive-going sinusoidal wave and its complex representation. At any time t, the complex vector rotates counterclockwise with increasing x. At any position x, it rotates clockwise with increasing t, with the angular velocity ω.

real axis is the physical wave (1.3.11). The relation between the real sinusoidal wave (1.3.11) and its complex representation (1.3.10) is nicely illustrated in Fig. 1.3.4, which shows a "snapshot" of the physical wave, i.e., the shape of the string, and a series of vector diagrams of the complex amplitude, with varying phase, of the complex wave. The magnitude of the complex wave is independent of x, but its orientation with respect to the x axis rotates counterclockwise as its position advances along the x axis. The behavior of the wave as time increases can be visualized by giving the amplitude vector in each of the circles the clockwise angular velocity ω.

(5) The addition (or superposition) of two or more waves of the same frequency traveling in the same direction but having differing amplitudes and phases is easily carried out by adding their complex amplitudes. The addition can be performed either algebraically (by adding separately the real and imaginary parts of their amplitudes) or graphically (by treating the complex amplitudes as two-dimensional vectors in the complex plane). The *resultant* complex amplitude, obtained in either way, gives the amplitude and phase constant of a single wave equivalent to the sum of the original waves. We make use of this vector addition of component waves in Chaps. 9 to 11.

Problems

1.3.1   The two waves $\eta_1 = 6 \cos(\kappa x - \omega t + \frac{1}{6}\pi)$ and $\eta_2 = 8 \sin(\kappa x - \omega t + \frac{1}{6}\pi)$ are traveling on a stretched string. (*a*) Find the complex representation of these waves. (*b*) Find the complex wave equivalent to their sum $\eta_1 + \eta_2$ and the physical (real) wave that it represents. (*c*) Endeavor to combine the two waves by working only with trigonometric identities in the real domain.

1.3.2   Prove the following trigonometric identities by expressing the left side in complex form:

(*a*) $\cos(x + y) = \cos x \cos y - \sin x \sin y$

(*b*) $\sin(x + y) = \sin x \cos y + \cos x \sin y$

(*c*) $\displaystyle\sum_{n=1}^{N} \cos n\theta = \frac{\cos\frac{1}{2}(N+1)\theta \, \sin\frac{1}{2}N\theta}{\sin\frac{1}{2}\theta}$

(*d*) $\displaystyle\sum_{n=1}^{N} \sin n\theta = \frac{\sin\frac{1}{2}(N+1)\theta \, \sin\frac{1}{2}N\theta}{\sin\frac{1}{2}\theta}$

Many other trigonometric identities can be similarly proved.

1.3.3   Investigate the multiplication and division of two complex numbers. Show that in general the product (quotient) of their real parts does not equal the real part of the product (quotient) of the two numbers.

1.4   Standing Sinusoidal Waves

Let us continue our study of sinusoidal waves by supposing that two such waves of identical amplitude and frequency travel simultaneously in *opposite* directions on a stretched string. For convenience in discussing the resulting wave pattern, we use the complex-exponential formalism, representing the waves by the real parts of (assume A to be real)

$$\eta_1 = \tfrac{1}{2}A\,e^{i(\kappa x - \omega t)}$$
$$\eta_2 = -\tfrac{1}{2}A\,e^{-i(\kappa x + \omega t)}, \tag{1.4.1}$$

where the minus sign preceding the second wave ensures that the wave amplitude will vanish at the origin. The factors $\tfrac{1}{2}$ make the maximum amplitude of the combined wave equal to A. When both waves are present,

$$\eta = \eta_1 + \eta_2 = \tfrac{1}{2}A\left(e^{i\kappa x} - e^{-i\kappa x}\right)e^{-i\omega t} = iA\,\sin\kappa x\,e^{-i\omega t}, \tag{1.4.2}$$

using the identity $\sin\theta = (e^{i\theta} - e^{-i\theta})/2i$. The actual disturbance of the string, given by

$$\eta = A\,\sin\kappa x\,\sin\omega t, \tag{1.4.3}$$

no longer has the space and time variables explicitly associated together as $x \pm ct$. Equation (1.4.3), nevertheless, is a valid solution of the wave equation (1.1.3), as direct substitution into the wave equation shows.

The wave disturbance we have found is called a *standing wave* since the wave pattern does not advance along the string. All elements of the string oscillate in phase; at any fixed position $x = x_1$ the string simply vibrates back and forth with the amplitude $A\,\sin\kappa x_1$. A plot of the standing wave at various times is shown in Fig. 1.4.1. *Nodes,* which are points of zero amplitude for all

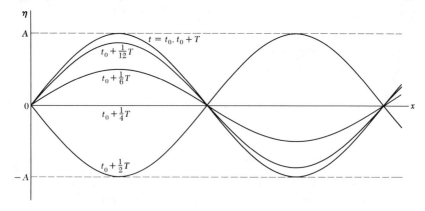

Fig. 1.4.1   A standing-wave pattern.

values of t, occur spaced half-wavelengths apart, with *loops* or *antinodes*, which are points of maximum amplitude, spaced halfway between the nodes. If rigid boundaries are introduced at any pair of nodes (in some way that preserves the tension in the string), the standing-wave pattern between the two boundaries is not disturbed. For example, the bridge and frets on a guitar constitute essentially rigid boundaries for the string segments used in playing particular tones.

Placing boundaries at nodes of a standing-wave pattern affords a means for finding all possible standing-wave patterns that can exist on a string stretched between rigid (nodal) supports. Let us place one such support at $x = 0$ and another at $x = l$. For the standing wave (1.4.3) to exist with nodes at these positions, it is necessary that the wave number (1.3.3) have one of the values

$$\kappa_n = n\frac{\pi}{l} \qquad n = 1, 2, 3, \ldots \tag{1.4.4}$$

which make $\sin\kappa x$ vanish at $x = l$. The corresponding *resonant* frequencies of the string segment are then

$$\omega_n = c\kappa_n = n\frac{\pi c}{l} = n\omega_1, \tag{1.4.5}$$

where $\omega_1 = \pi c/l$ (or $\nu_1 = c/2l$) is the frequency of the *fundamental*, or gravest, mode. The frequencies of higher modes, or *overtones*, are *harmonics* (integral multiples) of the fundamental, or first harmonic. Because the wave equation is linear, each permitted standing wave, or mode of oscillation, can exist with an arbitrary (small) amplitude and with an arbitrary phase constant simultaneously with, and independently of, the others.

Musical instruments are the most familiar practical application of standing waves on a string.[†] In this context, the pitches corresponding to the sequence of resonant frequencies (1.4.5) are, for example,

Harmonic number	1	2	3		4	5	6		7	8	9	10	11	12	13	14	15	16

In this illustration the fundamental is two octaves below middle C, which happens to be the tuning of the lowest string of the cello. The four harmonic pitches shown in parentheses depart

[†] For an interesting account of the connection between physics and music, see J. J. Josephs, "The Physics of Musical Sound," D. Van Nostrand Company, Inc., Princeton, N.J., 1967.

significantly from the usual chromatic scale; for instance, the eleventh harmonic falls almost exactly halfway between F and F♯ on the tempered scale. Most nonstring musical instruments have the same harmonic series. For example, a bugler, by controlled buzzing of his lips, excites one or another of the third to sixth harmonics of the bugle's fundamental.

The tone quality, or *timbre*, of a musical tone is determined largely by the relative strength of the various harmonics. In the case of a string instrument, this harmonic spectrum depends on how the string is plucked, struck, or bowed and also on the reinforcement of some of the harmonics by a sounding board or acoustic resonator built into the instrument. Thus the piano, harpsichord, violin, and guitar all have distinctive tone qualities. By locating the piano hammer approximately one-seventh of the way along the string, the "off-key" seventh harmonic can be weakened.

The stiffness of a real-life string, which we have ignored in our analysis, causes the wave velocity c to depend somewhat upon wavelength. This phenomenon, known as *dispersion*, progressively shifts the higher overtone frequencies away from strict integral multiples of the fundamental and consequently affects the quality of the resulting musical tone. An estimate of the upward shift in frequency is made in Sec. 4.4.

Problems

1.4.1   The two waves $\eta_1 = A \cos(\kappa x - \omega t)$ and $\eta_2 = A \sin(\kappa x + \omega t + \frac{1}{3}\pi)$ travel together on a stretched string. Find the resulting wave disturbance and make sketches of it at several different times, carefully locating the position of nodes on the x axis.

1.4.2   The two waves $\eta_1 = A \cos(\kappa x - \omega t)$ and $\eta_2 = \frac{1}{2}A \cos(\kappa x + \omega t)$ travel together on a stretched string. Make a sketch of the resulting wave pattern at several different times. Note that it has an interesting envelope. *Hint:* First express the two waves in complex form and perform their addition vectorially at a series of positions.

1.4.3   The two waves $\eta_1 = A_1 \cos(\kappa x - \omega t)$ and $\eta_2 = A_2 \cos(\kappa x + \omega t)$ travel together on a stretched string. Show how the wave amplitude ratio A_1/A_2 can be found from a measurement of the so-called *standing-wave ratio* S of the resulting wave pattern, which is defined as the ratio of the maximum amplitude to the minimum amplitude of the envelope of this pattern. Note that S and the position and amplitude of the minima (or maxima) wave disturbance uniquely determine the two traveling waves that give rise to the standing-wave pattern.

1.5   Solving the Wave Equation by the Method of Separation of Variables

In the preceding sections we have seen that the wave equation has traveling-wave solutions of the form $f(x \pm ct)$, of which sinusoidal functions $\cos(\kappa x \pm \omega t)$ are of particular interest. We now consider a powerful general technique, known as the method of *separation of variables*, for solving the wave equation. Although traveling-wave solutions can also be found by this method, it is especially useful for finding standing-wave, or normal-mode, solutions. This method can be used for solving a wide variety of the partial differential equations occurring in physical theory. If a partial differential equation cannot be solved by the method of separation of variables, very often the method can be used to solve a simplified form of the equation. The solution of the simplified equation then forms the basis for an approximate solution of the nonseparable equation.

The method starts by assuming that a solution of the partial differential equation can be found which consists of the product of functions of the independent variables taken individually. In the case of the wave equation (1.1.3) we thus assume a solution to exist of the form

$$\eta(x,t) = f(x) \cdot g(t), \tag{1.5.1}$$

where $f(x)$ is a function of x alone and $g(t)$ is similarly a function of t alone. On substituting (1.5.1) in (1.1.3) and dividing through by $\eta = fg$, we find that

$$\frac{c^2}{f}\frac{d^2f}{dx^2} = \frac{1}{g}\frac{d^2g}{dt^2}. \tag{1.5.2}$$

Equation (1.5.2) requires that a function of x be equal to a function of t for all values of x and t. Since x and t are independent variables, this can be true only if both sides of the equation equal the same constant, which (with a modest amount of foresight) we choose to designate as $-\omega^2$. The equation then separates into the two *ordinary* differential equations

$$\frac{d^2f}{dx^2} + \frac{\omega^2}{c^2}f = 0$$
$$\frac{d^2g}{dt^2} + \omega^2 g = 0. \tag{1.5.3}$$

The constant $-\omega^2$, for obvious reasons, is called the *separation constant*. We recognize the two equations as those of simple harmonic motion, with the general solutions

$$f(x) = A\cos\kappa x + B\sin\kappa x \qquad \kappa = \omega/c$$
$$g(t) = C\cos\omega t + D\sin\omega t. \tag{1.5.4}$$

The constants of integration A, B, C, D are arbitrary, and though ω and κ are related by $\omega = c\kappa$, we are free to choose either ω or κ to suit our needs. For each choice of ω or κ there is, in general, a different set of constants of integration. One can therefore regard the constants as being functions of ω, or κ, or of some parameter that determines ω and κ. The two equations (1.5.4) for f and g may now be combined to give the solution (1.5.1) as

$$\eta(x,t) = a\sin\kappa x\cos\omega t + b\sin\kappa x\sin\omega t + c\cos\kappa x\cos\omega t + d\cos\kappa x\sin\omega t, \tag{1.5.5}$$

where $a = BC$, etc., and each term is recognized as a standing wave of frequency ω and wavelength $\lambda = 2\pi/\kappa = 2\pi c/\omega$, the four terms differing only in the phases of the sinusoidal space and time dependence.

Let us apply the result just found to a finite segment of string stretched between supports separated a distance l. If we choose the x origin at the left end of the string, then necessarily $c = d = 0$, to ensure that η is zero there at all times. The condition that $\eta = 0$ at $x = l$ is met by choosing values of κ from the set (1.4.4), as in the earlier treatment of this problem. Hence (1.5.5) reduces to

$$\eta(x,t) = \sin\kappa_n x (a_n \cos\omega_n t + b_n \sin\omega_n t) \tag{1.5.6}$$

for one of the possible modes of oscillation of the string segment. Although the present result is basically the same as that found earlier, the method of variable separation has led us directly to the functions needed to express the various modes of vibration of a finite string segment. This economy in analysis depends on separating, in the beginning, the time dependence of the wave from its space dependence.

It is customary to call the set of functions $\sin\kappa_n x$ the *normal-mode functions*, or the *eigenfunctions* (German *eigen*, characteristic), pertaining to the wave motion that can exist on the finite string segment having fixed ends. The values of $\kappa_n = n\pi/l$ are the *eigenvalues* of the wave motion: only for this set of wave numbers do eigenfunctions exist that satisfy the boundary conditions of the problem. The corresponding frequencies, $\omega_n = c\kappa_n$, are often called *eigenfrequencies*. When the string is vibrating in one of its normal modes, all parts of it oscillate in phase, with a common time dependence.

The present application of the method of variable separation has not led to any essentially new results. In more complicated cases of waves in two- and three-dimensional media limited by material boundaries, it constitutes the chief method for finding the normal modes of oscillation that can exist for the case considered. It also constitutes the chief method for solving other important partial differential equations of theoretical physics, such as Laplace's equation of potential theory, the heat flow (or diffusion) equation, Schrödinger's equation of quantum theory, etc. The importance of the method far transcends its use in the present instance.

Problems

1.5.1   Show that all possible traveling-sinusoidal-wave solutions of the wave equation can be obtained from (1.5.3) by assuming they have solutions of the form $f = e^{\alpha x}$ and $g = e^{\beta t}$ and finding α and β.

1.5.2   Show how the solution (1.5.5) of the wave equation can be transformed into a general traveling-wave solution.

1.5.3 Express (1.5.6) for one of the possible modes of oscillation of the string segment in complex form such that its real part becomes (1.5.6).

1.6 The General Motion of a Finite String Segment

The most general solution of the wave equation for transverse waves on a segment of string of length l, supported at the ends, appears to consist of an infinite sum of the various normal-mode oscillations (1.5.6) just found. We may write such a solution

$$\eta(x,t) = \sum_{n=1}^{\infty} \sin\kappa_n x (a_n \cos\omega_n t + b_n \sin\omega_n t) \tag{1.6.1}$$

where the doubly infinite set of constants a_n, b_n are the amplitudes of the standing waves of frequency ω_n having $\cos\omega_n t$ and $\sin\omega_n t$ as time factors, respectively. The frequencies ω_n and the wave numbers κ_n are related by the equation $\omega_n = c\kappa_n$, with the κ_n determined by the boundary conditions that give the set of eigenvalues $\kappa_n = n\pi/l$ $(n = 1, 2, 3 \ldots)$.

The remarkable, and by no means obvious, fact now emerges that (1.6.1) represents *the most general (arbitrary) motion of the string segment that is consistent with the end constraints.* The proof of this assertion depends on showing that it is possible to find the values of the constants a_n and b_n from a knowledge of the initial shape $\eta_0(x)$ and velocity $(\partial\eta/\partial t)_0 \equiv \dot\eta_0(x)$ of the string segment. The subsequent motion is thus expressed as the superposition of an infinite set of normal-mode oscillations. Let us now see how the coefficients can be found.

As a simple example, consider the case for which the segment is given an initial displacement and released with no initial velocity. The coefficients in (1.6.1) must be chosen so that

$$\eta_0(x) = \sum_{n=1}^{\infty} a_n \sin\kappa_n x$$

$$\dot\eta_0(x) = 0 = \sum_{n=1}^{\infty} b_n \omega_n \sin\kappa_n x. \tag{1.6.2}$$

The standard procedure for finding the values of a_n and b_n consists in multiplying each equation by $\sin\kappa_n x$ and integrating the equation over the length of the string. By virtue of the definite integrals that are readily established (see Prob. 1.6.1)

$$\int_0^l \sin\frac{m\pi x}{l} \sin\frac{n\pi x}{l}\, dx = 0 \qquad m \neq n$$

$$\int_0^l \sin^2\frac{n\pi x}{l}\, dx = \frac{l}{2}, \tag{1.6.3}$$

we find that the coefficients are

$$a_n = \frac{2}{l} \int_0^l \eta_0(x) \sin \frac{n\pi x}{l}\, dx \qquad n = 1, 2, 3 \ldots \qquad (1.6.4)$$

$$b_n = 0.$$

Equations (1.6.2), with these coefficients, express the initial condition of the string, and (1.6.1) then tells us the shape of the string at all future times. The trigonometric series (1.6.2) with the coefficients found as in (1.6.4) are examples of *Fourier series*, some of whose properties are explained in the next section. The property of sinusoidal functions that an infinite sum (1.6.2) can be found for any function $\eta_0(x)$ is known as *completeness*.

To see how the method works out in detail for a particular case, suppose that the string is pulled aside at its center a small distance A $(A \ll l)$, and released at time $t = 0$. The initial velocity $\dot{\eta}_0(x)$ is zero, and the initial shape is given by

$$\eta_0(x) = \begin{cases} \dfrac{2A}{l}\, x & 0 \le x \le \tfrac{1}{2}l \\[2ex] \dfrac{2A}{l}\, (l - x) & \tfrac{1}{2}l \le x \le l. \end{cases} \qquad (1.6.5)$$

In this example it is necessary to write the initial shape in two parts. It is not necessary in developing a function in a Fourier series to require that the function be expressible by a *single* continuous function of x.

We have already established that the b_n in (1.6.1) all vanish because the string has no initial velocity. Equation (1.6.4) for the a_n becomes

$$a_n = \frac{4A}{l} \left[\int_0^{l/2} \frac{x}{l} \sin \frac{n\pi x}{l}\, dx + \int_{l/2}^l \left(1 - \frac{x}{l}\right) \sin \frac{n\pi x}{l}\, dx \right]. \qquad (1.6.6)$$

A close inspection of this expression shows that $a_n = 0$ when n is *even*. When n is *odd*, the two integrals are equal, so that

$$a_n = \frac{8A}{l^2} \int_0^{l/2} x \sin \frac{n\pi x}{l}\, dx$$

$$= \frac{8A}{\pi^2} \int_0^{\pi/2} \theta \sin n\theta\, d\theta \qquad \theta \equiv \frac{\pi x}{l}$$

$$= \frac{8A}{\pi^2} (-1)^{(n-1)/2} \frac{1}{n^2} \qquad n = 1, 3, 5, \ldots \qquad (1.6.7)$$

Hence we find that the Fourier series expressing the initial shape of the string is

$$\eta_0(x) = \frac{8A}{\pi^2} \left(\frac{1}{1^2} \sin \frac{\pi x}{l} - \frac{1}{3^2} \sin \frac{3\pi x}{l} + \cdots \right) \qquad (1.6.8)$$

and that the shape of the string at any later time is given by

$$\eta(x,t) = \frac{8A}{\pi^2}\left(\frac{1}{1^2}\sin\frac{\pi x}{l}\cos\omega_1 t - \frac{1}{3^2}\sin\frac{3\pi x}{l}\cos3\omega_1 t + \cdots\right) \tag{1.6.9}$$

where $\omega_1 = c\kappa_1 = \pi c/l$ is the fundamental frequency. Since all the cosine time factors in (1.6.9) return to their initial values with this frequency, $\eta(x,t)$ is a periodic function of time, with the period $T_1 = 2\pi/\omega_1$. Figure 1.6.1 shows the initial shape of the string and a sequence of curves computed using one, two, and three terms of the Fourier series (1.6.8). Figure 1.6.2 shows the actual shape of the string, as represented by (1.6.9), at several values of t (see Prob. 1.6.4).

We note the following additional points of interest in the result just found.

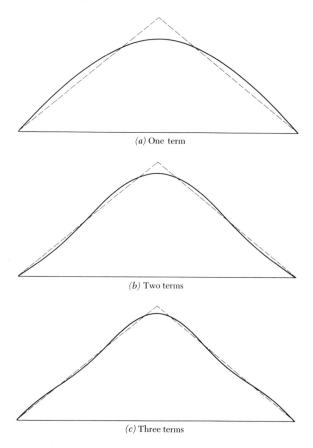

(a) One term

(b) Two terms

(c) Three terms

Fig. 1.6.1   Initial shape of string (*dashed curve*) and approximations obtained by truncating the Fourier series after one, two, or three terms.

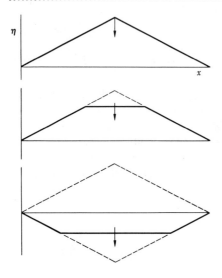

Fig. 1.6.2   Motion of string.

(1)   Since $\eta_0(l/2)$ must equal A, (1.6.8) shows that

$$\frac{\pi^2}{8} = \frac{1}{1^2} + \frac{1}{3^2} + \frac{1}{5^2} + \cdots,$$

which is a well-known series.

(2)   Only the odd harmonics of the string are excited. This is a reasonable result, since if n is even, $\sin(n\pi x/l)$ has a node at the center of the string, and with respect to this point as origin, it is an *odd* function of x, whereas the initial shape of the string about the center as origin is an *even* function of x. It can be shown that when a string is plucked, struck, or bowed at some position along the string, harmonics that have a node at that point are not excited.

(3)   The odd harmonics excited in the present example fall off as $1/n^2$ in amplitude.

Problems

1.6.1   Establish the definite integrals

$$\left.\begin{array}{l} \displaystyle\int_0^{\pi} \sin m\theta \, \sin n\theta \, d\theta = 0 \\[2mm] \displaystyle\int_0^{\pi} \cos m\theta \, \cos n\theta \, d\theta = 0 \end{array}\right\} \quad m \neq n$$

$$\int_0^{\pi} \sin n\theta \, \cos n\theta \, d\theta = 0$$

$$\int_0^{\pi} \cos^2 n\theta \, d\theta = \int_0^{\pi} \sin^2 n\theta \, d\theta = \frac{\pi}{2}$$

by making a substitution based on the identities $\sin x = (e^{ix} - e^{-ix})/2i$ and $\cos x = (e^{ix} + e^{-ix})/2$ and then carrying out the integration. What is the value of the integral $\int_0^\pi \sin m\theta \cos n\theta \, d\theta$?

1.6.2 Fill in the missing analysis between (1.6.6) and (1.6.7).

1.6.3 Find formulas for the coefficients a_n and b_n in (1.6.1) when the string segment has both an initial shape $\eta = \eta_0(x)$ and is given an initial velocity $\partial\eta/\partial t = \dot{\eta}_0(x)$ when it is released at $t = 0$.

1.6.4 Two transverse symmetrical sawtooth waves, each having the form at $t = 0$ shown in the figure are traveling in opposite directions on a stretched string. Investigate the resulting disturbance and plot the wave pattern at several times in the time interval $0 \le t \le l/c$.

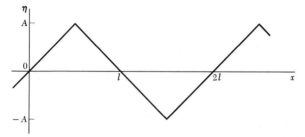

Prob. 1.6.4   Symmetrical sawtooth wave.

Does the pattern in the range $0 \le x \le l$ correspond to Fig. 1.6.2, which is a plot of $\eta(x,t)$ as given by (1.6.9)?

1.6.5 A string segment under tension with fixed supports at $x = 0$ and $x = l$ is pulled aside a small distance A at a point a distance d from the origin $(d < l)$ and released with no initial velocity. Find an expression for $\eta(x,t)$ analogous to (1.6.9), to which it should reduce when $d = l/2$.

1.6.6 Find two waves traveling in opposite directions whose superposition gives the motion of the string segment discussed in Prob. 1.6.5.

1.7   Fourier Series

It would be out of place here to attempt a discussion of Fourier series that makes any pretense to rigor.† Fourier analysis is of such great importance in

† See, for example, M. L. Boas, "Mathematical Methods in the Physical Sciences," chap. 6, John Wiley & Sons, Inc., New York, 1966, for an excellent introductory account; H. S. Carslaw, "Introduction to the Theory of Fourier's Series and Integrals," Dover Publications, Inc., New York, 1930, for an impressive account by a mathematician; and A. Sommerfeld, "Partial Differential Equations," chap. 1, Academic Press Inc., New York, 1949, for a short account by an eminent theoretical physicist.

physics, however, that it appears appropriate to discuss some of the properties of Fourier series and to give a few practical hints for finding the Fourier series of functions needed in the solution of problems of physical interest.

Let $f(\theta) = f(\theta - 2\pi)$ be a well-behaved periodic function defined for all θ. The graph of such a function, when shifted along the θ axis by 2π, coincides with the original graph. The modifier "well-behaved" permits $f(\theta)$ to have a finite number of discontinuities and turning points (maxima and minima) in the range $0 \leq \theta \leq 2\pi$, between which $f(\theta)$ is monotonic and continuous. It also permits $f(\theta)$ to become infinite, provided $\int_0^{2\pi} f(\theta) \, d\theta$ converges absolutely. These conditions on $f(\theta)$ are called *Dirichlet conditions*. Evidently such functions include any that are likely to arise in solving problems of physical interest.

The representation of a well-behaved periodic function $f(\theta)$ by the trigonometric series

$$f(\theta) = a_0 + \sum_{n=1}^{\infty} a_n \cos n\theta + \sum_{n=1}^{\infty} b_n \sin n\theta \tag{1.7.1}$$

is termed a *Fourier series*. The Fourier constants or coefficients a_0, a_n, b_n ($n = 1, 2, 3, \ldots$) are found by the equations

$$a_0 = \frac{1}{2\pi} \int_0^{2\pi} f(\theta) \, d\theta$$

$$a_n = \frac{1}{\pi} \int_0^{2\pi} f(\theta) \cos n\theta \, d\theta \tag{1.7.2}$$

$$b_n = \frac{1}{\pi} \int_0^{2\pi} f(\theta) \sin n\theta \, d\theta,$$

which are obtained from (1.7.1) by multiplying through by unity, $\cos m\theta$, and $\sin m\theta$, respectively, and then integrating over the range 0 to 2π. This procedure was used in the previous section to find the amplitudes of the various normal modes required to express the initial shape of a string segment.

Prior to Fourier's work,† mathematicians had used the series (1.7.1) for discussing certain problems where it was evident on other grounds that such a series should exist. It apparently came as a complete surprise to them when Fourier showed that arbitrary functions, occurring in his discussion of transient heat flow, could be expressed in this manner. Dirichlet later established what constitutes a well-behaved periodic function and showed that the sum of the series representing $f(\theta)$ is $\lim_{\epsilon \to 0} \frac{1}{2}[f(\theta - \epsilon) + f(\theta + \epsilon)]$ for every value of θ. That is, the series sums to $f(\theta)$ at points where $f(\theta)$ is continuous and to a value

---

† J. Fourier, "Theorie Analytique de la Chaleur," 1822, trans. as "The Analytical Theory of Heat," Dover Publications, Inc., New York, 1955.

midway between the two limiting values of $f(\theta)$ on each side of a discontinuity. Because of the latter property, it is often stated that the Fourier series of a function represents the function almost everywhere.

We present now a number of ideas that aid in finding a Fourier series for a given function.

(**1**) Because we consider $f(\theta)$ to be a periodic function, of period 2π, it is immaterial whether the limits in the integrals (1.7.2) for the Fourier coefficients extend from 0 to 2π, or from $-\pi$ to $+\pi$.

(**2**) The periodic function $f(\theta)$, of period 2π, can be changed to a periodic function $f(x)$, of period $2l$, by the substitution $\theta = \pi x/l$, with the limits of the integrals (1.7.2) changed to 0 to $2l$ (or $-l$ to $+l$).

(**3**) In many applications of Fourier series to a physical problem, a function $f(x)$ is specified as to functional form over a certain range of x, which we take to be from 0 to l (or from 0 to π in the variable θ). Outside this range, either from l to $2l$ or equivalently from $-l$ to 0, we are free to define *arbitrarily* the form of $f(x)$ such that we end up with a periodic function $f(x)$ of period $2l$ (or of period 2π in the variable θ). This procedure is possible since the problem is initially undefined outside the interval 0 to l. It often leads to an important simplification in the corresponding Fourier series when the additional part of the function is chosen to give certain symmetry properties to the entire function over the range $2l$. The effect of symmetry in eliminating terms from the Fourier series (1.7.1) is discussed presently.

(**4**) It can be shown (see Prob. 1.7.2) that the Fourier coefficients (1.7.2) give the best least-squares fit of a periodic function to a *finite* trigonometric series that approximates it [(1.7.1), with the upper summation limits of ∞ replaced by N]. It is found that increasing N improves the degree of approximation, by reducing the least-squares residual, and furthermore the values of a_n and b_n *previously calculated* are unaltered. When $N \to \infty$, the representation becomes a Fourier series whose sum equals the function (except at points of discontinuity).

Let us now examine how the symmetry properties of $f(\theta)$ control the form that its Fourier expansion takes.

(**1**) If $f(\theta) = -f(-\theta)$, as suggested in Fig. 1.7.1a, $f(\theta)$ is called an *odd* function of θ. The origin is a *center of symmetry* of the graph of such a function. It is evident that the coefficients a_0 and a_n all vanish for an odd function, and the

Fourier series becomes a *pure sine* series. For such a series

$$b_n = \frac{1}{\pi} \int_{-\pi}^{\pi} f(\theta) \sin n\theta \, d\theta = \frac{2}{\pi} \int_{0}^{\pi} f(\theta) \sin n\theta \, d\theta \tag{1.7.3}$$

since the integrand is an even function of θ. Shifting the origin to $\theta = \pi$ also gives rise to a pure sine series, but for other positions of the origin both sine and cosine terms are needed.

(2) If $f(\theta) = f(-\theta)$, as suggested in Fig. 1.7.1*b*, $f(\theta)$ is called an *even* function of θ. The f axis is a *line of mirror symmetry* of the graph of such a function. It is now evident that the coefficients b_n all vanish, and the Fourier series becomes a *pure cosine* series. For such a series

$$a_0 = \frac{1}{\pi} \int_{0}^{\pi} f(\theta) \, d\theta$$
$$a_n = \frac{1}{\pi} \int_{-\pi}^{\pi} f(\theta) \cos n\theta \, d\theta = \frac{2}{\pi} \int_{0}^{\pi} f(\theta) \cos n\theta \, d\theta \tag{1.7.4}$$

since the integrand is again an even function of θ. Note that if the θ origin is moved to $\theta = \pi$, the series is again a pure cosine series, but for other positions of the origin both sine and cosine terms are needed.

(3) An arbitrary periodic function $f(\theta)$ can always be separated into an odd and even function in the following way:

$$f(\theta) = \tfrac{1}{2}[f(\theta) - f(-\theta)] + \tfrac{1}{2}[f(\theta) + f(-\theta)], \tag{1.7.5}$$

where the first bracket is an odd function and the second is an even function of θ.

(4) The coefficient a_0 is simply the average value of $f(\theta)$. Its value can be made zero, if desired, by shifting the position of the origin on the f axis.

(5) There exists another type of symmetry that $f(\theta)$ can have, as suggested in Fig. 1.7.1*c*, namely, that $f(\theta) = -f(\theta - \pi)$. The θ axis is now an axis of *screw symmetry*, meaning that if the graph of the function is rotated about the θ axis 180° and advanced a distance π, it again coincides with the original graph. We now show that for symmetry of this sort all the even Fourier coefficients vanish. As a proof, consider the integral

$$\int_{0}^{2\pi} f(\theta) \cos n\theta \, d\theta = \int_{0}^{\pi} f(\theta) \cos n\theta \, d\theta + \int_{\pi}^{2\pi} f(\theta) \cos n\theta \, d\theta. \tag{1.7.6}$$

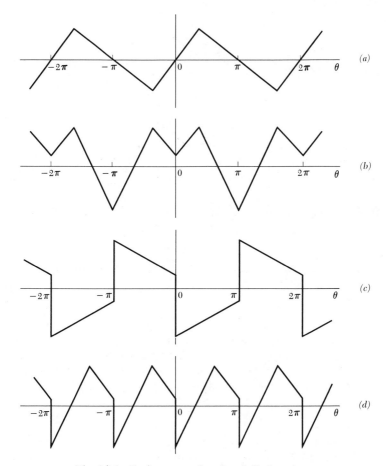

Fig. 1.7.1   Basic symmetries of periodic functions.

If we replace $f(\theta)$ by $-f(\theta - \pi)$ in the second integral on the right and change to the variable $\phi = \theta - \pi$, the integral becomes

$$- \cos n\pi \int_0^\pi f(\phi) \cos n\phi \, d\phi, \tag{1.7.7}$$

which just cancels the first integral when n is even. When n is odd, the two integrals are equal, so that

$$a_n = \frac{2}{\pi} \int_0^\pi f(\theta) \cos n\theta \, d\theta \qquad n \text{ odd.} \tag{1.7.8}$$

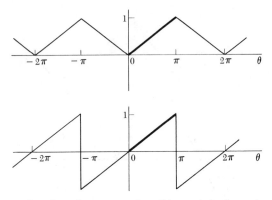

Fig. 1.7.2   Two functions that are equal to $f(\theta) = \theta/\pi$ in the region $0 \leq \theta \leq \pi$.

A similar calculation shows that the b_n vanish for n even and that

$$b_n = \frac{2}{\pi} \int_0^\pi f(\theta) \sin n\theta \, d\theta \qquad n \text{ odd.} \tag{1.7.9}$$

The Fourier series therefore contains only odd harmonics of $\cos\theta$ and $\sin\theta$. If, in addition, it is an even function of θ, then only odd harmonics of $\cos\theta$ occur, or if it is an odd function of θ, only odd harmonics of $\sin\theta$ occur, as in the expansion (1.6.8). The existence of only odd harmonics is independent of where the origin is placed.

(6)   An expansion containing only even harmonics of $\cos\theta$ and $\sin\theta$ occurs if $f(\theta) = f(\theta - \pi)$, since $f(\theta)$ now has a period π rather than 2π. A function having this symmetry is illustrated in Fig. 1.7.1d.

(7)   An arbitrary periodic function $f(\theta)$ having a period 2π can be separated into two functions, one having only odd harmonics and the other having only even harmonics, in the following way:

$$f(\theta) = \tfrac{1}{2}[f(\theta) - f(\theta - \pi)] + \tfrac{1}{2}[f(\theta) + f(\theta - \pi)]. \tag{1.7.10}$$

Each of these functions can in turn be separated into odd and even functions of θ, so that their Fourier series contain only sines or only cosines, respectively. If an origin on the θ axis can be chosen so that $f(\theta)$ becomes an odd or an even function, this choice will simplify the Fourier expansion. It is often helpful to make a sketch of $f(\theta)$ and rough plots of the first few terms that will occur in the Fourier expansion before undertaking the evaluation of the Fourier coefficients.

(8) Fourier series of functions that contain discontinuities converge more slowly than those of continuous functions. For example, if one needs a series to represent $f(\theta) = \theta/\pi$ in the range from 0 to π, it is much better to piece it out to have a period 2π by supplying a section that decreases linearly from $f(\pi) = 1$ to $f(2\pi) = 0$ than by supplying a linear section from $f(-\pi) = -1$ to 0 at the origin. The two possibilities are shown in Fig. 1.7.2. As might be expected intuitively, the more nearly $f(\theta)$ has the appearance of a sinusoidal function, the more rapidly the series converges.

(9) Fourier series must be differentiated with caution! For example, the derivative of the Fourier series of a function that has a discontinuity will not converge. Integration, however, improves the rate of convergence of a Fourier series, since it increases the rate at which the coefficients get small with increasing n.

Problems

1.7.1 Obtain the Fourier coefficients (1.7.2) from the Fourier series (1.7.1).

1.7.2 Show that if the periodic function $f(\theta)$ is approximated by the trigonometric series

$$f(\theta) = a_0 + \sum_{n=1}^{N} a_n \cos n\theta + \sum_{n=1}^{N} b_n \sin n\theta + \epsilon_N(\theta)$$

by the method of least squares, the coefficients given by (1.7.2) are obtained. *Hint:* Choose the coefficients so as to minimize the mean square error

$$\int_0^{2\pi} \epsilon_N^2(\theta)\, d\theta.$$

1.7.3 Show that the Fourier series for the two functions given in Fig. 1.7.2. are

$$f(\theta) = \frac{1}{2} - \frac{2}{\pi^2}\left(\cos\theta + \frac{1}{3^2}\cos 3\theta + \frac{1}{5^2}\cos 5\theta + \cdots\right)$$

$$f(\theta) = \frac{2}{\pi}\left(\sin\theta - \frac{1}{2}\sin 2\theta + \frac{1}{3}\sin 3\theta - \cdots\right).$$

1.7.4 It is desired to express the function $y(x) = lx - x^2$ in the interval $0 \le x \le l/2$ by a Fourier series. Investigate how to piece it out so that it becomes a periodic function, of period $2l$, so that the Fourier series converges as rapidly as possible. Obtain the Fourier series and plot curves showing the sum of one, two, and three terms of the series, as well as the function $y(x)$ that it represents.

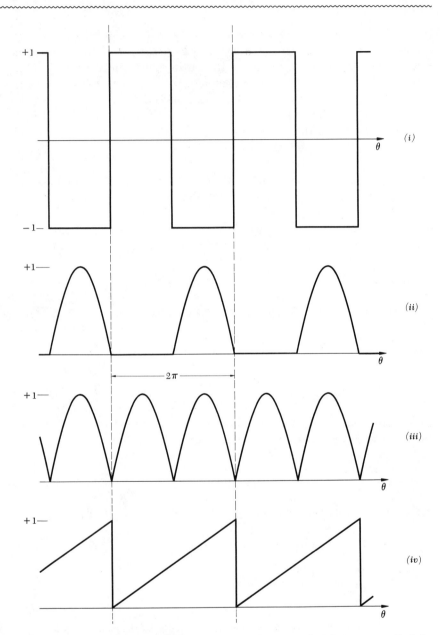

Prob. 1.7.4 (*i*) Square wave; (*ii*) half-wave rectified sine wave; (*iii*) full-wave rectified sine wave; (*iv*) asymmetrical sawtooth wave.

1.7.5 In the figure are given several periodic waveforms. (*a*) For each waveform, show where to place the θ origin so that the waveform has one or more types of symmetry. Identify the symmetry and indicate which, if any, of the Fourier coefficients vanish. (*b*) Obtain the Fourier coefficients, with the θ origin placed where the greatest simplification of the Fourier series results.

1.7.6 The function $y(x)$ in Prob. 1.7.4 is closely related to one of the functions plotted in the figure for Prob. 1.7.5. Find this relation and show that the two Fourier series are consistent with each other.

1.7.7 Show that the Fourier series (1.7.1) can be expressed in the complex form

$$f(\theta) = \sum_{n=-\infty}^{+\infty} \breve{A}_n e^{in\theta}, \tag{1.7.11}$$

where the complex Fourier coefficients are given by

$$\breve{A}_n = \frac{1}{2\pi} \int_{-\pi}^{\pi} f(\theta) e^{-in\theta} \, d\theta. \tag{1.7.12}$$

Here, the entire series (1.7.11), not just its real part, represents $f(\theta)$. If $f(\theta)$ is real, this implies that the imaginary part of the series sums to zero. Relate the $\breve{A}_n$ to the coefficients a_0, a_n, b_n given by (1.7.2). We make considerable use of Fourier series in the complex form in later chapters.

1.7.8 Obtain the real form of the Fourier series (1.7.1) and (1.7.2) from (1.7.11) and (1.7.12) by combining the latter equations with their complex conjugates. The complex conjugates constitute an alternative complex form for a Fourier series and are useful if the terms in the series must appear with a negative sign in the argument of the exponential to make the development consistent with an established sign convention.

1.8 Energy Carried by Waves on a String

The concept of energy in its various forms (including mass as a form of energy) occupies a key position in physical theory. Its importance rests primarily on the fact that energy is *conserved* in one form or another, and presumably the energy content of the universe is a constant. Nevertheless, keeping track of the location in space and time of a given amount of energy is often difficult and subtle, and sometimes it is meaningless. Energy can easily change form, move about in space, become irreversibly lost to the surroundings as heat, etc. In view of the theoretical, as well as practical, significance of energy, its transport by waves constitutes an important part of the theory of wave motion.

We shall consider that a string under tension has zero potential energy in the absence of a transverse wave. When a wave is present, it gains both potential and kinetic energy (in equal amounts, as we shall see). Since we are neglecting frictional effects and presume that no energy is supplied to the string along its length, the conservation of mechanical energy requires that the rate at which energy enters a section of the string through its ends, or boundaries, equal the rate of increase of energy present in the section. If the section has ends that are fixed to immovable supports, the energy content of the section must remain constant and equal to that expended in setting up its motion.

The kinetic energy dK of an element dx of the string having a velocity $\partial\eta/\partial t$ is evidently

$$dK = \tfrac{1}{2}\lambda_0\, dx \left(\frac{\partial\eta}{\partial t}\right)^2$$

where $\lambda_0\, dx$ is the mass of the element. The *kinetic energy density* of the string is therefore

$$K_1 = \frac{dK}{dx} = \tfrac{1}{2}\lambda_0 \left(\frac{\partial\eta}{\partial t}\right)^2 \tag{1.8.1}$$

and, in general, is a function of both time and position.

The calculation of the potential energy dV associated with the element dx, or the *potential energy density* $V_1 = dV/dx$ (the potential energy density per unit length), is a more subtle matter. The string as a whole must possess potential energy, since external work would have to be done to give it the non-equilibrium shape it momentarily has when a wave is present. Just where is this energy stored? Evidently it must exist in the form of elastic energy in the string, which is slightly longer when a wave is present. We expect, therefore, that the elastic properties of the string should somehow enter into the calculation of the potential energy. If true, we expect in turn that the velocity of a transverse wave should depend on the elastic "stretch" constant of the string, as well as on the static tension τ_0. We avoided this possibility in the derivation of the wave equation by assuming that τ_0 is independent of the presence of a wave.

Although it is common to regard the constancy of τ_0 simply as a good approximation, we may simplify the discussion of potential energy by idealizing the string to be such that the tension remains constant *no matter how much it is stretched.* Such an idealized string has a negligible Young's modulus (see Sec. 3.1) and therefore cannot transmit longitudinal waves along its length. Without this idealization, any local stretching caused by the passage of a transverse wave must increase the local tension. The increased tension, in turn, must excite longitudinal waves that run away with some of the energy of the transverse wave. The *coupling* between the two sorts of waves, transverse and longitudinal, is expressed by nonlinear terms that were neglected in the derivation of the wave equation in Sec. 1.1 but which can be included, if desired, in the more rigorous derivation of the wave equation for strings given in Sec. 1.10. We shall accordingly base the present discussion of potential energy on an idealized string that has a zero Young's modulus. (This idealization is consistent with the assumption of perfect flexibility, which requires a zero Young's modulus if the string has a small, but finite, cross-sectional area.)

To find an expression for V_1, let us first find how much a finite segment of string is stretched when it suffers a small transverse displacement consistent with fixed ends separated a distance l. If we recall that $ds = (dx^2 + d\eta^2)^{1/2}$ is the length of an element of the curve giving the shape of the displaced string, the segment of string must have been stretched the amount

$$\Delta s = \int ds - l = \int_0^l \left[1 + \left(\frac{\partial \eta}{\partial x} \right)^2 \right]^{1/2} dx - l$$

$$= \int_0^l \left[1 + \frac{1}{2} \left(\frac{\partial \eta}{\partial x} \right)^2 + \cdots \right] dx - l \approx \frac{1}{2} \int_0^l \left(\frac{\partial \eta}{\partial x} \right)^2 dx. \qquad (1.8.2)$$

Just as in the derivation of the wave equation for transverse waves, we have treated $\partial \eta / \partial x$ as a small quantity.

The work done to stretch the string an amount Δs against a constant tension τ_0 is $\tau_0 \, \Delta s$, which therefore equals the gain in potential energy stored in the displaced string. Hence the total potential energy of the finite segment is

$$V = \tfrac{1}{2}\tau_0 \int_0^l \left(\frac{\partial \eta}{\partial x} \right)^2 dx. \qquad (1.8.3)$$

Equation (1.8.3) implies that the localized potential energy density is given by

$$V_1 = \tfrac{1}{2}\tau_0 \left(\frac{\partial \eta}{\partial x} \right)^2 \qquad (1.8.4)$$

for our idealized string (having zero Young's modulus). When (1.8.4) is used to calculate the potential energy of a finite string segment with fixed ends, the result should apply fairly closely to a real string (or reasonably flexible wire), since the fixed ends block the escape of energy via longitudinal waves. In contrast, when used to discuss traveling waves on a string of indefinite length, any conclusions reached must apply less accurately to waves traveling on a real string or wire. This state of affairs is not unusual in theoretical physics: it is often necessary to make a simplified or idealized model of some aspect of the physical world in order to be able to discuss it theoretically with any ease. This process is also *physically* useful in that it serves to identify which features of a complex physical situation are essential to the main properties of the actual behavior.

Let us now investigate the energy transported by various waves. Consider first the arbitrary wave $\eta = f(u) = f(x - ct)$ proceeding in the positive x direction. At any position and time, (1.8.1) shows that the kinetic energy density in the wave is

$$K_1 = \tfrac{1}{2}\lambda_0 c^2 f'^2 = \tfrac{1}{2}\tau_0 f'^2, \tag{1.8.5}$$

where f' is the ordinary derivative $df(u)/du$ and use has been made of $c = (\tau_0/\lambda_0)^{1/2}$. At the same position and time, (1.8.4) shows that the potential energy density in the wave is

$$V_1 = \tfrac{1}{2}\tau_0 f'^2 = K_1. \tag{1.8.6}$$

Since the wave moves with the velocity c, the instantaneous rate of transport of the total energy density $E_1 = K_1 + V_1$ (i.e., the instantaneous power passing any position x) is

$$P = cE_1 = c\tau_0 f'^2. \tag{1.8.7}$$

A similar relation is found to hold for waves in other media, provided the wave velocity is independent of frequency.

If the traveling wave is the sinusoidal wave

$$\eta = A \cos(\kappa x - \omega t), \tag{1.8.8}$$

then

$$K_1 = V_1 = \tfrac{1}{2}\lambda_0 \omega^2 A^2 \sin^2(\kappa x - \omega t). \tag{1.8.9}$$

For a sinusoidal wave it is convenient to calculate the average energy densities

$$\bar{K}_1 = \bar{V}_1 = \tfrac{1}{4}\lambda_0 \omega^2 A^2 \tag{1.8.10}$$

by making use of the fact that

$$\overline{\sin^2(\kappa x - \omega t)} = \tfrac{1}{2},$$

where the bar signifies averaging over an integral number of periods. The average power transmitted by the wave is then

$$\bar{P} = c\bar{E}_1 = c(\bar{K}_1 + \bar{V}_1) = \tfrac{1}{2}\lambda_0 c\omega^2 A^2. \tag{1.8.11}$$

The energy density and power transmitted by a sinusoidal wave depend on the product of the square of the displacement amplitude and the square of the frequency, i.e., on the square of the amplitude ωA of the transverse "particle" velocity of the string. For the power transmitted, the properties of the string enter as $\lambda_0 c = (\lambda_0\tau_0)^{1/2}$, which is the so-called *characteristic impedance* of the string for transverse waves. A similar dependence on particle velocity and characteristic impedance exists for all elastic waves.

Another way of arriving at the energy transport by waves is to deduce a general expression for the power that passes by any position x along the string.

The power is given by the product of the transverse force that the left-hand portion of the string exerts on the right-hand portion and the transverse velocity of the string at x. From Fig. 1.1.1, we see that the force is

$$F = -\tau_0 \frac{\partial \eta}{\partial x}, \tag{1.8.12}$$

and since the transverse velocity is $v = \partial \eta / \partial t$, the (instantaneous) power transmitted is

$$P = Fv = -\tau_0 \frac{\partial \eta}{\partial x} \frac{\partial \eta}{\partial t}. \tag{1.8.13}$$

In using this expression, a positive value of P means power is flowing toward the positive x direction, and a negative value means it is flowing toward the negative x direction.

The *ratio* of force to velocity is of interest for the case of a traveling wave. Using the general form $f(x - ct)$, we find

$$\frac{F}{v} = -\tau_0 \frac{\partial \eta / \partial x}{\partial \eta / \partial t} = \frac{\tau_0}{c} = \lambda_0 c; \tag{1.8.14}$$

this ratio is called the *characteristic impedance* of the string. It turns out that the ratio of force to displacement velocity equals the characteristic impedance for all wave-transmitting mechanical systems. Electric power and impedance are given respectively by the *product* and *quotient* of voltage and current, which are the electrical analogs of mechanical force and displacement velocity.

Applying (1.8.13) to calculate the power transported by the sinusoidal wave (1.8.8), we find that

$$P = \tau_0 \omega \kappa A^2 \sin^2(\kappa x - \omega t) = \lambda_0 c \omega^2 A^2 \sin^2(\kappa x - \omega t). \tag{1.8.15}$$

Hence the average power carried by the wave computed here agrees with that found earlier, (1.8.11). In the case of standing waves, the absence of a velocity $\partial \eta / \partial t$ at a node means that no power is transmitted past such a point. Stated another way, the two traveling waves of equal amplitude and frequency going in opposite directions, which are equivalent to the standing wave, carry the same amounts of power in the positive and negative directions.

Before proceeding further, we note that we must not use the complex wave representation (1.3.7) in computing instantaneous energy or power in a wave, since the square of a complex number does not equal the square of its real part. There exists, however, a convenient way for obtaining the *average value* of the square of the real part of a complex sinusoidal oscillation or wave directly from the complex representation. If the wave has the form

$$\psi = \check{A} e^{i(\kappa x - \omega t)}, \tag{1.8.16}$$

then it is easily seen that

$$\overline{[\mathrm{Re}(\psi)]^2} = \tfrac{1}{2}|\check{A}|^2 = \tfrac{1}{2}\psi^*\psi, \tag{1.8.17}$$

where ψ^* is the *complex conjugate* of ψ, found from ψ by changing the sign of its imaginary part, $\psi^* = \psi_r - i\psi_i$.

Next let us find the total energy present on a vibrating string segment stretched between fixed supports; i.e., we consider a standing rather than a traveling wave. First suppose that the string segment is vibrating in one of its normal modes, as given by

$$\eta = A \sin\kappa_n x \sin\omega_n t. \tag{1.8.18}$$

The kinetic energy density (1.8.1) is then

$$K_1 = \tfrac{1}{2}\lambda_0\omega_n{}^2 A^2 \sin^2\kappa_n x \cos^2\omega_n t, \tag{1.8.19}$$

and the potential energy density (1.8.4) is

$$V_1 = \tfrac{1}{2}\tau_0\kappa_n{}^2 A^2 \cos^2\kappa_n x \sin^2\omega_n t. \tag{1.8.20}$$

The total kinetic energy of the string segment is

$$K = \int_0^l K_1 \, dx = \tfrac{1}{4}l\lambda_0\omega_n{}^2 A^2 \cos^2\omega_n t, \tag{1.8.21}$$

and the total potential energy is

$$V = \int_0^l V_1 \, dx = \tfrac{1}{4}l\tau_0\kappa_n{}^2 A^2 \sin^2\omega_n t. \tag{1.8.22}$$

Since $\tau_0\kappa_n{}^2 = \lambda_0\omega_n{}^2$, the total energy

$$E = K + V = \tfrac{1}{4}l\lambda_0\omega_n{}^2 A^2 \tag{1.8.23}$$

is constant in time, as required by the conservation-of-energy principle.

The examples just discussed show that a difference in the division of total energy into kinetic and potential exists between a traveling and a standing sinusoidal wave. In a traveling wave, V_1 and K_1 have an identical dependence on x and t (are in phase) and, in fact, are everywhere equal. In a standing wave, or normal-mode oscillation, V_1 and K_1 are out of phase by 90° both in x and in t. The total potential and kinetic energies of a string segment vibrating in a normal mode bear the same relationship to each other that the corresponding energies do in simple harmonic motion. Their sum is constant and their averages equal. This behavior turns out to be generally true for normal-mode oscillations of a linear system of whatever kind.

Let us next consider that the vibrating string segment has the arbitrary motion as specified by (1.6.1)

$$\eta(x,t) = \sum_{n=1}^{\infty} \sin\kappa_n x(a_n \cos\omega_n t + b_n \sin\omega_n t). \tag{1.8.24}$$

To find the total potential energy we must substitute this expression for η into (1.8.3) and carry out the integration over the length of the string segment. In so doing we obtain an expression for the integrand that contains many terms involving squares of $\cos\kappa_n x$ and products $\cos\kappa_n x \cos\kappa_m x$ ($m \neq n$), with factors involving the squares and products of the time factors ($a_n \cos\omega_n t + b_n \sin\omega_n t$). If we recall that $\kappa_n = n\pi/l$, the space integrals are either (see Prob. 1.6.1)

$$\int_0^l \cos^2 \frac{n\pi x}{l}\, dx = \frac{l}{2} \tag{1.8.25a}$$

or

$$\int_0^l \cos \frac{m\pi x}{l} \cos \frac{n\pi x}{l}\, dx = 0 \qquad m \neq n. \tag{1.8.25b}$$

Hence the expression for V becomes

$$V(t) = \frac{l\tau_0}{4} \sum_{n=1}^{\infty} \kappa_n^2 (a_n \cos\omega_n t + b_n \sin\omega_n t)^2 \tag{1.8.26}$$

since all cross-product terms involving $\cos\kappa_m x$ have disappeared.

Next we compute the time average of the potential energy,

$$\overline{V} = \frac{1}{T_1} \int_0^{T_1} V(t)\, dt, \tag{1.8.27}$$

where $T_1 = 2\pi/\omega_1 = 2l/c$ is the fundamental period. The integrand now contains terms that involve $\cos\omega_n t$ and $\sin\omega_n t$ squared, as well as various cross-product terms. The integrals now fall into several classes

$$\int_0^{T_1} \binom{\sin^2 n\omega_1 t}{\cos^2 n\omega_1 t}\, dt = \frac{T_1}{2}$$

$$\int_0^{T_1} \sin n\omega_1 t \cos n\omega_1 t\, dt = 0 \tag{1.8.28}$$

$$\int_0^{T_1} \binom{\sin m\omega_1 t}{\cos m\omega_1 t}\binom{\sin n\omega_1 t}{\cos n\omega_1 t}\, dt = 0 \qquad m \neq n$$

of which all are zero except the first, in which sine or cosine squared appears.†
As a result, the expression for the average potential energy is greatly simplified, becoming

$$\overline{V} = \frac{l\tau_0}{8} \sum_{n=1}^{\infty} \kappa_n^2 (a_n^2 + b_n^2). \tag{1.8.29}$$

---

† If the integral of the product of two functions vanishes when taken over their common fundamental range (period), as in the last two equations of (1.8.28), the functions are said to be *orthogonal*. Each member of an *orthogonal set* of functions, such as $\sin n\omega_1 t$, is orthogonal to every other member.

Since $a_n^2 + b_n^2$ is the square of the amplitude of the nth normal mode, the average potential energy is simply the sum of the average potential energies of the individual normal modes whose superposition describes the motion of the string segment.

A similar calculation (Prob. 1.8.2) for the average kinetic energy shows that it is given by

$$\bar{K} = \frac{l\lambda_0}{8} \sum_{n=1}^{\infty} \omega_n^2(a_n^2 + b_n^2). \tag{1.8.30}$$

Since $\lambda_0 \omega_n^2 = \tau_0 \kappa_n^2$ for each normal mode, the total energy of the string may be written

$$\bar{E} = \bar{V} + \bar{K} = \tfrac{1}{4}l\lambda_0 \sum_{n=1}^{\infty} \omega_n^2(a_n^2 + b_n^2). \tag{1.8.31}$$

Therefore the total energy of vibration of the string segment is made up of independent contributions from each of the normal-mode oscillations into which the vibration can be subdivided. This is a fundamental property of the normal-mode oscillations into which the general vibration of a conservative system can be analyzed when it is characterized by linear mathematical properties.

Problems

1.8.1 A steel wire 1.5 m long is stretched to a tension of 700 newtons, which is a typical length and tension of a string on a piano. It is vibrating in its fundamental mode with an amplitude of 2 mm. What is its total energy of vibration? Make some reasonable assumptions and decide whether or not a person playing the piano is likely to deliver this amount of energy to one of the three strings struck by one of the hammers.

1.8.2 Derive the expression for the average kinetic energy (1.8.30) of the string segment having the motion (1.8.24).

1.8.3 A string segment is vibrating with its motion given by (1.6.9). Find its total energy and show that it equals the work done in pulling the string aside to establish the initial shape of the string.

★1.8.4 Any physical quantity that is conserved must satisfy the so-called *equation of continuity*, which states that the rate of increase of the physical quantity per unit volume (or unit length or area) must equal the rate at which the quantity enters the unit volume (or unit length or area). When applied to the energy density in an elastic medium, the equation of continuity takes the form

$$\frac{\partial E_1}{\partial t} + \mathbf{\nabla} \cdot \mathbf{P} = 0 \tag{1.8.32}$$

where $E_1 = K_1 + V_1$ is the total energy density and $\mathbf{\nabla} \cdot \mathbf{P}$ is the divergence of the directed power flow per unit area. Show that the expressions for K_1, V_1, and P, as given by (1.8.1), (1.8.4), and (1.8.13), satisfy (1.8.32).

1.9   The Reflection and Transmission of Waves at a Discontinuity

We are here concerned with transverse waves on a string consisting of two parts, as shown in Fig. 1.9.1. The left part has a linear mass density λ_1 and the right part a different linear mass density λ_2, with both parts under the same tension τ_0. For convenience we place the x origin at the discontinuity. We suppose that a source of sinusoidal waves on the negative x axis is sending waves toward the discontinuity and that the waves continuing past it are absorbed with no reflection by a distant sink. We wish to examine how the abrupt change in properties of the string affects the passage of waves down the string.

Our first task is to find the so-called *boundary conditions* that the wave motion must satisfy at the discontinuity. Evidently there must exist two independent conditions, reflecting the fact that the differential wave equation is of second order. One of these is obviously the continuity of the string, i.e., of its displacement, or, equivalently, the continuity of its transverse velocity. The other is the continuity of the transverse force in the string, as given by (1.8.12). This boundary condition is basically a consequence of Newton's third law. If the force is not continuous at the boundary, an infinitesimal mass there would be subject to a finite force, resulting in an infinite acceleration. Accordingly for all times at $x = 0$, we require that

$$\eta_{\text{left}} = \eta_{\text{right}}$$

$$\left(-\tau_0 \frac{\partial \eta}{\partial x} \right)_{\text{left}} = \left(-\tau_0 \frac{\partial \eta}{\partial x} \right)_{\text{right}} . \tag{1.9.1}$$

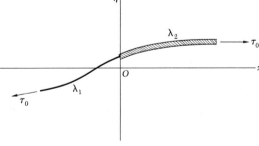

Fig. 1.9.1   Wave on string having a discontinuity in mass density at the origin.

Let us take the incident wave coming from the left to be the real part of

$$\eta_1 = A_1 e^{i(\kappa_1 x - \omega t)} \qquad -\infty < x < 0, \tag{1.9.2}$$

which has a specified amplitude A_1 and the velocity $c_1 = \omega/\kappa_1 = (\tau_0/\lambda_1)^{1/2}$. The wave transmitted past the discontinuity is assumed to be the real part of

$$\eta_2 = \check{A}_2 e^{i(\kappa_2 x - \omega t)} \qquad 0 < x < \infty, \tag{1.9.3}$$

which has a complex amplitude $\check{A}_2$ yet to be determined and a velocity and wave number that differ from those of the first wave, $c_2 = \omega/\kappa_2 = (\tau_0/\lambda_2)^{1/2}$. Both waves must necessarily have the same frequency.

We now discover that it is impossible with only these two waves to satisfy the boundary conditions (1.9.1), since the first condition would require that $A_1 = \check{A}_2$ and the second that $\kappa_1 A_1 = \kappa_2 \check{A}_2$. Necessarily, then, there must exist a *third* wave that is reflected from the boundary, in order that the boundary conditions (1.9.1) be satisfied.

We assume that the reflected wave traveling to the left is the real part of

$$\eta_1' = \check{B}_1 e^{i(-\kappa_1 x - \omega t)} \qquad -\infty < x < 0, \tag{1.9.4}$$

where $\check{B}_1$ is to be determined and the wave number is that appropriate to the string on the left side of the boundary.

The boundary conditions now require that

$$A_1 + \check{B}_1 = \check{A}_2$$
$$\kappa_1 A_1 - \kappa_1 \check{B}_1 = \kappa_2 \check{A}_2, \tag{1.9.5}$$

which are sufficient to determine $\check{B}_1$ and $\check{A}_2$ in terms of A_1, the amplitude of the incident wave. Solving for the amplitude ratios $\check{B}_1/A_1$ and $\check{A}_2/A_1$, which are defined as the complex *amplitude reflection coefficient* $\check{R}_a$ and *amplitude transmission coefficient* $\check{T}_a$, respectively, we find that

$$\check{R}_a \equiv \frac{\check{B}_1}{A_1} = \frac{\kappa_1 - \kappa_2}{\kappa_1 + \kappa_2} = \frac{Z_1 - Z_2}{Z_1 + Z_2}$$
$$\check{T}_a \equiv \frac{\check{A}_2}{A_1} = \frac{2\kappa_1}{\kappa_1 + \kappa_2} = \frac{2Z_1}{Z_1 + Z_2}, \tag{1.9.6}$$

where we have expressed the results in terms of the characteristic impedances $Z_1 [= \lambda_1 c_1 = (\lambda_1 \tau_0)^{1/2}]$ and Z_2 of the two parts of the string. The fact that $\check{R}_a$ and $\check{T}_a$ turn out to be real indicates that the reflected and transmitted waves are not shifted in phase, except for a possible 180° phase shift for the reflected wave, when $\check{R}_a$ is negative. We note that if $\lambda_1 > \lambda_2$, $\check{R}_a$ is positive, which implies that the reflected wave has the same phase as the incident wave, whereas if $\lambda_1 < \lambda_2$, $\check{R}_a$ is negative, showing that the reflected and incident waves are 180° out of phase. Since $\check{T}_a$ is always positive, the transmitted wave has the same phase as the incident wave.

It is also customary to define a *power reflection coefficient* R_p and a *power transmission coefficient* T_p to express the reflection and transmission of waves at a boundary. The power carried by a traveling sinusoidal wave is given by (1.8.11). Hence for the power reflection coefficient

$$R_p = \frac{\frac{1}{2}\lambda_1 c_1 \omega^2 B_1{}^2}{\frac{1}{2}\lambda_1 c_1 \omega^2 A_1{}^2} = \frac{B_1{}^2}{A_1{}^2} = \left(\frac{Z_1 - Z_2}{Z_1 + Z_2}\right)^2, \tag{1.9.7}$$

and for the power transmission coefficient

$$T_p = \frac{\frac{1}{2}\lambda_2 c_2 \omega^2 A_2{}^2}{\frac{1}{2}\lambda_1 c_1 \omega^2 A_1{}^2} = \frac{\lambda_2 c_2 A_2{}^2}{\lambda_1 c_1 A_1{}^2} = \frac{4 Z_1 Z_2}{(Z_1 + Z_2)^2}. \tag{1.9.8}$$

The fact that the incident power equals the reflected power plus the transmitted power is expressed by $R_p + T_p = 1$. Since (1.9.6) to (1.9.8) depend only on properties of the medium (the string) and not on the frequency of the waves, they must hold for waves of arbitrary shape. Reflection and transmission coefficients for *plane* waves of any sort incident normally on a plane boundary between two media have the same form as those found here when expressed in terms of the characteristic impedances of the media.

Problems

1.9.1 Obtain the boundary conditions (1.9.5) from (1.9.1) and show that they lead to (1.9.6).

1.9.2 A uniform string of linear mass density λ_0 and under a tension τ_0 has a small bead of mass m attached to it at $x = 0$. Find expressions for the complex amplitude and the power reflection and transmission coefficients for sinusoidal waves brought about by the mass discontinuity at the origin. Do these coefficients hold for a wave of arbitrary shape?

1.9.3 Three long identical strings of linear mass density λ_0 are joined together at a common point forming a symmetrical Y. Thus they lie in a plane 120° apart. Each is given the same tension τ_0. A distant source of sinusoidal waves sends transverse waves, with motion perpendicular to the plane of the strings, down one of the strings. Find the reflection and transmission coefficients that characterize the junction.

1.9.4 A long string under tension τ_0 having a linear mass density λ_1 is tied to a second string with linear mass density $\lambda_2 \ll \lambda_1$. Transverse waves on the heavy string are incident on the junction. Find what happens to them.

*1.10   Another Derivation of the Wave Equation for Strings

We now take a more rigorous look at the derivation of the wave equation for waves on a flexible string under tension. In the absence of a wave, the string coincides with the x axis. When a wave is present, a point on the string originally at x is displaced to some point A specified by the displacement vector

$$\varrho = \mathbf{i}\xi + \mathbf{j}\eta + \mathbf{k}\zeta. \tag{1.10.1}$$

We have now included the possibility that the displacement may have components ξ and ζ in the x and z directions, and not just the η component assumed earlier. Similarly,

$$\varrho + d\varrho = \mathbf{i}(\xi + d\xi) + \mathbf{j}(\eta + d\eta) + \mathbf{k}(\zeta + d\zeta) \tag{1.10.2}$$

is the vector displacement of a neighboring point B on the string.

The net vector force acting on the element AB is the vector sum of the force $-\boldsymbol{\tau}$ at A and the force $\boldsymbol{\tau} + (\partial\boldsymbol{\tau}/\partial x)\,dx$ at B, that is, $(\partial\boldsymbol{\tau}/\partial x)\,dx$. If the string has stiffness, the two forces will in general not be tangent to the string at these points. The mass of the element AB is $\lambda_0\,dx$, the mass that it had in the absence of the wave. An application of Newton's second law to the element, with no approximations having been made in the analysis so far, requires that

$$\frac{\partial\boldsymbol{\tau}}{\partial x} = \lambda_0\frac{\partial^2\varrho}{\partial t^2}, \tag{1.10.3}$$

where dx has been canceled from both sides of the equation.

Further progress rests with expressing the force $\boldsymbol{\tau}$ in terms of the tension τ_0 of the undisplaced string, the elastic constant of the string, and factors of geometrical origin. The displaced string element AB is the vector

$$\begin{aligned} d\mathbf{s} &= \mathbf{i}(dx + d\xi) + \mathbf{j}\,d\eta + \mathbf{k}\,d\zeta \\ &= \left[\mathbf{i}\left(1 + \frac{\partial\xi}{\partial x}\right) + \mathbf{j}\frac{\partial\eta}{\partial x} + \mathbf{k}\frac{\partial\zeta}{\partial x}\right]dx \end{aligned} \tag{1.10.4}$$

having the magnitude

$$ds = \left[\left(1 + \frac{\partial\xi}{\partial x}\right)^2 + \left(\frac{\partial\eta}{\partial x}\right)^2 + \left(\frac{\partial\zeta}{\partial x}\right)^2\right]^{1/2}dx. \tag{1.10.5}$$

We now assume that the string is flexible, so that the direction of $\boldsymbol{\tau}$ is given by the unit vector $d\mathbf{s}/ds$ tangent to the string. If it were not flexible, it would be necessary to take into account a force component at right angles to the string, depending in some way on the curvature and stiffness of the string. (We discuss this effect in Sec. 4.4.) The magnitude of $\boldsymbol{\tau}$ at any position is the tension τ_0 in the undisplaced string, increased by the incremental tension needed to stretch it the fractional amount

$$\frac{ds - dx}{dx} = \frac{ds}{dx} - 1.$$

If S is the constant cross-sectional area of the string and Y is *Young's modulus* (see Sec. 3.1), then

$$|\tau| = \tau_0 + SY \left(\frac{ds}{dx} - 1 \right),$$

so that

$$\tau = \left[\tau_0 + SY \left(\frac{ds}{dx} - 1 \right) \right] \frac{d\mathbf{s}}{ds}, \tag{1.10.6}$$

where no *geometrical* limitations have yet been put on the magnitude of ϱ and $\partial\varrho/\partial x$ (other than to prevent the string from being stretched beyond its elastic limit).

The four equations (1.10.3) to (1.10.6) can be combined into a single *vector* wave equation that expresses the propagation of waves on the string having three displacement components. The resulting equation is too complicated to be worth working out in detail, and its solution would present formidable difficulties. In particular, the equation is nonlinear, so that superposition no longer holds, and furthermore the three displacement components of the wave are coupled together. To obtain a wave equation that is sufficiently simple to solve, it is necessary to require that the *strains* $\partial\xi/\partial x$, $\partial\eta/\partial x$, and $\partial\zeta/\partial x$ be small compared with unity. Such an assumption is customarily made in the theory of elasticity and of elastic waves.

In Prob. 1.10.1 it is established that when (1.10.6) is developed in powers of the strains and only the first powers of the strains are retained, then

$$\tau = \mathbf{i} \left(\tau_0 + SY \frac{\partial\xi}{\partial x} \right) + \mathbf{j}\tau_0 \frac{\partial\eta}{\partial x} + \mathbf{k}\tau_0 \frac{\partial\zeta}{\partial x}. \tag{1.10.7}$$

On using this approximate equation for the force τ, it is found that (1.10.3) separates into the three wave equations

$$\frac{\partial^2\xi}{\partial x^2} = \frac{1}{c_b^2} \frac{\partial^2\xi}{\partial t^2} \tag{1.10.8}$$

$$\frac{\partial^2\eta}{\partial x^2} = \frac{1}{c_t^2} \frac{\partial^2\eta}{\partial t^2} \tag{1.10.9}$$

$$\frac{\partial^2\zeta}{\partial x^2} = \frac{1}{c_t^2} \frac{\partial^2\zeta}{\partial t^2} \tag{1.10.10}$$

where $c_b = (YS/\lambda_0)^{1/2}$ and $c_t = (\tau_0/\lambda_0)^{1/2}$ are wave velocities. The second and third of these equations govern the propagation of small-amplitude transverse waves, the second being identical with the wave equation (1.1.3) previously derived. The first equation governs the propagation of *longitudinal* (or *compressive*) waves on the string that have the wave velocity

$$c_b = \left(\frac{YS}{\lambda_0}\right)^{1/2} = \left(\frac{Y}{\rho_0}\right)^{1/2}. \tag{1.10.11}$$

in which $\rho_0 \equiv \lambda_0/S$ is the volume density of the material of the string. Waves of this sort are considered in more detail in Chap. 4. Each of the three sorts of waves on the string can thus exist independently of the others, provided it is possible to neglect powers of $\partial\xi/\partial x$, $\partial\eta/\partial x$, $\partial\zeta/\partial x$ of degree higher than the first. When this is not the case, each of the wave equations contains nonlinear terms, some of which represent a coupling between the three sorts of waves. In such event the analysis becomes too difficult for consideration here.†

Problems

1.10.1   Expand $d\mathbf{s}/ds$ and ds/dx in powers of $\partial\xi/\partial x$, $\partial\eta/\partial x$, and $\partial\zeta/\partial x$, where $d\mathbf{s}$ and ds are given by (1.10.4) and (1.10.5). Keep only terms through the second power. Then find the expansion of τ, as given by (1.10.6), again keeping only terms involving the second power. Show that this expression reduces to (1.10.7) when second-power terms are neglected. *Answer:* Let

$$\Delta \equiv \frac{1}{2}\left[\left(\frac{\partial\xi}{\partial x}\right)^2 + \left(\frac{\partial\eta}{\partial x}\right)^2 + \left(\frac{\partial\zeta}{\partial x}\right)^2\right];$$

then

$$\tau = i\left\{\tau_0 + SY\frac{\partial\xi}{\partial x} + SY\Delta - \tau_0\left[\left(\frac{\partial\xi}{\partial x}\right)^2 + \Delta\right]\right\} + j\left(\tau_0\frac{\partial\eta}{\partial x} + SY\frac{\partial\xi}{\partial x}\frac{\partial\eta}{\partial x}\right) + k(\cdots).$$

1.10.2   What units should be used when computing the wave velocities given by (1.1.4) and (1.10.11) in the cgs, mks, and English systems? Look up values of Y and ρ_0 for steel and compute c_b. Compare with the wave velocity $c_t = (\tau_0/\lambda_0)^{1/2}$ when the tension is such that the wire has static strain of 10^{-3}, about the largest it can have without permanent deformation. Does this result depend on the diameter of the wire?

1.10.3   Investigate the motion of a string having transverse waves on it specified by the equations $\eta = A\sin(\kappa x - \omega t)$ and $\zeta = A\cos(\kappa x - \omega t)$. If the string, instead of being infinitely long, is attached to a wall, so that reflections occur, find what sort of standing-wave pattern exists. Does the motion have a connection with rope jumping?

† For an analysis of the nonlinear vibration of a string segment with fixed ends, see G. F. Carrier, *Quart. Appl. Math.*, **3**: 157 (1945).

*1.11   Momentum Carried by a Wave

In Sec. 1.8 we found expressions for the energy that is carried by a transverse wave on a string. In later chapters we show that energy transport is a characteristic feature of all wave motion. Indeed, much of the transport of energy in the universe takes place by wave motion of some sort or another; the only competing mode of energy transport involves the bodily motion of matter, and even in this case, the motion of matter is described at a fundamental level by wave (quantum) mechanics. From the study of ordinary mechanics we know that momentum and momentum conservation have as fundamental a significance in physics as energy and energy conservation. We shall now show that there exists a close connection between energy transport by a wave traveling on a string and the transport of linear momentum by the wave.

We can arrive at an expression for the momentum density by examining the small longitudinal motion of the string, specified by the displacement component ξ, that occurs when a transverse wave is present. As a model we suppose that the string has a constant tension τ_0 and a negligible Young's modulus, as discussed in Sec. 1.8 in connection with the localization of potential energy. In Sec. 1.10 we found, as the result of a more rigorous discussion of the motion of an element of the string, the equation of motion (1.10.3), which is equivalent to the two component equations

$$\frac{\partial \tau_x}{\partial x} = \lambda_0 \frac{\partial^2 \xi}{\partial t^2} \tag{1.11.1}$$

$$\frac{\partial \tau_y}{\partial x} = \lambda_0 \frac{\partial^2 \eta}{\partial t^2} \tag{1.11.2}$$

when motion is restricted to the xy plane and where τ_x and τ_y are the components of the force

$$\boldsymbol{\tau} = \tau_0 \frac{d\mathbf{s}}{ds}. \tag{1.11.3}$$

We can evaluate $\partial \tau_x/\partial x$ and $\partial \tau_y/\partial x$ from (1.11.3) using (1.10.4) and (1.10.5) for $d\mathbf{s}/ds$ and keeping only the lowest-order term for each component (see Probs. 1.10.1 and 1.11.1). We find in this way that

$$\frac{\partial \tau_x}{\partial x} = -\tau_0 \frac{\partial \eta}{\partial x} \frac{\partial^2 \eta}{\partial x^2} + \text{higher-order terms} \tag{1.11.4}$$

$$\frac{\partial \tau_y}{\partial x} = \tau_0 \frac{\partial^2 \eta}{\partial x^2} + \text{higher-order terms.} \tag{1.11.5}$$

We suppose that $\partial \eta/\partial x \ll 1$, so that the higher-order terms in each expression can be safely neglected; moreover, the right-hand side of (1:11.5) is much larger than (1.11.4). Thus (1.11.5) substituted in the equation of motion (1.11.2) describes the dominant process, the usual wave equation for η

$$\tau_0 \frac{\partial^2 \eta}{\partial x^2} = \lambda_0 \frac{\partial^2 \eta}{\partial t^2}, \tag{1.11.6}$$

on which we have based our discussion of transverse waves on a flexible string under tension. Meanwhile (1.11.4), substituted in (1.11.1), describes a secondary process, the associated wave motion in the longitudinal direction

$$-\tau_0 \frac{\partial \eta}{\partial x} \frac{\partial^2 \eta}{\partial x^2} = \lambda_0 \frac{\partial^2 \xi}{\partial t^2} \tag{1.11.7}$$

which results from the small amount the string has to stretch in order to accommodate the distortion resulting from a transverse wave. We previously ignored this effect, except in calculating the potential energy density

$$V_1 = \tfrac{1}{2}\tau_0 \left(\frac{\partial \eta}{\partial x}\right)^2. \tag{1.11.8}$$

Let us now endeavor to find the momentum in the x direction of a transverse wave described by the transverse-wave equation (1.11.6) by an integration of (1.11.7), the equation giving the longitudinal acceleration accompanying the transverse wave. For simplicity we integrate with respect to time from t_0, a time when no wave is present on the string, to an arbitrary later time t and with respect to x over a finite string segment lying between x_1 and x_2. The result should be the momentum G_x acquired by the string segment as the result of transverse wave motion. This space-time integral of (1.11.7) is explicitly

$$G_x \equiv \int_{x_1}^{x_2} \lambda_0 \frac{\partial \xi}{\partial t} \, dx = -\int_{t_0}^{t} \int_{x_1}^{x_2} \tau_0 \frac{\partial \eta}{\partial x} \frac{\partial^2 \eta}{\partial x^2} \, dx \, dt, \tag{1.11.9}$$

where the time integration of $\partial^2 \xi / \partial t^2$ has been carried out. Before attempting the integration of the right side of (1.11.9), we must replace $\tau_0 \, \partial^2 \eta / \partial x^2$ by $\lambda_0 \, \partial^2 \eta / \partial t^2$, in order to bring into the calculation the fact that we are dealing with a transverse wave obeying (1.11.6) and not simply with a static state of sideways displacement of the string brought about by some external system of forces applied to the string in the y direction. If we then integrate the resulting equation by parts, we find that

$$G_x = -\lambda_0 \int_{x_1}^{x_2} \frac{\partial \eta}{\partial x} \frac{\partial \eta}{\partial t} \, dx + \int_{t_0}^{t} [K_1(x_2,t) - K_1(x_1,t)] \, dt, \tag{1.11.10}$$

where the kinetic energy density (1.8.1) has been introduced to simplify the form of the second integral.

Equation (1.11.10) for the momentum G_x has the following interpretation: the second integral on the right clearly represents momentum delivered to the string segment by impulses at the two boundaries at x_1 and x_2. If these boundaries are very remote, so that a wave disturbance initiated on the string segment has not yet had time to reach them, this integral vanishes. We are thus left with the first integral, whose form suggests that the quantity

$$g_x(x,t) \equiv -\lambda_0 \frac{\partial \eta}{\partial t} \frac{\partial \eta}{\partial x} \tag{1.11.11}$$

may be interpreted as a localized momentum density in the x direction associated with a transverse wave.

The close relationship of energy flow and momentum density is revealed on comparing (1.11.11) with (1.8.13) expressing the flow of wave energy along the string. We find that the energy flow is

$$P = -\tau_0 \frac{\partial \eta}{\partial x} \frac{\partial \eta}{\partial t} = c^2 g_x \tag{1.11.12}$$

where $c = (\tau_0/\lambda_0)^{1/2}$ is the wave velocity. A relation of this form connecting energy flow and momentum density holds in general for plane waves traveling in linear isotropic media.

When a wave traveling on a string is reflected or absorbed in some manner, the momentum it carries is either reversed in its direction of flow or is transferred from the wave to the external bodies that serve to absorb the energy of the wave. We therefore expect, from Newton's second law, that in either event the wave must exert a longitudinal force on its surroundings equal to the rate of change of momentum of the wave in its direction of propagation. Let us now see how this aspect of wave motion can be discussed theoretically, with the expectation that an analogous treatment can be applied to waves of other sorts.

A rather fundamental method of attack is to multiply the wave equation (1.11.6) on each side by $\partial \eta/\partial x$. We note that the left side then becomes

$$\tau_0 \frac{\partial^2 \eta}{\partial x^2} \frac{\partial \eta}{\partial x} = \frac{\partial}{\partial x} \left[\frac{1}{2} \tau_0 \left(\frac{\partial \eta}{\partial x} \right)^2 \right] = \frac{\partial V_1}{\partial x}, \tag{1.11.13}$$

where V_1 is the potential energy density (1.11.8). The right side is found to be

$$\lambda_0 \frac{\partial^2 \eta}{\partial t^2} \frac{\partial \eta}{\partial x} = \frac{\partial}{\partial t} \left(\lambda_0 \frac{\partial \eta}{\partial x} \frac{\partial \eta}{\partial t} \right) - \frac{\partial}{\partial x} \left[\frac{1}{2} \lambda_0 \left(\frac{\partial \eta}{\partial t} \right)^2 \right]$$

$$= -\frac{\partial g_x}{\partial t} - \frac{\partial K_1}{\partial x}, \tag{1.11.14}$$

where g_x is the momentum density (1.11.11) and K_1 is the kinetic energy density (1.8.1). Accordingly the wave equation implies that momentum density and total energy density $E_1 = K_1 + V_1$ are related by the basic equation

$$\frac{\partial g_x}{\partial t} = - \frac{\partial E_1}{\partial x},$$ (1.11.15)

which holds when the wave is not interacting with external bodies. If such an interaction takes place, conservation of total momentum in the x direction requires that (1.11.15) include an additional term $\partial g_{xe}/\partial t$ representing the rate at which momentum density is transferred to external bodies. This term arises from an external force density term in the wave equation, such as the ones considered in Probs. 1.1.2 and 1.1.3. With this term (1.11.15) becomes

$$\frac{\partial g_{xe}}{\partial t} + \frac{\partial g_x}{\partial t} = - \frac{\partial E_1}{\partial x}.$$ (1.11.16)

If we take the time average of this equation for a traveling wave having a sinusoidal time dependence, we find that the average force density on the surroundings is given by

$$\bar{F}_{1x} \equiv \overline{\frac{\partial g_{xe}}{\partial t}} = - \overline{\frac{\partial E_1}{\partial x}} = - \frac{d\bar{E}_1}{\partial x}$$ (1.11.17)

since $\overline{\partial g_x/\partial t} = 0$ for a periodic wave.

Let us now suppose that such a force exists to the right of $x = 0$. For example, the string could enter a viscous liquid medium at $x = 0$, so that the wave dies out slowly with distance beyond this point. The total *time-averaged* force exerted on the liquid is

$$\bar{F}_x = \int_0^\infty \bar{F}_{1x}\, dx = - \int_0^\infty \frac{d\bar{E}_1}{dx}\, dx = - \int_{\bar{E}_{10}}^0 d\bar{E}_1 = \bar{E}_{10}$$ (1.11.18)

where the bar signifies a time average and $\bar{E}_{10}$ is the average energy density of the wave at $x = 0$. For simplicity we have assumed that no reflection occurs where the wave enters the liquid. On combining this result with (1.8.11) and (1.11.12), we see that the average force exerted by the wave may be variously expressed as

$$\bar{F}_x = \bar{E}_{10} = \frac{\bar{P}_0}{c} = c\bar{g}_{x0}$$ (1.11.19)

where the subscript 0 refers to the value of the various quantities at $x = 0$. The average force $\bar{F}_x$ exists in addition to the tension τ_0 present in the string.

A result similar to (1.11.19) holds for waves of other types. In particular, the pressure exerted by electromagnetic (light) waves is of considerable theoretical and practical interest. The pressure exerted by an acoustic wave in a fluid affords a means for measuring the wave intensity.

Problems

1.11.1 Obtain the expansions (1.11.4) and (1.11.5), using the results found in Prob. 1.10.1, and supply the missing steps leading to (1.11.10).

1.11.2 Show that the force on a fixed support due to a reflected sinusoidal wave is equal to the average total energy density of the incident and reflected wave.

1.11.3 Show that energy flow and total energy are related by the continuity equation (see Prob. 1.8.4)

$$\frac{\partial P}{\partial x} = -\frac{\partial E_1}{\partial t} \qquad\qquad (1.11.20)$$

by multiplying each side of the wave equation (1.11.6) by $\partial\eta/\partial t$ and carrying out a development analogous to that leading to (1.11.15). Using (1.11.15) and (1.11.20), show that P, g_x, and E_1 are all solutions of the usual wave equation, with the wave velocity $c = (\tau_0/\lambda_0)^{1/2}$.

two

Waves on a Membrane

A vibrating membrane is a simple, easily visualized system from which we can learn much that also applies to more interesting three-dimensional waves, without the mathematical and geometrical complexity of full three-dimensional systems. For instance, we here introduce such topics as Bessel functions, vector wave numbers, and cutoff wavelengths and evanescent modes in waveguides.

An obvious common example of waves on a two-dimensional surface is given by water waves, particularly under conditions when surface tension is important. However, since the wave motion is not confined strictly to the water-air interface, the formal treatment of water waves requires a basic knowledge of hydrodynamics. We return to this question in Chap. 6. In the present chapter, attention is confined to an idealized membrane, exemplified by a soap film or a drumhead.

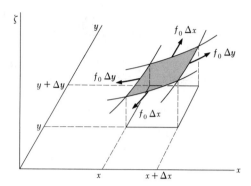

Fig. 2.1.1 Displaced element of a membrane.

2.1 The Wave Equation for a Stretched Membrane

A stretched flexible membrane is the two-dimensional counterpart of the stretched flexible string considered in Chap. 1. In the absence of a wave, the membrane is flat, lying in the xy coordinate plane. Across any straight line on the membrane, regardless of its orientation, we suppose that there exists a *surface tension* f_0 per unit length of the line, the two-dimensional analog of the linear tension existing in the string. As with the string, we idealize the elastic properties of the membrane to ensure that f_0 remains constant for small deflections normal to its flat equilibrium position. We use $\zeta(x,y,t)$ to specify the normal displacement at any position and time. The membrane has a mass density per unit area σ_0. We neglect the effect of gravity and the loading effect of the surrounding air.

The wave equation for small-amplitude transverse waves is derived by a method closely paralleling that adopted for the string in Sec. 1.1. Let us focus our attention on a square element $\Delta x \, \Delta y$ of the membrane, as illustrated in Fig. 2.1.1. When the membrane is displaced, there arises a net force in the z direction from each of the two pairs of tensile forces $f_0 \, \Delta x$ and $f_0 \, \Delta y$ acting on the four edges of the displaced square element. Each pair may be thought of as being equivalent to the tension in a hypothetical string consisting of a strip of the membrane of width Δx extending in the y direction, or of width Δy extending in the x direction, as the case may be. For the first strip, the net force in the z direction is

$$f_0 \, \Delta x \left[\left(\frac{\partial \zeta}{\partial y} \right)_{y+\Delta y} - \left(\frac{\partial \zeta}{\partial y} \right)_{y} \right] = f_0 \, \Delta x \, \frac{\partial^2 \zeta}{\partial y^2} \, \Delta y,$$

whereas for the second strip, the net force is

$$f_0 \, \Delta y \left[\left(\frac{\partial \zeta}{\partial x} \right)_{x+\Delta x} - \left(\frac{\partial \zeta}{\partial x} \right)_{x} \right] = f_0 \, \Delta y \, \frac{\partial^2 \zeta}{\partial x^2} \, \Delta x.$$

These forces correspond to (1.1.1) in the derivation of the wave equation on a string. Their sum must equal the mass $\sigma_0 \, \Delta x \, \Delta y$ of the element times its acceleration, that is,

$$\Delta x \, \Delta y \, f_0 \left(\frac{\partial^2 \zeta}{\partial x^2} + \frac{\partial^2 \zeta}{\partial y^2} \right) = \Delta x \, \Delta y \, \sigma_0 \frac{\partial^2 \zeta}{\partial t^2}.$$

On dividing out $\Delta x \, \Delta y$ we have the wave equation

$$\frac{\partial^2 \zeta}{\partial x^2} + \frac{\partial^2 \zeta}{\partial y^2} = \frac{1}{c_m^2} \frac{\partial^2 \zeta}{\partial t^2}, \tag{2.1.1}$$

where

$$c_m \equiv \left(\frac{f_0}{\sigma_0} \right)^{1/2} \tag{2.1.2}$$

turns out to be the wave velocity for transverse waves on the membrane. The wave equation (2.1.1) is a two-dimensional generalization of the one-dimensional wave equation (1.1.3) we found for the string.

Let us tackle the solution of (2.1.1) by the method of separation of variables, explained in Sec. 1.5. Since ζ is a function of x, y, and t, we assume that a solution exists of the form

$$\zeta(x,y,t) = X(x) \cdot Y(y) \cdot T(t), \tag{2.1.3}$$

where $X(x)$ is a function of x alone, etc. On substituting (2.1.3) into (2.1.1) and dividing through by XYT/c_m^2, we find that

$$\frac{c_m^2}{X} \frac{d^2 X}{dx^2} + \frac{c_m^2}{Y} \frac{d^2 Y}{dy^2} = \frac{1}{T} \frac{d^2 T}{dt^2} = -\omega^2. \tag{2.1.4}$$

Since the two sides of this equation are functions of different independent variables, we equate them to a constant, $-\omega^2$, as before. We then find that

$$\frac{d^2 T}{dt^2} + \omega^2 T = 0 \tag{2.1.5}$$

and that the other equation may be rearranged to read

$$\frac{1}{X} \frac{d^2 X}{dx^2} = -\frac{1}{Y} \frac{d^2 Y}{dy^2} - \frac{\omega^2}{c_m^2} = -\kappa_x^2 \tag{2.1.6}$$

where we have chosen to call the second separation constant $-\kappa_x^2$. This equation thus separates into the two equations

$$\begin{aligned} \frac{d^2 X}{dx^2} + \kappa_x^2 X &= 0 \\ \frac{d^2 Y}{dy^2} + \kappa_y^2 Y &= 0, \end{aligned} \tag{2.1.7}$$

where for the sake of symmetry we have introduced $\kappa_y{}^2$, related to the two separation constants by the equation

$$\kappa_x{}^2 + \kappa_y{}^2 = \frac{\omega^2}{c_m{}^2}. \tag{2.1.8}$$

Each of the three separated equations has the form of the equation for simple harmonic motion, and we are free to construct a general solution of the wave equation (2.1.1) by combining, with arbitrary constants, the three respective pairs of independent solutions, in any way we like. The solution is characterized by a single frequency ω and by the constants κ_x and κ_y, limited only by (2.1.8). In particular, we may choose real functions, such as $\cos\kappa_x x$, $\sin\kappa_y y$, $\cos\omega t$, etc., or complex functions such as $e^{i\kappa_x x}$, $e^{i\kappa_y y}$, $e^{-i\omega t}$, etc., or some of each. The constants of integration may be either real or complex.

Since we have not yet established that c_m is indeed the velocity of waves on the membrane, let us first consider for detailed study the particular solution

$$\zeta(x,y,t) = A e^{i(\kappa_x x + \kappa_y y - \omega t)}, \tag{2.1.9}$$

which we suspect should represent a traveling wave of frequency ω. The constants κ_x and κ_y are related to ω by (2.1.8). To discover the significance of (2.1.9), we examine loci of constant displacement ζ. Denote the phase of the wave by

$$\phi \equiv \kappa_x x + \kappa_y y - \omega t. \tag{2.1.10}$$

Then at some instant of time t_1, a particular locus or *wavefront* is specified by a particular value ϕ_1 of the phase. The points on this locus are all those with coordinates x_1, y_1 satisfying the equation

$$\kappa_x x_1 + \kappa_y y_1 - \omega t_1 = \phi_1. \tag{2.1.11}$$

That is, the wavefront is a straight line in the xy plane of the membrane. Equation (2.1.9) may be said to represent a (two-dimensional) *plane* wave. The geometry involved, as shown in Fig. 2.1.2, is made clearer by writing (2.1.11) in vector form. Thus let $\mathbf{r}_1 = \mathbf{i}x_1 + \mathbf{j}y_1$, and $\boldsymbol{\kappa} = \mathbf{i}\kappa_x + \mathbf{j}\kappa_y$, so that (2.1.11) can be written

$$\boldsymbol{\kappa} \cdot \mathbf{r}_1 = \phi_1 + \omega t_1. \tag{2.1.12}$$

The wavefront specified by ϕ_1 at the time t_1 is perpendicular to the constant vector $\boldsymbol{\kappa}$.

If we increase the phase ϕ_1 by 2π, keeping t_1 fixed, we have an identical parallel wavefront, the equation for which is

$$\boldsymbol{\kappa} \cdot \mathbf{r}_2 = \phi_1 + \omega t_1 + 2\pi. \tag{2.1.13}$$

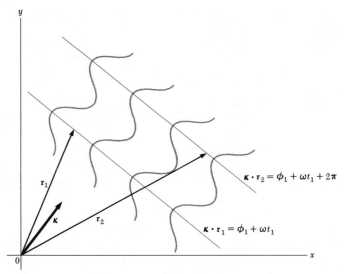

Fig. 2.1.2   Wavefronts on a membrane.

The wavelength λ is the perpendicular separation of the two parallel wavefronts; hence, subtracting the equations for the two wavefronts, we have

$$(\mathbf{r}_2 - \mathbf{r}_1) \cdot \mathbf{\kappa} = \lambda \kappa = 2\pi, \tag{2.1.14}$$

where by (2.1.8)

$$\kappa = |\mathbf{\kappa}| = (\kappa_x{}^2 + \kappa_y{}^2)^{1/2} = \frac{\omega}{c_m}. \tag{2.1.15}$$

Thus $\kappa = 2\pi/\lambda$ is the wave number of the plane wave (2.1.9).

Alternatively, if we let s be the perpendicular distance from the origin to the wavefront defined by (2.1.12), then, since s is along the line indicated by $\mathbf{\kappa}$ in Fig. 2.1.2,

$$\kappa s = \phi_1 + \omega t_1. \tag{2.1.16}$$

If we replace t_1 by the variable time t, we find that the velocity of the wavefront in the direction of $\mathbf{\kappa}$ is

$$\frac{ds}{dt} = \frac{\omega}{\kappa} = c_m. \tag{2.1.17}$$

It is $\mathbf{\kappa}$, *not* the wave velocity c_m, that is the mathematical quantity having a vector character which specifies the direction in which the wave is progressing.

The kinetic energy density per unit area of the membrane in the presence of a wave is evidently

$$K_1 = \tfrac{1}{2}\sigma_0 \left(\frac{\partial \zeta}{\partial t}\right)^2, \tag{2.1.18}$$

and the potential energy density per unit area is

$$V_1 = \tfrac{1}{2}f_0 \left[\left(\frac{\partial \zeta}{\partial x}\right)^2 + \left(\frac{\partial \zeta}{\partial y}\right)^2\right]. \tag{2.1.19}$$

These equations correspond to (1.8.1) and to (1.8.4) for a string in tension. We can establish (2.1.19) by noting that the increase in area of the displaced membrane is

$$\Delta S = \iint \left[1 + \left(\frac{\partial \zeta}{\partial x}\right)^2 + \left(\frac{\partial \zeta}{\partial y}\right)^2\right]^{1/2} dx\, dy - S_0, \tag{2.1.20}$$

where S_0 is the area of the undisplaced membrane. When the displacement derivatives $\partial \zeta/\partial x$ and $\partial \zeta/\partial y$ are small, (2.1.20) can be approximated by

$$\Delta S = \frac{1}{2} \iint \left[\left(\frac{\partial \zeta}{\partial x}\right)^2 + \left(\frac{\partial \zeta}{\partial y}\right)^2\right] dx\, dy. \tag{2.1.21}$$

Since the work done in stretching the surface is $f_0\,\Delta S$, (2.1.19) is the work done per unit area.

The localization of potential energy density on the membrane, of course, requires that the membrane be idealized so that the surface tension f_0 remains constant, no matter how much it is stretched. A similar requirement was discussed in some detail with regard to the potential energy density of the displaced flexible string in Sec. 1.8. A soap film possesses the property of a constant surface tension. Membranes of thin layers of fairly flexible solid materials do not have this property exactly, so that the theory of the idealized membrane can be considered only a good approximation for them.

The expression for energy flow across a line of unit width has a number of points of interest. Consider an element $\Delta x\, \Delta y$ of the membrane, as shown in Fig. 2.1.1. The rate at which energy passes into the element in the positive x direction across the edge Δy is the force component $-f_0\,\Delta y(\partial \zeta/\partial x)$ times the displacement velocity $\partial \zeta/\partial t$. Hence the energy flow per unit time in the positive x direction, per unit distance in the y direction, is

$$P_{1x} = -f_0 \frac{\partial \zeta}{\partial t}\frac{\partial \zeta}{\partial x}. \tag{2.1.22}$$

This expression is the membrane counterpart of (1.8.13) for waves on a string. A similar expression holds for the power-flow component in the positive y

direction

$$P_{1y} = -f_0 \frac{\partial \zeta}{\partial t} \frac{\partial \zeta}{\partial y}. \tag{2.1.23}$$

Now $\partial \zeta / \partial x$ and $\partial \zeta / \partial y$ are the components of the two-dimensional gradient vector

$$\nabla_2 \zeta \equiv \mathbf{i} \frac{\partial \zeta}{\partial x} + \mathbf{j} \frac{\partial \zeta}{\partial y} \tag{2.1.24}$$

so that the vector

$$\mathbf{P}_1 = -f_0 \frac{\partial \zeta}{\partial t} \nabla_2 \zeta \tag{2.1.25}$$

expresses both the direction in which the energy flow is taking place and the energy per unit time passing across a line of unit width perpendicular to the direction of flow. The application of these results to the plane wave (2.1.9) is made in Prob. 2.1.1.

Problems

2.1.1   Compute the total energy density $E_1 = K_1 + V_1$ for the wave (2.1.9). Show that the magnitude of the rate of energy flow (2.1.25) is the total energy density times the wave velocity c_m.

2.1.2   A distributed force per unit area $F(x,y,t)$ acts on a flat stretched membrane in a direction normal to its surface. Show how to modify the wave equation (2.1.1) to include the presence of this force density.

2.1.3   Find the form of the wave equation (2.1.1) with respect to new axes x' and y', which are rotated an angle θ with respect to the xy axes. *Hint:* $x' = x \cos\theta + y \sin\theta$, $y' = -x \sin\theta + y \cos\theta$.

2.1.4   Prove that the two-dimensional gradient $\nabla_2 \zeta(x,y)$ of the surface $\zeta = \zeta(x,y)$ at any point has a magnitude giving the maximum slope of the surface at that point and a direction giving the direction in which the maximum slope occurs.

2.1.5   From the arguments leading to the wave equation (2.1.1), show that in *static equilibrium* the displacement $\zeta(x,y)$ of an elastic membrane obeys the two-dimensional *Laplace equation*

$$\nabla_2^2 \zeta \equiv \frac{\partial^2 \zeta}{\partial x^2} + \frac{\partial^2 \zeta}{\partial y^2} = 0.$$

[Thus, given a uniformly stretched membrane, one may place blocks or clamps so as to fix some desired displacement along a closed boundary contour of arbitrary geometry. The apparatus is then an analog computer for solving Laplace's equation (in two dimensions) for geometries too complicated to handle analytically. This scheme is particularly useful in electrostatic problems, where the displacement ζ represents the electrostatic potential V. Moreover, the trajectories of electrons in vacuum tubes can be plotted by rolling small ball bearings on such a model made from a rubber membrane.]

★2.1.6 Show that

$$g_1 = -\sigma_0 \frac{\partial \zeta}{\partial t} \, \nabla_2 \zeta = c_m{}^2 \mathbf{P}_1 \tag{2.1.26}$$

is the momentum density associated with a transverse wave on a membrane. *Hint:* Generalize the results of Sec. 1.11.

2.2   Standing Waves on a Rectangular Membrane

The vibrations that can exist on a membrane having a rectangular shape afford a simple introduction to normal-mode, or eigenvalue, problems in a space of more than one dimension. We suppose that the membrane is attached to fixed supports along the x and y coordinate axes and along the lines $x = a$ and $y = b$. Accordingly, we need a solution of (2.1.5) and (2.1.7) that vanishes on the four straight boundaries. The tentative solution

$$\zeta(x,y,t) = A \, \sin\kappa_x x \, \sin\kappa_y y \, \cos\omega t \tag{2.2.1}$$

satisfies the condition of no displacement at the edges along the coordinate axes. If, in addition,

$$\begin{aligned} \kappa_x a = l\pi \qquad & l = 1, 2, 3, \ldots \\ \kappa_y b = m\pi \qquad & m = 1, 2, 3, \ldots, \end{aligned} \tag{2.2.2}$$

then the condition of no displacement is also satisfied along the other two edges. In view of (2.1.8), the frequencies of the various normal-mode vibrations are therefore

$$\omega_{lm} = \pi c_m \left[\left(\frac{l}{a}\right)^2 + \left(\frac{m}{b}\right)^2 \right]^{1/2}. \tag{2.2.3}$$

If a^2 and b^2 are incommensurable (meaning that a^2/b^2 cannot be expressed as the ratio of two integers), the frequencies are all *distinct*. Otherwise there can exist two or more identical frequencies for different pairs of *mode numbers l, m*. In such cases the normal-mode vibrations are said to be *degenerate*.

For a particular normal-mode vibration, the *nodal lines* on the membrane consist of two sets of straight lines parallel to the coordinate axes. They evidently divide the rectangular membrane into l by m small rectangles. Except for obvious alternations in the phase of the motion in adjacent rectangles, the motion of the membrane in each small rectangle is the same.

If no degeneracy exists and the membrane is vibrating at one of the normal-mode frequencies (2.2.3), the nodal pattern is that just described. If, however, there exists a degeneracy, such that two or more normal-mode vibrations are excited at the same frequency, the nodal pattern depends on the relative amplitudes and phases of the normal-mode vibrations that are present. Some of the nodal patterns of a square membrane, for which many of the normal-mode vibrations are degenerate, are dealt with in Prob. 2.2.1.

The most general vibration of the rectangular membrane consists of a superposition of the normal-mode vibrations (2.2.1), with both a cosine and sine time factor. Thus, corresponding to (1.6.1) for the most general vibration of a string segment, we now have

$$\zeta(x,y,t) = \sum_{l=1}^{\infty} \sum_{m=1}^{\infty} \sin\frac{l\pi x}{a} \sin\frac{m\pi y}{b} (A_{lm} \cos\omega_{lm}t + B_{lm} \sin\omega_{lm}t), \qquad (2.2.4)$$

where the ω_{lm} are given by (2.2.3). The amplitude coefficients A_{lm} and B_{lm} can be found from the initial displacement $\zeta = \zeta_0(x,y)$ and the initial velocity $\partial\zeta/\partial t = \dot{\zeta}_0(x,y)$ of the membrane by an extension of the procedure described in Sec. 1.6 for the one-dimensional case of the string. Here we find that

$$A_{lm} = \frac{4}{ab} \int_0^a \int_0^b \zeta_0 \sin\frac{l\pi x}{a} \sin\frac{m\pi y}{b}\, dx\, dy$$
$$B_{lm} = \frac{4}{ab\omega_{lm}} \int_0^a \int_0^b \dot{\zeta}_0 \sin\frac{l\pi x}{a} \sin\frac{m\pi y}{b}\, dx\, dy. \qquad (2.2.5)$$

Problems

2.2.1 Investigate the nodal pattern of a square membrane for the degenerate (12)–(21) modes that vibrate at a common frequency ω_{12}. Assume that the vibration has the form

$$\zeta(x,y,t) = \left(A \sin\frac{2\pi x}{a} \sin\frac{\pi y}{a} + B \sin\frac{\pi x}{a} \sin\frac{2\pi y}{a} \right) \cos\omega_{12}t.$$

Find the nodal lines for the four special cases $A = 0$, $B = 0$, $A - B = 0$, $A + B = 0$ and for some other case, such as $A = 2B$. Extend the analysis to one or more higher sets of degenerate modes, such as the (13)–(31) modes.

2.2.2 Investigate the normal modes of a triangular membrane consisting of half of a square. Find several of the lowest normal-mode frequencies and the corresponding nodal patterns. *Hint:* Make use of the results of Prob. 2.2.1.

2.2.3 Establish the expressions (2.2.5) for the amplitude coefficients for the most general vibration of a rectangular membrane (2.2.4).

2.2.4   Find expressions for the average kinetic, potential, and total energies of the vibration (2.2.4).

2.3   *Standing Waves on a Circular Membrane*

We have investigated, in Chap. 1, the normal-mode vibrations of a string segment with fixed ends and, in the preceding section, of a rectangular membrane with fixed boundaries. In each case, we based the analysis on a partial differential equation that described wave motion on the structure. We found that the method of variable separation leads to certain functions of position and of time, from which we are able to construct an infinity of normal-mode functions that satisfy the spatial boundary conditions of the problem, with sinusoidal time factors that can be adjusted to satisfy initial conditions. In particular, we found that the spatial boundary conditions inevitably lead to a discrete set of values (eigenvalues) of the separation constants, telling us the wave numbers and the frequencies of the normal-mode vibrations.

In the case of the two-dimensional rectangular membrane, we are able to satisfy the boundary condition of no displacement on all four edges of the rectangle because the edges coincide with the lines of constant x, or of constant y. That is, the method of variable separation, using cartesian coordinates, automatically gives functions that are suited to fitting boundary conditions along these coordinate lines.

To fit the boundary condition of no displacement on other than rectangular boundaries requires the use of an appropriate two-dimensional *orthogonal curvilinear coordinate system* such that the boundary of the membrane coincides with coordinate lines in this system. Furthermore, it is necessary that the variables of the wave equation be separable in the new system. It turns out that the choice of curvilinear coordinate systems is severely limited, and it is impossible, except in an approximate way, to analyze the vibrations of a membrane having an arbitrarily shaped boundary. A circular boundary, however, is a coordinate line of a polar coordinate system, and, as we shall see, it is possible to separate the variables of the wave equation in polar coordinates. The solution of one of the separated equations consists of Bessel functions; it is primarily to introduce these functions that we have chosen to investigate the vibrations on a circular membrane as a second example of two-dimensional normal-mode vibrations.

Our first task is to change the wave equation (2.1.1) from xy coordinates to polar coordinates r and θ, with the origin at the center of a circular membrane of radius a. According to Prob. 2.3.2, the wave equation then becomes

$$\frac{\partial^2 \zeta}{\partial r^2} + \frac{1}{r}\frac{\partial \zeta}{\partial r} + \frac{1}{r^2}\frac{\partial^2 \zeta}{\partial \theta^2} = \frac{1}{c_m{}^2}\frac{\partial^2 \zeta}{\partial t^2}. \tag{2.3.1}$$

We next assume that (2.3.1) has a solution of the form

$$\zeta(r,\theta,t) = R(r) \cdot \Theta(\theta) \cdot T(t) \tag{2.3.2}$$

and find, after multiplication through by $c_m{}^2/R\Theta T$, that

$$\frac{c_m{}^2}{R}\left(\frac{d^2R}{dr^2} + \frac{1}{r}\frac{dR}{dr}\right) + \frac{c_m{}^2}{r^2\Theta}\frac{d^2\Theta}{d\theta^2} = \frac{1}{T}\frac{d^2T}{dt^2} = -\omega^2. \tag{2.3.3}$$

As before, we have introduced the separation constant $-\omega^2$, and again we find the differential equation (2.1.5) for the time function. The spatial part of (2.3.3) can now be rearranged to read

$$\frac{r^2}{R}\left(\frac{d^2R}{dr^2} + \frac{1}{r}\frac{dR}{dr}\right) + \left(\frac{\omega}{c_m}\right)^2 r^2 = -\frac{1}{\Theta}\frac{d^2\Theta}{d\theta^2} = m^2, \tag{2.3.4}$$

where we have chosen to denote the second separation constant by m^2. Equation (2.3.4) thus separates into the two ordinary differential equations

$$\frac{d^2\Theta}{d\theta^2} + m^2\Theta = 0 \tag{2.3.5}$$

$$\frac{d^2R}{dr^2} + \frac{1}{r}\frac{dR}{dr} + \left[\left(\frac{\omega}{c_m}\right)^2 - \frac{m^2}{r^2}\right]R = 0. \tag{2.3.6}$$

The equation for $\Theta(\theta)$ has the independent complex solutions $e^{\pm im\theta}$, or the independent real solutions $\cos m\theta$ and $\sin m\theta$. We see that the separation constant m must be either zero, which makes Θ a constant, or a (positive) integer, which makes Θ a single-valued function of θ. In effect, we are making use of a boundary condition along a radial line on the circular membrane, namely, that the displacement and its θ derivative be continuous functions across this hypothetical boundary. For the vibrations of a sector-shaped membrane, m could have other than integral values.

The differential equation for $R(r)$ can be put in the standard form

$$\frac{d^2R}{du^2} + \frac{1}{u}\frac{dR}{du} + \left(1 - \frac{m^2}{u^2}\right)R = 0 \tag{2.3.7}$$

by changing to the dimensionless independent variable

$$u = \kappa r = \frac{\omega}{c_m}\,r. \tag{2.3.8}$$

Equation (2.3.7) is known as *Bessel's equation*. Since it is of second order, it must have two linearly independent solutions for each value of the parameter m, which in the present instance we know to be a positive integer or zero. The two solutions of (2.3.7) are normally designated by $J_m(u)$ and $N_m(u)$. They are tabulated functions, just as $\cos\theta$ and $\sin\theta$ are two independent tabulated functions.†

The solution $J_m(u)$ is called the Bessel function (of the first kind) of order m, and it remains finite over the entire range of u from 0 to ∞. The other solution, $N_m(u)$, is called the Neumann function (or the Bessel function of the second kind) of order m, and it becomes infinite at $u = 0$ though it is finite elsewhere. Since $N_m(u)$ cannot represent a possible displacement of a circular membrane, we need only examine the properties of the functions $J_m(u)$. Neumann functions, however, are needed in discussing problems with other boundary conditions, such as the vibrations of an annular membrane.

The function J_m can be expressed by the infinite series

$$J_m(u) = \frac{u^m}{2^m m!}\left[1 - \frac{u^2}{1!2^2(m+1)} + \frac{u^4}{2!2^4(m+1)(m+2)} - \cdots\right],$$

$$(2.3.9)$$

found by assuming a series solution for (2.3.7) expanded about the origin. The numerical coefficient $1/2^m m!$ is purely conventional. A plot of $J_m(u)$ for $m = 0$, 1, 2 is given in Fig. 2.3.1. All the Bessel functions but J_0 vanish at the origin,

† For a brief account of Bessel functions, see M. L. Boas, "Mathematical Methods in the Physical Sciences," pp. 559–577, John Wiley & Sons, Inc., New York, 1966.

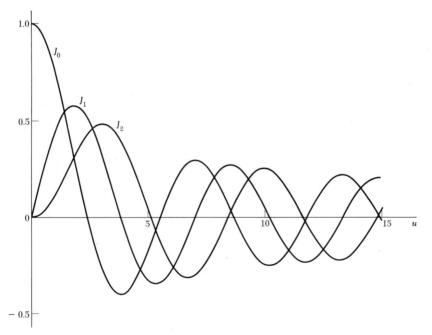

Fig. 2.3.1   Bessel functions of the first kind of order zero, one, and two.

TABLE 2.1   The nth Roots of $J_m(u) = 0$

n \ m	0	1	2	3
1	2.405	3.832	5.136	6.380
2	5.520	7.016	8.417	9.761
3	8.654	10.173	11.620	13.015
4	11.792	13.324	14.796	16.223

and $J_0(0) = 1$. Each Bessel function is seen to alternate in sign with increasing u, with its amplitude slowly dropping off (ultimately as $u^{-1/2}$), and with the spacing of its zeros becoming more nearly uniform (approaching π). The behavior reminds one of a damped sine wave. A few of the roots of $J_m(u) = 0$ are listed in Table 2.1. The roots of Bessel functions of adjacent orders interlace each other.

The Bessel functions obey *recursion relations*, such as

$$J_{m+1}(u) = \frac{2m}{u} J_m(u) - J_{m-1}(u) \tag{2.3.10}$$

$$\frac{dJ_m(u)}{du} = -\frac{m}{u} J_m(u) + J_{m-1}(u). \tag{2.3.11}$$

These relations may be established directly from the infinite series (2.3.9). They show that it is necessary to have numerical tables for only J_0 and J_1. The values of all higher-order Bessel functions, as well as all first derivatives, can then be calculated from the recursion relations.

Let us now see what the normal-mode vibrations of a circular membrane are like. When $m = 0$, Θ is independent of θ, so that

$$\zeta(r,t) = A\, J_0(\kappa r)\, \cos\omega t \tag{2.3.12}$$

is a possible solution of the wave equation. To satisfy the boundary condition that $\zeta(r,t) = 0$ at $r = a$, the value of $\kappa = \omega/c_m$ must be chosen to make

$$\kappa_{0n} a = u_{0n} \qquad n = 1, 2, 3, \ldots, \tag{2.3.13}$$

where u_{0n} is one of the roots of $J_0(u) = 0$, some of which are listed in the first column of Table 2.1. The frequencies of these radially symmetric $(0,n)$ modes are therefore

$$\omega_{0n} = \frac{c_m u_{0n}}{a}, \tag{2.3.14}$$

the lowest frequency being $\omega_{01} = 2.405(c_m/a)$.

When $m = 0$ and $n = 2$, there is a single nodal circle at the radius

$$r = \frac{u_{01}}{\kappa_{02}} = \frac{u_{01}}{u_{02}} a. \qquad (2.3.15)$$

There are evidently $n - 1$ nodal circles when the nth root of $J_0(u) = 0$ coincides with the fixed boundary. They are at the radii

$$r = \frac{u_{0p}}{u_{0n}} a \qquad p = 1, 2, \ldots, n - 1. \qquad (2.3.16)$$

Next suppose that $m = 1$ and that we choose $\cos\theta$ for the Θ function. A solution of the wave equation is then

$$\zeta(r,\theta,t) = A\, J_1(\kappa r)\, \cos\theta \cos\omega t. \qquad (2.3.17)$$

To satisfy the boundary condition at $r = a$ we must now have

$$\kappa_{1n}a = u_{1n} \qquad n = 1, 2, 3, \ldots, \qquad (2.3.18)$$

where u_{1n} are the roots of $J_1(u) = 0$ appearing in the second column of Table 2.1. The frequencies of these normal modes are evidently

$$\omega_{1n} = \frac{c_m u_{1n}}{a}. \qquad (2.3.19)$$

A nodal diameter exists at the angles $\theta = \pi/2,\ 3\pi/2$, as well as nodal circles at the radii

$$r = \frac{u_{1p}}{u_{1n}} a \qquad p = 1, 2, \ldots, n - 1. \qquad (2.3.20)$$

We could just as well have used $\sin\theta$ for $\Theta(\theta)$, or any linear combination of $\cos\theta$ and $\sin\theta$. That is, the nodal diameter can have any orientation, and its orientation depends on how the vibration is set up. We thus have a type of degeneracy which can be removed by stabilizing the orientation of the diameter by applying a *constraint* to the membrane at some point other than the center. The constraint forces the nodal diameter to pass through that point. We lose no generality by choosing $\cos\theta$ so long as we are free to pick the θ origin (polar axis) appropriately.

Our discussion of the various normal modes of a circular membrane can be readily extended to arbitrary values of m, with the outer boundary at $r = a$ such that $\kappa_{mn}a = u_{mn}$, where u_{mn} is the nth root of $J_m(u) = 0$. Evidently there are m nodal diameters, and $n - 1$ nodal circles. Figure 2.3.2 shows some of the possible modes of vibration of the circular membrane for small values of m and n.

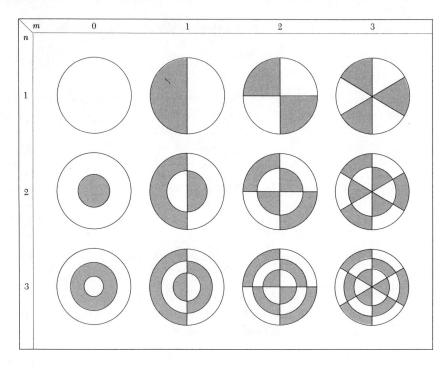

Fig. 2.3.2   Normal modes of the circular membrane.

Problems

2.3.1   Explain, in physical terms, why the method of separation of variables applied to the wave equations that we have considered always leads to a *sinusoidal* time function, though the spatial functions may take a variety of forms.

2.3.2   Use the relations $x = r \cos\theta$, $y = r \sin\theta$ connecting rectangular and polar coordinates to transform the wave equation in cartesian coordinates (2.1.1) to that in polar coordinates (2.3.1).

2.3.3   Assume a trial solution $R(u) = u^p \sum\limits_{0}^{\infty} a_n u^n$ for the Bessel equation (2.3.7) and establish the series solution (2.3.9), except for the arbitrary numerical factor $1/2^m m!$.

2.3.4   Establish the recursion relations (2.3.10) and (2.3.11) from the series (2.3.9) or directly from the differential equation (2.3.7). Note the special case $dJ_0(u)/du = -J_1(u)$. Can you develop a recursion relation for the second derivative, $d^2 J_m(u)/du^2$?

2.3.5   Show how to find the normal-mode frequencies of a membrane in the form of a semi-circle with fixed boundaries along its edges.

2.3.6 If the frequency of the (01) mode of circular membrane in Fig. 2.3.2 is 100 Hz, calculate the frequencies of the other modes shown in the figure.

2.4 Interference Phenomena with Plane Traveling Waves

In the one-dimensional case of transverse waves on a string, we found that the superposition of two sinusoidal waves of identical frequency and amplitude traveling in opposite directions gives rise to a pattern of standing waves. We were then able to put fixed boundaries at any pair of nodes and in this way arrive at a description of the vibration of a string segment with fixed ends. In the two-dimensional case of transverse waves on a membrane, the interference of two sinusoidal plane waves of the same frequency and amplitude is considerably richer in phenomena, since now the waves can travel in different directions across the membrane. To avoid the complicating effect of boundaries, let us assume, for the time being, that the membrane is of indefinitely great extent.

Let us suppose that, at a great distance, a line source of sinusoidal waves of frequency ω produces the plane wave

$$\zeta_1(x,y,t) = \tfrac{1}{2}A\ e^{i(\kappa_1 \cdot r - \omega t)}, \tag{2.4.1}$$

which travels across the membrane and continues on toward infinity in its direction of travel. At a great distance in some other direction, a second line source produces the plane wave

$$\zeta_2(x,y,t) = \tfrac{1}{2}A\ e^{i(\kappa_2 \cdot r - \omega t)}, \tag{2.4.2}$$

having the same frequency and amplitude. Each wave travels at the wave velocity

$$c_m = \frac{\omega}{|\kappa_1|} = \frac{\omega}{|\kappa_2|} = \left(\frac{f_0}{\sigma_0}\right)^{1/2} \tag{2.4.3}$$

characteristic of the membrane, with the direction of travel given by the vector wave number κ_1 or κ_2, as the case may be. The two waves exist independently of each other and do not interact in any way since they are of small amplitude and consequently are described by linear equations. The angle α between their directions of travel may be found from

$$\cos\alpha = \frac{\kappa_1 \cdot \kappa_2}{\kappa^2}, \tag{2.4.4}$$

where $\kappa = |\kappa_1| = |\kappa_2|$.

It is convenient to choose the x axis such that it bisects the angle α between the two waves. With respect to this axis, and the related y axis, the vector wave numbers may then be written

$$\kappa_1 = \mathbf{i}\kappa_x + \mathbf{j}\kappa_y$$
$$\kappa_2 = \mathbf{i}\kappa_x - \mathbf{j}\kappa_y, \tag{2.4.5}$$

with

$$\frac{\omega^2}{c_m{}^2} = \kappa^2 = \kappa_x{}^2 + \kappa_y{}^2. \tag{2.4.6}$$

Evidently the two waves travel in directions that make angles of $\alpha/2 = \tan^{-1}(\kappa_y/\kappa_x)$ with respect to the x axis.

We are now prepared to discuss the combined wave disturbance. Superposing (2.4.1) and (2.4.2) and writing out the wave numbers by (2.4.5), we have

$$\zeta(x,y,t) = \zeta_1 + \zeta_2 = A \frac{e^{i\kappa_y y} + e^{-i\kappa_y y}}{2} e^{i(\kappa_x x - \omega t)} = A \cos\kappa_y y \, e^{i(\kappa_x x - \omega t)}. \tag{2.4.7}$$

Clearly this disturbance is a sinusoidal wave traveling in the $+x$ direction. However it differs from either of the component waves in three important respects: (1) it travels in a new direction, bisecting the directions of the component waves, (2) its amplitude is *modulated* in the y direction by the factor $\cos\kappa_y y$, (3) the wave velocity is increased to

$$c_x = \frac{\omega}{\kappa_x} = \frac{\kappa c_m}{\kappa_x} = \frac{c_m}{\cos\frac{1}{2}\alpha}, \tag{2.4.8}$$

which depends on both the properties of the membrane c_m and the angle α between the wave-number vectors of the two component waves. A snapshot of the combined wave (Fig. 2.4.1) shows a periodic pattern of alternating hollows and hills in both the x and y directions. However, a movie shows that this disturbance is a traveling wave in the x direction but a standing wave in the y direction.

Along the set of lines

$$y_n = (n - \tfrac{1}{2})\frac{\pi}{\kappa_y} \qquad n = 0, \pm 1, \pm 2, \dots \tag{2.4.9}$$

parallel to the x axis, the cosine factor in (2.4.7) vanishes for all values of x and t. These constitute nodal lines in the moving interference pattern. We can discover some interesting aspects of wave behavior if we place a rigid boundary along one of the nodal lines. Such a boundary in effect divides the membrane into two regions that have no communication with each other. Nevertheless the boundary in no way disturbs the wave pattern that existed before it was put in place. Our interpretation of the pattern with the boundary in place is quite different, however, for we can remove the membrane on one side of the boundary without changing the interference pattern on the other side at all! What

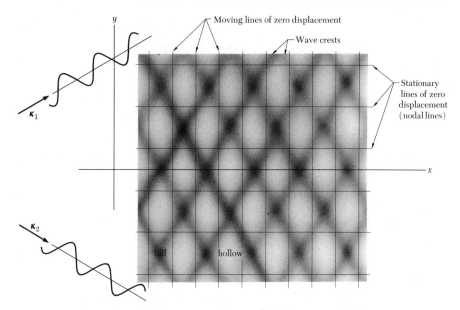

Fig. 2.4.1   Interference of two plane waves on a membrane.

we now have is a single distant line source that is sending plane sinusoidal waves toward a rigid straight boundary along which the wave amplitude is necessarily zero at all times. As a result the waves are being completely reflected, with the angle of reflection equal to the angle of incidence. The resulting interference pattern is precisely the one produced by the two line sources in the absence of the boundary. We can think of one source as being the *image* of the other in the rigid boundary and the boundary as being a *plane mirror*.

Let us next place rigid boundaries along a *pair* of the nodal lines and investigate the propagation of waves in the resulting channel, or *waveguide*. In so doing we can forget about the membrane outside the channel since the exterior part of the membrane is no longer relevant. The source of the waves in the channel can continue to be the portions of the two distant line sources that lie within the channel, or the membrane at some position x_0 in the channel can be given the motion specified by (2.4.7). The waves continue indefinitely toward infinity in the channel, or one can assume that some sort of sink absorbs them with no reflection.

To be definite, let us place rigid boundaries along the two nodal lines $y_0 = -\pi/2\kappa_y$ and $y_n = (n - \frac{1}{2})\pi/\kappa_y$, with n a positive integer. These lines have the separation $b = y_n - y_0 = n\pi/\kappa_y$, so that κ_y has the value

$$\kappa_y = \frac{n\pi}{b}. \tag{2.4.10}$$

If we regard the channel width b as fixed, then (2.4.10) assigns a set of possible values to the y component of the wave numbers κ_1 and κ_2 (2.4.5). Their component κ_x can then have only those values permitted by (2.4.6),

$$\frac{\omega^2}{c_m^2} = \kappa^2 = \kappa_x^2 + \left(\frac{n\pi}{b}\right)^2. \tag{2.4.11}$$

Evidently for κ_x to be real, it is necessary that $\kappa = \omega/c_m$ be greater than $n\pi/b$. The nature of this condition becomes clearer if we express (2.4.11) in terms of wavelength,

$$\frac{1}{\lambda^2} = \frac{1}{\lambda_x^2} + \left(\frac{n}{2b}\right)^2. \tag{2.4.12}$$

Hence for waves with $n - 1$ nodal lines to propagate in the channel it is necessary that the "open-membrane" wavelength satisfy the condition

$$\lambda < \frac{2b}{n} \equiv (\lambda_c)_n, \tag{2.4.13}$$

where $(\lambda_c)_n$ is termed the *cutoff wavelength* of the nth mode of propagation. Evidently the channel width must be greater than n half-wavelengths $\lambda/2$. Now if λ lies in the range

$$b < \lambda < 2b, \tag{2.4.14}$$

only waves with $n = 1$ can propagate along the channel. This mode is called the *dominant mode* in waveguide theory.

The velocity of the wave

$$\zeta(x,y,t) = A_n \cos \frac{n\pi y}{b} \, e^{i(\kappa_x x - \omega t)} \tag{2.4.15}$$

propagating in the nth mode in a channel of width b is

$$c_x = \frac{\omega}{\kappa_x} = \frac{\kappa c_m}{[\kappa^2 - (n\pi/b)^2]^{1/2}} \tag{2.4.16}$$

when expressed in terms of the wave number $\kappa = \omega/c_m$. In terms of wavelengths, the wave velocity becomes

$$c_x = \frac{c_m}{[1 - (\lambda/\lambda_c)^2]^{1/2}}, \tag{2.4.17}$$

where $\lambda_c = 2b/n$ is the cutoff wavelength (2.4.13); in terms of frequency

$$c_x = \frac{c_m}{[1 - (\omega_c/\omega)^2]^{1/2}}, \tag{2.4.18}$$

where

$$\omega_c \equiv (\omega_c)_n \equiv \frac{n\pi c_m}{b} \tag{2.4.19}$$

is the corresponding cutoff frequency. Thus, the wave velocity depends both on mode number and frequency and is always greater than c_m, which it approaches when $\omega \gg \omega_c$. At the other limit, the wave velocity becomes infinite as $\omega \to \omega_c$.

When the frequency is less than the cutoff frequency for a particular mode of propagation, the wave number κ_x becomes pure imaginary

$$\kappa_x = \left[\kappa^2 - \left(\frac{n\pi}{b}\right)^2 \right]^{1/2} = i\left[\left(\frac{n\pi}{b}\right)^2 - \kappa^2 \right]^{1/2} \tag{2.4.20}$$

since now $\kappa < n\pi/b$. The expression for the wave then becomes

$$\zeta(x,y,t) = A_n \cos\frac{n\pi y}{b} e^{-(n^2\pi^2/b^2 - \kappa^2)^{1/2}x} e^{-i\omega t}. \tag{2.4.21}$$

The wave disturbance is no longer periodic in x but decreases exponentially with x. Such an *evanescent* wave quickly dies out with increasing distance from its source.

We can see now what happens if we endeavor to send a wave down the channel in the dominant mode [that is, $n = 1$ and $(\omega_c)_1 < \omega < (\omega_c)_2$] by exciting it in some way that is not perfectly consistent with the spatial membrane motion implied by

$$\zeta(x,y,t) = A_1 \cos\frac{\pi y}{b} e^{i[(\kappa^2 - \pi^2/b^2)^{1/2}x - \omega t]}. \tag{2.4.22}$$

Not only is the wave (2.4.22) excited, but so are an infinite number of evanescent waves with mode numbers $n = 2, 3, \ldots$,

$$\zeta(x,y,t) = A_n \cos\frac{n\pi y}{b} e^{-(n^2\pi^2/b^2 - \kappa^2)^{1/2}x} e^{-i\omega t}. \tag{2.4.23}$$

The evanescent waves, however, rapidly die out in space, leaving only the wave in the dominant mode continuing down the channel.

Although transverse waves on membranes have little practical use other than in musical instruments of the drum family, they are simple to discuss theoretically and have given us a chance to introduce a number of important ideas that pertain to wave motion in general. More specifically, the qualitative ideas developed in this section, and indeed most of the quantitative analysis, are directly applicable to the propagation of electromagnetic waves (microwaves) in conducting hollow-pipe waveguides (Sec. 8.7). This latter problem, of great technological importance, is more complicated in its formal details since the wave motion involves vector electric and magnetic fields in three dimensions. For instance, two index numbers are needed to specify the three-dimensional electromagnetic modes, in contrast to the single mode index n needed for our two-dimensional membrane waves. However, the basic features of traveling-wave propagation in narrow channels or waveguides—discrete modes, cutoff wavelengths, and mode- and frequency-dependent velocities—are identical for the two cases.

Problems

2.4.1   The wave (2.4.15) is traveling down a channel of width b, and an identical wave is traveling in the opposite direction. Show how to obtain the various standing waves discussed in Sec. 2.2 from this model. Interpret this result in terms of four traveling waves having a wave number κ.

2.4.2   A plane wave of the form $\zeta(x,y,t) = F(y)e^{i(\kappa_x x - \omega t)}$ is traveling in the x direction on a membrane. Investigate what restrictions the wave equation (2.1.1) puts on the form that the function $F(y)$ may take.

★2.4.3   Two membranes have different areal mass densities, σ_1 and σ_2. They are joined together along a straight line, the x axis, and are stretched to a common surface tension f_0. Investigate the reflection and refraction of plane waves incident obliquely on the boundary.

three

Introduction to the Theory of Elasticity

A study of elastic waves and vibrations in fluid and solid media must start with a brief mathematical description of strain, stress, and the relation between the two under static conditions. An elastic medium is *homogeneous* when its physical properties are not a function of position. It is *isotropic* when the properties are independent of direction. It is said to obey Hooke's law when the distortion, or *strain*, in the medium is linearly related to the magnitude of the applied force, or *stress*, that causes the distortion. We shall find that a fluid is described by a single elastic constant relating stress and strain, whereas a linear homogeneous isotropic solid medium is described by two independent constants. In addition to developing an elementary description of stress and strain in an extended medium, we need to examine the static elastic deformation of a few simple structures, such as the stretching, bending, and twisting of a rod. In

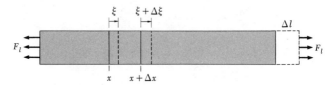

Fig. 3.1.1   The elongation of a rod. The lateral shrinkage specified by Poisson's ratio is not indicated.

Chap. 4 we investigate the traveling and standing waves that can occur on such structures. A more detailed account of stress and strain as tensor quantities is given in Chap. 7.

3.1   The Elongation of a Rod

Suppose that a rod of length l and constant cross-sectional area S is stretched to a length $l + \Delta l$ by opposing axial forces F_l applied uniformly over its ends, as in Fig. 3.1.1. The *tension stress*, or simply the *tension*, acting to stretch the rod is

$$f_l \equiv \frac{F_l}{S}. \tag{3.1.1}$$

The fractional increase in length of the rod,

$$\epsilon_l \equiv \frac{\Delta l}{l}, \tag{3.1.2}$$

is termed the *tension strain*, or *extension*, in the direction of the rod axis. Normally ϵ_l is very small, much less than unity. If we designate the rod axis to be the x axis and let ξ be the displacement of the rod at any position x due to the strain, the extension (3.1.2) can be written in the form

$$\epsilon_l = \frac{\Delta \xi}{\Delta x} \longrightarrow \frac{\partial \xi}{\partial x}, \tag{3.1.3}$$

where $\Delta \xi$ is the amount an element Δx has been stretched. Although for a uniform rod the extension ϵ_l is independent of position, the definition (3.1.3) continues to hold when the extension in the x direction varies with position. The extensions in the y and z directions may be defined similarly.

Hooke's law applied to the elongation of a uniform rod by axial tension states that

$$f_l = Y \epsilon_l, \tag{3.1.4}$$

where Y is known as *Young's modulus*. Representative values of Y, which has the dimensions of a force per unit area, are given in Table 3.1.

TABLE 3.1   Typical Elastic Constants at Room Temperature†

Material	Young's modulus Y, N/m²	Shear modulus μ, N/m²	Bulk modulus B, N/m²	Poisson's ratio σ	Density ρ_0, kg/m³
Aluminum	7.1×10^{10}	2.65×10^{10}	7.4×10^{10}	0.34	2.70×10^3
Brass	10.4	3.8	13	0.37	8.53
Copper	12.8	4.7	15	0.36	8.90
Iron	20	7.8	16	0.29	7.85
Nickel	21	8.0	18	0.31	8.9
Tungsten	36	13.4	37	0.34	19
Fused silica	7.3	3.1	3.7	0.17	2.2
Pyrex glass	6.2	2.5	4.0	0.24	2.32
Lucite	0.40	0.14	0.66	0.4	1.182
Water			2.2		1.0
Mercury			29		13.6

† Adapted from Dwight E. Gray (ed.), "American Institute of Physics Handbook," 2d ed., McGraw-Hill Book Company, 1963, with the values of B for the solids computed from Y and σ. Elastic constants for actual samples of the metals may differ significantly from those given, depending on the thermal and mechanical history of the sample and on the presence of small amounts of alloying elements.

A strictly linear relation between stress and strain, such as that expressed by (3.1.4), is a good approximation to the behavior of many materials. Clearly the tension must be kept below some limit, the so-called elastic limit, where a significant permanent distortion sets in. Even for smaller values of maximum tension, the rod, when cyclically loaded and unloaded, is likely to exhibit *elastic hysteresis*. In such event a plot of the stress-strain relation is not precisely a straight line but a slightly open loop whose area represents an irreversible loss of energy. Elastic vibrations die out rapidly in a material exhibiting appreciable elastic hysteresis. We shall ignore the effect of elastic hysteresis in treating wave phenomena and deal with idealized materials for which Hooke's law holds in a strict sense.

There is one additional important feature involved in the elongation of a rod. When a rod is stretched, it is found to shrink laterally in proportion to its increase in length. If we suppose the rod to be made of isotropic material, so that the lateral extension is independent of orientation, the lateral extension is then

$$\epsilon_D = \frac{\Delta D}{D},$$ (3.1.5)

where D is any lateral dimension of the rod. The negative ratio of the lateral extension to the axial extension,

$$\sigma \equiv -\frac{\epsilon_D}{\epsilon_l},$$ (3.1.6)

is known as *Poisson's ratio*. Its value is one of the elastic constants of an isotropic material. Representative values of σ are also given in Table 3.1. For most materials, Poisson's ratio is of the order of $\frac{1}{3}$.

It is useful to introduce a more general notation, in order that we may consider the simultaneous application of tension stresses in more than one dimension. We may generalize the tension stress (3.1.1) with the notation

$$f_l \rightarrow f_{xx}, \ f_{yy}, \ f_{zz}. \tag{3.1.7}$$

The first subscript designates the normal to the plane across which the force is applied, and the second subscript designates the direction in which the force acts.† A *pressure* is simply a negative tension. A *hydrostatic pressure* p is equivalent to three equal negative tensions in the three coordinate directions,

$$f_{xx} = f_{yy} = f_{zz} = -p. \tag{3.1.8}$$

Similarly we may generalize the tension strain (3.1.3) by making the definitions

$$\epsilon_{xx} \equiv \frac{\partial \xi}{\partial x} \qquad \epsilon_{yy} \equiv \frac{\partial \eta}{\partial y} \qquad \epsilon_{zz} \equiv \frac{\partial \zeta}{\partial z}. \tag{3.1.9}$$

In terms of this more general notation, the simple stress-strain relation (3.1.4) is written

$$f_{xx} = Y\epsilon_{xx} \qquad f_{yy} = f_{zz} = 0, \tag{3.1.10}$$

and the lateral strains (3.1.5) are

$$\epsilon_{yy} = \epsilon_{zz} = -\sigma\epsilon_{xx}. \tag{3.1.11}$$

The virtue of the double-subscript notation is that a general set of stress-strain equations for an isotropic elastic medium can be written down at this point. All we need are the results just obtained and the principle of superposition. Thus we may write that

$$\epsilon_{xx} = \frac{1}{Y} f_{xx} - \frac{\sigma}{Y} f_{yy} - \frac{\sigma}{Y} f_{zz}. \tag{3.1.12}$$

For if $f_{yy} = f_{zz} = 0$, then (3.1.12) reduces to (3.1.4), defining Young's modulus. However, if $f_{xx} = f_{zz} = 0, f_{yy}$ alone causes the extension $\epsilon_{yy} = f_{yy}/Y$, and (3.1.12) reduces to (3.1.6), defining Poisson's ratio. An analogous situation holds if $f_{xx} = f_{yy} = 0$. When all three tensions are present, the total extension in the x direction is that given by (3.1.12), namely, the sum of the three extensions

† In Sec. 3.3 and again in Chap. 7, we shall need the stress and strain components with mixed subscripts, which describe *shear*. The shearing strain components involve the mixed derivatives such as $\partial \eta / \partial x$ and $\partial \xi / \partial y$.

produced by the tensions acting singly. The complete set of elastic equations is therefore

$$\epsilon_{xx} = + \frac{1}{Y} f_{xx} - \frac{\sigma}{Y} f_{yy} - \frac{\sigma}{Y} f_{zz}$$

$$\epsilon_{yy} = - \frac{\sigma}{Y} f_{xx} + \frac{1}{Y} f_{yy} - \frac{\sigma}{Y} f_{zz} \tag{3.1.13}$$

$$\epsilon_{zz} = - \frac{\sigma}{Y} f_{xx} - \frac{\sigma}{Y} f_{yy} + \frac{1}{Y} f_{zz}.$$

These equations can be solved simultaneously to express the tensions (stresses) as linear functions of the extensions (strains). In Chap. 7 it is shown that (3.1.13) holds for a general state of strain in an isotropic elastic medium provided that we choose the orientation of the axes x, y, z to make them *principal axes* of the stress-strain system.

Later on when we discuss longitudinal waves on a rod, we shall need an expression for the elastic potential energy density of a stretched rod. We may find such an expression by noting that the work required to stretch the rod an amount Δl by the force F_l is

$$W = \tfrac{1}{2} F_l \, \Delta l, \tag{3.1.14}$$

since the increase in length grows linearly with the force. The elastic potential energy density per unit volume is therefore

$$V_1 = \frac{W}{Sl} = \frac{1}{2} \frac{F_l}{S} \frac{\Delta l}{l} = \tfrac{1}{2} f_l \epsilon_l. \tag{3.1.15}$$

With the understanding that the tensions in the y and z directions are zero, we may write

$$V_1 = \tfrac{1}{2} f_{xx} \epsilon_{xx} = \tfrac{1}{2} Y \epsilon_{xx}^2 = \frac{f_{xx}^2}{2Y}. \tag{3.1.16}$$

A general expression for elastic energy density, when other components of stress are present, is developed in Chap. 7.

Problems

3.1.1   Justify the factor $\tfrac{1}{2}$ in (3.1.14).

3.1.2   A uniform tensile stress f_{xx} is applied to the ends of a rectangular flat plate of thickness t. Concurrently the sides of the plate are constrained by applying a uniform tensile stress f_{yy}

of such magnitude that the width w of the plate remains constant. Show that the *plate Young's modulus* $Y_p \equiv f_{xx}/\epsilon_{xx}$ is $Y/(1 - \sigma^2)$ and that the *plate Poisson's ratio* $\sigma_p \equiv -\epsilon_{zz}/\epsilon_{xx}$ is $\sigma/(1 - \sigma)$.

3.1.3   Now suppose that the plate of Prob. 3.1.2 is further constrained by applying a tensile stress f_{zz} so that its thickness t remains constant as well as its width. Show then that the modulus for *pure linear strain* (meaning that $\epsilon_{xx} \neq 0$ but that $\epsilon_{yy} = \epsilon_{zz} = 0$) is $Y_B = Y(1 - \sigma)/[(1 - 2\sigma)(1 + \sigma)]$. (The two moduli Y_p and Y_B are needed in discussing longitudinal waves in plates and in extended media, where the inertia of the medium prevents sideways motion and thereby supplies the stresses assumed here.)

3.2   Volume Changes in an Elastic Medium

One of the simplest sorts of distortion that a homogeneous fluid or solid can experience is a change in its volume v, with or without an accompanying change in shape. We define the volume strain, or *dilatation*, to be

$$\theta \equiv \frac{\Delta v}{v}. \tag{3.2.1}$$

The physical properties of liquids and solids ensure that θ has a small magnitude, except under extremely high pressures. Even in gases, θ is small for sound waves of normal intensity (as distinct from blast or shock waves).

The tension strains discussed in Sec. 3.1 involve a change in volume of the medium, so that there must exist a connection between the dilatation, just defined, and the three extensions ϵ_{xx}, ϵ_{yy}, and ϵ_{zz}. To find this connection, let us consider what happens to a cubical volume element $\Delta x \, \Delta y \, \Delta z$ in the undistorted medium when the length Δx becomes $\Delta x + (\partial \xi/\partial x) \, \Delta x$, with corresponding changes in Δy and Δz. Evidently the dilatation is

$$\theta = \frac{\left(\Delta x + \frac{\partial \xi}{\partial x} \Delta x\right)\left(\Delta y + \frac{\partial \eta}{\partial y} \Delta y\right)\left(\Delta z + \frac{\partial \zeta}{\partial z} \Delta z\right) - \Delta x \, \Delta y \, \Delta z}{\Delta x \, \Delta y \, \Delta z}$$

$$= \frac{\partial \xi}{\partial x} + \frac{\partial \eta}{\partial y} + \frac{\partial \zeta}{\partial z} = \epsilon_{xx} + \epsilon_{yy} + \epsilon_{zz}, \tag{3.2.2}$$

where the three extensions have been treated as small quantities. In terms of the displacement vector

$$\varrho = \mathbf{i}\xi + \mathbf{j}\eta + \mathbf{k}\zeta, \tag{3.2.3}$$

the dilatation can be written

$$\theta = \mathrm{div}\,\varrho = \nabla \cdot \varrho. \tag{3.2.4}$$

That is, the dilatation is the divergence of the displacement vector.† The dilatation accompanying the stretching of a rod of isotropic material is evidently

$$\theta_{\text{rod}} = \epsilon_{xx}(1 - 2\sigma), \tag{3.2.5}$$

since $\epsilon_{yy} = \epsilon_{zz} = -\sigma\epsilon_{xx}$.

A fluid or a homogeneous elastic solid medium when subjected to an incremental hydrostatic pressure p suffers a decrease in volume specified by the strain $-\theta$. Hooke's law in this case may be written

$$p = -B\theta, \tag{3.2.6}$$

where B is the *bulk modulus*. Like Y, B has the dimensions of a force per unit area. Some representative values of B are given in Table 3.1 in Sec. 3.1, page 73. The bulk modulus of a gas for specified conditions can be calculated from the equations that summarize its thermodynamic properties. A brief account is given in Sec. 5.2; see also Prob. 3.2.2.

In an isotropic medium, the bulk modulus B can be related to Young's modulus Y and Poisson's ratio σ very simply by making use of the elastic equations (3.1.13). In these equations set $f_{xx} = f_{yy} = f_{zz} = -p$, so that the stress system becomes that of hydrostatic pressure. On adding the three equations, we find that

$$\theta = -\frac{3p}{Y}(1 - 2\sigma). \tag{3.2.7}$$

In view of the definition of B in (3.2.6), it follows that

$$Y = 3B(1 - 2\sigma). \tag{3.2.8}$$

This relation among the elastic constants shows that of the three elastic constants defined so far, only two are independent.

Finally, we derive an expression for the elastic potential energy density in a medium that is subject to a hydrostatic pressure p. We shall always regard p to be the pressure in *excess* of that normally present in the equilibrium state. Some care in this matter is necessary, since we shall later wish to use the result in computing the potential energy density accompanying a sound wave in a gas, which has a static equilibrium pressure P_0 that greatly exceeds the incremental pressure of the sound wave. If the (incremental) pressure p causes the small change in volume Δv of a medium having the equilibrium volume v, the work done is clearly

$$W = -\tfrac{1}{2}p\,\Delta v, \tag{3.2.9}$$

† A brief review of vector calculus is given in Appendix A, where a discussion of the divergence will be found.

since Δv decreases linearly with p. The potential energy density is therefore

$$V_1 = \frac{W}{v} = -\tfrac{1}{2}p\,\frac{\Delta v}{v}$$

$$= -\tfrac{1}{2}p\theta = \tfrac{1}{2}B\theta^2 = \frac{p^2}{2B}, \tag{3.2.10}$$

on using the definition of B (3.2.6). Equation (3.2.10) gives all the elastic potential energy in the distortion of a gas or liquid. It can be used for a solid only when the distortion is that produced by hydrostatic pressure alone, in other words, for *uniform dilatation*.

Problems

3.2.1   Show that the dilatation θ is equal to $-\Delta\rho/\rho$, where ρ is the density of a homogeneous elastic medium.

3.2.2   Show that an ideal gas has the isothermal bulk modulus $B = P$ when it obeys Boyle's law $Pv = \text{const}$. For rapid changes in volume, the temperature changes in the gas do not have time to equalize. The ideal gas then obeys the *adiabatic* equation $Pv^\gamma = \text{const}$, where γ is a constant. Show then that the adiabatic bulk modulus is $B = \gamma P$.

3.3   Shear Distortion in a Plane

We are primarily concerned here with distortions of an elastic medium for which no change in volume occurs. A simple example is illustrated in Fig. 3.3.1, which shows the end view of a cubical element $\Delta x\,\Delta y\,\Delta z$ that is *sheared* sideways into the form of a rhomboid by the action of applied stresses. The magnitude of the shearing strain for the distortion illustrated is defined to be $\gamma = \tan\alpha$,

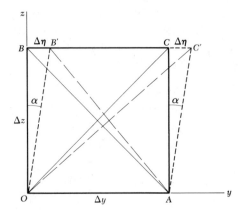

Fig. 3.3.1   A cubical element $\Delta x\,\Delta y\,\Delta z$ sheared in the yz plane.

where α is the angle shown in the figure. Since α is an extremely small angle for highly elastic materials under stresses that cause no permanent deformation, there is a negligible distinction between γ and α. The value of γ is given by

$$\gamma = \tan\alpha = \frac{\Delta\eta}{\Delta z} \rightarrow \frac{\partial\eta}{\partial z}. \tag{3.3.1}$$

Now the distortion illustrated in Fig. 3.3.1 is unchanged if the rhomboid is bodily rotated counterclockwise through the small angle α. We would then conclude that the shearing strain is given by

$$\gamma = \frac{\partial\zeta}{\partial y}. \tag{3.3.2}$$

With the rhomboid at other angular positions differing slightly from that illustrated, the shearing strain is given by

$$\gamma_{yz} = \frac{\partial\eta}{\partial z} + \frac{\partial\zeta}{\partial y} \tag{3.3.3}$$

where the subscripts yz identify the plane in which the shearing strain occurs.

Our discussion suggests that a small rotation of the elastic medium is likely to accompany shear distortion. Only if $\partial\zeta/\partial y = \partial\eta/\partial z$ does the cubical element become sheared into a rhomboid without an accompanying rotation. Otherwise the elastic medium is turned about the x axis through the small angle

$$\phi_x = \frac{1}{2}\left(\frac{\partial\zeta}{\partial y} - \frac{\partial\eta}{\partial z}\right). \tag{3.3.4}$$

Although relations similar to (3.3.3) and (3.3.4) hold for the other two coordinate axes, we shall defer a general consideration of them until Chap. 7 and limit our discussion here to the simple case of *plane shear*. We therefore need to introduce next the shearing stresses responsible for the shearing strain illustrated in Fig. 3.3.1. At first glance it would appear that a force in the y direction acting uniformly over the top face of the cubical element, with its equal but oppositely directed counterpart on the bottom face, is all that is needed. However, such a pair of shearing forces would result in a net torque, or couple, acting on the element, thus violating the condition for rotational equilibrium. Hence along with this pair of forces there must be an equal second pair acting on the side faces of the element, as shown in Fig. 3.3.2. We define shearing stress to be the tangential force acting across a plane of unit area and designate it by quantities such as f_{zy}. The first subscript indicates the direction of the normal of the plane, and the second, the direction of the force. For rotational equilibrium of the parts of the medium, evidently $f_{zy} = f_{yz}$.

The shearing stress f_{zy} ($= f_{yz}$) and the shearing strain γ_{zy} ($= \gamma_{yz}$) that it

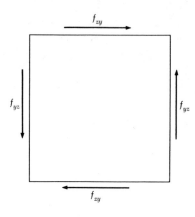

Fig. 3.3.2 Shearing-force system necessary for rotational equilibrium.

produces serve to define the *shear modulus,* or *modulus of rigidity* μ. Hooke's law in this case takes the form

$$f_{zy} = \mu\gamma_{zy}. \tag{3.3.5}$$

Like B and Y, μ has the dimensions of force per unit area. Representative values of μ for various solids are given in Table 3.1 in Sec 3.1, page 73. The shear modulus of an isotropic medium is not an independent elastic constant, for it is established in Prob. 3.3.3 that

$$Y = 2\mu(1 + \sigma). \tag{3.3.6}$$

Hence of the four elastic constants that have been defined for an isotropic medium (Y, σ, B, and μ), only two are independent.

We need to find the potential energy density associated with plane shear. If we assume that neither $\partial\zeta/\partial y$ nor $\partial\eta/\partial z$ vanishes, then the work done by the shearing stresses f_{zy} and f_{yz} in bringing about the shearing strain γ_{zy} in the cubical element $\Delta x\,\Delta y\,\Delta z$ amounts to

$$\Delta W = \tfrac{1}{2}(f_{zy}\,\Delta x\,\Delta y)\left(\frac{\partial\eta}{\partial z}\,\Delta z\right) + \tfrac{1}{2}(f_{yz}\,\Delta x\,\Delta z)\left(\frac{\partial\zeta}{\partial y}\,\Delta y\right). \tag{3.3.7}$$

In (3.3.7) we recognize $f_{zy}\,\Delta x\,\Delta y$ as the force in the y direction on the top face of the cubical element and $(\partial\eta/\partial z)\,\Delta z$ as the displacement of this face relative to the bottom face, which is considered as stationary. The other term pertains in similar fashion to the side faces. The factor of $\tfrac{1}{2}$ reflects the fact that the displacements grow linearly with the forces. Since $f_{zy} = f_{yz}$ and $\gamma_{zy} = \partial\eta/\partial z + \partial\zeta/\partial y$, the potential energy density per unit volume is

$$V_1 = \frac{\Delta W}{\Delta x\,\Delta y\,\Delta z} = \tfrac{1}{2}f_{zy}\gamma_{zy} = \tfrac{1}{2}\mu\gamma_{zy}{}^2 = \frac{f_{zy}{}^2}{2\mu}, \tag{3.3.8}$$

where use has been made of (3.3.5).

Problems

3.3.1   Suppose that all the stress components are zero except $f_{yz} = f_{zy}$, which therefore act to produce the state of shear illustrated in Fig. 3.3.1. Show that this shearing stress is equivalent to a positive tensile stress $f_{y'y'} = f_{yz}$ acting along a y' axis parallel to the diagonal OC, together with a negative tensile stress $f_{z'z'} = -f_{yz}$ acting along a z' axis parallel to the diagonal BA.

3.3.2   Show that the shearing strain illustrated in Fig. 3.3.1 is equivalent to an elongation $\epsilon_{y'y'} = \frac{1}{2}\gamma_{yz}$ along the y' axis and an elongation $\epsilon_{z'z'} = -\frac{1}{2}\gamma_{yz}$ along the z' axis, where these axes are defined in Prob. 3.3.1.

3.3.3   Use the results of Probs. 3.3.1 and 3.3.2 in conjunction with (3.3.5) defining μ and with the elastic equations (3.1.13) to prove that $Y = 2\mu(1 + \sigma)$.

3.3.4   Show that $Y = 9\mu B/(\mu + 3B)$ connects the three elastic moduli Y, μ, and B. By requiring that the three moduli be positive, show that Poisson's ratio is constrained to the range $-1 < \sigma < \frac{1}{2}$. Does this result imply that a rod increases its volume on being stretched?

3.3.5   Soft rubber deforms primarily by shear when acted on by ordinary forces. Show that this property is equivalent to considering B as infinite, $\sigma = \frac{1}{2}$, and $Y = 3\mu$.

3.3.6   Show that the modulus for pure linear strain, $Y_B = Y(1 - \sigma)/(1 - 2\sigma)(1 + \sigma)$, as derived in Prob. 3.1.3, may be expressed more simply in the form $Y_B = B + \frac{4}{3}\mu$.

★3.3.7   Write expressions analogous to (3.3.4) for the small rotations that take place about the y and z axes when a general state of strain exists. Show that when ϕ_x, ϕ_y, and ϕ_z are small (compared with unity), they may be considered as the components of a vector $\boldsymbol{\phi}$ related to the displacement $\boldsymbol{\varrho}$ by $\boldsymbol{\phi} = \frac{1}{2} \operatorname{curl} \boldsymbol{\varrho} = \frac{1}{2}\nabla \times \boldsymbol{\varrho}$.

3.4   The Torsion of Round Tubes and Rods

We are interested here in calculating the torque needed to twist a round tube (or rod) of length l through an angle ϕ. By considering first the torsion of a shell of radius r and thickness dr, we can find by integration how tubes with finite walls and solid rods behave.

In Fig. 3.4.1 is shown such a shell. We suppose that a pair of equal and opposing torques dM exist at the two ends of the shell, as the result of a constant circumferential shearing stress f_ϕ acting across the end areas $dS = 2\pi r\,dr$. The moment of the shearing force is

$$dM = rf_\phi\,dS = 2\pi r^2 f_\phi\,dr. \tag{3.4.1}$$

The shearing strain in the shell is measured by the angle α (strictly, by $\tan\alpha$) between a line on the shell parallel to the axis of the shell and the direction

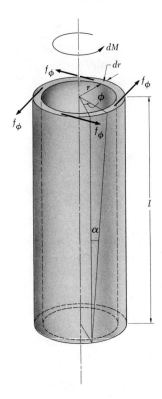

Fig. 3.4.1   The torsion of a thin cylindrical shell.

such a line takes with shear. It is evident from the geometry of the figure that $l\alpha = r\phi$. Hence the shearing strain is

$$\gamma = \tan\alpha \approx \alpha = \frac{r\phi}{l}. \tag{3.4.2}$$

From the definition of the shear modulus μ,

$$f_\phi = \mu\gamma. \tag{3.4.3}$$

Equations (3.4.1) to (3.4.3) show that

$$dM = \frac{2\pi\mu\phi}{l} r^3 \, dr. \tag{3.4.4}$$

On integrating (3.4.4) from an inner radius r_1 to an outer radius r_2, we find that

$$M = \frac{\pi}{2} \mu \frac{r_2{}^4 - r_1{}^4}{l} \phi. \tag{3.4.5}$$

For a solid rod $r_1 = 0$.

The coefficient of ϕ in (3.4.5) is called the *torsion constant* of the tube, i.e., the torque that is necessary to twist the tube through an angle of one radian. The torsion constant of a fiber ($r_1 = 0$) used in a torsion balance and in other sensitive measuring apparatus is proportional to the *fourth* power of its diameter. The torsion of a rod or tube of circular section, or the related standing torsional wave that can be set up on a rod or tube of finite length, constitutes one of the best means for measuring the shear modulus of an isotropic material. The wave equation for torsional waves is taken up in Sec. 4.5.

The elastic energy stored in unit length of the twisted tube is evidently

$$V_1 = \frac{1}{2}\frac{M\phi}{l} = \frac{\pi}{4}\mu\frac{r_2^4 - r_1^4}{l^2}\phi^2 = \frac{M^2}{\pi\mu(r_2^4 - r_1^4)}, \tag{3.4.6}$$

where use has been made of (3.4.5).

Problems

3.4.1 Devise a way of measuring the unknown moment of inertia of a motor armature by hanging it on a steel wire so that it becomes a torsion pendulum.

3.4.2 A Cavendish torsional pendulum for measuring the gravitational constant G is to be designed using a quartz fiber of length L as a torsion element. Assume that a safe tensile load for the fiber is proportional to its cross-sectional area. Investigate the relative merits of making a large or a small apparatus so far as getting the largest possible deflection is concerned.

3.4.3 The elastic potential energy density of a medium distorted by shear in a plane is given by (3.3.8). Use this expression to account in detail for the total elastic potential energy (3.4.6) stored in a twisted tube or rod. *Hint:* Use (3.4.2) to find V_1 at any point in the tube then integrate V_1 over the volume of the tube.

3.4.4 Show that the dependence on tube geometry may be expressed in terms of I, the second moment of area of the tube section about the tube axis, so that (3.4.5) may be written

$$\frac{M}{I} = \mu\frac{\phi}{l}. \tag{3.4.7}$$

3.5 The Statics of a Simple Beam

Presently we shall be interested in discussing *lateral* waves and vibrations that can exist on a long slender rod in which the elastic restoring force arises from the *stiffness* of the member. As a first step, we need to develop a way of expressing the distribution of forces and torques that can exist as a function of position along the rod, which in this context is often called a *beam*.

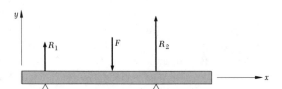

Fig. 3.5.1   A loaded simple beam.

We suppose that the beam is horizontal and is supported either by external forces at two points or by a torque and a force applied at one end (the so-called *cantilever beam*). All applied forces, or *loads*, are supposed to lie in a vertical plane and are normal to the axis of the beam, which we take to be the x axis of a coordinate frame. The positive y axis is taken vertically up, thus establishing a sign convention for forces. We distinguish *concentrated* loads, such as the forces R_1, R_2, and F in Fig. 3.5.1, and *distributed* loads $w(x)$, such as the weight of the beam. The conditions for equilibrium of the entire beam serve to determine R_1 and R_2 when the other applied loads are given.

Let us now investigate the implications arising from the condition that when the *entire* beam is in equilibrium, *any portion* of the beam is in equilibrium also. The two equations for equilibrium, force and torque, applied to a portion of the beam evidently require that a *shearing force* $V(x)$ and a torque or *bending moment* $M(x)$ be transmitted across a plane cutting the beam transversely. Let us agree to take $V(x)$ *positive* if the beam on the left of the plane exerts an *upward* force on the beam on the right of the plane. (By Newton's third law, the beam on the right of the plane exerts an equal downward force on the beam on the left of the plane.) The bending moment $M(x)$ is *positive* when the beam on the left of the plane exerts a *clockwise* torque on the beam on the right of the plane. Positive shearing force and bending moments are shown in Fig. 3.5.2 for an element dx of the beam.

Let us next find how $w(x)$, $V(x)$, and $M(x)$ are related by applying the conditions for equilibrium to the element dx of Fig. 3.5.2. For force equilibrium, $w(x)\,dx + V = V + dV$. Hence,

$$\frac{dV(x)}{dx} = w(x). \tag{3.5.1}$$

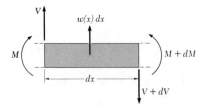

Fig. 3.5.2   An element dx of a loaded beam.

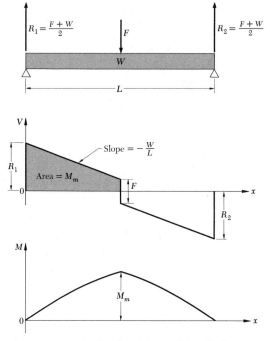

Fig. 3.5.3   Example of load, shearing-force, and bending-moment diagrams.

For torque equilibrium about the center of the element,

$$M + V \tfrac{1}{2}dx + (V + dV)\tfrac{1}{2}dx = M + dM.$$

Hence,

$$\frac{dM(x)}{dx} = V(x). \tag{3.5.2}$$

On combining (3.5.1) and (3.5.2),

$$\frac{d^2M(x)}{dx^2} = w(x). \tag{3.5.3}$$

These equations show that from a given distributed load, we can find both $V(x)$ and $M(x)$ by performing two successive indefinite integrations over x. The two constants of integration may be found by noting that (1) at a *free* end of the beam, both V and M must vanish; (2) at a positive (negative) concentrated load F, the $V(x)$ function has a discontinuity, ascending (descending) by $|F|$; (3) at a *clamped* end, such as occurs with a cantilever beam, V and M have the values needed to support the beam. A simple example is illustrated in Fig. 3.5.3, which shows a load diagram with shearing-force and bending-moment diagrams drawn beneath it. Several quantitative examples are included in Prob. 3.5.1.

Problems

3.5.1 Construct shearing-force and bending-moment diagrams for the cases shown in the figure. Find the equations for $V(x)$ and $M(x)$ in each case.

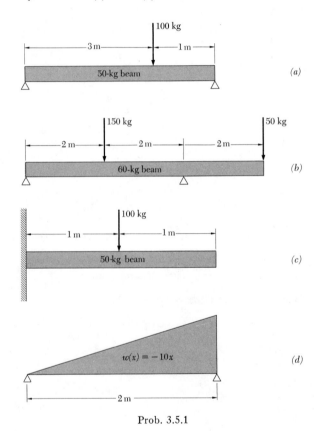

Prob. 3.5.1

3.5.2 Establish the relations (3.5.1) and (3.5.2) by computing the shearing force $V(x)$ and bending moment $M(x)$ acting at x that are necessary to ensure the equilibrium of the finite section of the beam between one end and a position x along the beam.

3.6   The Bending of a Simple Beam

In the preceding section we found that a simple beam is subject to a shearing force and a bending moment along its length, both of which tend to distort it

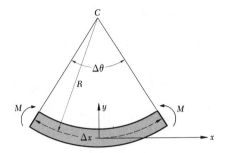

Fig. 3.6.1   Element Δx of beam, with bending greatly exaggerated.

from its straight unloaded shape. A calculation or an appeal to experience shows that the distortion due to shear in a beam whose length is much greater than its thickness is completely negligible compared with the distortion produced by the action of the bending moment.

As shown greatly exaggerated in Fig. 3.6.1, a positive bending moment M distorts an element Δx of the beam into a circular arc. The upper part of the element is under compression, and the lower part in tension. A curved surface of radius R separates the two regions of negative and positive tension, the *neutral surface*. A layer of the beam at a distance y above the neutral surface has a radius of curvature $R - y$ and therefore has been compressed the fractional amount

$$\epsilon_{xx} = \frac{(R - y)\,\Delta\theta - R\,\Delta\theta}{R\,\Delta\theta} = -\frac{y}{R}, \tag{3.6.1}$$

where $\Delta\theta$ is the angle subtended by Δx at the center of curvature C. In order to sustain this strain, there must exist the tension in the layer (often called the *fiber stress* in the beam) of the value

$$f_{xx} = -Y\frac{y}{R}, \tag{3.6.2}$$

where Y is Young's modulus. The area of an infinitesimal strip across the beam subject to this tensile stress is $dS = Z\,dy$, where Z is the width of the beam at a distance y from the neutral surface. Since no net force is assumed to exist along the beam,

$$\int f_{xx}\,dS = -\frac{Y}{R}\int y\,dS = 0, \tag{3.6.3}$$

where the integral is taken over the section of the beam. In order that this integral vanish, it is necessary that the y origin, i.e., the neutral surface, be situated at the *center of area*, or *centroid*, of the section of the beam. Equation (3.6.3) accordingly serves to define the position of the neutral surface.

The tensile force $f_{xx} dS$ acting at a distance y from the neutral surface constitutes a torque $yf_{xx} dS$ about a transverse axis lying in the neutral surface. When added up for the entire section, it exerts an elastic bending moment

$$M_{\text{elastic}} = \int yf_{xx} dS = -\frac{Y}{R} \int y^2 dS, \qquad (3.6.4)$$

which resists, or opposes, the applied bending moment M responsible for the curvature R. For equilibrium, $M_{\text{elastic}} + M = 0$, so that

$$M = \frac{Y}{R} \int y^2 dS. \qquad (3.6.5)$$

The so-called *section moment* I is defined as the second moment of area of the section of the beam about a transverse axis through the neutral surface,

$$I = \int y^2 dS, \qquad (3.6.6)$$

so that (3.6.5) becomes

$$M(x) = \frac{YI}{R(x)}. \qquad (3.6.7)$$

The notation I reminds us that (3.6.6) is formally the moment of inertia of a lamina of unit areal density. The fact that M is a function of x, and therefore that the radius of curvature of the beam R is likewise a function of x, has been indicated explicitly in writing (3.6.7).

We let $\eta(x)$ designate the displacement of the neutral surface, which is to be found from (3.6.7). In most practical applications of the present theory, the distortion of the beam is very slight, so that the up and down displacement η may be treated as a small quantity (strictly, $\partial\eta/\partial x$ is treated as a small quantity). The radius of curvature of the plane curve $\eta = \eta(x)$ is given by (see Prob. 3.6.1)

$$R = \frac{[1 + (d\eta/dx)^2]^{3/2}}{d^2\eta/dx^2} \rightarrow \frac{1}{d^2\eta/dx^2} \qquad (3.6.8)$$

when $d\eta/dx$ is small. Hence (3.6.7) becomes

$$M(x) = YI \frac{d^2\eta(x)}{dx^2}. \qquad (3.6.9)$$

Therefore if $M(x)$ is found by the methods of the previous section, the shape of the loaded beam can be found by integrating (3.6.9) twice and evaluating the two constants of integration.

The external work done in bending the element Δx in Fig. 3.6.1 to a radius of curvature R is evidently

$$\Delta W = \tfrac{1}{2}M \,\Delta\theta = \tfrac{1}{2}M \frac{\Delta x}{R} = \tfrac{1}{2}YI \left(\frac{d^2\eta}{dx^2}\right)^2 \Delta x, \tag{3.6.10}$$

in which R and M have been eliminated using (3.6.8) and (3.6.9). Hence the elastic potential energy density per unit length of beam is†

$$V_1 = \frac{\Delta W}{\Delta x} = \tfrac{1}{2}YI \left(\frac{d^2\eta}{dx^2}\right)^2 = \frac{M^2}{2YI}. \tag{3.6.11}$$

The total elastic potential energy of a loaded beam may be found by integrating (3.6.11) over the length of the beam.

Problems

3.6.1 Prove that the curvature at any point of the plane curve $y = y(x)$ having continuous first and second derivatives is given by

$$\frac{1}{R} = \frac{d\theta}{ds} = \frac{d^2y/dx^2}{[1 + (dy/dx)^2]^{3/2}}$$

where θ is the angle between a line tangent to the curve and the x axis and ds is an element of arc along the curve.

3.6.2 Find the shape $\eta(x)$ of the beams acted on by the loads of Prob. 3.5.1. (The product YI will appear as a scale factor in the answers.)

3.6.3 Derive formulas for the section moment I for the following cases: (*a*) a round rod; (*b*) a round tube; (*c*) a rectangular bar with neutral surface parallel to one face; (*d*) a triangular bar, two sides equal and base parallel to neutral section; (*e*) an I beam. (Why are I beams used in constructing buildings and bridges?)

3.6.4 A round rod used in constructing a piece of apparatus is found to be too flexible. How much must its diameter be increased to double its stiffness? If instead of a round rod, the member consists of a bar of rectangular section, how much must its thickness be increased to double its stiffness?

3.6.5 A horizontal beam of negligible weight and of length $2L$ is attached to a hinge at $x = 0$ and is supported at $x = L$. A downward force F_1 is applied at $x = L/2$, causing a small upward deflection D_2 at the end, $x = 2L$. The force F_1 is removed and a downward force F_2' is applied at the end, $x = 2L$, causing a small upward deflection D_1' at $x = L/2$. Show that

† The notation V_1 for potential energy density must not be confused with V for shearing force.

$F_1/D_2 = F_2'/D_1'$. Can you generalize this special case into a theorem, known as the *reciprocity theorem*, which applies to all linear elastic systems characterized by small strains?

★**3.6.6**　A slender beam of length L is placed under compression by a pair of opposing forces F applied at its ends parallel to its length (see figure). When the magnitude of F is increased above some critical value F_1, the beam starts to deform into an arc, with its ends remaining on the x axis (as the result of constraints not shown in the figure). Neglecting the weight of the beam, show that the bending moment at any position x is $M = -F_1\eta$, where η is the deflec-

F

F　　Prob. 3.6.6

tion at x, and that the equation governing small deflections is $d^2\eta/dx^2 + (F_1/IY)\eta = 0$. Show, therefore, that the critical value of F is $F_1 = (\pi/L)^2 IY$. (You may investigate this phenomenon by applying forces to the end of a thin steel ruler. The theory developed here is part of Euler's theory of struts and is important in estimating the gravitational load that a slender vertical column can support without *buckling*.)

★**3.6.7**　A uniform bar of rectangular section is elastically bent into a circular arc by applying a pair of equal and opposite couples M_0 to its two ends. By using the fact that each "fiber" under tension (or compression) shrinks (or expands) sideways an amount given by Poisson's ratio σ times the fiber elongation, show that the neutral surface does not remain plane but has a *lateral* radius of curvature R' given by

$$R' = \frac{R}{\sigma},$$

where R is the *longitudinal* radius of curvature given as usual by $R = YI/M_0$. If the bar is of thickness t, then on the top surface the lateral and longitudinal curvatures are $\rho' = R' + \frac{1}{2}t$, $\rho = R - \frac{1}{2}t$, and

$$\sigma = \frac{R}{R'} = \frac{\rho + \frac{1}{2}t}{\rho' - \frac{1}{2}t}.$$

(By polishing the surface of the undistorted bar to optical flatness, it is possible to use interference fringes, made with the help of an optically flat glass plate, to measure ρ and ρ', from which an accurate value of σ can be computed.)

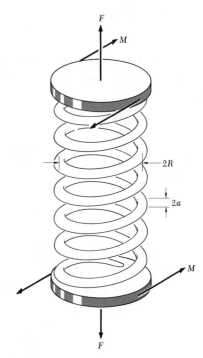

Fig. 3.7.1   A helical spring stretched by the forces
F or twisted by the torques M.

3.7 Helical Springs

A helical spring, Fig. 3.7.1, consists of a wire of highly elastic material that has
been given a permanent deformation by winding it tightly on a cylindrical
mandrel and then removing it. We suppose that the wire radius a is small
compared with the mean radius R of the spring and that the *pitch* of the spring
is also small compared with R.

A helical spring can be used either as a *linear* elastic member, which generates an axial restoring force proportional within limits to the amount of stretch,
or as a *torsional* elastic member, which generates an axial restoring torque proportional within limits to the angle of twist. In the first instance the deformation of the spring is brought about by a *twisting* of the wire of the helix,
whereas in the second instance the deformation is brought about by a *bending*
of the wire of the helix.

Let us first find how the familiar spring constant k for stretching depends
on the geometry of the spring and on the shear modulus μ. Consider that the
spring is pulled axially at each end by a pair of opposing forces F that stretch
the spring a small amount ξ. Since F and ξ are related linearly, the work done in

stretching the spring is

$$W = \tfrac{1}{2}F\xi, \tag{3.7.1}$$

and it is stored as elastic potential energy of torsion. According to (3.4.6), the elastic potential energy in a wire of length l twisted by a torque M is

$$V = \frac{l}{\pi\mu a^4} M^2 = \frac{l}{\pi\mu a^4} F^2 R^2, \tag{3.7.2}$$

since the twisting torque exerted by the axial force F on the wire is everywhere FR. Equating W and V, we find from (3.7.1) and (3.7.2) that the spring constant $k \equiv F/\xi$ is

$$k = \frac{\pi\mu a^4}{2R^2 l}. \tag{3.7.3}$$

The torsion constant k_α for twisting the helical spring can also be found conveniently by an energy method. Let us suppose that a pair of equal and opposite axial torques M are applied at the two ends of the spring, twisting one end through an angle α with respect to the other end. The work done,

$$W = \tfrac{1}{2}M\alpha, \tag{3.7.4}$$

is stored as elastic potential energy of bending. According to (3.6.11), the total elastic energy of bending a "beam" of length l by a constant bending moment M is

$$V = \frac{lM^2}{2IY} \tag{3.7.5}$$

where $I = (\pi/4)a^4$ is the section moment of the wire. On equating (3.7.4) and (3.7.5) and solving for the torsion constant $k_\alpha \equiv M/\alpha$, we find that

$$k_\alpha = \frac{2IY}{l} = \frac{\pi Y a^4}{l}. \tag{3.7.6}$$

Thus the stretching spring constant (3.7.3) depends on the shear modulus μ, while the twisting spring constant (3.7.6) depends on Young's modulus Y.

Problems

3.7.1 Design a practical helical spring that will be stretched 5 cm by a load of 500 g. Steel music wire having diameters of 0.020, 0.030, 0.040 and 0.050 in. is available.

3.7.2 Derive (3.7.3) and (3.7.6) by working directly with forces and torques instead of by the energy method used in the text.

four

One-dimensional Elastic Waves

In the last chapter, we introduced the elastic constants for various types of distortion of an elastic solid. We are now prepared to examine the wave motions that can exist in a solid body as a result of its elastic properties. Specifically, we pay most attention to the example of a long thin rod, which can vibrate longitudinally, transversely, and torsionally. As in the case of the stretched string of Chap. 1, we are concerned with both traveling waves and standing waves (normal-mode vibrations) and with energy relations. Furthermore, the examples of this chapter help us understand mechanical impedance, impedance matching, dispersion, and perturbation analysis—concepts whose importance and generality far transcend the particular examples at hand.

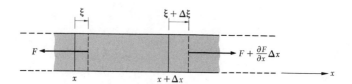

Fig. 4.1.1   Element of rod displaced by a longitudinal wave.

4.1   Longitudinal Waves on a Slender Rod

(a)   The Wave Equation

We have already found the wave equation for longitudinal waves (Sec. 1.10) as part of a more rigorous analysis of transverse waves on a string. Let us again consider its derivation, without the added complication of transverse motion. We suppose that the rod has a constant cross-sectional area S, a uniform volume density ρ_0, and a Young's modulus Y. We assume that it is sufficiently *slender* to permit us to neglect radial motion, arising from radial extension coupled through Poisson's ratio to the axial extension.

Referring to Fig. 4.1.1, we see that when a longitudinal wave is present, a plane at x moves to $x + \xi$ and a neighboring plane at $x + \Delta x$ moves to $x + \Delta x + \xi + \Delta \xi$. The element of the bar between these planes, of mass $S\rho_0\, \Delta x$, is acted on by the tensile force F at its left end and by the tensile force $F + (\partial F/\partial x)\, \Delta x$ at its right end, as shown. Since the net force on the element in the positive direction is $(\partial F/\partial x)\, \Delta x$, Newton's second law requires that

$$\frac{\partial F}{\partial x}\, \Delta x = S\rho_0\, \Delta x\, \frac{\partial^2 \xi}{\partial t^2}. \tag{4.1.1}$$

According to the definition of Young's modulus (3.1.4), the tensile force F and the strain $\partial \xi/\partial x$ are related by the equation

$$F = SY\, \frac{\partial \xi}{\partial x}. \tag{4.1.2}$$

On introducing this expression for F into (4.1.1) and canceling $S\, \Delta x$, we are left with the wave equation

$$\frac{\partial^2 \xi}{\partial x^2} = \frac{1}{c_b{}^2}\, \frac{\partial^2 \xi}{\partial t^2}, \tag{4.1.3}$$

where the wave velocity for longitudinal waves is

$$c_b \equiv \left(\frac{Y}{\rho_0}\right)^{1/2}. \tag{4.1.4}$$

The wave velocity (4.1.4) is often called the *bar velocity* to distinguish it from other wave velocities characterizing solid media. Let us find the magnitude of c_b for a common metal such as steel. Young's modulus for steel is about 2×10^{11} N/m², and the density is about 8,000 kg/m³. Hence the bar velocity is $c_b \approx (2 \times 10^{11}/8{,}000)^{1/2} = 5{,}000$ m/sec $\approx 11{,}000$ mph. This high velocity is easily measured by observing one or more normal-mode frequencies of a slender rod of known length. Measurement of c_b by this means affords one of the best methods for determining Young's modulus for the material of the rod.

(b) Standing Waves

It is virtually impossible to clamp the ends of a rod so that standing waves with nodes at the ends can occur, as with transverse waves on a string. The usual condition for standing waves is to have the ends "free," which implies that the force (4.1.2) vanishes at each end. Hence if we have a rod extending from $x = 0$ to $x = l$, the strain $\partial\xi/\partial x$ must vanish at these points. The vanishing of the strain, of course, does not mean that the displacement vanishes.

Now the wave equation (4.1.3) has a solution similar to (1.5.5), from which we select the standing wave

$$\xi = \cos\kappa x(c \cos\omega t + d \sin\omega t), \tag{4.1.5}$$

since for this particular solution the strain

$$\frac{\partial\xi}{\partial x} = -\kappa \sin\kappa x(c \cos\omega t + d \sin\omega t) \tag{4.1.6}$$

vanishes at $x = 0$. The strain (4.1.6) also vanishes at $x = l$ provided that

$$\kappa_n l = n\pi \qquad n = 1, 2, 3, \ldots, \tag{4.1.7}$$

which shows that there must be an integral number of half-waves in the distance l. The normal-mode frequencies of the bar are then given by $\omega_n = c_b\kappa_n$. When the wavelength $\lambda_n = 2\pi/\kappa_n = 2l/n$ begins to become comparable with the lateral dimensions of the bar, our assumption of slenderness fails and the normal-mode frequencies deviate increasingly from a harmonic progression. A first-order estimate of the effect of lateral motion on the wave velocity of longitudinal waves is given in Sec. 4.4.

How great an amplitude of longitudinal vibration can be given to a rod without damaging it? The upper limit is set basically by the maximum oscillating strain ϵ_m (or equivalently, the stress) that the material of the rod can endure without undue *elastic fatigue*, which will lead to its early failure. If the rod is vibrating in a particular mode, as given by (4.1.5) with $\omega = c_b\kappa = n\pi c_b/l$, the linear strain (4.1.6) has a maximum value $\kappa\xi_m$, where $\xi_m = (c^2 + d^2)^{1/2}$ is the amplitude of the vibration. Hence the maximum permitted amplitude is

$$\xi_m = \frac{\epsilon_m}{\kappa} = \frac{\epsilon_m c_b}{\omega}, \qquad (4.1.8)$$

which varies inversely with frequency. The maximum *velocity* amplitude

$$\left(\frac{\partial \xi}{\partial t}\right)_m = \omega \xi_m = \epsilon_m c_b \qquad (4.1.9)$$

depends only on the material properties of the rod. We have already computed that $c_b \approx 5{,}000$ m/sec for steel. An approximate value for ϵ_m is 10^{-3} for a high-strength steel. Hence an upper limit to the velocity amplitude (4.1.9) is about 5 m/sec. At a frequency of 16 kHz, ($\omega \approx 10^5$ sec^{-1}), which is a low *ultrasonic* frequency, the amplitude of motion is at most 5×10^{-5} m, or two-thousandths of an inch. A considerable ultrasonic technology has developed in which the very small motion just computed is put to practical use. Well-known applications include underwater communication, depth sounding, and object location (sonar). More recent applications include the use of high-powered ultrasonic vibrations as an aid in metal machining, forming, and joining operations.

(c)   *Energy and Power*

Let us suppose that the traveling wave

$$\xi = \xi_m \cos(\kappa x - \omega t) \qquad (4.1.10)$$

is present on a long slender rod. The kinetic energy density (per unit length) associated with the wave is

$$K_1 = \tfrac{1}{2}S\rho_0 \left(\frac{\partial \xi}{\partial t}\right)^2 = \tfrac{1}{2}S\rho_0 \omega^2 \xi_m^2 \sin^2(\kappa x - \omega t). \qquad (4.1.11)$$

According to (3.1.16), the potential energy density is

$$V_1 = \tfrac{1}{2}SY \left(\frac{\partial \xi}{\partial x}\right)^2 = \tfrac{1}{2}SY\kappa^2 \xi_m^2 \sin^2(\kappa x - \omega t)$$

$$= \tfrac{1}{2}S\rho_0 \omega^2 \xi_m^2 \sin^2(\kappa x - \omega t) \qquad (4.1.12)$$

where the substitution $Y\kappa^2 = \rho_0 \omega^2$ has been made. The equations for K_1 and V_1 are similar to (1.8.9) for transverse waves on a string in their dependence on x and t; little new can be added to the earlier discussion in Sec. 1.8. Note that if (4.1.11) and (4.1.12) are divided through by S, they become expressions for the *volume* density of energy.

The power transported by the wave (4.1.10) may be found from

$$P = -F \frac{\partial \xi}{\partial t} = -SY \frac{\partial \xi}{\partial x} \frac{\partial \xi}{\partial t}, \qquad (4.1.13)$$

an equation whose derivation parallels that of (1.8.13). We find that

$$P = S\rho_0 c_b \omega^2 \xi_m^2 \sin^2(\kappa x - \omega t), \tag{4.1.14}$$

which agrees with the sum of (4.1.11) and (4.1.12) multiplied by the wave velocity c_b. The *average* power transported is therefore

$$\bar{P} = \tfrac{1}{2} S\rho_0 c_b \omega^2 \xi_m^2, \tag{4.1.15}$$

which, as with the string, can be interpreted as the average total energy density moving with the velocity c_b. The *intensity* I of a wave is the power transported across a unit area normal to the direction of propagation. It has the value

$$I = \tfrac{1}{2}\rho_0 c_b \omega^2 \xi_m^2 \tag{4.1.16}$$

for longitudinal waves on a slender rod.

We have just seen that an upper practical limit to the particle velocity $\omega \xi_m$ is set by elastic-fatigue considerations. For special steels, this limit is about 5 m/sec. Hence a steel rod having a 1-cm² cross section can transport a maximum vibratory power by longitudinal waves amounting to

$$\bar{P} \approx \tfrac{1}{2} \times 10^{-4} \times 8,000 \times 5,000 \times 5^2 = 5,000 \text{ W}.$$

This result is independent of frequency, provided the lateral dimensions of the bar are considerably less than a wavelength. A metal rod therefore can be used as a *waveguide* to "conduct" ultrasonic power from a vibratory source to a "load" where some useful application of the vibratory power is to be made. Very respectable amounts of ultrasonic power can be transmitted by a waveguide of reasonable size.

(d) Momentum Transport

The mechanism of momentum transport by a longitudinal wave traveling on a slender rod is not difficult to discover. We note that when a traveling wave is present, the velocity of the medium $\partial \xi/\partial t$ is *in phase with* the increase in density $-\rho_0\, \partial \xi/\partial x$ resulting from the wave. Thus the portions of the rod moving forward (in the direction of the wave propagation) have a greater than average density, whereas those moving backward have a smaller than average density. Thus the rod possesses a net momentum associated with a longitudinal wave that can be calculated from the (linear) momentum density

$$g_x = -S\rho_0 \frac{\partial \xi}{\partial x} \frac{\partial \xi}{\partial t}. \tag{4.1.17}$$

Problems

4.1.1 How long must a slender steel rod be to resonate at fundamental frequencies of 440 Hz (an audio frequency); of 44 kHz (an ultrasonic frequency)?

4.1.2   A slender rod of area S_1, density ρ_1, and Young's modulus Y_1 is attached to a second rod of area S_2, density ρ_2, and Young's modulus Y_2. Longitudinal waves of frequency ω travel along the first rod and are partly reflected and partly transmitted at the discontinuity. Using the procedure followed in Sec. 1.9, find expressions for the complex amplitude reflection and transmission coefficients and the related power reflection and transmission coefficients.

4.1.3   The two rods of Prob. 4.1.2 have the lengths l_1 and l_2, respectively, with the outer ends free. Show that the equation

$$S_1\rho_1 c_1 \tan \frac{\omega l_1}{c_1} + S_2\rho_2 c_2 \tan \frac{\omega l_2}{c_2} = 0$$

gives the normal-mode frequencies of the system. *Hint:* Write expressions for a standing wave in each part and fit the two waves together using the boundary conditions that hold at the junction.

★4.1.4   Show that a longitudinal wave on a slender rod exerts an average force

$$\bar{F}_x = \bar{E}_1 = \frac{1}{c}\bar{P} = c\bar{g}_x$$

when it is absorbed. *Hint:* See Sec. 1.11.

4.2   The Impedance Concept

There are many similarities between the equations describing the propagation of elastic waves in one dimension and the equations describing the propagation of waves of potential and current on an electric transmission line, which are taken up in Sec. 8.1. This fact was noted in Sec. 1.8, in connection with transverse waves on a string. Here we develop the wave-impedance concept further, using the longitudinal waves on a slender rod as the basis for the discussion.

In electrical theory, the impedance of a linear circuit element having two terminals is the ratio of the complex sinusoidally varying voltage applied across the terminals to the complex current that flows in response to the voltage. (The actual voltage and current, of course, are given by the real parts of their complex representations.) The electrical impedance Z so defined is, in general, a frequency-dependent complex number, having a real part R and an imaginary part X. Its magnitude $(R^2 + X^2)^{1/2}$ gives the magnitude of the voltage-current ratio, and its phase angle $\phi = \tan^{-1}(X/R)$ gives the phase angle between the voltage and current. If the linear circuit element is a pure resistance, then Z is simply the ohmic resistance R of the element. The concept of electric impedance is discussed further in Sec. 8.1.

The impedance concept is a convenient way of dealing with various aspects of elastic-wave propagation. We suppose that a long slender rod extends along the positive x axis from the x origin. We consider, for the time being, that the rod is infinitely long. This assumption ensures that waves initiated at the origin do not return as reflected waves to complicate the analysis. Later we can take into account waves reflected from the far end.

In order to make the discussion of longitudinal waves on the rod waveguide formally analogous to the electrical case, it is necessary to regard the force that acts in the rod as a *compressive* force rather than a *tensile* force. Hence the relation between force and strain at any position and time is (4.1.2) with a negative sign, that is,

$$F(x,t) = -SY \frac{\partial \xi(x,t)}{\partial x}. \tag{4.2.1}$$

A sinusoidally varying compressive force

$$F = F_m e^{+j\omega t} \tag{4.2.2}$$

is now applied to the end of the waveguide at the origin. We depart here from our customary use of $e^{-i\omega t}$ as a time factor to make the results of this section follow the conventions of electrical theory, where the time factor $e^{+j\omega t}$ is universally employed.† Such a force excites longitudinal displacement waves having a frequency ω and a wave number $\kappa = \omega/c_b$, which we take to be

$$\xi(x,t) = \breve{A} e^{j(\omega t - \kappa x)}. \tag{4.2.3}$$

The complex amplitude $\breve{A}$ can be found by substituting (4.2.3) in (4.2.1) and requiring that it equal the force (4.2.2) at $x = 0$. We thus find that $\breve{A} = -jF_m/SY\kappa$, so that the displacement wave is in fact

$$\xi(x,t) = -j \frac{F_m}{SY\kappa} e^{j(\omega t - \kappa x)}. \tag{4.2.4}$$

Associated with this wave by (4.2.1) is the force wave

$$F(x,t) = F_m e^{j(\omega t - \kappa x)}. \tag{4.2.5}$$

The presence of the factor $-j$ in (4.2.4) signifies that the displacement lags the force by a phase angle of 90°. This interpretation is easily verified by taking real parts, that is,

$$\xi(x,t) = \frac{F_m}{SY\kappa} \sin(\omega t - \kappa x) \tag{4.2.6}$$

$$F(x,t) = F_m \cos(\omega t - \kappa x). \tag{4.2.7}$$

---

† We distinguish the two conventions by here using the alternative symbol j for $\sqrt{-1}$; the two notations are algebraically equivalent, with $-i \equiv +j$.

The relation between the force and the kinematical aspects of the wave is considerably simplified if we relate force to displacement velocity, $\dot{\xi} \equiv \partial \xi / \partial t$. From (4.2.4), we find that

$$\dot{\xi}(x,t) = \frac{F_m \omega}{S Y \kappa} e^{j(\omega t - \kappa x)}$$

$$= \frac{F_m}{S \rho_0 c_b} e^{j(\omega t - \kappa x)}. \tag{4.2.8}$$

We thus find that $\dot{\xi}$ and F are in phase. Furthermore, the relation between the displacement velocity amplitude $\dot{\xi}_m \equiv F_m / S \rho_0 c_b$ and the force amplitude F_m is frequency-independent. If we think of the force as being analogous to voltage and the displacement velocity as being analogous to current, we are led to define a *characteristic impedance* of the rod for longitudinal waves

$$Z_0 \equiv \frac{F(x,t)}{\dot{\xi}(x,t)} = S \rho_0 c_b = S(\rho_0 Y)^{1/2}. \tag{4.2.9}$$

In the mks system, Z_0 has the units of kilograms per second.

The power carried by the elastic wave, as given by (4.1.15), may now be expressed in several equivalent ways, namely,

$$\bar{P} = \tfrac{1}{2} S \rho_0 c_b \omega^2 \xi_m{}^2 = \tfrac{1}{2} Z_0 \dot{\xi}_m{}^2 = \tfrac{1}{2} \dot{\xi}_m F_m = \frac{F_m{}^2}{2 Z_0}, \tag{4.2.10}$$

where $\dot{\xi}_m = \omega \xi_m = F_m / Z_0$ is the amplitude of the displacement velocity. Equations (4.2.10) exhibit one use of the impedance concept, and, of course, they are exact analogs of familiar power relations in the electrical case. Normally in the latter case, the factor $\tfrac{1}{2}$ does not appear, since it is conventional to use *effective*, or *root-mean-square*, values of voltage and current, not *peak* values.

In the preceding discussion, we have assumed that the waveguide is infinitely long to avoid the complications of a reflected wave. Alternatively, if the waveguide has a finite length and is attached to some device (which we may term the *load*) that has an acoustic resistance $R_l = F_l / \dot{\xi}_l$ equal to the characteristic impedance of the waveguide Z_0, the waves are completely absorbed again with no reflected wave. Such a resistive loading device is often symbolized by a so-called *dashpot*, in which the viscous resistance of a fluid is used to produce a damping force proportional to the velocity of a loosely fitting piston. In the case of ultrasonic waves, the waveguide can often be loaded by immersing its end in a pot of molten lead. The design of a practical resistive load that accurately matches the impedance of an acoustic waveguide is a fairly difficult task. Nevertheless the concept of a resistive load that absorbs incident waves with no reflection is an important one, since all the power (4.2.10) delivered to the waveguide at the source end flows into the distant load. We then say that the load *matches* the impedance of the waveguide. It is as though the oscillating force (4.2.2) were directly applied to the load, without the waveguide's being present. The only effect of the waveguide, assuming that it is lossless, is to alter the phase of the force at the load with respect to that at the input end of the waveguide.

A more common, but more complicated, situation exists if the distant load has an impedance

$$\check{Z}_l = \frac{F_l}{\check{\xi}_l} \tag{4.2.11}$$

that does not equal the characteristic impedance of the waveguide. In such an event waves are reflected from the load, and they travel back to the source end, causing a change in the amplitude and phase of the displacement velocity there. When a steady state of wave motion has been established, there are then present sinusoidal waves of different amplitudes traveling in both directions on the waveguide. We looked at one aspect of this situation in Prob. 1.4.3, where it was established that a measurement of the standing-wave ratio and the position of the maximum or minimum amplitude of the envelope of the standing-wave pattern suffice to determine the equations specifying the two waves.

Suppose that we represent the two traveling displacement waves by

$$\xi(x,t) = (\check{A}_+ e^{-j\kappa x} + \check{A}_- e^{+j\kappa x})e^{+j\omega t}, \tag{4.2.12}$$

where $\check{A}_+$ is the complex amplitude of the positive-going (incident) wave and $\check{A}_-$ is the complex amplitude of the negative-going (reflected) wave. The wave impedance at any position x is now found to be

$$\check{Z}(x) = \frac{F}{\check{\xi}} = Z_0 \frac{\check{A}_+ e^{-j\kappa x} - \check{A}_- e^{+j\kappa x}}{\check{A}_+ e^{-j\kappa x} + \check{A}_- e^{+j\kappa x}}, \tag{4.2.13}$$

which is obtained by making use of (4.2.1) to calculate the compressive force at any point x and by writing Z_0 for $S\rho_0 c_b$.

The ratio of the complex amplitudes of the two waves can be found by using the fact that $\check{Z}(l)$ must equal the complex load impedance $\check{Z}_l$, whatever it is. Thus, if we solve

$$\frac{\check{A}_+ e^{-j\kappa l} - \check{A}_- e^{+j\kappa l}}{\check{A}_+ e^{-j\kappa l} + \check{A}_- e^{+j\kappa l}} = \frac{\check{Z}_l}{Z_0} \tag{4.2.14}$$

for this ratio, we find that

$$\check{R}_a \equiv \frac{\check{A}_- e^{+j\kappa l}}{\check{A}_+ e^{-j\kappa l}} = \frac{Z_0 - \check{Z}_l}{Z_0 + \check{Z}_l}. \tag{4.2.15}$$

Since $\breve{A}_-e^{+jkl}$ is the amplitude of the wave reflected by the load and $\breve{A}_+e^{-jkl}$ is the amplitude of the wave incident on the load, their ratio $\breve{R}_a$ is the complex *amplitude reflection coefficient* for the displacement.† Such a coefficient was used in Sec. 1.9.

Equation (4.2.15) can be used to eliminate the unknown wave amplitudes in the expression (4.2.13) for $\breve{Z}(x)$. We choose to do this here only for $x = 0$, since we can treat l as a variable telling us how far we are from the load $\breve{Z}_l$. After a modicum of algebra (Prob. 4.2.1), we find that

$$\breve{Z}(0) = Z_0 \frac{\breve{Z}_l + jZ_0 \tan\kappa l}{Z_0 + j\breve{Z}_l \tan\kappa l}. \tag{4.2.16}$$

This result directly relates the impedance load $\breve{Z}_l$ attached to the waveguide at $x = l$ to the impedance presented by the waveguide to a sinusoidal driving force applied to the end at $x = 0$. Equation (4.2.16) is basic in all waveguide theory.

Equation (4.2.16) tells us that $\breve{Z}(0)$ is a periodic function of the distance l separating the source and the load ends of the waveguide. Since the period of the tangent is π, $\breve{Z}(0)$ goes through a complete cycle of values when l covers the range of one-half wavelength. If, however, $\breve{Z}_l = Z_0$, we find that $\breve{Z}(0) = Z_0$, regardless of how long the waveguide is. We have already discussed this *matched loading* of the waveguide.

Suppose next that $\breve{Z}_l = 0$, indicating that the distant end is *free*, since a vanishing impedance permits a displacement velocity to exist with no accompanying force. The impedance at the source end is now a *pure reactance*,

$$\breve{Z}(0) = jZ_0 \tan\kappa l, \tag{4.2.17}$$

which can be made to have any value from $-j\infty$ to $+j\infty$ by choosing the value of l suitably. As a special case of (4.2.17), let $l \ll \lambda$. We can then approximate $\tan\kappa l$ by $\kappa l = \omega l/c_b$, so that

$$\breve{Z}(0) \to jS\rho_0 c_b \frac{\omega l}{c_b} = j\omega m, \tag{4.2.18}$$

where m is the mass $Sl\rho_0$ of the short section of the rod. The same result may be obtained directly by considering the force (4.2.2) applied to a *lumped* mass, i.e., dimensions much less than λ, which then undergoes an oscillatory motion described by Newton's second law

$$F = F_m e^{+j\omega t} = m \frac{d^2\xi}{dt^2}. \tag{4.2.19}$$

---

† The *force* (or pressure) reflection coefficient is the negative of $\breve{R}_a$, as defined by (4.2.15), but the *velocity* reflection coefficient is identical with $\breve{R}_a$. This distinction is important in relating the present case of acoustic waves to the analogous case of electromagnetic waves on an electric transmission line (Sec. 8.1). Voltage is usually considered the analog of force, current that of displacement velocity.

The displacement velocity, found by integration, is

$$\xi = \frac{F_m}{j\omega m} e^{+j\omega t}. \qquad (4.2.20)$$

Hence the small mass presents the impedance

$$\check{Z} = \frac{F}{\xi} = j\omega m \qquad (4.2.21)$$

to the force, in agreement with the value (4.2.18).

If $\kappa l = \pi/2$, that is, $l = \lambda/4$, the reactance (4.2.17) becomes infinite. A finite oscillating force can then produce no displacement velocity at the end of the rod where it is applied, although the rest of the segment has a standing wave on it.

As a final application of (4.2.16), suppose that we have a piece of waveguide one-quarter wavelength long with a load $\check{Z}_l$ at one end. We find that now

$$\check{Z}(0) = \frac{Z_0^2}{\check{Z}_l} \qquad (4.2.22)$$

is the impedance at the other end of the quarter-wave section. In particular, if $\check{Z}_l$ is real but greater than Z_0, then $\check{Z}(0)$ is also real but smaller than Z_0. Hence a quarter-wave section of waveguide acts much the way an electric transformer does in changing current, voltage, and impedance levels. This impedance-transforming property of the quarter-wave section is restricted to a narrow range of frequencies.

A quarter-wave section finds its greatest use in matching together two waveguides having different characteristic impedances. The matching section is chosen to have a characteristic impedance that is the geometrical mean of the two characteristic impedances of the waveguides being joined. Sinusoidal waves of correct frequency traveling down either waveguide can pass through the matching section to the other waveguide without reflection.

A useful application of the impedance-matching properties of a quarter-wave section is the layer put on glass to reduce surface reflections—the "coated" lenses that are now commonly used in optical equipment. The characteristic impedance of an optical medium for light waves is proportional to its refractive index. The coating layer is therefore chosen to have a refractive index that is the geometric mean of the index of air (unity) and that of the glass (≈ 1.5.) The optical thickness of the layer is made one-quarter of a wavelength of light in the middle of the visible spectrum, $\lambda \approx 550$ nm. The effectiveness of the coating in reducing reflections falls off toward the blue and red ends of the spectrum, since its optical thickness deviates progressively from a quarter-wavelength. Hence a coated lens appears to have a somewhat purple appearance when viewed in white light.

Problems

4.2.1   Verify the algebra leading to (4.2.13), (4.2.15), and (4.2.16).

4.2.2   A spring of negligible mass and of spring constant k has one end attached securely to a rigid support. What complex impedance does its free end present for axial motion? A small mass, whose impedance is given by (4.2.21), is attached to the spring at the free end. The combined impedance is the *sum of the two impedances*, since both the spring and the mass have the *same displacement velocity*. The force (4.2.2) is now applied to the mass-spring system. Find the resultant steady-state displacement velocity and the resonant frequency of the system.

4.2.3   Obtain the solution to Prob. 4.1.3 by requiring that the sum of the reactances of the two sections of waveguide vanish. Why does this condition give all the normal-mode frequencies?

4.2.4   Two long waveguides having different characteristic impedances Z_1 and Z_2 are joined by a quarter-wave section having the characteristic impedance $(Z_1Z_2)^{1/2}$. A wave $\xi_1 = A_1 e^{j(\omega t - \kappa_1 x)}$ is traveling toward the coupling section. Assume that there is a reflected wave $\xi_1 = \breve{B}_1 e^{j(\omega t + \kappa_1 x)}$, that there are waves going in both directions in the coupling section, and that there is a wave $\xi_2 = \breve{A}_2 e^{j(\omega t - \kappa_2 x)}$ transmitted into the other waveguide. By satisfying boundary conditions at the two junctions, show that $\breve{B}_1 = 0$ and that $\breve{A}_2 = A_1(Z_1/Z_2)^{1/2}$, which is the value one may get very easily from (4.2.10) by power considerations alone.

4.3   Rods with Varying Cross-sectional Area

We wish to see how the wave equation (4.1.3) is altered when the cross-sectional area of a rod is a slowly varying function of position x. The problem will not be treated exactly, since for ease in analysis we ignore the fact that the wavefronts in the rod deviate somewhat from being plane.

The wave equation can be obtained from (4.1.1) and (4.1.2) on the assumption that S is now a function of x. From (4.1.2) we find that

$$\frac{\partial F}{\partial x} = Y \frac{dS}{dx} \frac{\partial \xi}{\partial x} + YS \frac{\partial^2 \xi}{\partial x^2}. \tag{4.3.1}$$

Hence the wave equation becomes

$$\frac{\partial^2 \xi}{\partial x^2} + \frac{1}{S} \frac{dS}{dx} \frac{\partial \xi}{\partial x} = \frac{1}{c_b^2} \frac{\partial^2 \xi}{\partial t^2} \tag{4.3.2}$$

upon dividing through by SY and introducing the wave velocity (4.1.4).

The solution of (4.3.2) for most functions $S(x)$ is likely to be very difficult and have few features of general interest. It is instructive to choose forms for

$S(x)$ that permit the wave equation to be easily solved. Waveguides based on these forms have considerable practical importance.

We consider here only the simplest choice for $S(x)$, namely, the one that makes the coefficient of $\partial\xi/\partial x$ constant. If we call this constant 2α, it follows that

$$S(x) = S_0 e^{2\alpha x}, \tag{4.3.3}$$

where S_0 is the area of the waveguide at $x = 0$. A waveguide of this shape is called an *exponential horn*. The area of the waveguide increases by the factor e in a distance $1/2\alpha$.

Let us now turn to solving the wave equation for the horn,

$$\frac{\partial^2 \xi}{\partial x^2} + 2\alpha \frac{\partial \xi}{\partial x} = \frac{1}{c_b^2} \frac{\partial^2 \xi}{\partial t^2}, \tag{4.3.4}$$

by trying the function

$$\xi(x,t) = f(x)e^{-i\omega t}, \tag{4.3.5}$$

which depends sinusoidally on time. We find that $f(x)$ must satisfy the ordinary differential equation

$$\frac{d^2 f}{dx^2} + 2\alpha \frac{df}{dx} + \kappa_0^2 f = 0, \tag{4.3.6}$$

where $\kappa_0 = \omega/c_b$. This equation is closely related to that of damped simple harmonic motion and has the general solution

$$f(x) = A e^{-\alpha x} e^{i\kappa_1 x} + B e^{-\alpha x} e^{-i\kappa_1 x}, \tag{4.3.7}$$

in which $\kappa_1 \equiv (\kappa_0^2 - \alpha^2)^{1/2}$ and A and B are arbitrary constants. Hence the waves that can travel on the horn with a sinusoidal time dependence are of the form

$$\xi(x,t) = A e^{-\alpha x} e^{i(\kappa_1 x - \omega t)} + B e^{-\alpha x} e^{i(-\kappa_1 x - \omega t)}. \tag{4.3.8}$$

The first term represents a sinusoidal wave traveling in the $+x$ direction. Its amplitude *decreases* with the exponential factor $e^{-\alpha x}$ as it advances. The second term represents a similar sinusoidal wave traveling in the $-x$ direction. Its amplitude *increases* with the same exponential factor as it advances in the negative direction. For both waves, the wave velocity is

$$c_{\text{horn}} = \frac{\omega}{\kappa_1} = \frac{\omega}{(\kappa_0^2 - \alpha^2)^{1/2}} = \frac{c_b}{(1 - \alpha^2 c_b^2/\omega^2)^{1/2}} \tag{4.3.9}$$

which is greater than the wave velocity $c_b = \omega/\kappa_0$ on a straight section of the waveguide. Our analysis shows that the magnitude of α must be *less* than κ_0 for traveling waves to exist. In terms of frequency, ω must be *greater* than αc_b, which is termed the *cutoff frequency* of the horn. Exponential horns have considerable usefulness in increasing the efficiency of antique acoustic phonographs and modern "tweeter" loudspeakers. With negative α (decreasing cross section), the amplitude of ultrasonic vibrations can be increased up to the endurance or fatigue limit of the material of the horn, a technique of considerable technological importance.

Problems

4.3.1   Show that the resonant frequencies of an exponential horn of length l whose ends are free are given by $\omega_n = c_b(\alpha^2 + \pi^2 n^2/l^2)^{1/2}$.

4.3.2   A positive-going sinusoidal wave travels along an exponential horn. Show that the power transmitted is the same at all points.

4.3.3   Find the shape of a rod on which the amplitude of a traveling longitudinal wave varies inversely with distance x over some limited range.

4.3.4   Investigate what variation in rod area permits sinusoidal waves having a constant velocity of propagation on the rod. *Hint:* Assume $\xi = f(x)e^{i(\kappa x - \omega t)}$ and find what restriction this assumption puts on $S(x)$ in (4.3.2).

★4.3.5   An exponential horn whose area is given by (4.3.3) makes an excellent impedance transformer. Let us investigate the properties of such a horn whose length is an *integral number of half-wavelengths*, so that $\kappa_1 l = n\pi$. In the notation of Sec. 4.2, the wave on the horn is

$$\xi = e^{-\alpha x}(\breve{A}_+ e^{-j\kappa_1 x} + \breve{A}_- e^{j\kappa_1 x})e^{j\omega t}.$$

Compute the characteristic impedance $\breve{Z}(x) = F/\xi$ at any position x on the horn. Set it equal to the load impedance $\breve{Z}_l$ at $x = l = \pi/n\kappa_1$ and show that the impedance at $x = 0$ is then $\breve{Z}(0) = (S_0/S_l)\breve{Z}_l$, where S_0 is the area of the horn at $x = 0$ and S_l is the area at $x = l$.

★4.3.6   Consider the resonating exponential horn of Prob. 4.3.1, with $n = 1$. Let the ratio of diameters at the two ends be 10. Sketch the amplitude of the oscillatory strain along the horn and find the location and value of its maximum. Then evaluate (numerically) the "figure of merit"

$$\phi \equiv \frac{\text{velocity amplitude at small end}}{(c_b)(\text{maximum strain amplitude})}.$$

By (4.1.9), this quantity is unity for a uniform rod ($\alpha = 0$). The value of ϕ for a horn shows that a greater velocity amplitude can be attained with a given material than for a uniform rod. How big can the figure of merit get for horns of large diameter ratio? *Answer:* 30 percent of length from small end; $\phi = 1.99$; $\phi_{max} = 2.72$.

4.4 The Effect of Small Perturbations on Normal-mode Frequencies

The examples of normal-mode vibrations (standing waves) treated so far—especially transverse waves on a finite length of string and on a rectangular membrane, and longitudinal waves on a finite length of rod—have illustrated a general characteristic of such vibrations, namely, that the sum of the potential energy and the kinetic energy at any time is a constant and their average values are equal. This is a property of a normal-mode vibration of any conservative system characterized by linear elastic restoring forces. For each normal-mode vibration, then,

$$K(t) + V(t) = E = \text{const} \tag{4.4.1}$$

$$\overline{K(t)} = \overline{V(t)} = \tfrac{1}{2}E. \tag{4.4.2}$$

Furthermore, $\overline{V}$ and $\overline{K}$ depend on the amplitude A and frequency ω in the following way,

$$\overline{V} = \alpha A^2 \tag{4.4.3}$$

$$\overline{K} = \beta \omega^2 A^2, \tag{4.4.4}$$

where α and β are constants independent of A. Evidently α depends on the elastic properties of a vibrating system and β on the inertial properties of the system. Both α and β may depend on the integers specifying the normal mode, although β is often independent of mode numbers [see, for example, the expressions for V and K applying to a vibrating string segment (1.8.21) and (1.8.22)]. In view of (4.4.2), α and β are related to the normal-mode frequency by

$$\omega = \left(\frac{\alpha}{\beta}\right)^{1/2}. \tag{4.4.5}$$

In the course of analyzing various vibratory systems, it is often necessary to make certain approximations, or idealizations, in setting up a mathematical model that corresponds, nevertheless, fairly closely to the physical world. For example, we postulated a *flexible* string, whereas the stiffness of a real string or wire must make a small contribution to the linear restoring force produced primarily by tension in the string and hence modify the value of α slightly. Again, in the case of longitudinal waves on a rod, we postulated a *slender* rod to minimize a small error that we were making in β as a result of neglecting the inertia of the radial motion of the rod, related to the longitudinal motion by a finite Poisson's ratio. We may think of these neglected effects, which depend linearly on amplitude, as constituting small *perturbations* in the equation of motion for the normal-mode vibration of the idealized system. We now show that it is possible to make a first-order correction to the frequency of a normal-mode vibration by considering how the perturbation alters the *energy* of the vibration.

First suppose that we wish to correct the normal-mode frequency for a small additional *elastic restoring force* that was previously left out of consideration. When such an elastic perturbation is taken into account, the value of α in (4.4.3), which relates average potential energy to amplitude of vibration, is increased by a small amount $\delta\alpha$, with a negligible change in β, since no additional inertia is added to the system. We are assuming here that the perturbation is so small that it does not alter the functional form of the normal-mode motions significantly. If $\delta\overline{V}$ is the increase in potential energy due to the perturbation, then clearly

$$\frac{\delta\alpha}{\alpha} = \frac{\delta\overline{V}}{\overline{V}}. \tag{4.4.6}$$

When α changes by this fractional amount with no change in β, (4.4.5) shows that the fractional *increase* in frequency is

$$\frac{\delta\omega}{\omega} = \frac{1}{2}\frac{\delta\alpha}{\alpha} = \frac{1}{2}\frac{\delta\overline{V}}{\overline{V}} = \frac{\delta\overline{V}}{E}. \tag{4.4.7}$$

Note that for a given value of the amplitude A, an increase in α implies an increase in $\overline{V}$, which, of course, is always precisely equal to $\overline{K}$. An increase in $\overline{K}$ requires the increase in frequency specified by (4.4.7).

Next suppose that the system is perturbed by a small additional *mass loading* that does not alter its elastic properties. In such an event the value of β in (4.4.4) is increased by a small amount $\delta\beta$, with a negligible change in α. The fractional increase in β equals the fractional increase in average kinetic energy,

$$\frac{\delta\beta}{\beta} = \frac{\delta\overline{K}}{\overline{K}} \tag{4.4.8}$$

when both $\delta\overline{K}$ and $\overline{K}$ are calculated for the same value of ω. The fractional *decrease* in frequency brought about by the increase $\delta\overline{K}$ in average kinetic energy is therefore

$$\frac{\delta\omega}{\omega} = -\frac{1}{2}\frac{\delta\beta}{\beta} = -\frac{1}{2}\frac{\delta\overline{K}}{\overline{K}} = -\frac{\delta\overline{K}}{E}. \tag{4.4.9}$$

Note that now, for a given value of the amplitude A, a value of α unchanged by the perturbation implies that $\overline{V}$ and hence its equal $\overline{K}$ are also unchanged by the perturbation. Hence an increase in β necessitates the decrease in frequency specified by (4.4.9). No real (physical) increase $\delta\overline{K}$ in kinetic energy occurs: the value of $\delta\overline{K}$ in (4.4.9) is a virtual increase calculated on the assumption that no compensating decrease in frequency takes place.

The method used in establishing (4.4.7) and (4.4.9), though simple, is not entirely satisfactory. In particular, it gives no indication of how big a frequency correction $\delta\omega/\omega$ can be accommodated for a desired accuracy in the calculation; i.e., the method of derivation does not allow one to proceed to second and higher stages of approximation. A more satisfactory, but more involved, treatment of perturbations requires that the perturbation, whatever it is, be included as a term in the differential equation describing the wave motion. The perturbed equation is then solved by a sequence of approximations, with increased accuracy at each step in the sequence. The results obtained here turn out to be those given by a first-order perturbation calculation made in this way. In Sec. 4.9, a perturbation calculation based on the differential equation is carried out for longitudinal vibrations of a rod having a variable cross-sectional area.

Let us now apply the present results to find how stiffness alters the resonant frequencies of an idealized flexible-string segment stretched between fixed supports. The string, or wire as we may now call it, is under a steady tension τ_0 and has a mass λ_0 per unit length. We assume that the wire has a circular cross section of radius a, so that its section moment (3.6.6) is

$$I = \frac{\pi}{4} a^4. \tag{4.4.10}$$

Equation (3.6.11) shows that the potential energy density of bending is

$$V_1 = \tfrac{1}{2} I Y \left(\frac{\partial^2 \eta}{\partial x^2}\right)^2. \tag{4.4.11}$$

Now suppose that the normal-mode vibration, *neglecting stiffness*,

$$\eta = A \sin\kappa_n x \sin\omega_n t \tag{4.4.12}$$

is present on the wire, where $\omega_n = c\kappa_n = n\pi c/l$ is the nth harmonic frequency. The total average potential energy is then that found by averaging (1.8.22),

$$\overline{V} = \tfrac{1}{8} l \tau_0 \kappa_n^2 A^2. \tag{4.4.13}$$

When a small amount of stiffness contributes to the elastic restoring force, we can assume that the shape of the oscillating segment of wire is very little changed. Hence we can substitute (4.4.12) into (4.4.11) and integrate over the length of the wire to find the perturbation in the average potential energy due to stiffness. In this way we find that

$$\delta\overline{V} = \tfrac{1}{8} l I Y \kappa_n^4 A^2. \tag{4.4.14}$$

Therefore (4.4.7) shows that the fractional increase in frequency due to stiffness, when the increase is small, amounts to

$$\frac{\delta\omega}{\omega} = \frac{I Y \kappa_n^2}{2\tau_0}. \tag{4.4.15}$$

This result can be expressed in the form

$$\frac{\delta\omega}{\omega} = n^2 \frac{\pi^2}{8} \left(\frac{a}{l}\right)^2 \frac{SY}{\tau_0},\tag{4.4.16}$$

where $S = \pi a^2$ is the cross-sectional area of the wire and SY is the hypothetical tension that would stretch the wire to twice its normal length. For a typical midrange string of a piano, $SY/\tau_0 \approx 10^3$, $a/l \approx 10^{-3}$, so that the overtones are sharp by the fractional amount $\delta\omega/\omega \approx 10^{-3}n^2$. The human ear is accustomed to this small amount of inharmonicity in the overtones of a piano and can distinguish an actual piano tone from a synthetic tone that differs from it only in having precise harmonic overtones.

Next let us estimate how much the resonant frequencies of a vibrating rod of circular section are lowered by radial motion. The rod has a length l, a radius a, a Young's modulus Y, and a Poisson's ratio σ. The radius of the rod is assumed small compared with a wavelength, so that a knowledge of Poisson's ratio enables us to calculate the radial strain $\partial\eta/\partial r$ at any position along the rod where the longitudinal strain is $\partial\xi/\partial x$. Thus,

$$\frac{\partial\eta}{\partial r} = -\sigma \frac{\partial\xi}{\partial x}.\tag{4.4.17}$$

By integrating (4.4.17) from 0 to r, we find that the radial displacement η of the rod at a distance r from the axis is

$$\eta = -\sigma \frac{\partial\xi}{\partial x} r.\tag{4.4.18}$$

We take the unperturbed normal-mode vibration to be of the form

$$\xi = A \cos\kappa_n x \sin\omega_n t,\tag{4.4.19}$$

where $\omega_n = c_b\kappa_n = n\pi c_b/l$ is the nth harmonic frequency. The average kinetic energy of the unperturbed vibration is, from (4.1.11),

$$\bar{K} = \tfrac{1}{8}lS\rho_0\omega_n^2A^2,\tag{4.4.20}$$

in which $S = \pi a^2$. To find the average kinetic energy associated with the perturbation, first note that the contribution to it from the volume element $2\pi r\, dr\, dx$ is

$$d^2(\delta K) = \tfrac{1}{2}\rho_0 2\pi r\, dr\, dx \left(\frac{\partial\eta}{\partial t}\right)^2.\tag{4.4.21}$$

The radial velocity $\partial\eta/\partial t$ can be found from (4.4.18) and (4.4.19) on the assumption that the perturbation does not change the form of (4.4.19) appreciably. On substituting this expression for $\partial\eta/\partial t$ in (4.4.21), integrating over the entire rod, and averaging over one period, we find that

$$\delta\bar{K} = \tfrac{1}{16}\sigma^2 lS\rho_0 a^2\kappa_n^2\omega_n^2A^2.\tag{4.4.22}$$

Therefore the fractional decrease in frequency amounts to

$$\frac{\delta\omega}{\omega} = -\frac{\delta\overline{K}}{2\overline{K}} = -\tfrac{1}{4}a^2\sigma^2\kappa_n{}^2 = -n^2\frac{\pi^2}{4}\left(\frac{a}{l}\right)^2\sigma^2. \tag{4.4.23}$$

This estimate of the frequency correction begins to lose its accuracy when its magnitude exceeds more than a few percent.

Since the wave number κ_n in both examples is set by n and the length l of the vibrating member, we can interpret the fractional change in frequency as being caused by an equal fractional change in wave velocity. When the wave velocity depends on frequency (or wavelength), the medium is said to possess the property of *dispersion*. The consequences of dispersion are discussed in Sec. 4.7.

Problems

4.4.1   Carry out the algebra leading to (4.4.14) and (4.4.22).

4.4.2   A small bead of mass m is attached to a flexible string of linear mass density λ_0 and tension τ_0 stretched between fixed supports separated a distance l. Assume that $m \ll l\lambda_0$. Discuss how it changes the normal-mode frequencies of the string when it is placed at various positions along the string.

4.4.3   A simple pendulum consists of a particle of mass m hung on a massless string of length l in the earth's gravitational field. Its period for small-amplitude motion is then $T = 2\pi(l/g)^{1/2}$. Find by a perturbation calculation how the period is altered if the pendulum actually has a small moment of inertia I_0 about its center of mass, which we continue to take at a distance l from the point of support. The moment of inertia I_0 includes contributions from the mass of the string and from the finite dimensions of the particle.

4.4.4   A long slender rod is vibrating in its fundamental longitudinal mode. It is desired to raise its frequency 1 percent by drilling holes of equal depth in each end of the rod. The holes are to have an area equal to one-fourth the cross-sectional area of the rod. How deep should the holes be drilled? How will their presence affect the first few overtones?

4.4.5   Consider again the rod of Prob. 4.4.4. It is now desired to *lower* its fundamental frequency 1 percent by removing material from the middle section of the rod. How much material should be removed? How will its removal affect the first few overtones?

★4.4.6   Show that longitudinal waves on a rod of circular section whose radius a is small compared with a wavelength obey the wave equation

$$Y\frac{\partial^2\xi}{\partial x^2} = \rho_0\left(\frac{\partial^2\xi}{\partial t^2} - \frac{1}{2}\sigma^2 a^2\frac{\partial^4\xi}{\partial x^2\,\partial t^2}\right) \tag{4.4.24}$$

when radial motion is taken into account through Poisson's ratio σ. Generalize the equation to a rod of noncircular cross section. Show that (4.4.24) is consistent with (4.4.23).

4.5   Torsional Waves on a Round Rod

We now study the wave equation for *torsional* waves on a round rod of radius a, density ρ_0, and shear modulus μ. Figure 4.5.1 shows an element Δx of the rod, with the x axis along the axis of the rod. When a torsional wave is present, at any instant of time the end at x has been rotated through an angle ϕ by the wave, whereas the end at $x + \Delta x$ has been rotated through the angle $\phi + (\partial \phi / \partial x) \, \Delta x$. According to (3.4.5) the net twist of the element $\Delta \phi = (\partial \phi / \partial x) \, \Delta x$ requires a torque of magnitude

$$M = \frac{\pi}{2} \mu a^4 \frac{\Delta \phi}{\Delta x} \rightarrow \frac{\pi}{2} \mu a^4 \frac{\partial \phi}{\partial x}. \tag{4.5.1}$$

If $-M$ is the torque acting on the element at x and $M + (\partial M / \partial x) \, \Delta x$ the torque acting on it at $x + \Delta x$, not only is the torque condition (4.5.1) satisfied, but there is a net unbalanced torque $(\partial M / \partial x) \, \Delta x$ available to give the element an angular acceleration. The moment of inertia of the element about the axis is

$$\Delta I = \tfrac{1}{2} \Delta m \, a^2 = \frac{\pi}{2} a^4 \rho_0 \, \Delta x. \tag{4.5.2}$$

Hence we have that

$$\frac{\partial M}{\partial x} = \frac{\pi}{2} a^4 \rho_0 \frac{\partial^2 \phi}{\partial t^2}, \tag{4.5.3}$$

where Δx has been canceled. On evaluating $\partial M / \partial x$, using (4.5.1), we find that

$$\frac{\partial^2 \phi}{\partial x^2} = \frac{1}{c_r^2} \frac{\partial^2 \phi}{\partial t^2}, \tag{4.5.4}$$

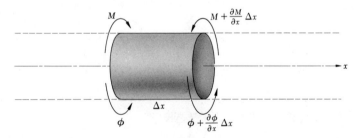

Fig. 4.5.1   Element of rod twisted by a torsional wave.

where

$$c_r \equiv \left(\frac{\mu}{\rho_0}\right)^{1/2}, \tag{4.5.5}$$

the velocity for torsional waves, is independent of the radius of the rod.

We have not had to make approximations in obtaining the wave equation for torsional waves on a circular rod. Therefore the normal-mode frequencies of a rod of length l with free ends vibrating in torsion constitute an accurate harmonic series. In view of the relation (3.3.6)

$$Y = 2\mu(1 + \sigma),$$

connecting Y, μ, and σ, the velocity of torsional waves is less than that of longitudinal waves on a slender rod of the same material by the factor $[2(1 + \sigma)]^{-1/2}$, which for most metals amounts to about 0.6.

The kinetic energy density associated with the traveling wave

$$\phi = \phi_m \cos(\kappa x - \omega t) \tag{4.5.6}$$

is

$$K_1 = \tfrac{1}{2} I_0 \omega^2 \phi_m^2 \sin^2(\kappa x - \omega t), \tag{4.5.7}$$

where

$$I_0 = \frac{\pi}{2} \rho_0 a^4 \tag{4.5.8}$$

is the moment of inertia of unit length of the rod. The potential energy density, from (3.4.6), is

$$V_1 = \frac{\pi}{4} a^4 \mu \kappa^2 \phi_m^2 \sin^2(\kappa x - \omega t)$$
$$= \tfrac{1}{2} I_0 \omega^2 \phi_m^2 \sin^2(\kappa x - \omega t) \tag{4.5.9}$$

upon introducing I_0 from (4.5.8) and substituting $\rho_0 \omega^2$ for $\mu \kappa^2$. The two energy densities are equal, and the average total energy density is

$$\bar{E}_1 = \tfrac{1}{2} I_0 \omega^2 \phi_m^2. \tag{4.5.10}$$

The power passing any point is evidently

$$P = -M \frac{\partial \phi}{\partial t}. \tag{4.5.11}$$

For the wave (4.5.6), we find, on using (4.5.1) for M, (4.5.8) for I_0, and (4.5.5) for c_r,

$$P = \frac{\pi}{2} \mu a^4 \kappa \phi_m^2 \sin^2(\kappa x - \omega t)$$
$$= I_0 c_r \omega^2 \phi_m^2 \sin^2(\kappa x - \omega t). \tag{4.5.12}$$

The average power is therefore

$$\bar{P} = \tfrac{1}{2}I_0 c_r \omega^2 \phi_m{}^2 = c_r \bar{E}_1. \tag{4.5.13}$$

That is, the power transmission may be visualized as the total energy per unit length flowing at the wave velocity c_r.

Problems

4.5.1 Develop the theory of the normal-mode torsional vibrations of a rod of length l having free ends. Show how to express an arbitrary vibration as a sum of normal-mode vibrations. Show how to evaluate the normal-mode amplitudes from an initial configuration $\phi(x)$ and an initial velocity $\partial \phi(x)/\partial t$.

4.5.2 Apply the results of Prob. 4.5.1 to find the subsequent vibration if the rod is initially twisted by equal and opposite torques applied to its two ends and released at $t = 0$. *Hint:* Use the methods developed in Sec. 1.6.

4.5.3 Show that the expressions for energy densities (4.5.7) and (4.5.9) and for power flow (4.5.13) also hold for a hollow cylindrical tube provided I_0 is properly interpreted.

4.5.4 Show how to define a characteristic impedance for torsional waves on a round rod or tube and express the power (4.5.13) in terms of this impedance.

4.5.5 A round rod has a radius that is a gradual function of position x. Derive the wave equation for torsional waves on the rod, which corresponds to (4.3.2) for longitudinal waves on a rod having a varying cross-sectional area. *Hint:* Introduce I, as defined in Prob. 3.4.4, to simplify your equations.

4.6   Transverse Waves on a Slender Rod

(a)   The Wave Equation

A more complicated case of wave motion is that of *transverse* waves on a slender rod, whose static bending is described by the beam equation (3.6.9). The x axis is taken to be along the rod, in equilibrium, at the centroid of its cross section. The y axis is taken to be in the transverse direction of the wave displacement, with η specifying the displacement of the neutral surface. We ignore the effect of gravity, as well as the presence of supports necessary to carry the weight of the rod. Although the rod must be uniform along its length, it may have any shape of section, with an area S and section moment I (3.6.6) about the z axis.

For a continuous static load $w(x)$, (3.5.3) shows that the bending moment

$M(x)$ is related to the load by

$$\frac{d^2M}{dx^2} = w(x). \tag{4.6.1}$$

On combining this equation with (3.6.9), which relates the bending moment to the displacement of the neutral surface, we have that

$$IY\frac{d^4\eta}{dx^4} = w(x). \tag{4.6.2}$$

When a transverse wave is present, the load $w(x)$ becomes the *kinetic reaction* of the rod in opposing the acceleration that the wave gives to it at any position and time. Most of this equivalent load comes from the transverse acceleration $\partial^2\eta/\partial t^2$, which is equivalent to the load (note the negative sign!)

$$w_t(x,t) = -\rho_0 S \frac{\partial^2\eta}{\partial t^2} \tag{4.6.3}$$

acting on unit length of the rod. There is a second, normally much smaller, contribution to the equivalent load, brought about by the fact that any element Δx of the rod has a moment of inertia $I\rho_0\,\Delta x$ about a transverse axis lying in the neutral surface at the element. The transverse wave turns the element through a small angle $\partial\eta/\partial x$ with an angular acceleration $\partial^3\eta/(\partial x\,\partial t^2)$. The torque required for the rotary acceleration comes from a shearing force V_r, drawn as a positive shearing force in Fig. 4.6.1. In the case of a static loading of the rod, the shearing force V_r would give rise to a difference in bending moment at the two ends of the element Δx. Here the torque, or couple, $-V_r\,\Delta x$ is responsible for the rotary acceleration of the element,

$$-V_r\,\Delta x = I\rho_0\,\Delta x\,\frac{\partial^3\eta}{\partial x\,\partial t^2}. \tag{4.6.4}$$

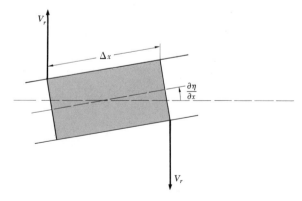

Fig. 4.6.1   Element of rod rotated by transverse wave.

By (3.5.1), the existence of a shearing force V_r would indicate a static load dV_r/dx on the rod. Since such a load is not present, dV_r/dx becomes an equivalent kinetic-reaction load per unit length (note the change in sign)

$$w_r(x,t) = -\frac{\partial V_r}{\partial x} = I\rho_0 \frac{\partial^4 \eta}{\partial x^2 \, \partial t^2} \tag{4.6.5}$$

that must be added to $w_t(x,t)$, as given by (4.6.3), to give the total equivalent kinetic reaction per unit length of the rod in "resisting" accelerations produced by the passage of a transverse wave. On substituting the sum of (4.6.3) and (4.6.5) for $w(x)$ into (4.6.2) and dividing by $S\rho_0$, we thus arrive at the wave equation

$$-\frac{YI}{\rho_0 S}\frac{\partial^4 \eta}{\partial x^4} + \frac{I}{S}\frac{\partial^4 \eta}{\partial x^2 \, \partial t^2} = \frac{\partial^2 \eta}{\partial t^2}. \tag{4.6.6}$$

If the rod is reasonably slender or thin in the y direction, the term in (4.6.6) due to rotary inertia is very small compared with the other two terms. It is customarily omitted in discussing transverse waves on a slender rod. Therefore we attempt to solve only the somewhat simpler wave equation

$$\frac{\partial^4 \eta}{\partial x^4} = -\frac{\rho_0 S}{YI}\frac{\partial^2 \eta}{\partial t^2}, \tag{4.6.7}$$

which is *fourth* order in x, and *second* order in t.

(b)   Solution of the Wave Equation

A wave of arbitrary functional form $f(x \pm ct)$ does not satisfy the wave equation (4.6.7), though of course it satisfies the ordinary wave equation such as (1.1.3). This fact indicates that a wave of arbitrary shape cannot retain its shape as it travels along the rod. We next try a wave having a sinusoidal time dependence

$$\eta(x,t) = f(x)e^{-i\omega t}, \tag{4.6.8}$$

where $f(x)$ may be complex. On substituting this trial solution into (4.6.7), we find that $f(x)$ satisfies the ordinary differential equation

$$\frac{d^4 f}{dx^4} - \kappa^4 f = 0, \tag{4.6.9}$$

where

$$\kappa \equiv \left(\frac{\rho_0 S \omega^2}{YI}\right)^{1/4} \tag{4.6.10}$$

has the dimensions of reciprocal length.

To find the solutions of (4.6.9), let us try $f(x) = e^{\alpha x}$. We find that $e^{\alpha x}$ is a solution provided

$$\alpha^4 = \kappa^4, \tag{4.6.11}$$

which is satisfied by the four roots $\alpha = \kappa, -\kappa, i\kappa, -i\kappa$. Hence a general solution of (4.6.7) having $e^{-i\omega t}$ as a time factor is

$$\eta(x,t) = (A e^{\kappa x} + B e^{-\kappa x} + C e^{i\kappa x} + D e^{-i\kappa x}) e^{-i\omega t}, \tag{4.6.12}$$

where A, B, C, D are arbitrary constants, which may be complex. The actual vibration of the rod is represented by the real part of (4.6.12).

(c) Traveling Waves

By choosing the values of the constants in (4.6.12) suitably, we can construct a sinusoidal wave traveling in the positive direction having as a real part

$$\eta(x,t) = \eta_m \cos(\kappa x - \omega t). \tag{4.6.13}$$

The velocity of such a wave is evidently

$$c_t \equiv \frac{\omega}{\kappa} = \left(\frac{YI\omega^2}{\rho_0 S}\right)^{1/4}. \tag{4.6.14}$$

If we introduce the *radius of gyration* of the section $b \equiv (I/S)^{1/2}$ and the *bar velocity* $c_b \equiv (Y/\rho_0)^{1/2}$ and eliminate ω, the transverse velocity (4.6.14) becomes

$$c_t = \kappa b c_b = \frac{2\pi b}{\lambda} c_b. \tag{4.6.15}$$

Thus we can say either that the transverse wave velocity varies with the square root of the frequency or, equivalently, that it varies inversely with the wavelength. A slender rod thus possesses *dispersion* to a marked degree so far as transverse waves are concerned.

The kinetic energy density associated with the wave (4.6.13) is

$$K_1 = \tfrac{1}{2}\rho_0 S \left(\frac{\partial \eta}{\partial t}\right)^2 = \tfrac{1}{2}\rho_0 S \omega^2 \eta_m^2 \sin^2(\kappa x - \omega t). \tag{4.6.16}$$

It has the average value

$$\overline{K}_1 = \tfrac{1}{4}\rho_0 S \omega^2 \eta_m^2. \tag{4.6.17}$$

The potential energy density is given by (3.6.11),

$$V_1 = \tfrac{1}{2}YI \left(\frac{\partial^2 \eta}{\partial x^2}\right)^2 = \tfrac{1}{2}YI\kappa^4 \eta_m^2 \sin(\kappa x - \omega t). \tag{4.6.18}$$

It has the average value

$$\overline{V}_1 = \tfrac{1}{4}YI\kappa^4 \eta_m^2 = \tfrac{1}{4}\rho_0 S \omega^2 \eta_m^2 \tag{4.6.19}$$

on replacing κ^4 by its value (4.6.10). Hence the kinetic and potential energy densities are equal, as in the case of the other elastic waves we have been studying. The total average energy density is

$$\bar{E}_1 = \bar{K}_1 + \bar{V}_1 = \tfrac{1}{2}\rho_0 S \omega^2 \eta_m^2. \tag{4.6.20}$$

The power carried by the transverse wave is a bit more difficult to compute. Power is transferred past any position both by the shearing force† V, acting in conjunction with the displacement velocity $\partial\eta/\partial t$, and by the bending moment M, acting in conjunction with the angular velocity $\partial^2\eta/(\partial x\,\partial t)$. The sign conventions for V and M, as defined in Sec. 3.5, are such that a positive V and positive $\partial\eta/\partial t$ constitute an energy flow in the positive direction, whereas a positive M and positive $\partial^2\eta/(\partial x\,\partial t)$ constitute an energy flow in the negative direction. Hence the total energy flow past any position in the positive direction is

$$P = V\frac{\partial\eta}{\partial t} - M\frac{\partial^2\eta}{\partial x\,\partial t}. \tag{4.6.21}$$

In view of the fact that

$$M = YI\frac{\partial^2\eta}{\partial x^2}$$

$$V = \frac{\partial M}{\partial x} = YI\frac{\partial^3\eta}{\partial x^3},$$

(4.6.21) becomes

$$P = YI\left(\frac{\partial^3\eta}{\partial x^3}\frac{\partial\eta}{\partial t} - \frac{\partial^2\eta}{\partial x^2}\frac{\partial^2\eta}{\partial x\,\partial t}\right). \tag{4.6.22}$$

The power carried by the traveling wave (4.6.13) is then

$$P = YI[\kappa^3\omega\eta_m^2\,\sin^2(\kappa x - \omega t) + \kappa^3\omega\eta_m^2\,\cos^2(\kappa x - \omega t)]$$

$$= YI\kappa^3\omega\eta_m^2 = \rho_0 S c_t \omega^2\eta_m^2, \tag{4.6.23}$$

after eliminating κ by its definition (4.6.10) and introducing the transverse wave velocity $c_t = \omega/\kappa$. Our calculation shows that each of the terms in (4.6.22) contributes the same average amount to P. Since at any position and time the two terms are out of phase, the instantaneous power is a constant independent of x and t! This behavior differs from that of other waves we have investigated, as exemplified by (4.1.11) giving the power carried by a longitudinal wave on a slender rod.

We now call attention to a puzzling conclusion. Our calculation of energy

---

† The notation V for shearing force must not be confused with V_1 for potential energy density.

flow (4.6.23) and average total energy density (4.6.20) shows that

$$P = \bar{P} = (2c_t)\bar{E}_1, \tag{4.6.24}$$

whereas for the other traveling waves we have studied, the average flow of energy turned out to be the average energy density times the wave velocity. An extra factor of 2 has appeared! An explanation is given in Sec. 4.7, where the concept of *group velocity* is discussed.

(d)   Normal-mode Vibrations

An alternative way of writing the solution (4.6.12) is in the real form

$$\eta(x,t) = (A \sinh\kappa x + B \cosh\kappa x + C \sin\kappa x + D \cos\kappa x) \cos\omega t. \tag{4.6.25}$$

The solution in this form constitutes a more convenient starting point for treating normal-mode vibrations of a finite length of rod, i.e., standing waves. Hyperbolic sines and cosines are related to the exponential function by the real form of Euler's relation,

$$e^{\pm\theta} = \cosh\theta \pm \sinh\theta. \tag{4.6.26}$$

This equation reveals the intimate connection between the hyperbolic and circular functions. For example, when θ is replaced by $i\theta$, it follows that $\cosh i\theta = \cos\theta$ and $\sinh i\theta = i \sin\theta$.

Three distinct sorts of boundary conditions can hold at the end of a transversely vibrating rod.

FREE END

The bending moment and the shearing force both vanish at a free end. In view of (3.6.9) and (3.5.2), these conditions are equivalent to

$$\frac{\partial^2\eta}{\partial x^2} \equiv \eta'' = 0 \qquad \frac{\partial^3\eta}{\partial x^3} \equiv \eta''' = 0. \tag{4.6.27}$$

HINGED END

The bending moment and the displacement both vanish at a hinged end, that is,

$$\frac{\partial^2\eta}{\partial x^2} \equiv \eta'' = 0 \qquad \eta = 0. \tag{4.6.28}$$

CLAMPED END

The displacement and the slope both vanish at a clamped end, that is,

$$\eta = 0 \qquad \frac{\partial\eta}{\partial x} \equiv \eta' = 0. \tag{4.6.29}$$

Accordingly there are six possible cases for standing waves on a slender rod, depending on which of the three sets of boundary conditions are applied at the ends of the rod. We shall consider here only the so-called free-free case, for which at both $x = 0$ and $x = l$, $\eta'' = \eta''' = 0$.

The displacement $\eta(x)$ and its first three space derivatives become

$$\begin{aligned}
\eta &= (A \sinh\kappa x + B \cosh\kappa x + C \sin\kappa x + D \cos\kappa x) \cos\omega t \\
\eta' &= \kappa(A \cosh\kappa x + B \sinh\kappa x + C \cos\kappa x - D \sin\kappa x) \cos\omega t \\
\eta'' &= \kappa^2(A \sinh\kappa x + B \cosh\kappa x - C \sin\kappa x - D \cos\kappa x) \cos\omega t \\
\eta''' &= \kappa^3(A \cosh\kappa x + B \sinh\kappa x - C \cos\kappa x + D \sin\kappa x) \cos\omega t.
\end{aligned} \tag{4.6.30}$$

At $x = 0$, the condition $\eta'' = 0$ requires that $B = D$, and the condition $\eta''' = 0$ that $A = C$. The same conditions at $x = l$ give the two equations

$$\begin{aligned}
A(\sinh\kappa l - \sin\kappa l) + B(\cosh\kappa l - \cos\kappa l) &= 0 \\
A(\cosh\kappa l - \cos\kappa l) + B(\sinh\kappa l + \sin\kappa l) &= 0.
\end{aligned} \tag{4.6.31}$$

These homogeneous equations give the ratio B/A provided their determinant vanishes, i.e., provided

$$(\sinh\kappa l - \sin\kappa l)(\sinh\kappa l + \sin\kappa l) = (\cosh\kappa l - \cos\kappa l)^2.$$

This transcendental equation for κl readily simplifies into the form

$$\cosh m \cos m = 1. \tag{4.6.32}$$

The roots m_n of (4.6.32) give the values, i.e., the eigenvalues, of $\kappa_n = m_n/l$ for the normal-mode vibrations of the free-free rod. By numerical methods, the roots of (4.6.32) are found to be

$$\begin{aligned}
m_1 &= 4.730 = \tfrac{3}{2}\pi(1.0037) \\
m_2 &= 7.853 = \tfrac{5}{2}\pi(0.9999) \\
m_3 &= 10.996 = \tfrac{7}{2}\pi(1.0000) \\
m_{n>3} &= (n + \tfrac{1}{2})\pi \qquad \text{to 5 significant figures.}
\end{aligned} \tag{4.6.33}$$

The frequencies of the various normal-mode vibrations follow from (4.6.15),

$$\omega_n = \kappa_n c_t = \kappa_n^2 b c_b = m_n^2 \frac{b c_b}{l^2}. \tag{4.6.34}$$

The overtones are not in a harmonic progression. For example, the frequency ratio of the first overtone to the fundamental is $(m_2/m_1)^2 \approx 2.76$. The two frequencies are separated two octaves and about $5\tfrac{1}{2}$ semitones on the musical scale. Hence the lateral vibration of a free-free bar when many overtones are excited, as for instance by striking it, has a characteristic discordant sound.

Let us return to the shape of the standing wave on the rod. To find how η depends on x for one of the normal modes, say the nth mode, the value of m_n from (4.6.33) is assigned to κl in one or the other of the equations (4.6.31) and the ratio B/A found. If the first of these equations is used, then

$$\eta(x,t) = a[(\cosh m_n - \cos m_n)(\sinh \kappa_n x + \sin \kappa_n x) \\ - (\sinh m_n - \sin m_n)(\cosh \kappa_n x + \cos \kappa_n x)] \cos \omega_n t, \quad (4.6.35)$$

where a is arbitrary. Figure 4.6.2 shows curves depicting the dependence of η on x for $n = 1, 2$, and 3. The fundamental mode has two nodes, at a distance $0.22418l$ from the ends. The nth mode has $n + 1$ nodes. The functions of x multiplying $\cos \omega_n t$ in (4.6.35) constitute the set of normal-mode functions, or eigenfunctions, of the vibrating bar.

We need to investigate the magnitude of the error made in neglecting the rotary inertia term in the wave equation (4.6.6). It is fairly laborious to evaluate the kinetic-energy integrals for the normal-mode functions of the free-free case to make the perturbation calculation described in Sec. 4.4 possible. We shall be content here to estimate the order of magnitude of the frequency correction by examining how much the velocity of traveling waves is altered when rotary inertia is taken into account. According to Prob. 4.6.2, the transverse velocity is then

$$c_t = \kappa b c_b (1 + b^2 \kappa^2)^{-1/2}, \qquad (4.6.36)$$

to be compared with (4.6.15). The fractional error made in neglecting rotary inertia is therefore

$$\frac{\delta c_t}{c_t} \approx -\tfrac{1}{2} b^2 \kappa^2 = -\frac{1}{2}\left(\frac{2\pi b}{\lambda}\right)^2 \qquad (4.6.37)$$

when it is small. Hence a rod may be considered "slender" for transverse waves when the wavelength is large compared with the thickness of the rod in the plane of vibration. The fractional error made in computing the normal-mode

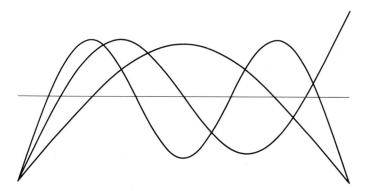

Fig. 4.6.2   The first three normal modes of a free-free vibrating bar.

frequencies of a free-free rod by (4.6.34) will be approximately given by (4.6.37) if λ is interpreted as twice the distance between adjacent nodes on the vibrating rod.

Problems

4.6.1   Derive the wave equation for transverse waves on a slender rod under longitudinal tension. The equation should reduce to (4.6.6) for zero tension and to (1.1.3) for zero section moment I.

4.6.2   Show that the transverse wave velocity for a traveling sinusoidal wave, when rotary inertia is taken into account, is given by $c_t = c_b b \kappa (1 + b^2 \kappa^2)^{-1/2}$. *Hint:* Make use of (4.6.13) and (4.6.6).

4.6.3   Show that the frequencies for the transverse vibrations of a bar of length l clamped at one end and free at the other may be found from the equation $\cosh m \cos m = -1$, in which $m = \kappa l$. Verify that the smallest root is $m_1 = 1.875$.

4.6.4   A tuning fork consists of two clamped-free bars, clamped to each other at the stalk and free at their other ends. From the results of Prob. 4.6.3 show that a tuning fork made with rectangular tines l cm long and t cm thick from steel for which $c_b = 5,237$ m/sec has a frequency $\nu = 84,590 t / l^2$. Design a practical tuning fork for $\nu = 440$ Hz (standard pitch for A on the musical scale).

★4.6.5   The roots of the transcendental equation (4.6.32) are conveniently found by making the substitution $m = (n + \frac{1}{2})\pi + (-1)^{n-1}\alpha$, where α is a small positive number. Show that this leads to a transcendental equation for α whose roots may be found by expanding the functions of α in power series. By this means verify the roots (4.6.33).

4.7   Phase and Group Velocity

We have found that when wave propagation in a medium is described by the usual wave equation

$$\frac{\partial^2 \eta}{\partial x^2} = \frac{1}{c^2} \frac{\partial^2 \eta}{\partial t^2}, \tag{4.7.1}$$

where c is a constant independent of frequency, a wave of arbitrary shape keeps its form and travels with the wave velocity c. For many media, a more complicated linear wave equation describes wave propagation, e.g., the fourth-order wave equation (4.6.7) for lateral waves on a rod. In such cases only sinusoidal waves can propagate without change in form. Their wave velocity, however, depends on frequency or, equivalently, on wavelength. A medium is then said to possess the property of *dispersion*.

In both dispersive and nondispersive media, the wave velocity for sinusoidal waves of a particular frequency is termed the *phase velocity*, and it is related to frequency and wave number by

$$c_{\text{phase}} = \frac{\omega}{\kappa}. \tag{4.7.2}$$

In the case of dispersive media, if c_{phase} decreases with frequency in some frequency range, the medium is said to possess *normal* dispersion in this range. If it increases with frequency in some frequency range, it is said to possess *anomalous* dispersion in this range.†

An arbitrarily shaped wave traveling in a dispersive medium changes its shape as it progresses. If we pick out some identifiable aspect of the wave and follow it along, the wave packet, or wave group, is found to travel with a speed that depends on its shape. To treat this situation analytically, in linear media we can decompose the arbitrary wave mathematically into a *spectrum* of sinusoidal waves by means of Fourier analysis. The phase velocity of each of the sinusoidal components is known from the differential equation that describes wave propagation in the medium. At a later time the shape of the wave can be found by adding up the component waves at that time. If the arbitrary wave is periodic, the Fourier analysis involves only the Fourier series discussed briefly in Sec. 1.7. If the arbitrary wave is nonperiodic, it is necessary to employ Fourier integrals, which are a generalization of Fourier series. Fourier integrals are discussed briefly in Chap. 12; here we shall discuss the motion of wave packets from a more elementary point of view. We return to a more complete discussion in Sec. 12.5.

Let us therefore consider the simplest possible wave packet consisting of two sinusoidal waves of equal amplitude but of slightly different frequencies. The wavelengths of the two waves differ somewhat. If the waves are instantaneously in phase at certain positions in the medium, there are other positions where they are out of phase and therefore cancel each other. A snapshot of the wave disturbance reveals the pattern of *beats* illustrated in Fig. 4.7.1. If the medium is dispersive, the two waves travel along with their slightly differing wave velocities. The pattern of beats, however, travels with a so-called *group*

---

† The terms *normal* and *anomalous* dispersion are borrowed from optics. The normal behavior of an optically transparent material, such as glass, is for the velocity of light to decrease with increasing frequency. A prism made from a material having normal dispersion "disperses" a light beam into a spectrum with a greater angle of refraction for the spectral components having higher frequency or shorter wavelength.

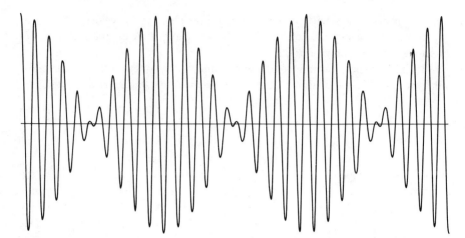

Fig. 4.7.1   The superposition of two sinusoidal waves having identical amplitudes but different frequencies.

velocity, which may be quite different from the average of the phase velocities of the two waves.

To obtain a formula for calculating the group velocity, analogous to (4.7.2) for the phase velocity, let us superpose the two sinusoidal waves

$$\eta_1 = A \exp i[(\kappa + \tfrac{1}{2}\Delta\kappa)x - (\omega + \tfrac{1}{2}\Delta\omega)t]$$
$$\eta_2 = A \exp i[(\kappa - \tfrac{1}{2}\Delta\kappa)x - (\omega - \tfrac{1}{2}\Delta\omega)t], \tag{4.7.3}$$

which have identical amplitudes but have wave numbers differing by $\Delta\kappa$ and frequencies differing by $\Delta\omega$. The mean wave number is κ, and the mean frequency is ω, implying an average phase velocity ω/κ for the two waves. The sum of the two waves may be written

$$\eta = \eta_1 + \eta_2 = A\{\exp[i(\tfrac{1}{2}\Delta\kappa x - \tfrac{1}{2}\Delta\omega t)] + \exp[-i(\tfrac{1}{2}\Delta\kappa x - \tfrac{1}{2}\Delta\omega t)]\}e^{i(\kappa x - \omega t)}. \tag{4.7.4}$$

The physical wave is represented by the real part of (4.7.4),

$$\eta = 2A \cos(\tfrac{1}{2}\Delta\kappa x - \tfrac{1}{2}\Delta\omega t) \cos(\kappa x - \omega t), \tag{4.7.5}$$

which is shown graphically in Fig. 4.7.1. The fine-scaled wave has a wave number κ that is the average of the wave numbers of the two component waves and has a wavelength $\lambda = 2\pi/\kappa$. The amplitude of this wave is modulated by a coarse-scaled wave envelope that has a wave number $\Delta\kappa/2$, which is half the difference of the wave numbers of the two component waves. Although the wavelength of the modulating wave is $4\pi/\Delta\kappa$, the separation of successive beats is half this distance, namely, $2\pi/\Delta\kappa$.

At any position x, a plot of η against time also has the appearance shown in Fig. 4.7.1 if the x axis is now interpreted as the time axis. The fine-scaled wave has a frequency ω, which is the average of the frequencies of the two component waves, and the coarse-scaled wave envelope has a frequency $\Delta\omega/2$, which is half the difference of the frequencies of the two component waves. The beats, however, occur with the frequency $\Delta\omega$ and are separated in time by $2\pi/\Delta\omega$. The pattern of beats travels along the x axis with the *group velocity*

$$c_{\text{group}} = \frac{\Delta\omega}{\Delta\kappa} \rightarrow \frac{d\omega}{d\kappa}. \tag{4.7.6}$$

If it is known how phase velocity depends on wave number, (4.7.2) expresses ω as a function of κ, that is, $\omega = \kappa c_{\text{phase}}(\kappa)$, from which (4.7.6) makes it possible to calculate the group velocity. In a medium having no dispersion, c_{phase} is independent of κ, and (4.7.6) shows that $c_{\text{group}} = c_{\text{phase}}$. Although (4.7.6) applies strictly to a wave packet consisting of only two waves separated slightly in frequency and wave number, it continues to be useful for more complicated wave packets of sinusoidal waves involving a range of frequencies, provided that in this range ω is a smoothly varying function of κ.

An alternative way of expressing the group velocity is

$$c_{\text{group}} = c_{\text{phase}} - \lambda \frac{dc_{\text{phase}}}{d\lambda}, \tag{4.7.7}$$

which may be readily established from (4.7.2) and (4.7.6). Equation (4.7.7) has the interesting geometrical interpretation shown in Fig. 4.7.2. A further geometrical interpretation of the properties of a dispersive medium is a graph of frequency ω against wave number κ (Fig. 4.7.3). Such a plot is often called an *ω-β diagram* in electrical-engineering terminology, or sometimes a *Brillouin diagram*. Its distinctive feature is that the phase and group velocities are given, respectively, by the slope of the line from the origin to a particular point on the curve and by the slope of the tangent to the curve at the point, in accordance with (4.7.2) and (4.7.6). The equation connecting ω with κ is often called the *dispersion relation*, even when the relation is linear so that no dispersion exists.

To illustrate the distinction between phase and group velocity, let us compute the group velocity for transverse waves on a slender rod. For such a medium the dispersion relation has the form (4.6.15)

$$\omega = \kappa^2 b c_b, \tag{4.7.8}$$

so that

$$c_{\text{group}} = \frac{d\omega}{d\kappa} = 2\kappa b c_b = 2c_{\text{phase}}. \tag{4.7.9}$$

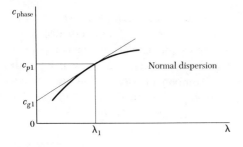

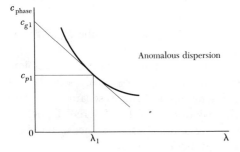

Fig. 4.7.2 The relation between phase and group velocities and wavelength.

Hence, when rotary inertia can be neglected, the group velocity is twice the phase velocity. When rotary inertia is taken into account, Prob. 4.7.4 shows that with decreasing wavelength the phase velocity approaches the group velocity and both converge on the wave velocity c_b for longitudinal waves.

The distinction between group and phase velocity serves to clear up the puzzle uncovered in attempting to account for the energy carried by a trans-

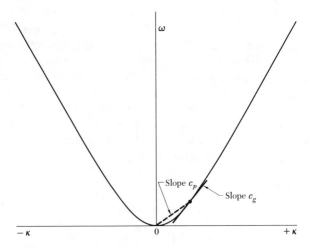

Fig. 4.7.3 The relation between frequency and wave number for a dispersive medium.

verse sinusoidal wave on a slender rod. Equation (4.6.24) shows that the rate of energy transport is $2c_t$ times the average total energy density $\bar{E}_1$ on the rod. We now recognize that $2c_t$ is the group velocity, so that (4.6.24) becomes

$$\bar{P} = c_{\text{group}}\bar{E}_1. \tag{4.7.10}$$

A study of wave propagation in dispersive media shows that in general the rate of energy transport by a sinusoidal wave is given by the group velocity times the average total energy density.

Problems

4.7.1 Consider a medium in which $c_{\text{phase}} = A\omega^n$, where A and n are constants. Show that $c_{\text{group}} = c_{\text{phase}}/(1 - n)$. For what values of n is the dispersion normal? Anomalous?

4.7.2 The condition for anomalous dispersion is $dc_{\text{phase}}/d\lambda < 0$. What does this condition mean geometrically on an ω-κ diagram similar to Fig. 4.7.3? There exist wave-propagation systems for which $d\omega/d\kappa$ has the opposite sign from ω/κ; that is, the group and phase velocities are in opposite directions. How would this situation show up on an ω-κ diagram? (Such *backward waves* occur in the backward-wave oscillator, a type of vacuum tube for generating microwaves. Have you ever noticed a caterpillar crawling along with ripples running backward from head to tail?)

4.7.3 There exist media in which the product of phase velocity and group velocity is a constant. If such a medium possesses normal dispersion, how does the phase velocity depend on wavelength? (Electromagnetic wave propagation in ionized gases and in hollow-pipe waveguides is an example of this situation.)

4.7.4 Use the wave velocity for transverse waves on a rod found in Prob. 4.6.2 to find how group velocity depends on wavelength when rotary inertia is taken into account. Make a sketch showing both phase velocity and group velocity as a function of wavelength. Note that Fig. 4.7.3 is plotted for this case.

4.8   *Waves on a Helical Spring*

A mass hung on a spring is a favorite example of a system that exhibits simple harmonic motion. Ordinarily the mass m of the spring is considered to be negligible compared with that of the suspended mass M. If ξ_M measures the downward displacement of the suspended mass from its equilibrium position, the equation of motion is

$$-k\xi_M = M\frac{d^2\xi_M}{dt^2}, \tag{4.8.1}$$

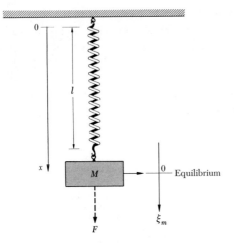

Fig. 4.8.1   Mass-spring oscillator.

where k is the spring constant of the spring. Equation (4.8.1) has the general solution

$$\xi_M = a \cos\omega_0 t + b \sin\omega_0 t, \tag{4.8.2}$$

where

$$\omega_0 = \left(\frac{k}{M}\right)^{1/2} \tag{4.8.3}$$

is the frequency of oscillation.

Let us now reexamine the theory of the mass-spring oscillator (Fig. 4.8.1), taking into account the distributed mass of the spring. We assume that the spring is uniform, so that its linear mass density is $m_1 = m/l$, where l is the equilibrium length of the spring, with M in place. If a static tensile force F is applied to the spring, stretching it a small amount Δl, then by the definition of the spring constant

$$F = k\,\Delta l = kl\,\frac{\Delta l}{l} = kl\,\frac{\partial\xi}{\partial x}, \tag{4.8.4}$$

an equation that continues to hold when the local extension $\partial\xi/\partial x$ varies with position. Evidently kl is the hypothetical force that would double the length of the spring, no matter how long a piece of spring is involved. The wave equation for longitudinal waves on the spring is easily found to be

$$\frac{\partial^2\xi}{\partial x^2} = \frac{1}{c_s^2}\frac{\partial^2\xi}{\partial t^2}, \tag{4.8.5}$$

where

$$c_s \equiv \left(\frac{kl}{m_1}\right)^{1/2} = \left(\frac{k}{m}\right)^{1/2} l \tag{4.8.6}$$

is the wave velocity.

We are now ready to consider that the mass M is moving up and down in simple harmonic motion. Evidently there then exists a longitudinal standing wave on the spring having a node at the point of support, $x = 0$. The standing wave is a solution of (4.8.5) of the form

$$\xi = A \sin\kappa x \cos\omega t \tag{4.8.7}$$

with a frequency ω, as yet undetermined, related to the wave number κ by $\omega = \kappa c_s$. The motion of the mass M at $x = l$ is therefore

$$\xi_M = A \sin\kappa l \cos\omega t. \tag{4.8.8}$$

The boundary condition at the junction of the spring and mass is that the force exerted by the spring on the mass, as given by (4.8.4) with a minus sign, shall equal M times its acceleration. Accordingly, we have that

$$-kl \left(\frac{\partial \xi}{\partial x}\right)_{x=l} = M \frac{d^2 \xi_M}{dt^2}. \tag{4.8.9}$$

After making substitutions involving (4.8.6) to (4.8.8), we find that the equation

$$\kappa l \tan\kappa l = \frac{m}{M} \tag{4.8.10}$$

determines the wave number κ of the standing wave and hence the frequency of the simple harmonic motion.

We can solve the transcendental equation (4.8.10) for κl by a sequence of approximations. Let $\theta \equiv \kappa l$, so that

$$\theta \tan\theta = \frac{m}{M}. \tag{4.8.11}$$

The Taylor series expansion for $\tan\theta$ is

$$\tan\theta = \theta + \tfrac{1}{3}\theta^3 + \tfrac{2}{15}\theta^5 + \cdots . \tag{4.8.12}$$

ZEROTH–ORDER APPROXIMATION

Suppose that $m/M \ll 1$. Approximate $\tan\theta$ by θ. Then $\theta = \kappa_0 l = (m/M)^{1/2}$, and

$$\omega_0 = \kappa_0 c_s = \frac{1}{l}\left(\frac{m}{M}\right)^{1/2} \left(\frac{k}{m}\right)^{1/2} l = \left(\frac{k}{M}\right)^{1/2}, \tag{4.8.13}$$

which agrees with (4.8.3).

FIRST–ORDER APPROXIMATION

Now keep one more term in the expansion of $\tan\theta$. Then

$$\theta(\theta + \tfrac{1}{3}\theta^3) = \theta^2(1 + \tfrac{1}{3}\theta^2) = \frac{m}{M}.$$

Since $\tfrac{1}{3}\theta^2$ is small compared with unity, we replace it by its value according to the zeroth-order approximation. Hence

$$\theta^2\left(1 + \frac{1}{3}\frac{m}{M}\right) = \frac{m}{M}$$

$$\theta = \kappa_1 l = \left(\frac{m}{M + \tfrac{1}{3}m}\right)^{1/2},$$

and

$$\omega_1 = \kappa_1 c_s = \left(\frac{k}{M + \tfrac{1}{3}m}\right)^{1/2}. \tag{4.8.14}$$

Thus the finite mass of the spring can be taken into account to a first approximation by adding one-third of its mass to M. It is not profitable to carry the iteration further. When m/M is not small, the frequency can be found by solving (4.8.11) numerically. The first-order approximation is equivalent to assuming that the strain along the spring is uniform, so that any element of the spring suffers a displacement proportional to the distance x from the fixed end.

Problems

4.8.1 Investigate the roots of $\theta \tan\theta = m/M$ by sketching $y = \tan\theta$ and $y = m/M\theta$ for small m/M. What is the physical significance of the roots slightly in excess of $\theta = \pi$, $2\pi, \ldots$?

4.8.2 Show that in the first-order approximation, the relative shift in period $T = 2\pi/\omega$ due to the mass of the spring is

$$\frac{T_1 - T_0}{T_0} = \frac{1}{6}\frac{m}{M} = \frac{\pi^2}{6}\left(\frac{t_l}{T_0}\right)^2,$$

where $t_l = 2l/c_s$ is the time required for a longitudinal wave in the spring to travel from M to the point of support and back again. Thus, show that the familiar limit of negligible spring mass is equivalent to the limit $t_l^2 \ll T_0^2$.

4.8.3 Make a perturbation calculation, as described in Sec. 4.4, to show how the frequency of a mass-spring oscillator is altered when the small mass m of the spring is taken into account.

4.8.4 A disk of moment of inertia I is hung on a wire so as to constitute a torsion pendulum. Investigate how the inertial properties of the wire contribute to the period of the pendulum.

*4.9 Perturbation Calculations

In Sec. 4.4 we developed a simple method for estimating how much the frequencies of normal-mode vibrations are altered by small perturbations. The method described there is based on calculating the small change in kinetic (or potential) energy of a normal-mode vibration caused by the perturbation, assuming that the normal-mode motion is unchanged except for frequency, and then relating the fractional change in energy to the fractional change in frequency. Here we consider another, more accurate, technique for taking into account a small perturbing term in the differential equation for the normal-mode vibrations. We illustrate the method by showing how a small variation in the cross-sectional area of a rod alters its normal-mode longitudinal vibrational frequencies.

The wave equation (4.3.2) describes the propagation of longitudinal waves on a slender rod when the cross-sectional area varies with position. If we are interested in the normal-mode frequencies of a free-free rod whose area varies with position, we can start by looking for a solution of (4.3.2) of the form

$$\xi(x,t) = f(x) \cos\omega t. \tag{4.9.1}$$

We then find that $f(x)$ must satisfy the ordinary differential equation

$$\frac{d^2f}{dx^2} + \frac{\omega^2}{c_b{}^2}f = -F(x)\frac{df}{dx}, \tag{4.9.2}$$

where

$$F(x) \equiv \frac{1}{S(x)}\frac{dS(x)}{dx} \tag{4.9.3}$$

is a prescribed function of x and where we have put the small perturbing term on the right-hand side of the equation.

A rough approximation to the normal-mode functions and frequencies of the rod consists in neglecting completely the perturbing term. The normal-mode functions are then the ones discussed in Sec. 4.1, namely,

$$f_n(x) = \cos\kappa_n x, \tag{4.9.4}$$

where the wave numbers are

$$\kappa_n l = n\pi \qquad n = 1, 2, 3, \ldots \tag{4.9.5}$$

and the normal-mode frequencies are

$$\omega_n = c_b \kappa_n = \frac{n\pi c_b}{l}. \tag{4.9.6}$$

When the perturbing term is taken into account, the normal-mode functions must continue to satisfy the requirement that df/dx vanish at $x = 0$ and

$x = l$. Although we need these functions only in the range $0 < x < l$, it is evident that they may be considered to be periodic functions of period $2l$, thus reducing to $\cos\kappa_n x$ when the perturbing term is negligible. Furthermore they must be *even* functions of x, to ensure that their derivatives df/dx be *odd* functions vanishing at $x = 0$ and l. We know, from our discussion of Fourier series in Sec. 1.7, that any well-behaved function of this type can be expanded in a cosine series. Accordingly, we attempt to solve (4.9.2) by an expansion of the perturbed normal-mode functions in a Fourier series of the normal-mode functions of the unperturbed equation.

To simplify the notation, we find it convenient to write

$$\phi_n(x) = \left(\frac{2}{l}\right)^{1/2} \cos\kappa_n x, \tag{4.9.7}$$

with the κ_n given by (4.9.5) for the unperturbed normal-mode functions (eigenfunctions). The coefficient $(2/l)^{1/2}$ is called a *normalizing factor*, making the integral $\int_0^l \phi_n{}^2(x) \, dx$ equal unity. We then have

$$\int_0^l \phi_m(x)\phi_n(x) \, dx = \delta_{mn}, \tag{4.9.8}$$

where the Kronecker delta function is defined by the properties

$$\delta_{mn} = \begin{cases} 0 & m \neq n \\ 1 & m = n. \end{cases} \tag{4.9.9}$$

Suppose we wish to find the normal-mode frequency of the fundamental mode, $n = 1$. We start by writing the Fourier series expansion of $f_1(x)$ in the form

$$f_1(x) = \phi_1(x) + \sum_{n=2}^{\infty} a_n \phi_n(x) \tag{4.9.10}$$

where the coefficients a_n are to be found by requiring that $f_1(x)$ satisfy the differential equation (4.9.2). If we substitute (4.9.10) into (4.9.2) and endeavor to find the a_n by multiplying the resulting equation by $\phi_m(x)$ and integrating from 0 to l, we obtain an infinite set of simultaneous algebraic equations for the unknown a_n and the unknown frequency ω_1. The equations cannot, in general, be solved in closed form, so that it is necessary to resort to an iterative procedure for finding increasingly more accurate values of ω_1 and of the a_n.

The rationale of the procedure is based on the premise that the perturbing term in (4.9.2) produces only a relatively small change in the normal-mode functions and in the normal-mode frequencies. Hence the a_n in (4.9.10) are all much less than unity, so that to a *zeroth-order* approximation $f_{01}(x) = \phi_1(x)$ and the zeroth-order frequencies are simply $\omega_{0n} = c_b\kappa_n = n\pi c_b/l$. (We find it convenient to designate the order of the approximation by a subscript preceding the normal-mode number.)

A *first-order* approximation is then made by finding ω_{11} and the a_{1n} using the zeroth-order approximation for $f_1(x)$ in the perturbing term. The perturbation calculations of Sec. 4.4 were of this character.

A *second-order* approximation is next made by finding ω_{21} and the a_{2n}, using the first-order approximation for $f_1(x)$ in the perturbing term, and so on. Usually the second-order approximation is sufficiently accurate, making the labor of carrying the procedure to a higher order of iteration pointless.

The procedure for carrying out each of these iterations consists in substituting (4.9.10) in the left side of (4.9.2), multiplying the equation through by ϕ_m ($m = 1, 2, 3, \ldots$), and integrating from 0 to l, using (4.9.8) for the integrals on the left-hand side of the equation and noting that $d^2\phi_n/dx^2 = -\kappa_n^2\phi_n$. The rth-order iteration is then found from the $(r-1)$st by

$$\omega_{r1}^2 = \omega_{01}^2 - c_b^2 \int_0^l \phi_1 F \frac{df_{r-1,1}}{dx} \, dx \tag{4.9.11}$$

$$a_{rn} = \frac{c_b^2 \int_0^l \phi_n F \dfrac{df_{r-1,1}}{dx} \, dx}{\omega_{0n}^2 - \omega_{r-1,1}^2}. \tag{4.9.12}$$

The first-order approximation consists in replacing $f_{01}(x)$ in the integrals by its zeroth-order value, $\phi_1(x)$. If we let

$$F_{nm} \equiv \int_0^l \phi_n F \frac{d\phi_m}{dx} \, dx, \tag{4.9.13}$$

we find to first order that

$$\omega_{11}^2 = \omega_{01}^2 - c_b^2 F_{11} \tag{4.9.14}$$

$$a_{1n} = \frac{c_b^2 F_{n1}}{\omega_{0n}^2 - \omega_{01}^2}. \tag{4.9.15}$$

The second-order approximation is found by substituting the first-order normal-mode function

$$f_{11}(x) = \phi_1(x) + \sum_{n=2}^{\infty} \frac{c_b^2 F_{n1} \phi_n(x)}{\omega_{0n}^2 - \omega_{01}^2} \tag{4.9.16}$$

into (4.9.11) and (4.9.12) to find better values for the frequency ω_1 and the coefficients a_n. We shall find only the second-order estimate of the frequency here, since we shall not attempt to use the a_{2n} to carry the iteration for the frequency to a third order. We thus substitute (4.9.16) into (4.9.11) and find that

$$\omega_{21}^2 = \omega_{01}^2 - c_b^2 F_{11} - c_b^4 \sum_{n=2}^{\infty} \frac{F_{1n} F_{n1}}{\omega_{0n}^2 - \omega_{01}^2}. \tag{4.9.17}$$

This result can be put in a slightly neater form by dividing through by $\omega_{01}{}^2$ and using (4.9.6) for the unperturbed normal-mode frequencies,

$$\left(\frac{\omega_{21}}{\omega_{01}}\right)^2 = 1 - \left(\frac{l}{\pi}\right)^2 F_{11} - \left(\frac{l}{\pi}\right)^4 \sum_{n=2}^{\infty} \frac{F_{1n}F_{n1}}{n^2 - 1}. \tag{4.9.18}$$

We have thus obtained a second-order correction to the frequency of the fundamental mode. The result is found to be amazingly good provided the variation in area of the rod with position is not too great. For example, (4.9.18) happens to give exactly the fundamental frequency of the exponential horn with free ends. The mathematical technique described here is of great importance in other areas of theoretical physics, particularly in quantum mechanics, which abounds in such eigenvalue problems.

Problems

4.9.1   Carry out the algebra of finding the first- and second-order estimates of the frequency, as given by (4.9.14) and (4.9.17).

4.9.2   For the free-free exponential horn of length l, for which $F(x) = 2\alpha$, show that (4.9.18) gives exactly the result obtained in Prob. 4.3.1 for the fundamental frequency. *Note:*

$$\sum_{n=1}^{\infty} \frac{n^2}{(4n^2 - 1)^3} = \frac{\pi^2}{256}.$$

★4.9.3   Show that (4.9.18) for the second-order frequency calculation can be put in the form

$$\left(\frac{\omega_{21}}{\omega_{01}}\right)^2 = 1 - G_2 - \frac{1}{4} \sum_{n=2}^{\infty} n \left(\frac{n+1}{n-1} G_{n+1}^2 - \frac{n-1}{n+1} G_{n-1}^2\right),$$

where

$$G_n = \frac{2}{l} \int_0^l \ln S(x) \cos \frac{n\pi x}{l} \, dx.$$

This form is more suitable for practical calculations.

five

Acoustic Waves in Fluids

A relatively simple class of three-dimensional waves comprises those which can propagate in a nonviscous fluid. A familiar example is sound waves in air. As in previous chapters, we use this specific model as a vehicle for discussing several important features of wave motion in general. In particular we examine spherically symmetric traveling waves and standing waves in a rectangular enclosure. A brief account of shock waves in a gas serves to introduce the properties of large-amplitude (nonlinear) waves.

5.1 The Wave Equation for Fluids

The only stress component in a fluid when viscous effects are assumed negligible is the hydrostatic pressure P. In the absence of wave motion, P has an equilibrium value P_e, which may vary with position, because of gravitational effects. When a compressional wave is passing through the fluid but the fluid is otherwise at rest, the local pressure differs from its equilibrium value by the amount†

$$p(\mathbf{r},t) = P(\mathbf{r},t) - P_e(\mathbf{r}). \tag{5.1.1}$$

We shall find it more convenient to use the incremental pressure p, rather than the total pressure P, as one of the dependent variables characterizing compressional waves in the fluid.

The incremental pressure $p(\mathbf{r},t)$ is a *scalar* function of position and time, whereas the related displacement of the fluid $\varrho(\mathbf{r},t)$ is a *vector* function of position and time. Since it is much easier to work with a scalar wave equation than with a vector equation, we regard p, rather than ϱ, as the basic dependent variable in the wave equation for compressional waves in the fluid.

The volume strain, or dilatation θ, which is associated with the incremental pressure, is related to the displacement ϱ by the equation

$$\theta = \frac{\Delta V}{V} = \frac{\partial \xi}{\partial x} + \frac{\partial \eta}{\partial y} + \frac{\partial \zeta}{\partial z} = \mathbf{\nabla} \cdot \varrho, \tag{5.1.2}$$

as discussed in Sec. 3.2. The incremental pressure and the accompanying dilatation are linearly related through the bulk modulus B,

$$p = -B\theta = -B\,\mathbf{\nabla} \cdot \varrho, \tag{5.1.3}$$

which expresses Hooke's law for a fluid. We assume for the waves considered here that $\theta \ll 1$.

Let us apply Newton's second law to the motion of the fluid contained in the cubical element $\Delta x \, \Delta y \, \Delta z$ shown in Fig. 5.1.1. As a first step, we need to find the net force on the element arising from wave-induced pressure variations on its six faces. The force in the x direction on the face $ABCD$ is $p \, \Delta y \, \Delta z$, whereas the force in the x direction on the face $EFGH$ is $[p + (\partial p/\partial x)\Delta x] \, \Delta y \, \Delta z$. Hence the net force in the positive x direction is

$$\Delta F_x = -\frac{\partial p}{\partial x} \Delta x \, \Delta y \, \Delta z.$$

Similarly the net forces in the positive y and z directions are

$$\Delta F_y = -\frac{\partial p}{\partial y} \Delta x \, \Delta y \, \Delta z \qquad \Delta F_z = -\frac{\partial p}{\partial z} \Delta x \, \Delta y \, \Delta z.$$

---

† The notation $p(\mathbf{r},t)$ denotes a dependence on three spatial coordinates and on time without implying the specific choice of cartesian coordinates x, y, z.

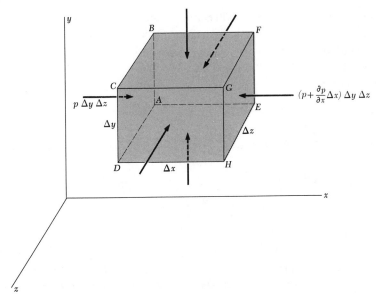

Fig. 5.1.1   Forces acting on a fluid element.

The net *vector* force on the cubical element is therefore

$$\Delta \mathbf{F} = -\nabla p \, \Delta x \, \Delta y \, \Delta z, \tag{5.1.4}$$

where

$$\nabla p \equiv \mathbf{i} \frac{\partial p}{\partial x} + \mathbf{j} \frac{\partial p}{\partial y} + \mathbf{k} \frac{\partial p}{\partial z} \tag{5.1.5}$$

is the *pressure gradient*. Since the vector acceleration of the fluid in the element is $\partial^2 \varrho / \partial t^2$ and its mass is $\rho_0 \, \Delta x \, \Delta y \, \Delta z$, where ρ_0 is the density of the fluid,[†] Newton's second law requires that

$$\Delta \mathbf{F} = -\nabla p \, \Delta x \, \Delta y \, \Delta z = \rho_0 \, \Delta x \, \Delta y \, \Delta z \, \frac{\partial^2 \varrho}{\partial t^2}.$$

On dividing through by $\Delta x \, \Delta y \, \Delta z$,

$$-\nabla p = \rho_0 \frac{\partial^2 \varrho}{\partial t^2}. \tag{5.1.6}$$

At this stage in deriving earlier wave equations, we expressed the elastic force, on the left side of the equations corresponding to (5.1.6), in terms of the

[†] The customary use of ρ_0 for density and our systematic use of ϱ for the displacement vector should cause no confusion. Note also that we are ignoring a possible variation in ρ_0 due to variations in equilibrium hydrostatic pressure with depth.

appropriate strain component and elastic modulus. Since we now wish to regard the scalar pressure p, rather than the vector displacement ϱ, as the basic dependent variable, we need to use Hooke's law in the form (5.1.3) to eliminate ϱ from the right side of (5.1.6). Let us therefore take the divergence of (5.1.6) to obtain an equation containing $\nabla \cdot \varrho$,

$$-\nabla \cdot \nabla p = \rho_0 \frac{\partial^2 (\nabla \cdot \varrho)}{\partial t^2}. \tag{5.1.7}$$

By eliminating $\nabla \cdot \varrho$ using (5.1.3), we obtain the three-dimensional scalar wave equation

$$\nabla^2 p = \frac{1}{c_f^2} \frac{\partial^2 p}{\partial t^2}, \tag{5.1.8}$$

where

$$c_f \equiv \left(\frac{B}{\rho_0} \right)^{1/2} \tag{5.1.9}$$

is the wave velocity in the fluid. The left side of (5.1.8) is the *three-dimensional laplacian* of the pressure p. The *laplacian operator* in three dimensions may be written variously as

$$\nabla^2 \equiv \nabla \cdot \nabla \equiv \text{div } \mathbf{grad} \equiv \frac{\partial^2}{\partial x^2} + \frac{\partial^2}{\partial y^2} + \frac{\partial^2}{\partial z^2}, \tag{5.1.10}$$

where the last form is its expression in rectangular coordinates.

In deciding to use the scalar wave equation (5.1.8) to describe compressional waves in fluids, we have not given up the possibility of finding the displacement ϱ for a pressure wave that is the solution of (5.1.8). We need to compute the pressure gradient ∇p, substitute it in (5.1.6), and obtain the displacement in the wave by making two integrations with respect to time. If the wave has a sinusoidal time dependence, the two integrations amount to dividing the expression being integrated by $-\omega^2$. Hence, for a sinusoidal wave of frequency ω,

$$\varrho(\mathbf{r},t) = \frac{1}{\rho_0 \omega^2} \nabla p_\omega(\mathbf{r}) \, e^{-i\omega t}, \tag{5.1.11}$$

where $p_\omega(\mathbf{r})$ is the spatial part of the pressure wave. Evidently $p_\omega(\mathbf{r})$ satisfies the time-independent wave equation

$$\nabla^2 p_\omega + \frac{\omega^2}{c_f^2} p_\omega = 0 \tag{5.1.12}$$

obtained by substituting

$$p(\mathbf{r},t) = p_\omega(\mathbf{r}) e^{-i\omega t} \tag{5.1.13}$$

in (5.1.8). Equation (5.1.12) is known as the *scalar Helmholtz equation.*

Equation (5.1.11) has an interesting consequence for wave motion in a fluid. We can easily show that the curl of the gradient of any scalar function of position is identically zero (see Prob. 5.1.2). Hence for a compressional wave in a fluid satisfying (5.1.8) and having a sinusoidal time dependence the curl of the displacement vector vanishes, that is,

$$\nabla \times \varrho = 0. \tag{5.1.14}$$

Accordingly the strain accompanying such a wave is free from rotation, which we showed to be $\phi = \frac{1}{2}\nabla \times \varrho$ in Sec. 3.3. A vector function of position whose curl vanishes in some region is said to be *irrotational* in that region.

By superposing many waves having the form of (5.1.11), it is possible to construct three-dimensional waves having a great variety of forms. All of these waves thus have the property of being irrotational.

Problems

5.1.1 Show that the most general solution of the wave equation (5.1.8) consisting of plane waves traveling in the direction of the unit vector $\mathbf{n}$ is given by

$$p = f_1(\mathbf{n} \cdot \mathbf{r} - c_f t) + f_2(\mathbf{n} \cdot \mathbf{r} + c_f t),$$

where f_1 and f_2 are arbitrary functions.

5.1.2 Show that $\nabla \times \mathbf{V} \equiv 0$ when $\mathbf{V}$ is the gradient of a scalar function of position, $\mathbf{V} = \nabla \phi$.

⋆5.2 The Velocity of Sound in Gases

Let us examine some of the consequences of the expression (5.1.9) for the velocity of compressional waves in a fluid before discussing solutions of the three-dimensional wave equation. It is usual to call the wave velocity in a fluid the *velocity of sound*. In the present section we wish to show how the sound velocity in a gas is related to other properties of the gas, without going into great detail regarding how the actual thermodynamic calculations are made.

A gas of molecular weight μ is a thermodynamic substance whose equilibrium state is determined by its pressure P and absolute temperature T. A definite volume V of the gas of mass m has the density $\rho_0 = m/V$ and contains $n = m/\mu$ moles. The so-called *equation of state* of the gas is a functional relation between P, V, and T for a specified amount of gas, such as 1 mole. Under equilibrium conditions, two of the state variables determine the third. The equation of state of real gases, for conditions remote from those causing liquefaction, is expressed to good approximation by that of an *ideal gas*

$$PV = nRT, \tag{5.2.1}$$

where $R = 8{,}314$ J/kg mole-deg is the universal gas constant, one of the fundamental constants of nature.

Thermodynamics teaches that we need to know more than the equation of state to specify completely the thermodynamic properties of a simple substance having the state variables P, V, and T. The additional information can consist of a knowledge of a specific heat, such as the specific heat at constant pressure C_p or the specific heat at constant volume C_v, as a function of temperature. With this information, and that of the equation of state, thermodynamics enables us to compute many other properties of the substance, including its bulk modulus B and therefore its velocity of sound.

Gases that have an equation of state closely approximated by (5.2.1) usually have nearly constant specific heats C_p and C_v over a limited temperature range. When such a gas is expanded (or compressed) in such a way that no heat flows into or out of the gas, a thermodynamic calculation shows that the ideal gas then obeys the *adiabatic* equation

$$PV^\gamma = \text{const} \quad \text{(adiabatic)}, \tag{5.2.2}$$

where $\gamma \equiv C_p/C_v$ is the (constant) ratio of the specific heats. When heat can flow so as to maintain constant temperature, the ideal gas obeys Boyle's law

$$PV = \text{const} \quad \text{(isothermal)}. \tag{5.2.3}$$

Both these equations become more and more accurate for real gases in the limit of vanishing pressure. The first gives the *adiabatic* bulk modulus, $B_{ad} = \gamma P$, and the second gives the *isothermal* bulk modulus, $B_{iso} = P$, as found earlier in Prob. 3.2.2. The corresponding sound velocities (5.1.9) are then

$$c_{ad} = \left(\gamma \frac{P}{\rho_0}\right)^{1/2} = \left(\gamma \frac{RT}{\mu}\right)^{1/2} = \left(\gamma \frac{kT}{m_0}\right)^{1/2} \tag{5.2.4}$$

$$c_{iso} = \left(\frac{P}{\rho_0}\right)^{1/2} = \left(\frac{RT}{\mu}\right)^{1/2} = \left(\frac{kT}{m_0}\right)^{1/2}. \tag{5.2.5}$$

The final forms given in (5.2.4) and (5.2.5) express the velocities in microscopic (molecular) rather than macroscopic terms by substituting Boltzmann's constant $k = R/N$ and the mass of an individual molecule $m_0 = \mu/N$, where N is Avogadro's number.

It is the adiabatic velocity (5.2.4) that gives more nearly the correct value for the velocity of sound in gases, because the compressions and rarefactions are so widely separated that a negligible transfer of heat takes place. The isothermal formula (5.2.5) is chiefly of historical interest. Newton proposed it in 1686 on theoretical grounds, before the distinction between isothermal and adiabatic processes was recognized. The newtonian formula for the velocity of sound in air gives a value considerably less than the experimental value, since

TABLE 5.1   Acoustic Properties of Common Gases (STP)†

Gas		Velocity c, m/sec	Density ρ_0, kg/m³	Impedance $\rho_0 c$, kg/m²-sec	$\gamma = \dfrac{C_p}{C_v}$
Hydrogen	H_2	1270	0.090	114	1.41
Helium	He	970	0.178	173	1.66
Neon	Ne	435	0.900	385	1.64
Nitrogen	N_2	337	1.25	421	1.40
Air		331	1.29	428	1.40
Argon	Ar	319	1.78	569	1.67
Oxygen	O_2	317	1.43	453	1.40
Carbon dioxide	CO_2	258	1.96	508	1.30

† Adapted from Dwight E. Gray (ed.), "American Institute of Physics Handbook," 2d ed., McGraw-Hill Book Company, New York, 1963, which contains more extensive tables, with references.

$\gamma \approx 1.4$ for air. Laplace, in 1816, pointed out the need to regard the process as adiabatic. Undoubtedly, a very small heat flow does take place. However, it is easy to show that such a heat flow causes a dying out, or *attenuation*, of a traveling sound wave. Since everyday experience indicates that sound can travel great distances with little attenuation (other than that associated with the inverse-square law), the amount of heat flow must be extremely small, implying that the adiabatic assumption is an extremely good one.

In Table 5.1 are listed some values of the velocity of sound for a number of gases at standard conditions, together with measured values of γ. It may be recalled that the classical value of γ for a monatomic gas is $\frac{5}{3} = 1.67$, for a diatomic gas $\frac{7}{5} = 1.4$, and, in general, $(\nu + 2)/\nu$, where ν is the number of classical degrees of freedom of the gas molecules.

Real gases, except in the limit of vanishing pressure, obey equations of state that differ more or less from that of the theoretical ideal gas. Furthermore, the specific heat ratios of real gases can vary with pressure and temperature. A number of semiempirical equations of state for gases have been proposed, with constants adjusted to make the equations fit accurate experimental data. From these equations of state it is possible to derive expressions for the sound velocity differing somewhat from (5.2.4) but containing the specific heat ratio γ as an unknown quantity. It is thus possible to find an accurate value of γ from sound-velocity measurements. Since a thermodynamic calculation based on the equation of state can give the difference $C_p - C_v$ of the molar specific heats (for example, $C_p - C_v = R$ for an ideal gas), values of both C_p and C_v can then be found. This example illustrates the fact that the accurate measurement of sound velocities under carefully controlled conditions constitutes a good way to obtain useful information about some of the thermal properties of gases.

TABLE 5.2   Acoustic Properties of Representative Liquids
(Room Temperature)†

Liquid		Velocity c, m/sec	Density ρ_0, kg/m³	Impedance $\rho_0 c$, kg/m²-sec
Water (4°C)	H_2O	1,418	1,000.0	1.4183×10^6
(25°C)		1,493	997.1	1.4888
Mercury	Hg	1,450	13,600	19.70
Kerosene		1,315	810	1.06
Benzene	C_6H_6	1,300	870	1.13
Ethyl alcohol	C_2H_6O	1,210	790	0.96
Methyl alcohol	CH_4O	1,130	790	0.89
Carbon disulfide	CS_2	1,150	1,260	1.45
Carbon tetrachloride	CCl_4	930	1,600	1.48

† Selected from Dwight E. Gray (ed.), "American Institute of Physics Handbook," 2d ed.,
McGraw-Hill Book Company, New York, 1963, which contains more extensive tables.

The velocity of sound in a liquid depends on the bulk modulus B and density ρ_0. No satisfactory theory for the liquid state exists that is comparable with that for the gaseous state. Hence, except so far as order of magnitude is concerned, it is necessary to depend on empirical values of the velocity of sound in liquids. Pertinent data for several common liquids are given in Table 5.2.

Problems

5.2.1 Express the velocity of sound in a gas in terms of the kinetic-theory rms average velocity of the atoms. Is the result physically reasonable?

5.2.2 Compute the velocity of sound in hydrogen, the lightest gas, and in UF_6, a heavy gas, both at standard conditions. Assume $\gamma = 1.3$ for the latter gas.

5.2.3 Helium, with a few percent oxygen, is used instead of air for supporting the life of deep-sea divers. Estimate how much the nasal and throat resonances, which contribute to the intelligibility of speech, are increased in frequency by the change in gas. Express the result as so many notes on the piano.

5.3   Plane Acoustic Waves

We now investigate the solutions of the wave equation (5.1.8) representing plane waves of sound in a uniform gaseous medium. Sound waves of almost this character can be obtained by moving a piston at one end of a long tube of sufficient bore. The perturbation of the walls of the tube through frictional effects, heat flow from the gas, and a lack of perfect rigidity of the walls can ordinarily be neglected and the tube considered an ideal waveguide for plane compressional waves of sound whose wavelength is much greater than the lateral dimensions of the tube.

Let x be measured in the direction perpendicular to the plane wavefronts, so that the pressure in the sound wave is a function of x and t but not of y and z. The wave equation (5.1.8) then simplifies to

$$\frac{\partial^2 p}{\partial x^2} = \frac{1}{c^2}\frac{\partial^2 p}{\partial t^2},$$

(5.3.1)

where we have dropped the subscript from the wave velocity c_f.

(a) Traveling Sinusoidal Waves

The pressure in a sinusoidal sound wave of frequency ω traveling in the positive x direction has the form (real part understood)

$$p(x,t) = p_m e^{i(\kappa x - \omega t)}.$$

(5.3.2)

To find the displacement in the wave, we substitute the pressure gradient†

$$\nabla p = \mathbf{i}\frac{\partial p}{\partial x} = \mathbf{i} i\kappa p_m e^{i(\kappa x - \omega t)}$$

in (5.1.11), which gives, after noting that $\kappa = \omega/c$,

$$\varrho(x,t) = \mathbf{i}\,\xi(x,t) = \mathbf{i}\,\frac{i}{\omega \rho_0 c}\,p_m e^{i(\kappa x - \omega t)}.$$

(5.3.3)

The displacement velocity in the wave is

$$\frac{\partial \varrho}{\partial t} = \mathbf{i}\frac{\partial \xi}{\partial t} = \mathbf{i}\,\frac{p_m}{\rho_0 c}\,e^{i(\kappa x - \omega t)}.$$

(5.3.4)

We find that the pressure and displacement velocity are in phase. Their (real) frequency-independent ratio

$$Z_c = \frac{p}{\partial \xi/\partial t} = \rho_0 c = (B\rho_0)^{1/2}$$

(5.3.5)

is the characteristic wave impedance of the medium for plane sound waves. Values of $\rho_0 c$ are given in Table 5.1 for several common gases.

The kinetic energy density in the wave (5.3.2) can be computed from

$$K_1 = \tfrac{1}{2}\rho_0 |\mathrm{Re}(\dot{\varrho})|^2,$$

(5.3.6)

† The x-direction unit vector $\mathbf{i}$ must not be confused with the imaginary symbol $i = \sqrt{-1}$.

where $|\text{Re}(\dot{\varrho})|$ is the magnitude of the real part of the complex vector displacement velocity (5.3.4). Since

$$|\text{Re}(\dot{\varrho})| = \frac{p_m}{\rho_0 c} \cos(\kappa x - \omega t),$$

we find that

$$K_1 = \frac{1}{2} \frac{p_m{}^2}{\rho_0 c^2} \cos^2(\kappa x - \omega t)$$
$$= \tfrac{1}{2}\rho_0\dot{\xi}_m{}^2 \cos^2(\kappa x - \omega t), \tag{5.3.7}$$

where $\dot{\xi}_m = p_m/\rho_0 c = p_m/Z_c$ is the magnitude of the displacement velocity.

The potential energy density in the sound wave can be found from (3.2.10),

$$V_1 = \frac{1}{2} \frac{p^2}{B} = \frac{1}{2} \frac{p^2}{\rho_0 c^2}, \tag{5.3.8}$$

where use has been made of (5.1.9). Since the pressure in real form is

$$p = p_m \cos(\kappa x - \omega t),$$

we have that

$$V_1 = \frac{1}{2} \frac{p_m{}^2}{\rho_0 c^2} \cos^2(\kappa x - \omega t)$$
$$= \tfrac{1}{2}\rho_0\dot{\xi}_m{}^2 \cos^2(\kappa x - \omega t). \tag{5.3.9}$$

Evidently the kinetic and potential energy densities in the wave are equal and depend on position and time in an identical manner. We have found this to be true in nondispersive media for all the traveling elastic waves we have investigated. The power carried by a sound wave through unit area perpendicular to the x axis is easily shown to be

$$\mathbf{P}_1 = p \frac{\partial \varrho}{\partial t} = \mathbf{i} p \frac{\partial \xi}{\partial t}. \tag{5.3.10}$$

For the wave under discussion

$$\mathbf{P}_1 = \mathbf{i} p_m \dot{\xi}_m \cos^2(\kappa x - \omega t) = \mathbf{i}(K_1 + V_1)c = \mathbf{i} E_1 c. \tag{5.3.11}$$

Hence, as with other elastic waves showing no dispersion, the power flow per unit area equals the wave (phase) velocity times the sum of the (equal) kinetic and potential energy densities.

The intensity I of a sound wave is a scalar quantity expressing the average power flow per unit area. We thus find for a traveling sinusoidal sound wave that

$$I = \tfrac{1}{2} p_m \dot{\xi}_m = \frac{p_m{}^2}{2Z_c} = \tfrac{1}{2} Z_c \dot{\xi}_m{}^2. \tag{5.3.12}$$

We have come to expect that the stress variable (here p_m), the velocity variable (here $\dot{\xi}_m$), and the characteristic wave impedance (here $Z_c = p_m/\dot{\xi}_m = \rho_0 c$) give the average energy flow in a wave by a formula like (5.3.12).

In the case of sound waves, the expression for the intensity has some simple, but important, implications in the recording, transmission, and reproduction of speech and music. In terms of the displacement amplitude, $\xi_m = \dot{\xi}_m/\omega$, the expression for I may be written

$$I = \tfrac{1}{2}\rho_0 c\omega^2\xi_m{}^2. \tag{5.3.13}$$

For a given level of sound intensity (sound intensity corresponds closely to the subjective notion of *loudness*), the amplitude ξ_m of a sinusoidal wave, which may be a Fourier component of a complex wave, must vary inversely with frequency to keep $\omega^2\xi_m{}^2$, and therefore I, constant. Hence it would appear that extremely large amplitudes are required to record and reproduce sounds of low frequency. For example, the separation of grooves on a phonograph record would have to be impractically great. To avoid this difficulty in recording sounds having a wide frequency range, the intensity of the sinusoidal components is attenuated as $1/\omega^2$ below about 500 Hz. For a constant intensity this procedure keeps the amplitude of the displacement constant below this frequency. At high frequencies, in contrast, the displacement amplitude tends to become so small that the recorded signal is comparable with irregularities in the record surface (surface *noise*). Hence improved recording results if frequencies above 1,000 Hz are progressively amplified before recording. The *deemphasis* of low frequencies and the *preemphasis* of high frequencies in the recording is compensated for by electrical networks in the amplifier of the reproducing system.†

An expression for the momentum transported by a plane wave of sound is readily discovered. The particle velocity associated with a wave traveling in the x direction, $\partial\xi/\partial t$, is in phase with the increase in density, $-\rho_0\,\partial\xi/\partial x$, just as in the case of a longitudinal wave on a rod, discussed in Sec. 4.1d. Hence the fluid possesses a net momentum density (per unit volume)

$$g_x(x,t) = -\rho_0\frac{\partial\xi}{\partial x}\frac{\partial\xi}{\partial t} \tag{5.3.14}$$

in the direction of wave travel. Further results are dealt with in Prob. 5.3.6.

(b) *Standing Waves of Sound*

If the two traveling plane waves

$$p_1 = \tfrac{1}{2}p_m e^{i(\kappa x-\omega t)}$$
$$p_2 = \tfrac{1}{2}p_m e^{i(-\kappa x-\omega t)}$$

† For a technical treatment, see J. L. Bernstein, "Audio Systems," John Wiley & Sons, Inc., New York, 1966.

occur simultaneously in a gaseous medium, the combined wave

$$p = p_1 + p_2 = p_m \cos\kappa x \, e^{-i\omega t} \tag{5.3.15}$$

represents a one-dimensional standing wave. From (5.1.11) we find that the displacement in the wave is

$$\varrho = -i\frac{p_m}{\omega\rho_0 c} \sin\kappa x \, e^{-i\omega t}. \tag{5.3.16}$$

The standing wave is therefore given in real form by

$$p = p_m \cos\kappa x \cos\omega t$$

$$\varrho = -i\frac{p_m}{\omega\rho_0 c} \sin\kappa x \cos\omega t. \tag{5.3.17}$$

The pressure and related displacement in the standing wave are out of phase with respect to position, but not with respect to time. A displacement node occurs at the origin and at positions spaced integral half-wavelengths from the origin. The pressure nodes symmetrically interlace the displacement nodes, so that a displacement (pressure) node occurs at a pressure (displacement) antinode.

Let us now suppose that the standing wave (5.3.17) is present inside a long tube, ignoring the small perturbation of the smooth sidewalls. Rigid boundaries can be placed at any pair of displacement nodes without disturbing the standing wave. The resulting "organ pipe" of length l, closed at both ends, has the resonant frequencies

$$\omega_n = n\omega_1 = n\frac{\pi c}{l} \qquad n = 1, 2, 3, \ldots, \tag{5.3.18}$$

which form a harmonic sequence.

A tube with closed ends is sometimes used for measuring the velocity of sound in the gas that it contains. Standing waves in the container can be excited and detected by coupling a pair of electromagnetic transducers, such as miniature loudspeakers, to the cavity through small holes drilled in its walls. For accurate results it is necessary to correct the sound velocity computed from (5.3.18) for the perturbation of the walls. The correction may amount to several tenths of a percent. The theory of the wall correction is too difficult to consider here. It is, at best, only approximate, and the uncertainty in the wall correction appears to be one of the chief limitations in the accuracy of published sound velocities in gases, most of which have been measured in this way.

The boundary condition at an open end of a tube corresponds roughly to a pressure node. An approximate analysis,[†] confirmed by measurements, shows

† Lord Rayleigh, "The Theory of Sound," vol. 2, p. 201, Dover Publications, Inc., New York, 1945.

that the effective end of a tube of circular section is about $0.6a$ beyond the physical end, where a is the radius of the tube. It is assumed that the radius is much less than a wavelength.

If both ends of an organ pipe are open, the resonant frequencies again constitute ideally the harmonic sequence (5.3.18) with l increased to include the two end corrections. Since the end correction depends somewhat on a/λ, there is an increasing deviation from exact harmonicity as λ becomes comparable with the tube diameter.

The resonant frequencies of an organ pipe closed at one end but open at the other are given by

$$\omega_n = (2n - 1)\omega_1 = (n - \tfrac{1}{2}) \frac{\pi c}{l} \qquad n = 1, 2, 3, \ldots, \tag{5.3.19}$$

where l contains the correction for a single open end. The organ pipe is now an odd number of quarter-wavelengths long. Sound emitted by a musical instrument based on a pipe closed at one end and open at the other, so that only odd harmonics can be present, has a characteristic "hollow" sound. The lower tones of a clarinet are an example of this tonal quality.

Problems

5.3.1 Show that the sound wave specified by

$$p(\mathbf{r},t) = p_m \exp i(\mathbf{\kappa} \cdot \mathbf{r} - \omega t) \tag{5.3.20}$$

satisfies the wave equation (5.3.1) and constitutes a plane wave traveling in the direction of the vector wave number $\mathbf{\kappa}$. Find an expression for the particle displacement $\mathbf{\varrho}$ and velocity $\dot{\mathbf{\varrho}}$ in the wave. Verify that the wave is irrotational by computing $\nabla \times \mathbf{\varrho}$. Justify the use of the term "longitudinal" in referring to a plane sound wave, such as (5.3.20).

5.3.2 Show that the *average* kinetic energy density in a traveling sinusoidal sound wave may be written

$$\overline{K}_1 = \tfrac{1}{4}\rho_0 \dot{\mathbf{\varrho}} \cdot \dot{\mathbf{\varrho}}^* \tag{5.3.21}$$

where $\dot{\mathbf{\varrho}}^*$ is the complex conjugate of $\dot{\mathbf{\varrho}}$. Also establish that the average potential energy density and the average power flow are

$$\overline{V}_1 = \frac{1}{4} \frac{p p^*}{\rho_0 c^2}$$

$$\overline{\mathbf{P}}_1 = \tfrac{1}{2}p \dot{\mathbf{\varrho}}^* = \tfrac{1}{2}p^* \dot{\mathbf{\varrho}}.$$

5.3.3 A long tube of area S_1 is joined to a second tube of area S_2, as shown in the figure. Sound waves are sent down the first tube toward the junction, where they are partly reflected.

Prob. 5.3.3

What approximate boundary conditions exist at the junction? Find the amplitude and intensity coefficients of reflection and transmission at the junction. Confirm your results by showing that the incident power equals the sum of the reflected and transmitted power.

5.3.4 Let S be the cross-sectional area of a gas-filled tube through which plane sound waves can be sent. Instead of regarding p and ξ as the variables that describe the sound wave in the tube, it is sometimes more convenient to use p and $\dot{v} = S\dot{\xi}$ as the two variables, where $v = S\xi$ is volume of the air displaced by the passage of the sound wave. Discuss wave impedance and power flow using these variables.

5.3.5 Adapt the treatment in Sec. 4.3 devoted to rods with varying cross-sectional area to sound waves in air-filled tubes of slowly varying area. In particular examine the theory of the exponential horn used in certain loudspeakers.

★5.3.6 Establish the acoustic analogs of Eqs. (1.11.12) and (1.11.19) for a string. Here the equations relate average power density, energy density, momentum density, and the pressure exerted by a plane sound wave when it is absorbed or reflected.

5.4   The Cavity (Helmholtz) Resonator

Before electronic instrumentation simplified the study of sound waves, a number of nonelectrical devices were used which are now obsolete. For example, frequency on an absolute scale was established using an air-driven siren, whose rate of revolution could be timed with the aid of reducing gears. The cavity resonator constituted another convenient nonelectrical device for establishing the presence or absence of a particular frequency component in a complex sound wave. The resonator consists of a bottle almost entirely enclosing a volume of air, with a small opening, or *neck*, constituting a coupling between the air in the bottle and the external air of the room. The dimensions of the resonator are small compared with a wavelength of sound at which it will resonate. If the resonator is exposed to sound containing a frequency component at which it resonates and the external sound source is suddenly stopped, the resonator will continue to resonate, or "sing," as its vibration decays. Although this particular use of cavity resonators has long since disappeared, the theory of the resonator has points of continuing interest. Cavity resonators are often used as components in acoustic filters, the acoustic analog of electric wave filters. The ocarina

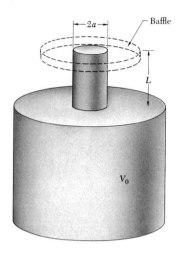

Fig. 5.4.1   A cavity resonator.

is a musical instrument based on the phenomenon of cavity resonance, and the tone that can be produced by blowing across the mouth of a soda bottle or jug is a familiar example.

Let us endeavor to compute the resonant frequency of the cavity resonator shown in Fig. 5.4.1. The resonator consists of a container of volume V_0 with a neck consisting of a tube of radius a and length L. The baffle B is not a necessary part of the resonator but enables a better theoretical estimate to be made of the end correction to be added to the length of the neck. We may regard the resonator, so far as its cavity resonance is concerned, as equivalent approximately to a mass m attached to a spring of spring constant k. The mass m includes the mass of the air in the tubular neck, plus a small additional amount equivalent to increasing the neck length L by 2α, where α is the end correction to a circular tube ending in a baffle. This end correction is somewhat greater than that for a tube without a baffle, and has been estimated to be $\alpha = 0.82a$. Hence the mass of air that moves in the neck during a vibration is

$$m = S(L + 2\alpha)\rho_0, \tag{5.4.1}$$

where $S = \pi a^2$ is the area of the neck and ρ_0 is the density of the air. We have tacitly assumed that $L \ll \lambda$, where λ is the wavelength of the sound corresponding to the resonant frequency that we are computing. Without this assumption, it would be incorrect to regard the air in the neck as moving as a unit.

Suppose that the plug of air in the neck moves out bodily a small amount ξ. The volume V_0 of air in the container then suffers an expansion $S\xi$ or a dilatation

$$\theta = \frac{S\xi}{V_0} \tag{5.4.2}$$

causing, according to (3.2.6), the pressure change

$$p = -B\theta = -\frac{\gamma PS}{V_0}\xi, \tag{5.4.3}$$

where we have chosen the adiabatic bulk modulus, $B = \gamma P$, because of the rapidity with which the pressure changes when an actual oscillation is taking place. The restoring force on the plug of air is thus

$$F = Sp = -\frac{\gamma PS^2}{V_0}\xi, \tag{5.4.4}$$

showing that the "spring constant" of the air in the container is

$$k = -\frac{F}{\xi} = \frac{\gamma PS^2}{V_0}. \tag{5.4.5}$$

We have again made the tacit assumption that the dimensions of the resonator, here of the container, are much less than λ, so that at each instant the dilatation of the enclosed air may be considered uniform. The volume V_0 should include a small additional volume to take into account the springiness of the air in the neck. This correction is about one-third the volume of the neck (Prob. 5.4.3). We are also ignoring frictional effects and loss of sound energy by radiation, both of which damp oscillations of the resonator but alter very little the resonant frequency that we are calculating.

If we now apply Newton's second law to the motion of the plug of air, we arrive at the equation

$$S(L + 2\alpha)\rho_0\frac{d^2\xi}{dt^2} + \frac{\gamma PS^2}{V_0}\xi = 0, \tag{5.4.6}$$

which becomes

$$\frac{d^2\xi}{dt^2} + \frac{c^2S}{(L + 2\alpha)V_0}\xi = 0 \tag{5.4.7}$$

on rearrangement and with the substitution $c = (\gamma P/\rho_0)^{1/2}$, from (5.2.4), for the velocity of sound. We recognize (5.4.7) as the equation for undamped simple harmonic motion having the frequency

$$\omega = c\left[\frac{S}{(L + 2\alpha)V_0}\right]^{1/2}. \tag{5.4.8}$$

There exist no overtones, harmonic or otherwise, related to this frequency, although, of course, at much higher frequencies there will occur unrelated resonances caused by standing waves in the container, in the neck, etc.

It is interesting to note that the vibration constituting the "cavity" resonance does not involve wave theory or standing waves, yet the velocity of sound c summarizes the properties of the air that control the resonant frequency. The other factor in (5.4.8) is essentially of geometrical origin.

A similar situation exists in the case of an electric resonant circuit consisting of an inductor of inductance L in parallel with a capacitor of capacitance C. The resonant frequency of the parallel combination is well known to be

$$\omega = (LC)^{-1/2}. \tag{5.4.9}$$

On closer analysis, it is found that (5.4.9) can be written

$$\omega = c_{\text{light}} \times \text{geometrical factor} \tag{5.4.10}$$

where the geometrical factor, having the dimensions of reciprocal length, depends on the geometry of the inductor and capacitor. In both cases the velocity of a wave—in the one case acoustic, in the other electromagnetic—can be found without recourse to wave techniques.

When it is possible to ignore wave theory and to assign separately the two properties of "matter" that are needed for waves—inertia and springiness in the case of elastic waves, inductance and capacitance in the case of electromagnetic waves—to elements that have dimensions small compared with a wavelength, we say that these elements constitute *lumped parameters*. In contrast, a continuous medium in which waves can propagate is said to have *distributed parameters*. Nevertheless, as we have seen in the case of the cavity resonator and have pointed out in the case of the LC circuit, the inherent wave nature of the basic phenomena involved shows up in a calculation of the resonant frequency of a simple lumped-parameter resonator.

Problems

5.4.1   Compute the resonant frequency of a cavity resonator having the form shown in Fig. 5.4.1 if $V_0 = 500$ cm³, $L = 15$ cm, and $a = 2$ cm for air at room temperature (20°C). *Answer:* 191 Hz.

5.4.2   A loudspeaker cone with a mass of 100 g and a diameter of 25 cm has a very flexible airtight support. It is mounted in a tight rigid box having a volume of 1 m³. Derive a formula for the "cavity" resonant frequency of the system and find its numerical value.

5.4.3   Show that the volume V_0 of the container of the cavity resonator of Fig. 5.4.1 should include one-third of the volume of the neck. *Hint:* See Sec. 4.4. Assume the dilatation varies linearly along the neck.

5.4.4   Develop an approximate expression for the complex impedance (actually a *reactance*) of a cavity resonator, $Z = p/\dot{v}$, where p is a sinusoidally varying external pressure applied at the opening of the neck and $\dot{v} = -S\dot{\xi}$ is the volume flow of air *into* the neck of resonator. Make a diagram showing Z as a function of frequency. *Hint:* See Prob. 4.2.2.

5.5   Spherical Acoustic Waves

We have examined some of the properties of plane waves of sound in Sec. 5.3. As a second example of sound waves in three dimensions, let us look at the equations for spherically symmetric sound waves, such as would be produced in air by a sphere whose surface moves in and out sinusoidally at some frequency ω. We assume that the pressure in such a wave is a function of r and t, where r is the radial distance from the origin at the center of the sound source. The wave equation (5.1.8) then takes the form

$$\frac{1}{r^2}\frac{\partial}{\partial r}\left(r^2\frac{\partial p}{\partial r}\right) = \frac{1}{c^2}\frac{\partial^2 p}{\partial t^2},\tag{5.5.1}$$

which follows from the laplacian in spherical coordinates, as established in Prob. 5.5.1. The well-known substitution

$$\psi = rp\tag{5.5.2}$$

changes (5.5.1) into the simpler equation

$$\frac{\partial^2\psi}{\partial r^2} = \frac{1}{c^2}\frac{\partial^2\psi}{\partial t^2}.\tag{5.5.3}$$

We recognize the wave equation for ψ to have the familiar form of a one-dimensional wave equation. From the discussion in Sec. 1.2 we know that it has the general solution

$$\psi(r,t) = f_1(r - ct) + f_2(r + ct),\tag{5.5.4}$$

so that the spherically symmetric pressure wave

$$p(r,t) = \frac{1}{r}f_1(r - ct) + \frac{1}{r}f_2(r + ct)\tag{5.5.5}$$

has an inverse r dependence. The first term in (5.5.5) represents waves of arbitrary shape traveling outward with their amplitude diminishing inversely with distance from the origin. The second term represents unrelated waves of arbitrary shape converging on the origin with their amplitude increasing inversely with distance from the origin.

Let us now examine in greater detail an outgoing sinusoidal wave of pressure,

$$p(r,t) = \frac{A}{r}\,e^{i(\kappa r - \omega t)},\tag{5.5.6}$$

where A, the pressure amplitude at unit radius, may be complex in order to express a shift in time origin or phase of the wave. We suppose that the wave continues outward indefinitely (or is absorbed at the walls of a room, as in an *anechoic chamber*).

To find the displacement velocity in the pressure wave (5.5.6), we need to compute the gradient of the pressure ∇p in spherical coordinates. The gradient operator in spherical coordinates is

$$\nabla = \mathbf{r}_1 \frac{\partial}{\partial r} + \boldsymbol{\theta}_1 \frac{1}{r \sin\theta} \frac{\partial}{\partial \theta} + \boldsymbol{\phi}_1 \frac{1}{r} \frac{\partial}{\partial \phi}, \tag{5.5.7}$$

where $\mathbf{r}_1$, $\boldsymbol{\theta}_1$, and $\boldsymbol{\phi}_1$ are unit vectors in the directions of increasing r, θ, and ϕ, respectively. Since p is not a function of θ or ϕ,

$$\nabla p = \mathbf{r}_1 \frac{\partial p}{\partial r} = \mathbf{r}_1 \left(-\frac{A}{r^2} + i \frac{A\kappa}{r} \right) e^{i(\kappa r - \omega t)}. \tag{5.5.8}$$

Hence from (5.1.11) we find, on taking a time derivative, that the displacement velocity in the wave is

$$\frac{\partial \boldsymbol{\varrho}}{\partial t} = \mathbf{r}_1 \frac{A}{\rho_0 c} \left(\frac{1}{r} + \frac{i}{\kappa r^2} \right) e^{i(\kappa r - \omega t)}. \tag{5.5.9}$$

The characteristic wave impedance for spherical waves (a scalar quantity)[†]

$$Z_c(r) = \frac{p}{\partial p / \partial t} = \frac{\rho_0 c}{1 + i/\kappa r} \tag{5.5.10}$$

approaches that of a plane wave for large values of κr, that is, when $\kappa r \gg 1$. For very small values of κr, it becomes almost pure imaginary; for intermediate values, it is a complex number, having a real (or *resistive*) part and an imaginary (or *reactive*) part.

We can discover most readily what a complex wave impedance implies by comparing the equations for the pressure and the displacement velocity in real form. If, for convenience, we assume that A is not complex, then

$$p = \frac{A}{r} \cos(\kappa r - \omega t) \tag{5.5.11}$$

$$\frac{\partial p}{\partial t} = \frac{A}{\rho_0 c} \left[\frac{1}{r} \cos(\kappa r - \omega t) - \frac{1}{\kappa r^2} \sin(\kappa r - \omega t) \right]. \tag{5.5.12}$$

Evidently (5.5.12) contains one term that is simply $p/\rho_0 c$ and a second that lags the first by 90° and contains the additional factor $1/\kappa r$. The imaginary term in the wave impedance is responsible for the 90° phase shift. When the second term in (5.5.12) is important, one speaks of the *near field* of the spherical wave.

[†] Note that Z_c cannot be defined so as to keep track of the spatial direction of $\partial \boldsymbol{\varrho}/\partial t$. If, however, one defines a *characteristic vector wave admittance*, $\mathbf{Y}_c \equiv (\partial \boldsymbol{\varrho}/\partial t)/p$, whose scalar (complex) magnitude is the reciprocal of Z_c, then $\mathbf{Y}_c$ gives the direction of the displacement velocity, as well as giving its magnitude and phase, for a specified sinusoidal pressure wave.

The rate at which energy passes through unit area in the radial direction is clearly $p(\partial \rho / \partial t)$, which is the form (5.3.10) takes for a spherical wave. The total power passing through a spherical surface of radius r, therefore, is

$$P = 4\pi r^2 p \frac{\partial \rho}{\partial t}$$

$$= \frac{4\pi A^2}{\rho_0 c} \left[\cos^2(\kappa r - \omega t) - \frac{1}{\kappa r} \cos(\kappa r - \omega t) \sin(\kappa r - \omega t) \right]. \qquad (5.5.13)$$

In the limit as $r \to \infty$, only the first term survives, giving

$$P = \frac{4\pi A^2}{\rho_0 c} \cos^2(\kappa r - \omega t). \qquad (5.5.14)$$

Even when the last term is not small, the power flow outward, when averaged over one period, comes entirely from the first term in (5.5.13). Hence the second term represents a local surging radially in and out of energy that does not, on the average, move outward. The displacement-velocity term that decreases as $1/r$ is solely responsible for the continuous outward transport of energy in a spherical wave. We can speak of this part of the wave either as the *far field* or the *radiative component* of the wave.

A simple physical source of spherical sound waves consists of a sphere of radius a whose surface oscillates sinusoidally in the radial direction with a small amplitude ρ_m. Since the gaseous medium remains in contact with the surface of the sphere, the sound wave produced has the displacement at the radius a

$$\rho(a,t) = \rho_m e^{-i\omega t} \qquad (5.5.15)$$

and the displacement velocity

$$\dot{\rho}(a,t) \equiv \frac{\partial \rho}{\partial t} = -i\omega \rho_m e^{-i\omega t}. \qquad (5.5.16)$$

If we equate (5.5.16) to the displacement velocity (5.5.9) of the spherical wave we have been discussing, we find that the constant A, hitherto arbitrary, is related to the amplitude ρ_m of the spherical sound source by the equation

$$A = -i\omega \rho_0 c \frac{ae^{-i\kappa a}}{1 + i/\kappa a} \rho_m. \qquad (5.5.17)$$

This result is in reality quite simple if we express it in a different form. The oscillating pressure at $r = a$, from (5.5.6), has the value

$$p(a,t) = \left(\frac{A}{a} e^{i\kappa a} \right) e^{-i\omega t} \qquad (5.5.18)$$

and the related displacement velocity is given by (5.5.16). Using the value (5.5.17) just found for A, we find that the complex impedance presented to the pulsating sphere is

$$\frac{p(a,t)}{\dot{\rho}(a,t)} = \frac{\rho_0 c}{1 + i/\kappa a} = Z_c(a),$$ (5.5.19)

where $Z_c(a)$ is the characteristic wave impedance (5.5.10) evaluated at $r = a$. This result shows again how useful the concept of characteristic wave impedance can be. Using (5.5.10), we could have immediately written down the relation between the (complex) pressure amplitude A and the (real) displacement amplitude ρ_m of the pulsating sphere.

Problems

5.5.1   By changing variables to spherical coordinates r, θ, ϕ, such that $x = r \sin\theta \cos\phi$, $y = r \sin\theta \sin\phi$, $z = r \cos\theta$, show that the laplacian of p becomes

$$\nabla^2 p = \frac{1}{r^2} \frac{\partial}{\partial r}\left(r^2 \frac{\partial p}{\partial r}\right) + \frac{1}{r^2 \sin\theta} \frac{\partial}{\partial \theta}\left(\sin\theta \frac{\partial p}{\partial \theta}\right) + \frac{1}{r^2 \sin^2\theta} \frac{\partial^2 p}{\partial \phi^2}.$$

5.5.2   Verify that the substitution $\psi = rp$ in the spherically symmetric wave equation (5.5.1) reduces the equation to the simpler form (5.5.3).

5.5.3   Find an expression for the total average power radiated by a pulsating sphere of radius a. Discuss the power radiated as a function of λ/a for a constant amplitude of the velocity displacement. What semiquantitative conclusions can be made with regard to the frequency dependence of the radiation of sound from ordinary paper-cone loudspeakers?

★5.5.4   Investigate the radially symmetric standing waves in a rigid spherical container of radius R filled with a gas in which the velocity of sound is c. In particular find an expression giving the frequency of the fundamental and its overtones.

5.6   *Reflection and Refraction at a Plane Interface*

Whenever waves traveling in a homogeneous medium reach an interface where the properties of the medium change abruptly, we expect both reflection and transmission (refraction) of the waves to occur. Here we are concerned with compressive waves in a fluid (sound waves) and choose to examine what happens to plane waves traveling in fluid I separated from fluid II by a plane interface, as shown in section in Fig. 5.6.1. For example, fluid I might be air and fluid II water. Let us suppose that the first fluid has a density ρ_{01} and a wave velocity c_1 and the second a density ρ_{02} and a wave velocity c_2.

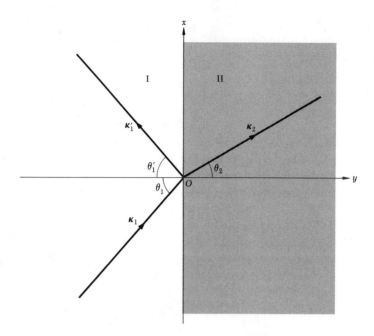

Fig. 5.6.1   Reflection and refraction at a plane interface.

In fluid I a plane wave of frequency ω and wave number $\kappa_1 = \omega/c_1$ is incident at an angle θ_1 on the interface, where θ_1 is the angle between $\mathbf{\kappa}_1$ and the normal to the interface. Let us take $\mathbf{\kappa}_1$ to be in the plane of the figure, which is the *plane of incidence* of the incoming wave. A reflected wave, if it exists, has the same frequency and, therefore, a wave number of the same magnitude as the incident wave. By symmetry the direction of the wave number $\mathbf{\kappa}_1'$ of the reflected wave must lie in the plane of incidence, making some angle θ_1' with the interface normal, as shown. The refracted plane wave also has the same frequency ω but a different wave number $\kappa_2 = \omega/c_2$. However, again by symmetry, the direction of $\mathbf{\kappa}_2$ lies in the plane of incidence and makes some angle θ_2 with the interface normal, as shown.

We can make further progress in describing what happens to the waves at the interface before it becomes necessary to apply boundary conditions that restrict the treatment to compressive waves in fluids. The general boundary conditions, holding for all sorts of waves, are kinematical in nature, involving only the geometry of the waves in space and time.

At the intersection of the plane of incidence and the interface, which we take to be the x direction, the number of waves per unit length at any instant in time must be common to the incident, reflected, and refracted waves (see Fig. 5.6.2). This requirement of geometry tells us that the x components of the three wave numbers must be equal, that is,

$$\kappa_{1x} = \kappa'_{1x} = \kappa_{2x} \qquad (5.6.1)$$

Since $\kappa_{1x} = \kappa_1 \sin\theta_1$, $\kappa'_{1x} = \kappa'_1 \sin\theta'_1$, and since $\kappa_1 = \kappa'_1$, (5.6.1) in turn tells us that the angle of reflection θ'_1 must be equal to the angle of incidence θ_1, that is,

$$\theta'_1 = \theta_1. \qquad (5.6.2)$$

Furthermore, since $\kappa_{2x} = \kappa_2 \sin\theta_2$, the equality (5.6.1) also shows that

$$\kappa_1 \sin\theta_1 = \kappa_2 \sin\theta_2, \qquad (5.6.3)$$

which may be written

$$\frac{\sin\theta_1}{\sin\theta_2} = \frac{\kappa_2}{\kappa_1} = \frac{c_1}{c_2}. \qquad (5.6.4)$$

This equation is usually known as *Snell's law*, especially when the waves involved are light waves. The ratio c_1/c_2 is then defined to be the *index of refraction* of the second medium relative to the first. Thus, from the kinematical aspects of the three waves we can determine the directions, but not the amplitudes, of the reflected and refracted waves.

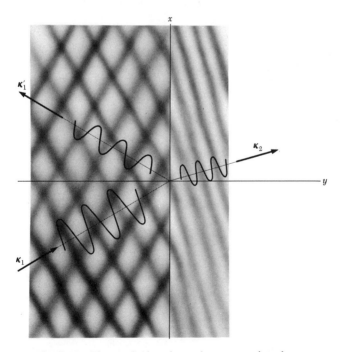

Fig. 5.6.2   The continuity of wavefronts at an interface.

Turning now to compressive waves in a fluid, we recognize that both the pressure in the wave and the component of the displacement normal to the interface must be continuous. Otherwise, respectively, Newton's laws would be violated or a physical discontinuity would occur at the interface.

Let the expressions for the pressure in the incident, reflected, and refracted waves be

$$p_1 = A_1 \exp i(\mathbf{\kappa}_1 \cdot \mathbf{r} - \omega t)$$
$$p_1' = A_1' \exp i(\mathbf{\kappa}_1' \cdot \mathbf{r} - \omega t) \tag{5.6.5}$$
$$p_2 = A_2 \exp i(\mathbf{\kappa}_2 \cdot \mathbf{r} - \omega t)$$

where it is most convenient to take the origin for $\mathbf{r}$ in the interface, as at point O in Fig. 5.6.1.

If we now take the y axis perpendicular to the interface, the three vector wave numbers have only x and y components, whereas the $\mathbf{r}$ vector, when in the interface, has only x and z components. Hence $\mathbf{\kappa}_1 \cdot \mathbf{r} = \kappa_{1x}x$, $\mathbf{\kappa}_1' \cdot \mathbf{r} = \kappa_{1x}'x$, and $\mathbf{\kappa}_2 \cdot \mathbf{r} = \kappa_{2x}x$. In view of (5.6.1), we then establish that

$$\mathbf{\kappa}_1 \cdot \mathbf{r} = \mathbf{\kappa}_1' \cdot \mathbf{r} = \mathbf{\kappa}_2 \cdot \mathbf{r}, \tag{5.6.6}$$

which in turn ensures that the three waves (5.6.5) have identical exponential wave factors at all points in the interface. The boundary condition for pressure at the interface

$$p_1 + p_1' = p_2 \tag{5.6.7}$$

thus leads to the equation

$$A_1 + A_1' = A_2 \tag{5.6.8}$$

connecting the pressure amplitudes.

A second equation among the A's can be found from the continuity condition on the normal displacement components at the interface,

$$\eta_1 + \eta_1' = \eta_2'. \tag{5.6.9}$$

In view of (5.1.11), (5.6.9) may be written

$$\frac{1}{\rho_{01}} \frac{\partial p_1}{\partial y} + \frac{1}{\rho_{01}} \frac{\partial p_1'}{\partial y} = \frac{1}{\rho_{02}} \frac{\partial p_2}{\partial y}. \tag{5.6.10}$$

On introducing the three pressure waves into this boundary condition, setting $\kappa_{1y} = -\kappa_1 \cos\theta_1$, $\kappa_{1y}' = \kappa_1 \cos\theta_1$, and $\kappa_{2y} = -\kappa_2 \cos\theta_2$ and making use of Snell's law, we find that

$$A_1 - A_1' = \frac{\rho_{01}c_1 \sec\theta_1}{\rho_{02}c_2 \sec\theta_2} A_2. \tag{5.6.11}$$

We can now solve (5.6.8) and (5.6.11) for the amplitude reflection coefficient, to find that

$$R_a \equiv \frac{A_1'}{A_1} = \frac{Z_2 \sec\theta_2 - Z_1 \sec\theta_1}{Z_2 \sec\theta_2 + Z_1 \sec\theta_1}, \tag{5.6.12}$$

where the characteristic wave impedances $Z_1 \equiv \rho_{01}c_1$ and $Z_2 \equiv \rho_{02}c_2$ have been introduced.

The reflection coefficient just found depends not only on the two wave impedances and the angle of incidence but also on the velocity-of-sound ratio c_1/c_2, through Snell's law, which gives the angle θ_2 for a given value of θ_1. We note that it is possible, under special conditions, for R_a to vanish. The angle of incidence at which this happens is known as *Brewster's angle* (see Prob. 5.6.4).

At normal incidence the amplitude reflection coefficient takes the simple form

$$R_a(\theta_1=0) = \frac{Z_2 - Z_1}{Z_2 + Z_1}, \tag{5.6.13}$$

which we have previously found to hold for various one-dimensional waves (Secs. 1.9 and 4.2). When the wave impedances Z_1 and Z_2 of the two media are equal, the reflection coefficient vanishes at normal incidence but not for oblique incidence. If the wave velocities c_1 and c_2 also happen to be equal, the two media have identical acoustical properties and no reflection occurs. A high reflection coefficient always occurs when Z_1 and Z_2 differ greatly. For example, at an air-water interface at normal incidence, the amplitude of a reflected sound wave is greater than 99.9 percent of the incident wave. The calculation of the amplitude transmission coefficient and of power relations is left to Probs. 5.6.2 and 5.6.3.

Problems

5.6.1 Supply convincing symmetry arguments establishing that the reflected and refracted plane waves at a plane interface have vector wave numbers lying in the plane of incidence. Why must all three waves have the same frequency?

5.6.2 Supply the missing algebra in the derivation of the reflection coefficient (5.6.12). Show that the amplitude transmission coefficient is

$$T_a \equiv \frac{A_2}{A_1} = \frac{2Z_2 \sec\theta_2}{Z_2 \sec\theta_2 + Z_1 \sec\theta_1} \tag{5.6.14}$$

5.6.3 Find the power incident on unit area of the interface. Show that it equals the sum of the power in the reflected and refracted waves leaving this area.

5.6.4 Show that Brewster's angle, the angle of incidence at which no reflection occurs, is given by

$$\cot^2(\theta_1)_{\text{Brewster}} = \frac{Z_1^2}{Z_2^2 - Z_1^2} \frac{c_1^2 - c_2^2}{c_1^2}. \tag{5.6.15}$$

Under what conditions is Brewster's angle real? *Answer:* $1 < c_1/c_2 < \rho_{02}/\rho_{01}$.

5.7 Standing Waves in a Rectangular Box

A rectangular box of dimensions a, b, c has smooth rigid walls and contains a gas of density ρ_0 and velocity of sound c_f. We wish to investigate the standing waves, or normal modes of oscillation, that can be set up in the gas in the box. The various standing waves that can occur satisfy the boundary condition that the displacement component perpendicular to a wall of the box vanishes. Motion parallel to a wall is not restricted, since we are ignoring viscosity. The results of the present analysis are useful in discussing the acoustic properties of rooms, and they are closely related to the problems of standing electromagnetic waves in a box and of elastic waves in a rectangular block of solid, both of which turn out to be of fundamental importance in quantum physics. It is one of the few normal-mode problems in three dimensions that involves elementary mathematical functions.

The appropriate normal-mode functions are easily found by the method of separation of variables developed in Sec. 1.5. Let us take rectangular axes along three edges of the box with a corner as origin. If we assume that the pressure in the box can be written

$$p(\mathbf{r},t) = X(x) \cdot Y(y) \cdot Z(z) \cdot T(t), \tag{5.7.1}$$

we find, on substituting in the wave equation (5.1.8), that the separated equations are

$$\frac{d^2X}{dx^2} + \kappa_x^2 X = 0$$

$$\frac{d^2Y}{dy^2} + \kappa_y^2 Y = 0$$

$$\frac{d^2Z}{dz^2} + \kappa_z^2 Z = 0 \tag{5.7.2}$$

$$\frac{d^2T}{dt^2} + \omega^2 T = 0,$$

where the four separation constants are related by

$$\kappa^2 \equiv \kappa_x^2 + \kappa_y^2 + \kappa_z^2 = \frac{\omega^2}{c_f^2}. \tag{5.7.3}$$

It is evident from our earlier treatment of two-dimensional waves on membranes that κ is the magnitude of the vector wave number $\boldsymbol{\kappa}$, which has components κ_x, κ_y, κ_z. We can construct possible standing waves of pressure by combining, in various ways, the solutions of (5.7.2), namely, $\cos\kappa_x x$, $\sin\kappa_x x$, $\cos\kappa_y y$, . . . , $\cos\omega t$, $\sin\omega t$. To be a valid standing wave in the box, the normal displacement component, computed from the pressure using (5.1.11), must vanish at the walls. Now (5.1.11) is equivalent to the three scalar equations

$$\xi = \frac{1}{\rho_0\omega^2}\frac{\partial p}{\partial x} \qquad \eta = \frac{1}{\rho_0\omega^2}\frac{\partial p}{\partial y} \qquad \zeta = \frac{1}{\rho_0\omega^2}\frac{\partial p}{\partial z}. \tag{5.7.4}$$

For the displacement components to vanish at $x = 0$, or $y = 0$, or $z = 0$, ξ, η, and ζ need to contain a sine function of x, or y, or z, respectively. It follows that the expression for the pressure can contain only cosine functions. We are therefore led to try the solution

$$p = p_m \cos\kappa_x x \, \cos\kappa_y y \, \cos\kappa_z z \, e^{-i\omega t}, \tag{5.7.5}$$

for which

$$\xi = -\frac{\kappa_x p_m}{\omega^2 c_f} \sin\kappa_x x \, \cos\kappa_y y \, \cos\kappa_z z \, e^{-i\omega t}$$

$$\eta = -\frac{\kappa_y p_m}{\omega^2 c_f} \cos\kappa_x x \, \sin\kappa_y y \, \cos\kappa_z z \, e^{-i\omega t} \tag{5.7.6}$$

$$\zeta = -\frac{\kappa_z p_m}{\omega^2 c_f} \cos\kappa_x x \, \cos\kappa_y y \, \sin\kappa_z z \, e^{-i\omega t}.$$

We see that the normal components of the displacement vanish at the three walls $x = 0$, $y = 0$, $z = 0$. They also vanish at the three walls $x = a$, $y = b$, $z = c$ provided that

$$\begin{array}{ll} \kappa_x a = \pi l & l = 0, 1, 2, \ldots \\ \kappa_y b = \pi m & m = 0, 1, 2, \ldots \\ \kappa_z c = \pi n & n = 0, 1, 2, \ldots . \end{array} \tag{5.7.7}$$

We note that any two (but not all three) of the mode numbers l, m, n can be zero simultaneously. From (5.7.3) and (5.7.7) we find that the normal-mode frequencies are

$$\omega_{lmn} = \pi c_f \left[\left(\frac{l}{a}\right)^2 + \left(\frac{m}{b}\right)^2 + \left(\frac{n}{c}\right)^2\right]^{1/2}. \tag{5.7.8}$$

An examination of (5.7.6) for various values of the mode numbers l, m, n shows that there are $(l-1)$ x-displacement nodal planes perpendicular to the x axis, $(m-1)$ such planes perpendicular to the y axis, and $(n-1)$ perpendicular to the z axis. These planes subdivide the box into lmn rectangular volume elements in which the standing waves differ only in phase, adjacent elements having opposite phase. The planes of no motion, including the walls of the box, are planes of maximum pressure. The pressure nodal planes lie midway between the displacement nodal planes.

It turns out to be of considerable theoretical interest to estimate the number of different standing waves that can exist in a box with frequencies less than some specified maximum frequency ω_{max}. We assume that ω_{max} is sufficiently great to ensure that the mode numbers in (5.7.8), or rather the quantities $u \equiv l/a$, $v \equiv m/b$, $w \equiv n/c$, can be treated as continuous, rather than discrete variables. Evidently each set of positive integers l, m, n defines a *lattice point* in u,v,w space. The number of possible standing waves is the number of these lattice points consistent with the inequality

$$u^2 + v^2 + w^2 \leq \left(\frac{\omega_{max}}{\pi c_f}\right)^2 \equiv R^2. \tag{5.7.9}$$

We see that all such points lie inside an octant of a sphere of radius R in u,v,w space. The lattice points are spaced a distance $1/a$ in the u direction, $1/b$ in the v direction, and $1/c$ in the w direction, so that there is one lattice point in a volume $1/abc$ in this space. Hence the total number of points is closely

$$N = \frac{\frac{1}{8}\left(\frac{4}{3}\pi R^3\right)}{(1/abc)} = \frac{\omega_{max}^3}{6\pi^2 c_f^3}\, abc. \tag{5.7.10}$$

Since N depends only on the total volume abc, and not on the dimensions a, b, c individually, it is meaningful to talk about the number of standing waves *per unit volume* in the box (with frequencies equal to or less than ω_{max})

$$n \equiv \frac{N}{abc} = \frac{\omega_{max}^3}{6\pi^2 c_f^3}. \tag{5.7.11}$$

A rough calculation indicates that there are over 100,000 possible standing waves per cubic meter in air in the audible frequency range (below 10 kHz).

This estimate of the number of standing waves is only a good approximation (valid in the limit $N \gg 1$). We expect it to continue to be almost as good an approximation for a box of any reasonable shape, provided the maximum frequency is sufficiently great. With the possible exception of a different numerical factor, we also can expect (5.7.11) to hold for other types of standing waves, e.g., standing elastic waves in a solid (such as a block of metal) and standing electromagnetic waves in an enclosure having walls of perfectly conducting material (see Probs. 8.7.13 and 8.7.17).

Rayleigh and Planck in 1900 made use of the latter interpretation of (5.7.11) (with its value doubled to accommodate the two states of polarization of electromagnetic waves) in the theory of thermal radiation, which gave birth to the quantum theory. Debye in 1912 considered that thermal energy in a solid can be described by many standing elastic waves, whose number per unit volume for frequencies less than ω_{max} is given by a slight modification of (5.7.11). By making use of the quantum theory, he was then able to account for the important features of the hitherto puzzling variation in the specific heats of solids with temperature.

A discussion of the use of (5.7.11) in technological acoustics can be found in Morse.[†] Its use in modern physics is covered in nearly all introductory accounts of quantum physics.

Problems

5.7.1 Show how to express an arbitrary oscillation of the pressure of the gas in a rectangular room by a Fourier development analogous to (1.6.1) holding for a finite string segment. Give formulas for evaluating the Fourier coefficients from the initial pressure distribution in the gas.

5.7.2 How many different standing waves can occur in a typical room of a house 15 by 20 by 8 ft for frequencies less than 10 kHz? What is the frequency of the lowest standing wave?

★5.7.3 Each possible normal mode of oscillation of a distributed system constitutes a mechanical degree of freedom of the system. According to the *classical equipartition-of-energy theorem*, each oscillatory degree of freedom is to be assigned an average thermal energy of kT, where k is Boltzmann's constant and T is the absolute temperature. Hence the internal thermal energy *per unit volume* of a monatomic gas should be $E_v = nkT$, with n given by (5.7.11). For a *mole* of monatomic gas, kinetic theory tells us that the internal energy is $E_\mu = \frac{3}{2}RT$, where R is the gas constant per mole. We can bring the two viewpoints into agreement by choosing a suitable value of ω_{max}. Compute this value and show that the minimum wavelength is of the order of the mean spacing of the atoms in the gas.

5.7.4 Derive a formula analogous to (5.7.11) giving the number of standing waves with frequency less than ω_{max} associated with unit length of a vibrating string segment. Do the same for a vibrating rectangular membrane, finding in this case the number of standing waves per unit area (with frequency less than ω_{max}). Express each of the three results in terms of the minimum wavelength λ_{min} and note how they differ.

★5.7.5 Formula (5.7.10) does not include the standing waves for which one or two of the mode numbers l, m, n are zero. Establish the more accurate formula[‡]

$$N = \frac{\omega_{max}^3}{6\pi^2 c_f^3} V + \frac{\omega_{max}^2}{16\pi c_f^2} A + \frac{\omega_{max}}{16\pi c_f} L,$$

where $V = abc$, $A = 2(ab + ac + bc)$, and $L = 4(a + b + c)$.

[†] P. M. Morse, "Vibration and Sound," 2d ed., chap. 8, McGraw-Hill Book Company, New York, 1948.
[‡] See Morse, *op. cit.*, p. 394.

5.8   The Doppler Effect

The shift in the observed frequency of a wave as the result of relative motion of source and observer is known as the *Doppler effect*. It is of particular interest in connection with fluids, since both the source and observer can readily move through the fluid. The effect is even more important for electromagnetic waves, e.g., the red shift of stellar radiation. Here our interest is principally in sound waves in fluids, since an analysis of the Doppler effect for light is complicated by relativistic considerations.

We have always regarded an elastic medium to be at rest in an inertial frame when deriving the wave equation for the medium. Otherwise the laws of physics, in particular Newton's laws, would not hold in their usual form. Accordingly the speed of the elastic wave in the wave equation is the speed *with respect to the medium*. With respect to other inertial frames a plane elastic wave would appear to have its wave velocity diminished by the component of the velocity of the moving frame in the direction of the wave travel. For example, imagine a boat ride across a lake covered with "plane" waves. Their wave velocity relative to the boat is reduced by the component of the boat's velocity in the direction of the wave travel. The component of the boat's velocity parallel to the wave crests does not change the speed with which the crests approach the boat but simply amounts to a steady sideways motion of the lake relative to the boat.

An elastic medium may be said to constitute a *preferred reference frame* for the elastic waves that can travel in it. This statement implies that one can discover the motion, if any, of an elastic medium by experiments with waves in the medium. For example, it is possible to measure the rate of flow of a liquid in a pipe using high-frequency sound waves. The statement takes on greater significance when we recognize that electromagnetic waves traveling in vacuo have the same wave speed with respect to all inertial frames, *regardless of their velocity*. This surprising behavior, verified by experiments such as that of Michelson and Morley, is a central feature of the special theory of relativity, where it is made one of the basic postulates. For electromagnetic waves, there evidently exists *no preferred reference frame*. We thus expect for elastic waves that the Doppler effect can depend on the relative velocity of the source and medium and independently on the relative velocity of observer and medium, whereas, for electromagnetic waves, only the relative velocity of observer and source are of significance. A calculation of the Doppler effect for light is taken up in Prob. 5.8.3 for the benefit of readers familiar with the elements of relativity theory.

Let us now examine the Doppler effect arising from the motion of a point source through a fluid at rest in an inertial frame. We suppose that the source moves with a velocity v_s along the x axis of the frame and emits sinusoidal waves at a frequency ω_s, which spread out as spherical wavefronts of sound. Any such wavefront is centered in the reference frame of the fluid on the po-

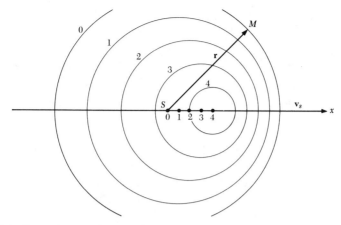

Fig. 5.8.1   Successive wavefronts emitted by a source moving with a velocity $v_s = \frac{1}{2}c$.

sition of the point source in the fluid at the time of emission. A series of such wavefronts is illustrated in Fig. 5.8.1. Evidently a wavefront emitted at time t_s reaches a point M in the medium at a distance r at a time $t_s + r/c$, and one emitted at a time $t_s + T_s$, one period later, reaches the same point at the time $t_s + T_s + r'/c$, where r' is the new distance between source and point M. The geometry is shown in Fig. 5.8.2. If r and r' are large compared with v_sT_s (the distance the source moves in one period), then

$$r - r' = v_s T_s \cos\theta_s, \tag{5.8.1}$$

where θ_s is the angle the line SM (or $S'M$) makes with the x axis. Hence the period of the wave reaching M is

$$T_M = \left(t_s + T_s + \frac{r'}{c}\right) - \left(t_s + \frac{r}{c}\right)$$

$$= T_s\left(1 - \frac{v_s}{c}\cos\theta_s\right), \tag{5.8.2}$$

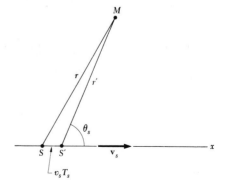

Fig. 5.8.2   A periodic sound source moving in a stationary medium.

and the frequency of the wave reaching M accordingly is

$$\omega_M = \frac{\omega_s}{1 - (v_s/c)\cos\theta_s}. \tag{5.8.3}$$

The waves being observed at any time t_M at M were emitted at the earlier time $t_M - r/c$, where r is the distance to the source *at this earlier time*. At the time t_M the waves appear to emanate from the position of the source at the earlier time.

Finally let us examine the Doppler effect arising from the motion of a (point) observer with respect to the inertial frame attached to the medium. We can suppose that a steady pattern of traveling sinusoidal waves of frequency ω_M and wave speed c moves past the position temporarily occupied by the observer. Usually the waves are so nearly plane in the vicinity of the point observer that they can be treated as plane waves. If we take the x' axis to be the direction of wave travel and let θ_o be the angle with respect to this axis of the line along which the observer travels with velocity v_o, then the speed of the wavefront relative to the moving observer is $c - v_o\cos\theta_o$. Since the actual spacing of wave crests in the medium is $\lambda_M = 2\pi c/\omega_M$, the point observer infers that their frequency is

$$\omega_o = 2\pi\frac{c - v_o\cos\theta_o}{\lambda_M} = \left(1 - \frac{v_o}{c}\cos\theta_o\right)\omega_M. \tag{5.8.4}$$

If both source and observer are moving with respect to the medium, then (5.8.3) enables the frequency of the wave to be found at any stationary point in the medium and (5.8.4) enables the frequency of the wave to be found with respect to a moving observer at this point.

Problems

5.8.1 A satellite passes across the sky sending out radio signals at a constant but inaccurately known frequency ω_s. Assume that the satellite altitude is small compared to the earth's radius and hence that the trajectory can be considered to be a straight line at constant altitude above a flat earth. By beating the signals against a known standard frequency and measuring the *difference* frequency, an observer on the earth can accurately measure the received frequency, with its Doppler shift, as a function of time, with $\cos\theta_s$ in (5.8.3) [or (5.8.5), with $\gamma = 1$, since $v \ll c$] ranging from 1 to -1. Show how the observations can be made to yield a value of v, the velocity of the satellite, and a value of D, the closest distance of approach of the satellite.†

5.8.2 Make a construction similar to Fig. 5.8.1 when $v_s > c$. Relate the diagram to the sonic boom that is heard when an airplane flies faster than the speed of sound. In what area

† See *Proc. IRE*, **45**: 1552–1555 (1957); **46**: 782–783 (1958).

of the sky should one look for the plane after hearing the boom? Find the half-angle of the conical envelope of the sound waves emitted by the airplane, noting that this angle is independent of frequency.

★5.8.3 To obtain the formula for the Doppler effect with electromagnetic waves (light waves), one needs to take into account the relativistic time dilatation of the clocks in a moving reference system, the so-called slowing down of a moving clock. The calculation is best made in the reference frame of the observer. Show that the relativistic Doppler effect for light of frequency ω_s (as measured when the source is at rest) is

$$\omega_0 = \frac{\omega_s}{\gamma(1 - \beta \cos\theta_s)} \tag{5.8.5}$$

where $\beta \equiv v_s/c$ and $\gamma \equiv 1/(1 - \beta^2)^{1/2}$. This equation corresponds to (5.8.3); there is no equation corresponding to (5.8.4). A relativistic *second-order* effect (depending on β^2) is superposed on the nonrelativistic *linear* effect (depending on β). The second-order effect is independent of θ_s.

★5.9   The Velocity Potential

In Sec. 5.1 we obtained the scalar wave equation for the (excess) pressure in a fluid from first principles and showed that we could obtain the vector displacement, and therefore the displacement velocity, for any solution of the wave equation for pressure. A more common procedure, particularly in more advanced treatments of acoustics or of hydrodynamics,† is to start with the basic equations of hydrodynamics and specialize them to irrotational wave motion in which the bulk modulus of the fluid is the pertinent elastic constant of the medium. Whenever one has an irrotational vector field, such as the displacement velocity $\partial\varrho/\partial t$ in a three-dimensional wavefield in a fluid (or, more familiarly, the electrostatic field **E**), it is always possible to express the field as the gradient of a scalar potential field (see Appendix A). Usually this procedure leads to a simplification, since the scalar field has but a single component whereas the vector field has three interrelated spatial components.

In Sec. 5.1 we accomplished an equivalent simplification by basing the discussion of waves in fluids on the scalar wave equation for pressure. Here we wish to establish a connection between the equations there and those based on the explicit use of a velocity potential.

We start by assuming, as is customary, that any wave encountered in a nonviscous fluid has an irrotational displacement velocity. As already pointed

---

† See, for example, R. W. B. Stephens, "Acoustics and Vibrational Physics," chap. 21, Edward Arnold (Publishers) Ltd., London, 1966; and H. Lamb, "Hydrodynamics," Dover Publications, Inc., New York, 1945.

out, we can then write that

$$\dot{\varrho} \equiv -\nabla\phi, \tag{5.9.1}$$

where $\phi(\mathbf{r},t)$ is the (scalar) *velocity potential*. The minus sign in (5.9.1) is not necessary but is often included so that the velocity field points in the direction of decreasing velocity potential.

We can relate the velocity potential just defined to the pressure $p(\mathbf{r},t)$ on the basis of (5.1.3),

$$p = -B\nabla \cdot \varrho. \tag{5.9.2}$$

If we take a time derivative of (5.9.2) and introduce (5.9.1) defining ϕ, then

$$\frac{\partial p}{\partial t} = -B\nabla \cdot \dot{\varrho} = B\nabla^2\phi. \tag{5.9.3}$$

If we now take a second time derivative and introduce the wave equation for p, (5.1.8), then

$$\frac{\partial^2 p}{\partial t^2} = c^2\nabla^2 p = B\nabla^2\frac{\partial\phi}{\partial t}. \tag{5.9.4}$$

The last equation implies that

$$p = \frac{B}{c^2}\frac{\partial\phi}{\partial t} = \rho_0\frac{\partial\phi}{\partial t}, \tag{5.9.5}$$

at least so far as the parts of p and of $\partial\phi/\partial t$ that are wavefunctions are concerned. Finally, if we take a time derivative of (5.9.5) and compare the result with (5.9.3), it is evident that

$$\nabla^2\phi = \frac{1}{c^2}\frac{\partial^2\phi}{\partial t^2}. \tag{5.9.6}$$

Hence the velocity potential satisfies the same scalar wave equation that the pressure does. Starting with a solution of the wave equation for the velocity potential, we can find the (scalar) pressure wave from (5.9.5) and the (vector) displacement velocity wave from (5.9.1).

The results obtainable from the velocity potential differ in no way from those based on the equations in Sec. 5.1. The importance of the method of the velocity potential rests primarily in its putting the treatment of irrotational waves on a common basis with the treatment of other irrotational vector fields. Scalar (and vector) potential theory constitutes an important part of theoretical physics. In developing the theory of any part of physics involving effects that occur in space and time, it is usually possible to achieve a simplification and unification by introducing ideas from potential theory.

Problems

5.9.1 Work out a treatment of plane waves of sound on the basis of the velocity potential covering the material in Sec. 5.3.

5.9.2 What boundary condition must the velocity potential satisfy (*a*) at a rigid boundary; (*b*) at a free surface of a fluid; (*c*) at a plane interface between two different fluids?

★5.9.3 Solve Prob. 5.5.4 using the velocity potential.

★5.10 Shock Waves

We have assumed throughout our discussion of sound waves in gases that we are dealing with waves of small amplitude, which means that the pressure amplitude is much less than the static pressure in the gas. Without this restriction, the equations governing the propagation of waves become very complicated. For example, the adiabatic bulk modulus of a gas, $B = \gamma P$, has significance only when a pressure change is small compared with the static pressure P.

When the pressure amplitude in a sound wave begins to approach the static pressure, new *nonlinear* phenomena become appreciable. Since there is a lower limit to the pressure in a gas, namely, zero, whereas no clearly defined upper limit exists, we expect an asymmetry in the phenomena associated with a traveling pulse of high compression, as compared with a traveling pulse of considerable rarefaction. In the blast wave leaving an explosion, for example, the forward displacement velocity of the gas in the highly compressed region may be so great that it cannot be neglected in comparison with the wave velocity. Consequently the compressive pulse changes shape as it goes along, and an analysis shows that the front edge of the pulse tends to become steeper and steeper. Hence it is not surprising to find that a pulse of high compression quickly forms, and maintains, a steep *shock front*, which propagates through the gas with a speed considerably in excess of normal sound velocity for the gas. In a pulse of rarefaction of large amplitude, the reverse state of affairs exists, causing the pulse to spread out in the backward direction and to weaken as it travels along. No shock front develops in this case.

The subject of shock waves and shock phenomena is an extensive one.† Here we give a simple derivation of the shock-wave speed in an ideal gas that is steadily driven forward by a moving piston and discuss some of the consequences of our findings. Let us therefore consider what happens to gas at rest in a tube of cross-sectional area S when a piston is driven down the tube at

---

† See, for example, R. Courant and K. O. Friedrichs, "Supersonic Flow and Shock Waves," Interscience Publishers, Inc., New York, 1948.

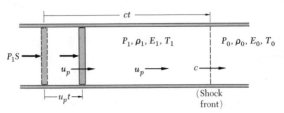

Fig. 5.10.1   A piston-driven shock front.

some constant speed u_p starting from rest at time $t = 0$. The tube is filled initially with gas of pressure P_0, density ρ_0, and internal energy per unit mass E_0, corresponding to a temperature T_0. The gas has a specific-heat ratio γ and the (normal) sound velocity (5.2.4), namely,

$$c_0 = \left(\gamma \frac{P_0}{\rho_0}\right)^{1/2}. \tag{5.10.1}$$

We shall use c without a subscript for the velocity of the shock front that travels down the tube when the piston is suddenly given the velocity u_p.

Figure 5.10.1 shows the state of affairs in the tube after the piston has been moving for a time t. We note that:

(**1**)   The piston has moved forward a distance $u_p t$.

(**2**)   The shock front has moved forward a distance ct.

(**3**)   The pressure in the gas behind the shock front has increased from P_0 to some new pressure P_1, with the density going from ρ_0 to ρ_1.

(**4**)   The gas between the piston and the shock front is moving as a unit with the velocity u_p of the piston.

(**5**)   The compressed gas has been heated from its original temperature T_0 to some new temperature T_1, with its internal energy per unit mass going from E_0 to E_1.

(**6**)   The external force on the piston required to keep it moving with the steady velocity u_p is $P_1 S$.

Conservation of mass implies that $\rho_0 c t S = \rho_1 (c - u_p) t S$, so that across the shock front the density ratio is

$$\frac{\rho_1}{\rho_0} = \frac{c}{c - u_p}. \tag{5.10.2}$$

Next let us apply Newton's second law to the gas between the piston and the shock front. At time t the gas involved has the mass $\rho_0 ctS$, and it has acquired the velocity u_p, so that it has acquired the mechanical momentum $\rho_0 ctSu_p$. The rate at which the gas is acquiring momentum must equal the net external force causing it. Since the pressure on the ever-increasing gas sample is P_0 at the right end and P_1 at the piston end, the net force is simply $(P_1 - P_0)S$. Accordingly, Newton's second law tells us that

$$P_1 - P_0 = \rho_0 c u_p. \tag{5.10.3}$$

Let us next equate the work done on the sample by the external forces to the total energy that the sample has acquired in the time t. Since the work is being done solely by the piston, we have that

$$P_1 Su_p t = \tfrac{1}{2}(\rho_0 ctS)u_p^2 + (\rho_0 ctS)(E_1 - E_0), \tag{5.10.4}$$

where the two terms on the right are, respectively, the kinetic energy of the sample and its increase in internal energy. Here the entire pressure P_1 acting on the piston is involved, not just the net pressure $P_1 - P_0$ responsible for the rate of increase in momentum of the gas. On dividing by S and t, we have that

$$P_1 u_p = \tfrac{1}{2}\rho_0 c u_p^2 + \rho_0 c(E_1 - E_0). \tag{5.10.5}$$

The three equations, (5.10.2), (5.10.3), and (5.10.5), which express, respectively, the conservation laws of mass, momentum, and energy as applied to a shock wave, are usually termed the *Rankine-Hugoniot relations*. These equations explicitly involve eight variables, namely, the three variables connected with the initial state of the gas, P_0, ρ_0, E_0; the three variables connected with the final state of the gas, P_1, ρ_1, E_1; and the two velocities, u_p and c.

Thermodynamics teaches that a definite quantity of a homogeneous substance of fixed composition, such as a gas, is usually characterized by three extensive variables: its volume, its (total) internal energy, and its entropy; and by two intensive variables, its pressure and its absolute temperature. Of these five variables, only two can be independently chosen. These two determine the so-called *thermodynamic state* of the substance, and the values of the other three are then definite. For example, the equation of state of the gas, connecting P, V, and T, permits T to be calculated given values of P and V (or the density $\rho = m/V$, where m is the mass of the sample). An expression for the internal energy per unit mass E as a function of T and ρ then permits E to be calculated for the given values of P and ρ. The entropy, which we shall not need, can be found from the equation of state and the expression for $E(T,\rho)$ using the laws of thermodynamics.

Evidently for a particular gas in the shock tube, the thermodynamic equations of the gas tell us E_0 and T_0 for given values of P_0 and ρ_0, so that we may regard all these as known quantities. If we further specify the piston velocity u_p, we are left the four variables P_1, ρ_1, E_1, and c as still unknown but with the three Rankine-Hugoniot equations connecting them with known quantities. The thermodynamic equations for the gas give us a fourth equation connecting P_1, ρ_1, and E_1, and, incidentally, give us the final temperature T_1.

In the case of small-amplitude sound waves, a reversible increase in the internal energy of a gas is associated with the adiabatic (elastic) compression of the gas. In the case of a shock wave, however, the increase in internal energy turns out to be greater than the reversible increase in potential energy that would occur if a gas sample were adiabatically compressed with the density ratio (5.10.2). Stated differently, the passage of a shock front involves an irreversible heating of the gas, as though friction were acting. Such an effect is not unexpected, for whenever a discontinuous jump in the velocity of mass elements of a mechanical system occurs, there is always an attendant loss in mechanical energy. Familiar examples include rain falling on a moving train or the flight of a harpoon that continuously lifts more and more of a rope initially at rest. Here the rapidly moving shock front continuously picks up gas at rest and includes it in the moving gas behind the shock front. It is incorrect to assume that the gas suffers an adiabatic compression in this process.

To make further progress in the discussion of shock waves, we need an equation relating the internal energy E of the gas to P and ρ. For simplicity, we take the gas to be an *ideal gas*. Either statistical mechanics or kinetic theory gives a simple expression for the internal energy of a unit mass of an ideal gas at the absolute temperature T, namely,

$$E = \frac{1}{\mu} C_v T, \tag{5.10.6}$$

where C_v is the (constant) molar specific heat at constant volume and μ is the molecular weight. For an ideal gas, E is independent of ρ. Furthermore, the molar specific heat of an ideal gas at constant pressure C_p is related to C_v by the simple equation

$$C_p = C_v + R, \tag{5.10.7}$$

where R is the universal gas constant. The equation of state of an ideal gas takes the form

$$\frac{P}{\rho} = \frac{R}{\mu} T \tag{5.10.8}$$

when density ρ, instead of volume, is used as a variable. If we introduce the specific-heat ratio $\gamma = C_p/C_v$ into (5.10.6) and eliminate T using (5.10.8), we find that the expression for E becomes

$$E = \frac{P}{(\gamma - 1)\rho}. \tag{5.10.9}$$

The difference in internal energy of a unit mass of gas between the initial and final states is therefore given by

$$E_1 - E_0 = \frac{1}{\gamma - 1}\left(\frac{P_1}{\rho_1} - \frac{P_0}{\rho_0}\right).$$ (5.10.10)

With the addition of this equation we are now prepared to solve the Rankine-Hugoniot equations for the shock speed and for the conditions of the gas behind the shock front.

The solution of the four equations can be carried out by eliminating $E_1 - E_0$ between (5.10.5) and (5.10.10), using (5.10.3) to eliminate P_1 and (5.10.2) to eliminate ρ_1. By this means we arrive at the equation

$$c^2 - \tfrac{1}{2}(\gamma + 1)u_p c - c_0^2 = 0,$$ (5.10.11)

where c_0 is the normal sound velocity (5.10.1). Solving for the shock velocity, we find that for an ideal gas,

$$c = \tfrac{1}{4}(\gamma + 1)u_p + [\tfrac{1}{16}(\gamma + 1)^2 u_p^2 + c_0^2]^{1/2}.$$ (5.10.12)

This expression shows clearly that the shock velocity always exceeds both c_0, and the piston velocity u_p (since $\gamma > 1$). When $u_p \ll c_0$, to first order,

$$c = c_0\left[1 + \tfrac{1}{4}(\gamma + 1)\frac{u_p}{c_0} + \cdots\right]$$ (5.10.13)

and the shock velocity approaches the normal sound velocity, which is its limiting value.

The shock velocity can be easily expressed in terms of the excess pressure $P_1 - P_0$ that exists in the gas behind the shock front. Using (5.10.3) to eliminate u_p from (5.10.11) and using (5.10.1) for $\rho_0 c_0^2$, we find that

$$c = c_0\left[1 + \tfrac{1}{2}\left(1 + \frac{1}{\gamma}\right)\left(\frac{P_1 - P_0}{P_0}\right)\right]^{1/2}.$$ (5.10.14)

In discussing shock phenomena, it is convenient to define the *Mach number* of the shock as the ratio of the shock velocity to the ordinary velocity of sound,

$$M \equiv \frac{c}{c_0}.$$ (5.10.15)

In terms of M, (5.10.14) can be rearranged to read

$$\frac{P_1}{P_0} = 1 + \frac{2\gamma}{\gamma + 1}(M^2 - 1).$$ (5.10.16)

TABLE 5.3   Some Shock Properties of an Ideal Monatomic Gas

M	$\dfrac{\rho_1}{\rho_0}$	$\dfrac{P_1}{P_0}$	$\dfrac{T_1}{T_0}$	M_p	Adiabatic $\dfrac{T_1}{T_0}$
$1 + \epsilon$	$1 + \frac{3}{2}\epsilon + \cdots$	$1 + \frac{5}{2}\epsilon + \cdots$	$1 + \epsilon + \cdots$	$\frac{3}{2}\epsilon$	$1 + \epsilon + \cdots$
2	2.28	4.75	2.08	1.13	1.74
5	3.57	31.0	8.68	1.44	2.34
10	3.88	124	31.9	7.43	2.47
$M \to \infty$	4.00	$\frac{5}{4}M^2$	$\frac{5}{16}M^2$	$\frac{3}{4}M$	2.50

The Mach number of the moving gas behind the shock front (i.e., of the piston) can be found from (5.10.11),

$$M_p \equiv \frac{u_p}{c_0} = \frac{2}{1 + \gamma}\left(M - \frac{1}{M}\right). \tag{5.10.17}$$

The density ratio, as given by (5.10.2), then becomes

$$\frac{\rho_1}{\rho_0} = \frac{c}{c - u_p} = \frac{M}{M - [2/(\gamma + 1)](M - 1/M)} = \frac{\gamma + 1}{\gamma - 1 + 2/M^2} \tag{5.10.18}$$

on introducing (5.10.17) for u_p/c_0. Finally, the temperature ratio is

$$\frac{T_1}{T_0} = \frac{P_1 \rho_0}{P_0 \rho_1} = \left[1 + \frac{2\gamma}{\gamma + 1}(M^2 - 1)\right]\left[1 - \frac{2}{\gamma + 1}\left(1 - \frac{1}{M^2}\right)\right]. \tag{5.10.19}$$

The results just obtained are based on the thermal properties assumed for an ideal gas. To see what they imply, the various ratios (5.10.16) to (5.10.19) have been computed for a monatomic gas ($\gamma = \frac{5}{3}$) for several values of M (Table 5.3). The last column shows the temperature ratio T_1/T_0 of a monatomic gas compressed adiabatically from a density ρ_0 to a density ρ_1, for the density ratio given in the second column.

Table 5.3 reveals a number of interesting properties of shocks:

(**1**)   The density ratio on the two sides of a shock front approaches an upper limit (depending on γ) with increasing M.

(**2**)   The pressure and temperature ratios go up rapidly with Mach number, ultimately as M^2.

(**3**)   The piston velocity, i.e., the velocity of the gas behind the shock front, is always less than M but ultimately goes up linearly with M.

(4) A hypothetical adiabatic (reversible) temperature rise, corresponding to the density ratio in the second column, approaches as a lower limit the temperature rise produced by a weak shock, but it rapidly falls behind as M increases; in fact it has a modest upper limit set by γ. In view of the close connection between temperature and internal energy, the difference between the two columns dealing with temperature indicates to what extent the transformation of mechanical energy into heat energy is irreversible. For higher Mach numbers, most of the heating of the gas is evidently irreversible.

For not too high Mach numbers, say a shock of Mach 5, the gas behind the shock front is heated to incandescence, and at still higher Mach numbers, ionization of the gas takes place. In such cases the simple theory given here needs considerable modification, since a constant specific-heat ratio no longer exists. The three Rankine-Hugoniot equations, coupled with thermal data for a particular gas, enable the shock-wave properties of the gas to be predicted.

Shock phenomena can be studied experimentally in so-called *shock tubes.*[†] In one form of shock tube a compressed gas contained behind a thin plastic barrier at one end of a long tube is suddenly released by rupturing the barrier. A shock front is generated as the pressure wave passes down the tube. In other shock tubes, the shock front is generated by an energetic electromagnetic discharge in the gas of the tube. Shock phenomena are important in connection with aircraft and missiles traveling at speeds exceeding Mach 1. In the laboratory, certain chemical reactions can be studied conveniently under the extremes of pressure and temperature that occur behind a shock front of high Mach number.

Problems

5.10.1 If we view the shock front from a reference frame moving with the velocity of the front, gas approaches from the right with the velocity $v_0 = -c$, suffers a compression as it moves through the front, and leaves toward the left with the velocity $v_1 = -(c - u_p)$. Show that $v_0 v_1 = (P_1 - P_0)/(\rho_1 - \rho_0)$, which is known as *Prandtl's relation*.

5.10.2 Construct a table similar to Table 5.3 for a diatomic gas for which $\gamma = 1.4$.

5.10.3 Derive the three Rankine-Hugoniot relations from the viewpoint of an observer in a reference frame traveling with a plane shock front so that the gas motion is steady.

[†] Ya. B. Zel'dovich and Yu. P. Raizer, "Physics of Shock Waves and High-temperature Hydrodynamic Phenomena," Academic Press Inc., New York, 1966.

six

⋆Waves on a Liquid Surface

Probably the most familiar type of wave motion is that which occurs at the interface between a liquid and a gas (or between two immiscible liquids). Most of the phenomena common to all types of wave motion can be demonstrated, or at least visualized in the mind's eye, by means of water waves in a lake or ripple tank.† For instance, tossing a stone into a lake excites an outgoing circular wave packet. Since the medium is dispersive, the distinction between phase and group velocities can be seen. The waves diffract around obstacles, refract when the depth changes abruptly, and show interference patterns when they recombine after having passed through two or more apertures. There

† A number of magnificant photographs of natural water waves, along with an interesting discussion, can be found in R. A. R. Tricker, "Bores, Breakers, Waves and Wakes," American Elsevier Publishing Company, New York, 1965.

are also more complicated processes, beyond the scope of this book. The bow wave of a fast ship is a direct analog of the Cerenkov radiation generated by fast subatomic particles passing through matter. The driving of waves by the wind is an analog of the traveling-wave tube, an important electromagnetic device used for amplifying microwave signals. The breaking of waves on a beach is an analog of complex processes by which magnetohydrodynamic waves heat ionized gases (plasmas) in current research work seeking to achieve controlled nuclear fusion.

These surface waves are two-dimensional in the sense of the available directions of propagation and so are very similar to the membrane waves discussed in Chap. 2. However, the wave motion necessarily penetrates into the fluid on each side of the interface. Thus the motion is in fact three-dimensional and involves the hydrodynamics of the fluid. The quantitative description is considerably more complex than that encountered in Chap. 2.

6.1 *Basic Hydrodynamics*

Before we can investigate wave motion on the surface of liquids, we need to develop the basic equations of the hydrodynamics of an ideal (nonviscous) fluid. These equations fall into three categories: (1) equations that are essentially kinematical in nature; (2) an equation that expresses the conservation of mass when a fluid moves, the *equation of continuity;* (3) a dynamical equation, *Bernoulli's equation*, that summarizes the application of Newton's second law to fluids, in particular to the irrotational motion of an incompressible liquid.

(a) *Kinematical Equations*

There exist two equivalent ways of describing a fluid when it is in motion. In the *eulerian method* we describe some aspect of the fluid in terms of what takes place at a fixed point $\mathbf{r}_0$, as the fluid flows by. We are thus continually observing new particles as they pass the point of observation. In the *lagrangian method* we pick out a single particle of the fluid and follow it along as it partakes of the fluid motion. We can then describe the same aspect of the fluid that was of interest before but now in terms of what happens to the fluid in the immediate vicinity of a specific moving particle. Let us see how the two viewpoints are related so far as time rates of change are concerned.

Let $F(x,y,z,t)$ stand for a (scalar) property of the fluid that may be of interest, e.g., its density, or one of the components of its velocity. If we stay at a fixed point $\mathbf{r}_0$, we can observe F as a function of time and can therefore calculate the rate of change $\partial F/\partial t$ at this point. From this limited knowledge of F we have no way of knowing, as yet, how F changes with time if we stay with a particular particle and follow it along as it passes through the point at $\mathbf{r}_0$. To find this lagrangian rate of change we need to relate the value of F at $\mathbf{r}_0$ at time t_0 to its value at a neighboring point $\mathbf{r}_0 + d\boldsymbol{\varrho}$ at time $t_0 + dt$, where $d\boldsymbol{\varrho} = \mathbf{v}\,dt$ is a small displacement along the flow line passing through the point at $\mathbf{r}_0$. Now F has the value $F_0 \equiv F(x_0, y_0, z_0, t_0)$ at $\mathbf{r}_0$ at time t_0. When a particle at this point at the time t_0 arrives at the neighboring point at the time $t_0 + dt$, the function F has the value

$$F(x_0 + d\xi, y_0 + d\eta, z_0 + d\zeta, t_0 + dt)$$
$$= F_0 + \left(\frac{\partial F}{\partial x}\right)_0 d\xi + \left(\frac{\partial F}{\partial y}\right)_0 d\eta + \left(\frac{\partial F}{\partial z}\right)_0 d\zeta + \left(\frac{\partial F}{\partial t}\right)_0 dt. \quad (6.1.1)$$

The total increment in F is therefore

$$dF = \left(\frac{\partial F}{\partial x}\right)_0 d\xi + \left(\frac{\partial F}{\partial y}\right)_0 d\eta + \left(\frac{\partial F}{\partial z}\right)_0 d\zeta + \left(\frac{\partial F}{\partial t}\right)_0 dt. \quad (6.1.2)$$

Hence the time rate of change of F from the lagrangian viewpoint is

$$\frac{dF}{dt} = \left(\frac{\partial F}{\partial x}\right)_0 \frac{\partial \xi}{\partial t} + \left(\frac{\partial F}{\partial y}\right)_0 \frac{\partial \eta}{\partial t} + \left(\frac{\partial F}{\partial z}\right)_0 \frac{\partial \zeta}{\partial t} + \frac{\partial F}{\partial t}, \quad (6.1.3)$$

which may be more compactly written

$$\frac{dF}{dt} = \mathbf{v} \cdot \nabla F + \frac{\partial F}{\partial t}, \quad (6.1.4)$$

where

$$\mathbf{v} = \mathbf{i}\frac{\partial \xi}{\partial t} + \mathbf{j}\frac{\partial \eta}{\partial t} + \mathbf{k}\frac{\partial \zeta}{\partial t}$$
$$\nabla F = \mathbf{i}\frac{\partial F}{\partial x} + \mathbf{j}\frac{\partial F}{\partial y} + \mathbf{k}\frac{\partial F}{\partial z}$$

are, respectively, the velocity of the fluid and the gradient of the function F and where $\partial F/\partial t$ is the eulerian time derivative observed at a fixed point. Although we calculated (6.1.4) at a particular point and at a particular time, both position and time can be arbitrarly chosen; thus (6.1.4) constitutes a general kinematical relation that always holds for a fluid.

Three different applications of (6.1.4) are of particular importance: (1) If F is the density ρ of a compressible fluid, then

$$\frac{d\rho}{dt} = \mathbf{v} \cdot \nabla \rho + \frac{\partial \rho}{\partial t} \quad (6.1.5)$$

relates the rate of change of density of a particular element of the fluid, as it moves along, to the time rate of change at a fixed point $\partial\rho/\partial t$ and to the gradient (or spatial rate of change) $\nabla\rho$. (2) If F is considered to be, in turn, each of the three rectangular velocity components v_x, v_y, v_z of the fluid and the three component scalar equations are combined into a single vector equation, then

$$\frac{d\mathbf{v}}{dt} = \mathbf{v} \cdot \nabla\mathbf{v} + \frac{\partial\mathbf{v}}{\partial t}. \tag{6.1.6}$$

Here $d\mathbf{v}/dt$ is the vector acceleration of the fluid at any point, which we need when we apply Newton's law to relate force fields to the acceleration of the fluid. If the fluid flow is *steady*, then $\partial\mathbf{v}/\partial t = 0$. However, there can still exist an acceleration of the fluid, $\mathbf{v} \cdot \nabla\mathbf{v}$, resulting from a dependence of $\mathbf{v}$ on position. (3) If $F(x,y,z,t) = 0$ is the equation of a surface moving with the fluid—of particular interest a boundary surface—then the differential equation

$$\frac{dF}{dt} = \mathbf{v} \cdot \nabla F + \frac{\partial F}{\partial t} = 0 \tag{6.1.7}$$

is satisfied by the moving surface.

A good example of the second application is afforded by liquid flowing steadily through a pipe whose diameter changes size. Clearly the liquid must suffer an acceleration as it passes into the smaller pipe, though it has a constant velocity at each position. The increase in fluid velocity is associated with a decrease in the pressure in the fluid.

A second kinematical equation, which we state without proof (see Prob. 6.1.5), relates the average angular velocity $\boldsymbol{\omega}$ of the fluid in the vicinity of any point to the linear velocity $\mathbf{v}$ of the fluid,

$$\boldsymbol{\omega} = \tfrac{1}{2}\,\mathbf{curl}\,\mathbf{v} = \tfrac{1}{2}\nabla \times \mathbf{v}. \tag{6.1.8}$$

In discussing waves on the surface of a liquid (as well as many other problems in hydrodynamics), it is customary to assume that the liquid motion is irrotational, which means that the curl of $\mathbf{v}$ everywhere vanishes. This assumption in effect means that we are not interested in eddies, or vortices, which, when once set up in an ideal fluid, persist forever to conserve angular momentum. (In real fluids vortices die out because of viscosity in the fluid, as illustrated by the behavior of smoke rings.) The vanishing of the curl of $\mathbf{v}$ is a necessary and sufficient condition that $\mathbf{v}$ can be derived from a scalar velocity potential ϕ,

$$\mathbf{v} = -\nabla\phi. \tag{6.1.9}$$

In Sec. 5.9 we briefly introduced the velocity potential as an alternative basis on which sound waves in gases can be discussed. Now, however, we shall find that the velocity potential is almost indispensable for discussing waves on a liquid surface.

(b)   The Equation of Continuity

We have already pointed out that the equation of continuity expresses the continued existence of the mass of a fluid as it moves about. To obtain an equation embodying this idea, let us compute the net entry of fluid mass into an element of volume $\Delta x \, \Delta y \, \Delta z$ in time Δt and equate it to the net increase in mass within the element. Now the quantity $\rho \mathbf{v}$ is the (vector) mass flow of the fluid, where $\mathbf{v} = \mathbf{i} v_x + \mathbf{j} v_y + \mathbf{k} v_z$ is the velocity of the fluid and ρ its density, both of which may be functions of position and time. The mass entering the element through the face $OBEC$ (see Fig. 6.1.1) in time Δt is $\rho v_x \, \Delta y \, \Delta z \, \Delta t$, and that leaving face $ADGF$ is

$$\left(\rho v_x + \frac{\partial}{\partial x} \rho v_x \, \Delta x \right) \Delta y \, \Delta z \, \Delta t.$$

Hence the net amount of mass entering the element in the x direction in time Δt is

$$- \frac{\partial}{\partial x} \rho v_x \, \Delta x \, \Delta y \, \Delta z \, \Delta t$$

with similar expressions for the y and z directions. The total influx of mass is

$$\Delta M = - \left(\frac{\partial}{\partial x} \rho v_x + \frac{\partial}{\partial y} \rho v_y + \frac{\partial}{\partial z} \rho v_z \right) \Delta x \, \Delta y \, \Delta z \, \Delta t. \tag{6.1.10}$$

On dividing by $\Delta x \, \Delta y \, \Delta z \, \Delta t$, passing to the limit $\Delta x \to 0$, etc., and identifying the quantity in parentheses in (6.1.10) as the divergence of the mass flow $\rho \mathbf{v}$, we obtain the *equation of continuity*

$$\frac{\partial \rho}{\partial t} = - \operatorname{div} \rho \mathbf{v} = - \mathbf{\nabla} \cdot \rho \mathbf{v}. \tag{6.1.11}$$

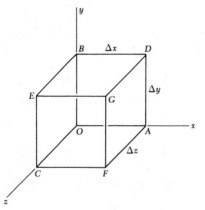

Fig. 6.1.1   Element of volume in a fluid.

In the case of an incompressible fluid, which is an accurate model for most liquids when the pressure changes are small, the density ρ is a constant. The equation of continuity then becomes

$$\mathbf{\nabla} \cdot \mathbf{v} = 0. \tag{6.1.12}$$

We recognize (6.1.12) as equivalent to the assumption that the liquid has a vanishing dilatation (3.2.4),

$$\theta = \mathbf{\nabla} \cdot \boldsymbol{\varrho} = 0, \tag{6.1.13}$$

since $\mathbf{v}$ is simply an alternative notation for $\dot{\boldsymbol{\varrho}}$, the displacement velocity in an elastic medium. As we shall always regard $\mathbf{v}$ as being derivable from a velocity potential, it is now evident that the velocity potential in an incompressible liquid must satisfy *Laplace's equation*

$$\mathbf{\nabla} \cdot \mathbf{\nabla}\phi = 0. \tag{6.1.14}$$

In the case of surface waves, the velocity potential ϕ necessarily depends on time, as well as on position, in some way that is in agreement with (6.1.14). Evidently ϕ does not itself satisfy what we have been calling a (differential) wave equation. The possibility of surface waves on a liquid arises when we relate wavelike solutions of (6.1.14) to the boundary condition that the pressure at the exposed surface of liquid is the constant atmospheric pressure. Hence we must not expect to find an obvious suggestion of waves in the hydrodynamic equations for incompressible fluids until we *actually seek* wavelike solutions of (6.1.14) that satisfy appropriate boundary conditions. For this reason the theory of surface waves is rarely used as an introduction to wave theory, though the waves themselves, which are so familiar in everyday experience, are often used as the example par excellence of waviness and wave behavior.

(c) *The Bernoulli Equation*

Let us now apply Newton's second law to the mass $\rho \, \Delta\tau$ in an element of volume $\Delta\tau$ of a fluid. The net force acting on the element to give it an acceleration $d\mathbf{v}/dt$ includes an external body force, which we shall assume to be due to gravity, and an internal force, which arises from a variation of the pressure from point to point in the fluid. In Sec. 5.1 we established that the latter force is given by

$$\Delta\mathbf{F}_i = -\mathbf{\nabla}P \, \Delta\tau, \tag{6.1.15}$$

where $\mathbf{\nabla}P$ is the pressure gradient. The force due to gravity is

$$\Delta\mathbf{F}_e = \rho \mathbf{g} \, \Delta\tau = -\rho \, \mathbf{\nabla}\Omega \, \Delta\tau, \tag{6.1.16}$$

where Ω is the gravitational potential energy per unit mass such that $\mathbf{g} = -\mathbf{\nabla}\Omega$. (For instance, if the y axis is directed upward, then $\Omega = gy$, and

$\mathbf{g} = -\mathbf{j}g$.) Newton's second law therefore states that

$$-\rho\,\nabla\Omega - \nabla P = \rho\frac{d\mathbf{v}}{dt}. \tag{6.1.17}$$

We now introduce (6.1.6) to express the (lagrangian) acceleration in a form suitable for studying the fluid motion by the more convenient eulerian method of observing it from an inertial frame. We then have that

$$-\nabla\Omega - \frac{1}{\rho}\nabla P = \frac{\partial\mathbf{v}}{\partial t} + \mathbf{v}\cdot\nabla\mathbf{v}. \tag{6.1.18}$$

A further simplification of the equation of motion is possible when we restrict the treatment to irrotational motion, so that $\nabla\times\mathbf{v} = 0$ and $\mathbf{v} = -\nabla\phi$. Now when $\nabla\times\mathbf{v} = 0$,

$$(\nabla\times\mathbf{v})\times\mathbf{v} = \mathbf{v}\cdot\nabla\mathbf{v} - \nabla\mathbf{v}\cdot\mathbf{v} = 0,$$

so that

$$\mathbf{v}\cdot\nabla\mathbf{v} = (\nabla\mathbf{v})\cdot\mathbf{v} = \nabla(\tfrac{1}{2}v^2). \tag{6.1.19}$$

Since

$$\frac{\partial\mathbf{v}}{\partial t} = -\nabla\frac{\partial\phi}{\partial t}, \tag{6.1.20}$$

(6.1.18) becomes

$$-\nabla\Omega - \frac{1}{\rho}\nabla P = -\nabla\frac{\partial\phi}{\partial t} + \nabla(\tfrac{1}{2}v^2). \tag{6.1.21}$$

We can integrate this equation, provided ρ is a function of P alone, by dotting $d\mathbf{r}$ into it and making use of the fact that

$$dF = d\mathbf{r}\cdot\nabla F, \tag{6.1.22}$$

where $F(x,y,z,t)$ is any scalar function of position and time and dF is the difference in its value at two neighboring points separated by $d\mathbf{r}$. We thus find that (6.1.21) becomes

$$-d\Omega - \frac{1}{\rho}dP = -d\frac{\partial\phi}{\partial t} + d(\tfrac{1}{2}v^2), \tag{6.1.23}$$

which, on integration, gives the relation

$$\int\frac{dP}{\rho} = \frac{\partial\phi}{\partial t} - \Omega - \tfrac{1}{2}v^2. \tag{6.1.24}$$

Finally, when the fluid can be considered incompressible, then

$$\frac{P}{\rho} + \Omega + \tfrac{1}{2}v^2 = \frac{\partial \phi}{\partial t} + F(t), \tag{6.1.25}$$

where the constant of integration, which may be a function of time, is often absorbed into $\partial \phi / \partial t$.

We refer to the integral (6.1.25) of the equation of motion as *Bernoulli's equation*, a name which is also used for somewhat different forms the equation takes when the limitations of irrotational motion and incompressibility are not placed on the fluid. We shall use Bernoulli's equation as a connecting link between the boundary condition of a constant pressure at the surface of a liquid and the variables that describe the kinematical aspects of a surface wave.

Problems

6.1.1 Prove that the equation of continuity (6.1.5) can be expressed in the alternative form

$$\frac{d\rho}{dt} + \rho \mathbf{\nabla} \cdot \mathbf{v} = 0. \tag{6.1.26}$$

★6.1.2 Construct an alternative derivation of the equation of continuity based on Gauss' theorem

$$\oint \mathbf{V} \cdot d\mathbf{S} = \int \mathbf{\nabla} \cdot \mathbf{V} \, d\tau.$$

Hint: Let $\mathbf{V} = \rho \mathbf{v}$ and interpret what Gauss' theorem is telling us.

6.1.3 For steady (irrotational) flow of an incompressible liquid, the right side of (6.1.25) is a constant. Using this equation, discuss the pressure and velocity variations in the steady streamline flow of a liquid through a pipe of varying cross section and varying elevation.

★6.1.4 Give a qualitative explanation, based on Bernoulli's equation, of how it is possible to throw a baseball in a "curve."

6.1.5 The fluid in the vicinity of an arbitrary point P is rotating with the angular velocity ω with respect to an inertial frame. Take an origin for a position vector $\mathbf{r}$ at P and show that the velocity $\mathbf{u}$ with respect to P of fluid in the vicinity of P, whose position vector is $\mathbf{r}$, is given by $\mathbf{u} = \omega \times \mathbf{r}$. If $\mathbf{v}_P$ is the velocity of transport of the point P (with respect to an inertial frame), then the velocity of the fluid in the vicinity of P with respect to the inertial frame is $\mathbf{v} = \mathbf{v}_P + \omega \times \mathbf{r}$. Show that $\omega = \tfrac{1}{2}\mathbf{\nabla} \times \mathbf{v}$ by taking the curl of $\mathbf{v}$.

6.2   Gravity Waves

We are now prepared to investigate the waves that can occur at a water-air interface under the influence of gravity. We may think of the water as the idealization of a lake or the ocean. For the time being we omit the effect of surface tension on the motion of surface waves. As with several cases of wave motion studied in earlier chapters, it is necessary to restrict the analysis to waves of small amplitude—here, in order that the nonlinear term $\frac{1}{2}v^2$ in Bernoulli's equation (6.1.25) can be neglected. In effect we are equating the eulerian acceleration $\partial v/\partial t$ to the lagrangian acceleration dv/dt by neglecting the term $\mathbf{v} \cdot \nabla \mathbf{v}$ in (6.1.6). This assumption was made implicitly in the derivation of (5.1.8) for sound waves in a fluid.

We also limit the discussion to plane (or line) waves traveling in one direction, which we designate as the x direction. If the y axis is taken vertically upward, the variables of the wave motion are independent of z. It is often convenient to consider that the body of water is confined to a channel, or canal, of unit width, with smooth vertical walls and a smooth level bottom. The equilibrium depth of the water is h, with the origin at the bottom of the canal. As mentioned earlier, we assume that the motion of the water associated with surface waves is irrotational, so that the displacement velocity of the water can be derived from a velocity potential ϕ, which satisfies the equation of continuity (6.1.14). Here this equation takes the form of the two-dimensional Laplace equation

$$\frac{\partial^2 \phi}{\partial x^2} + \frac{\partial^2 \phi}{\partial y^2} = 0. \tag{6.2.1}$$

In studying oscillations and waves on a liquid surface by the eulerian method, we need to distinguish clearly the coordinates of a definite point in an inertial frame, x, y, z, from the displacement of an element of liquid from its equilibrium position at this point. As in our discussions of elastic waves, we continue to use ξ, η, ζ to designate the components of the displacement. The displacement velocity of the water at any position and time is then

$$v_x = \frac{\partial \xi}{\partial t} = -\frac{\partial \phi}{\partial x}$$

$$v_y = \frac{\partial \eta}{\partial t} = -\frac{\partial \phi}{\partial y} \tag{6.2.2}$$

$$v_z = \frac{\partial \zeta}{\partial t} = -\frac{\partial \phi}{\partial z} = 0,$$

where v_z vanishes for the plane surface waves we are considering here.

At the bottom of the canal, the boundary condition of no vertical velocity is

$$v_y = -\frac{\partial \phi}{\partial y} = 0 \qquad\qquad (y = 0). \qquad\qquad (6.2.3)$$

At the surface of the water, the pressure is the (constant) atmospheric pressure P_0. We can turn this latter condition into a boundary condition on ϕ using Bernoulli's equation (6.1.25) but omitting the nonlinear term $\frac{1}{2}v^2$ for simplicity of analysis. The potential energy of unit mass of water at the surface when it has been displaced η_h in the vertical direction by the passage of a wave is evidently

$$\Omega = g(h + \eta_h). \qquad\qquad (6.2.4)$$

Hence Bernoulli's equation at the surface of the water becomes

$$P_0 + \rho_0 g(h + \eta_h) = \rho_0 \frac{\partial \phi}{\partial t} + \text{const} \qquad (y = h), \qquad\qquad (6.2.5)$$

where any time-dependent part of $F(t)$ has been absorbed in the as yet undetermined function $\partial \phi / \partial t$.

Equation (6.2.5), basically, constitutes a relation between the variation in pressure at $y = h$ arising from the vertical displacement of the surface, the term $\rho_0 g \eta_h$, and the time derivative of the velocity potential, the term $\rho_0 \, \partial \phi / \partial t$. A similar relation exists for compressional waves in a fluid between the excess pressure p and $\rho_0 \, \partial \phi / \partial t$, as given by (5.9.5). Here the variation in surface elevation gives rise to the (weak) elastic restoring force that replaces the (strong) elastic restoring force that exists for compressional waves in fluids.

If we now take a time derivative of (6.2.5), treating P_0 as a constant and replacing $\partial \eta_h / \partial t$ by its equal $-\partial \phi / \partial y$, evaluated at $y = h$,† we find that ϕ satisfies the boundary condition at the surface

$$v_y = -\frac{\partial \phi}{\partial y} = \frac{1}{g} \frac{\partial^2 \phi}{\partial t^2} \qquad\qquad (y = h). \qquad\qquad (6.2.6)$$

Once we have established how ϕ depends on position and time, Bernoulli's equation can be used to compute the pressure at any point in the water.

Our task is now to find a velocity potential function $\phi(x,y,t)$ that satisfies Laplace's equation (6.2.1) and the two boundary conditions (6.2.3) and (6.2.6). Although it is very easy to guess what form the function should take for sinusoidal waves in the x direction, we shall find it instructive to proceed by the more formal method of separation of variables. Accordingly assume that a

---

† Strictly speaking, we should equate the normal velocity component of fluid particles at the free surface, as computed from ϕ, to the normal velocity of the free surface. The error is negligible for small-amplitude waves, for which the inclination of the surface is small.

suitable solution of (6.2.1), satisfying the two boundary conditions, has the form

$$\phi(x,y,t) = X(x) \cdot Y(y) \cdot T(t). \tag{6.2.7}$$

Substitution of this trial solution into (6.2.1) gives no information about the time function $T(t)$, other than that it is satisfactory to have the time dependence of ϕ as a separate factored-out function $T(t)$. For the spatial functions, we find that

$$\frac{d^2X}{dx^2} + \kappa^2 X = 0 \tag{6.2.8}$$

$$\frac{d^2Y}{dy^2} - \kappa^2 Y = 0, \tag{6.2.9}$$

where the separation constant κ^2 has been given the sign that makes $X(x)$ a periodic function of x. General solutions of the separated equations are

$$X(x) = A e^{i\kappa x} + B e^{-i\kappa x}, \tag{6.2.10}$$
$$Y(y) = C e^{\kappa y} + D e^{-\kappa y}. \tag{6.2.11}$$

We could just as well have written these two solutions using trigonometric and hyperbolic functions, respectively.

We next need to introduce the boundary conditions at the bottom and at the top of the water. At the bottom, for all values of x and t, it is necessary that

$$v_y = -\frac{\partial \phi}{\partial y} = -XT \frac{dY}{dy} = 0. \tag{6.2.12}$$

This condition requires that $C = D$, so that the Y function becomes

$$Y(y) = 2C \cosh\kappa y. \tag{6.2.13}$$

For simplicity let us take $B = 0$, so that if $T(t)$ turns out to have the form $e^{-i\omega t}$, the solution we are constructing will represent sinusoidal waves traveling in the positive x direction.

Up to this point, we have established that the velocity potential has the form

$$\phi = A \cosh\kappa y \, e^{i\kappa x} \, T(t) \tag{6.2.14}$$

where A is an arbitrary constant. As a last step, let us substitute (6.2.14) in (6.2.6), which is the remaining boundary condition at $y = h$, to obtain the equation for $T(t)$

$$\frac{d^2T}{dt^2} + (g\kappa \tanh\kappa h) T = 0. \tag{6.2.15}$$

Hence the method of separation of variables has finally led us to a simple harmonic time dependence with the angular frequency

$$\omega = (g\kappa\ \tanh\kappa h)^{1/2}. \tag{6.2.16}$$

If we choose for T the solution $e^{-i\omega t}$, then

$$\phi = A\ \cosh\kappa y\ e^{i(\kappa x - \omega t)} \tag{6.2.17}$$

is the velocity potential for a simple harmonic surface wave traveling in the positive x direction with a wave velocity

$$c = \frac{\omega}{\kappa} = \left(\frac{g}{\kappa}\ \tanh\kappa h\right)^{1/2}. \tag{6.2.18}$$

As suggested by the notation, the separation constant κ has turned out to be the wave number. The dependence of the wave velocity on wave number indicates that dispersion exists, so that a distinction must be made between phase and group velocity (see Prob. 6.2.2). A plot of (6.2.18) showing c as a function of wavelength for several depths h is given in Fig. 6.2.1.

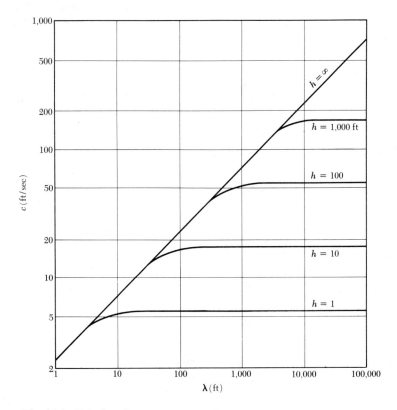

Fig. 6.2.1 Velocity of water waves as a function of wavelength and depth.

For deep water, say when $h \geq \lambda/2$, $\tanh \kappa h \geq \tanh \pi = 0.996$, so that the expression for the wave velocity (6.2.18) becomes, very closely,

$$c = \left(\frac{g}{\kappa}\right)^{1/2} = \left(\frac{\lambda g}{2\pi}\right)^{1/2} \qquad (\lambda < h). \qquad (6.2.19)$$

Waves commonly seen on the ocean and on most lakes are of this type. In contrast, when the wavelength is much greater than the depth, i.e., when $\kappa h \ll 1$, $\tanh \kappa h \approx \kappa h$, and (6.2.18) now becomes, very closely,

$$c = \left(\frac{g}{\kappa}\right)^{1/2} (h\kappa)^{1/2} = (gh)^{1/2} \qquad (\lambda \gg h). \qquad (6.2.20)$$

Waves of this type therefore travel with a speed that varies with the square root of the depth but without dispersion. This speed is that acquired by a body falling a distance $h/2$ from rest.

Let us now obtain expressions for the displacement components in a wave having the velocity potential (6.2.17), which has the real part (assume that A is real)

$$\phi = A \cosh \kappa y \cos(\kappa x - \omega t). \qquad (6.2.21)$$

The displacement velocity of the water, from (6.2.2) and (6.2.21), has the components

$$v_x = \frac{\partial \xi}{\partial t} = A\kappa \cosh \kappa y \sin(\kappa x - \omega t)$$

$$v_y = \frac{\partial \eta}{\partial t} = -A\kappa \sinh \kappa y \cos(\kappa x - \omega t). \qquad (6.2.22)$$

Integrating with respect to time and setting $\eta_m \equiv (A\kappa/\omega) \sinh \kappa h$ for the vertical amplitude of the wave at the surface gives

$$\xi = \eta_m \frac{\cosh \kappa y}{\sinh \kappa h} \cos(\kappa x - \omega t)$$

$$\eta = \eta_m \frac{\sinh \kappa y}{\sinh \kappa h} \sin(\kappa x - \omega t). \qquad (6.2.23)$$

These are the parametric equations of the ellipse

$$\frac{\xi^2}{a^2} + \frac{\eta^2}{b^2} = 1 \qquad (6.2.24)$$

having the major and minor semiaxes

$$a = \eta_m \frac{\cosh\kappa y}{\sinh\kappa h}$$

$$b = \eta_m \frac{\sinh\kappa y}{\sinh\kappa h}.$$

(6.2.25)

The axes of the ellipse diminish with depth, the minor axis b vanishing at the bottom. When the depth of water is of the order of one wavelength or greater, $e^{-\kappa h}$ is very small (so that $a \approx b$) and the ellipses become very nearly circles with radii that decrease exponentially with depth. These waves have the velocity (6.2.19).

When the depth is much less than a wavelength, i.e., when $\kappa h \ll 1$, then $\cosh\kappa y \approx 1$, $\sinh\kappa y \approx \kappa y$, and

$$a \approx \frac{\eta_m}{\kappa h}$$

$$b \approx \eta_m \frac{y}{h}.$$

(6.2.26)

For this case of shallow water, the horizontal motion of the water due to a wave is the same at all depths, with the vertical motion decreasing linearly with amplitude to the bottom where it vanishes. A wave of this type, whose velocity is given by (6.2.20), is usually termed a *tidal wave*. Some of the properties of tidal waves and tides are discussed in Sec. 6.4.

If we compare the equations for the displacement velocity (6.2.22) with those for the particle displacement (6.2.23), we see that the water making up a crest of the wave is moving in the direction in which the wave is traveling, whereas the water in a trough is moving in the backward direction. Although both the ξ and η displacements depend sinusoidally on $\kappa x - \omega t$ (the first with a cosine, the second with a sine factor), the actual profile of the water surface differs slightly from a sine wave. At the top of a crest, where $\eta = \eta_m$, ξ is zero; in the half-wavelength behind this point, ξ is positive, whereas in the half-wavelength in front of it, ξ is negative. Evidently the crests are somewhat narrower than those of a pure sine wave, whereas the troughs are correspondingly wider. For waves of finite amplitude, the sort of distortion just described is considerably greater than that predicted by the present linearized theory. Analytical treatment becomes very difficult when the amplitude is so large that the nonlinear term in Bernoulli's equation cannot be ignored. Figure 6.2.2 shows a highly nonlinear wave treated by numerical computation using a high-speed computer.

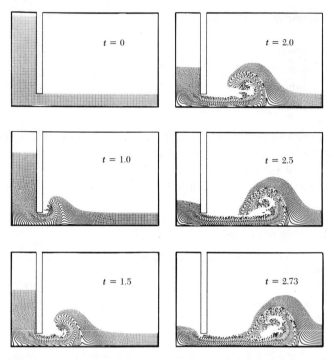

Fig. 6.2.2   Sequence of frames, produced by a computer, depicting the escape of water under a newly opened sluice gate into a tranquil pond. [*From F. H. Harlow, J. P. Shannon, and J. E. Welch, Group T-3 of Los Alamos Scientific Laboratory, pub. by Science,* **149**: 1092 (1965)].

Problems

6.2.1   A layer of oil of density ρ_0' is superposed on water of density ρ_0. Find the velocity of waves that can exist on their common boundary when the depth of each liquid considerably exceeds one wavelength. *Answer:* $c^2 = (g/\kappa)(\rho_0 - \rho_0')/(\rho_0 + \rho_0')$.

6.2.2   Show that the group velocity of the surface wave whose (phase) wave velocity is given by (6.2.18) is

$$c_g = \tfrac{1}{2}c\left(1 + \frac{2\kappa h}{\sinh 2\kappa h}\right) \qquad (6.2.27)$$

Discuss the variation of c_g with the ratio h/λ.

6.3   *Effect of Surface Tension*

The interface between two different material phases, such as a gas and liquid, or two liquids that do not mix, is characterized by a *surface energy density* that is related to the asymmetry of the intermolecular forces acting on the molecules in the interface. When the interfacial area is increased, external energy must be supplied to bring additional molecules into the interface. An equivalent measure of the interfacial energy density is by means of a *surface tension* τ_0, which acts in a direction tangent to the surface so as to minimize its area. By considering the work done in increasing (stretching) the area of the interface, it is easy to establish that τ_0, which is a force per unit length, is numerically and dimensionally equal to the interfacial energy density. The surface tension at an interface is very much like the tension in a uniformly stretched elastic membrane of the sort we considered in Chap. 2. It differs in that its value remains the same, no matter how much the interfacial surface area is increased by stretching. The measured surface tension of water at 18°C is 0.073 N/m.

When waves are present on a liquid surface, the surface area is increased, resulting in an increase in its surface energy. Such an increase adds to the gravitational potential energy associated with the waves, and therefore should be taken into account in the equations describing the wave motion. Let us first discuss waves where surface tension alone is responsible for the restoring force that tends to keep a water (or other liquid) surface flat, i.e., of minimal area, before considering the combined effect of gravity and surface tension.

We can conveniently use the notation of the last section and need only modify the boundary condition at the surface of the water by replacing the gravitational term with one depending on surface tension. When the surface has a small curvature as the result of waves traveling in the x direction, there exists a net upward force due to surface tension

$$\Delta F_y = \tau_0 \frac{\partial^2 \eta}{\partial x^2} \Delta x \qquad (y = h) \qquad (6.3.1)$$

acting on a strip of surface whose width is Δx and whose length in the z direction is unity. The derivation of (6.3.1) parallels that of the left member of (1.1.2) for the sideways force acting on unit length of a displaced string. The existence of an upward force component means that the pressure P just beneath the surface is *less* than atmospheric pressure by the amount $\Delta F_y/\Delta x$, so that the pressure there is

$$P = P_0 - \tau_0 \frac{\partial^2 \eta}{\partial x^2} \qquad (y = h). \qquad (6.3.2)$$

Hence, if we set $\Omega = 0$ in the Bernoulli equation (6.1.25), in order to leave out gravity, again omit the nonlinear term $\frac{1}{2}v^2$, and use (6.3.2) for the pressure term, the equation reads

$$P_0 - \tau_0 \frac{\partial^2 \eta}{\partial x^2} = \rho_0 \frac{\partial \phi}{\partial t} + \text{const} \qquad (y = h). \qquad (6.3.3)$$

If we then take a time derivative to eliminate the constant pressure term and replace $\partial\eta/\partial t$ by $-\partial\phi/\partial y$ as before, the boundary condition on the velocity potential at the surface becomes

$$\frac{\partial^3\phi}{\partial^2x\,\partial y} = \frac{\rho_0}{\tau_0}\frac{\partial^2\phi}{\partial t^2} \qquad (y = h), \tag{6.3.4}$$

which replaces (6.2.6) applying when gravity alone is acting.

In the preceding section we found that the boundary condition at the surface serves to establish the form of the time function $T(t)$ occurring as a factor in the velocity potential and that it does not influence the form of the spatial factors. Hence we can adopt the velocity potential (6.2.14) as suitable for pure surface-tension waves and substitute it in (6.3.4) to obtain the equation for $T(t)$. We find that

$$\frac{d^2T}{dt^2} + \left(\frac{\tau_0\kappa^3}{\rho_0}\tanh\kappa h\right)T = 0, \tag{6.3.5}$$

so that the time dependence again is simple harmonic, but now with the frequency

$$\omega = \left(\frac{\tau_0}{\rho_0}\kappa^3\tanh\kappa h\right)^{1/2}. \tag{6.3.6}$$

The wave velocity of pure surface tension waves is therefore

$$c = \frac{\omega}{\kappa} = \left(\frac{\tau_0}{\rho_0}\kappa\tanh\kappa h\right)^{1/2}. \tag{6.3.7}$$

The velocity is small compared with that of gravity waves unless κ is large, i.e., waves of short wavelength. In this limit we can set $\tanh\kappa h = 1$, so that

$$c = \left(\frac{\tau_0\kappa}{\rho_0}\right)^{1/2}. \tag{6.3.8}$$

The speed of waves controlled by surface tension therefore varies as $\kappa^{1/2}$, whereas the speed of waves in deep water controlled by gravity (6.2.19) varies as $\kappa^{-1/2}$. Hence we expect that surface tension is the controlling factor for waves of short wavelength (ripples), whereas gravity is the controlling factor for waves of long wavelength.

When both surface tension and gravity are acting, we must restore the gravitational term to Bernoulli's equation, giving for the boundary condition at $y = h$

$$\frac{P_0}{\rho_0} - \frac{\tau_0}{\rho_0}\frac{\partial^2\eta}{\partial x^2} + g(\eta_h + h) = \frac{\partial\phi}{\partial t} + \text{const} \qquad (y = h). \tag{6.3.9}$$

The boundary condition on ϕ, in turn, takes the form

$$\frac{\tau_0}{\rho_0}\frac{\partial^3 \phi}{\partial x^2 \,\partial y} - g\frac{\partial \phi}{\partial y} = \frac{\partial^2 \phi}{\partial t^2} \qquad (y = h). \tag{6.3.10}$$

The equation for $T(t)$, found by substituting (6.2.14) into (6.3.10), is now

$$\frac{d^2 T}{dt^2} + \left[\left(g\kappa + \frac{\tau_0 \kappa^3}{\rho_0}\right)\tanh\kappa h\right]T = 0, \tag{6.3.11}$$

so that the frequency of the wave is

$$\omega = \left[\left(g\kappa + \frac{\tau_0 \kappa^3}{\rho_0}\right)\tanh\kappa h\right]^{1/2}. \tag{6.3.12}$$

For deep water, $\tanh\kappa h = 1$, so that the expression for the wave velocity becomes

$$c = \frac{\omega}{\kappa} = \left(\frac{g}{\kappa} + \frac{\tau_0 \kappa}{\rho_0}\right)^{1/2}. \tag{6.3.13}$$

This result reduces, as it should, to (6.2.19) when $\tau_0 = 0$ and to (6.3.8) when $g = 0$. The velocity of waves on deep water as a function of wavelength is shown in Fig. 6.3.1. The dashed lines show how the wave velocity would vary either in the absence of gravity or in the absence of surface tension. With both

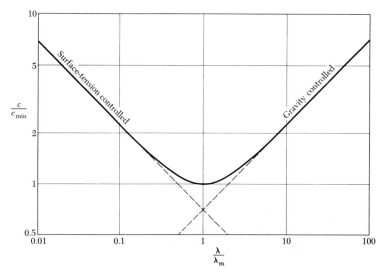

Fig. 6.3.1   Wave velocity of surface waves when both surface tension and gravity are acting ($h \gg \lambda$).

acting, there is a minimum wave velocity

$$c_{\min} = \left(\frac{4\tau_0 g}{\rho_0}\right)^{1/4} = 23.5 \text{ cm/sec } (18°C), \tag{6.3.14}$$

occurring at the wavelength

$$\lambda_m \equiv 2\pi \left(\frac{\tau_0}{\rho_0 g}\right)^{1/2} = 1.73 \text{ cm.} \tag{6.3.15}$$

The existence of a minimum wave velocity means that a wind will fail to disturb the surface of water and will not set up a system of waves unless its velocity exceeds the minimum value just calculated. Additional results related to waves controlled by gravity and surface tension may be found among the problems.

Problems

6.3.1 Derive (6.3.14) and (6.3.15) and verify the numerical results.

6.3.2 Show that (6.3.13), used in conjunction with (6.3.14) and (6.3.15), gives the relation

$$\left(\frac{c}{c_{\min}}\right)^2 = \frac{1}{2}\left(\frac{\lambda}{\lambda_m} + \frac{\lambda_m}{\lambda}\right). \tag{6.3.16}$$

This implies that for a given $c > c_{\min}$, the two corresponding wavelengths have the constant λ_m as their geometric mean.

6.3.3 Use the relation (4.7.7), namely, $c_{\text{group}} = c_{\text{phase}} - \lambda(dc_{\text{phase}}/d\lambda)$, in conjunction with Fig. 6.3.1 to describe how the group velocity of water waves depends on wavelength. Show that

$$c_{\text{group}} = c_{\text{phase}}\left(1 - \frac{1}{2}\frac{\lambda^2 - \lambda_m^2}{\lambda^2 + \lambda_m^2}\right), \tag{6.3.17}$$

by making use of (6.3.16).

6.3.4 Make an ω-κ diagram for water waves when both surface tension and gravity are taken into account. Assume that the water is deep, so that $\tanh \kappa h = 1$. Discuss the phase and group velocity of the water waves on the basis of this diagram.

6.3.5 Show how to modify the theory of pure surface tension waves when the plane interface is between two liquids of densities ρ_0 and ρ_0' and, in particular, show that $c = [\tau_0 \kappa/(\rho_0 + \rho_0')]^{1/2}$.

6.3.6 Extend the treatment of Prob. 6.3.5 to include gravity, as well as surface tension, and show that for $\kappa h \gtrsim 1$ (see Fig. 6.3.1),

$$c = \left(\frac{\rho_0 - \rho_0'}{\rho_0 + \rho_0'} \frac{g}{\kappa} + \frac{\tau_0 \kappa}{\rho_0 + \rho_0'} \right)^{1/2}. \tag{6.3.18}$$

6.4 Tidal Waves and the Tides

In Sec. 6.2 we noted that so-called tidal waves, which are gravity waves having a wavelength much greater than the water depth, are especially simple. We showed that the horizontal motion of the water is very nearly the same at all depths and that the wave velocity is independent of wavelength, though it depends on water depth. It is possible to give an elementary treatment of these waves, without use of the velocity potential. We then discuss briefly the cause of natural tides in the ocean, which are basically tidal waves generated by the moon and sun.

(a) Tidal waves

We suppose that water is confined to a long canal of depth h and unit width, as in our earlier treatment of gravity waves. The x axis is along the canal, and the y axis vertical, with the origin at the bottom of the canal. Since the water can be considered as incompressible, the equation of continuity requires that the dilatation vanish,

$$\nabla \cdot \varrho \to \frac{\partial \xi}{\partial x} + \frac{\partial \eta}{\partial y} = 0, \tag{6.4.1}$$

where, as usual, ξ and η are the x and y displacement components. We now suppose that the waves in the canal have a wavelength that is very great compared with the depth of water, so that we can assume that the horizontal displacement ξ is independent of y and much greater than any related vertical displacement. In this case we can immediately integrate the equation of continuity (6.4.1) to obtain

$$\eta = -y \frac{\partial \xi}{\partial x}. \tag{6.4.2}$$

The constant of integration has been set equal to zero, since the vertical displacement η must vanish at the bottom of the canal. The vertical displacement of the water at the surface is therefore proportional to the strain $\partial \xi / \partial x$,

$$\eta_h = -h \frac{\partial \xi}{\partial x}. \tag{6.4.3}$$

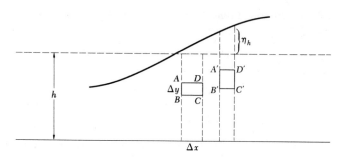

Fig. 6.4.1   Tidal wave in a canal.

Let us now apply Newton's second law to a displaced element of volume of the water $A'B'C'D'$, as shown in Fig. 6.4.1. The element has the equilibrium position $ABCD$ and has the dimensions Δx and Δy in the xy plane and unity in the z direction. As a result of the slope of the water surface, the pressure on the face $C'D'$ of the element is greater than the pressure on the face $A'B'$ by the amount

$$\Delta P = \rho_0 g \frac{\partial \eta_h}{\partial x} \Delta x. \tag{6.4.4}$$

We equate the net horizontal force on the element $-\Delta P\, \Delta y$ to the mass of the element $\rho_0\, \Delta x\, \Delta y$ times its horizontal acceleration $\partial^2 \xi / \partial t^2$, to obtain the equation of motion

$$-\rho_0 g \frac{\partial \eta_h}{\partial x} \Delta x\, \Delta y = (\rho_0\, \Delta x\, \Delta y) \frac{\partial^2 \xi}{\partial t^2}. \tag{6.4.5}$$

Since the motion of the water is very small in the vertical direction in comparison with that in the horizontal direction, we can neglect entirely the vertical components of force and acceleration. If we now eliminate η_h using (6.4.3), we arrive at the ordinary wave equation

$$\frac{\partial^2 \xi}{\partial x^2} = \frac{1}{c^2} \frac{\partial^2 \xi}{\partial t^2}, \tag{6.4.6}$$

where the wave velocity is given by

$$c \equiv (gh)^{1/2}, \tag{6.4.7}$$

in agreement with the value (6.2.20) found earlier. It is interesting that here, where we have made simplifying assumptions at the start, we are led to a differential wave equation, whereas in the earlier more exact treatment, no differential wave equation was found to exist. By taking a partial derivative of the wave equation (6.4.6) with respect to x and making use of (6.4.2), we see that the vertical displacement of waves in the canal satisfies an identical wave equation. We have discussed the principal features of these tidal waves in Sec. 6.2.

(b)   Tide-generating Forces

Before we can relate the wave motion we have been discussing to ocean tides, we need to see how a heavenly body, such as the moon or the sun, produces tide-generating forces on the earth. For simplicity, let us consider the forces produced just by the moon, of mass M, located at a distance R from the center of the earth. We shall find it convenient to take the earth's center to be the origin of a coordinate frame with the z axis directed away from the moon, as in Fig. 6.4.2. A unit mass at a point P, located on the earth, has its potential energy increased by

$$V_M = -\frac{GM}{r} = -\frac{GM}{[(R+z)^2 + x^2 + y^2]^{1/2}} \tag{6.4.8}$$

because of the presence of the moon, where r is the distance from P to the center of the moon and G is the gravitational constant. Since the coordinates of point P are small compared with R, we can expand V_M in powers of x/R, y/R, and z/R,

$$
\begin{aligned}
V_M &= -\frac{GM}{R}\frac{1}{[(1+z/R)^2 + (x/R)^2 + (y/R)^2]^{1/2}} \\
&= -\frac{GM}{R} + \frac{GM}{R^2}z - \frac{GM}{2R^3}(2z^2 - x^2 - y^2) + \cdots .
\end{aligned}
\tag{6.4.9}
$$

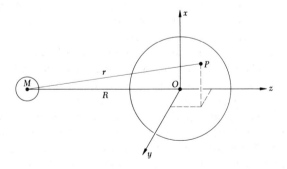

Fig. 6.4.2   The moon-earth system.

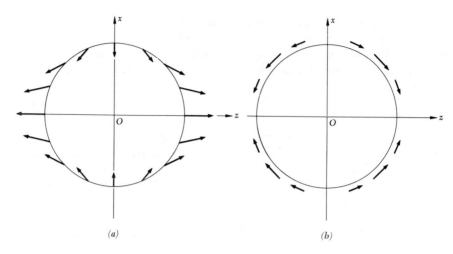

Fig. 6.4.3   Tide-generating forces in the xz plane.

Hence the moon's attraction for a unit mass at point P has the components

$$F_x = -\frac{\partial V_M}{\partial x} = -\frac{GM}{R^3}x + \cdots$$

$$F_y = -\frac{\partial V_M}{\partial y} = -\frac{GM}{R^3}y + \cdots \qquad (6.4.10)$$

$$F_z = -\frac{\partial V_M}{\partial z} = -\frac{GM}{R^2} + \frac{2GM}{R^3}z + \cdots.$$

The first term in F_z is considerably greater than the second, since the radius of the earth is about one-sixtieth the distance to the moon. The first term is evidently equal to the *average* force of the moon on each unit mass of the earth and is responsible for the centripetal acceleration in the earth-moon system. An observer on earth cannot observe this term as a force, however, for the same reason that an astronaut cannot observe the gravitational attraction of the earth: it is precisely cancelled by the free-fall acceleration that accompanies it. The remaining force components, whose averages vanish for the entire earth, have the interesting distribution at the earth's surface as shown in the xz plane in Fig. 6.4.3a. The component of these forces parallel to the earth's surface, shown in Fig. 6.4.3b, is responsible for the principal components of tides.

The sun produces a similar pattern of tide-generating forces, of about half the amplitude of those of the moon, symmetrical about the earth-sun axis. The much greater mass of the sun is more than offset by its greater distance away, which enters as an inverse cube in the expressions for the tide-generating force. The existence of two (not one) regions of high tide on the earth at any time, arises from the mirror symmetry that the force system illustrated in Fig. 6.4.3 possesses with respect to the xy plane.

(c) Equilibrium Theory of Tides

Newton proposed an explanation for the tides in the ocean known as the *equilibrium theory*. As a model he imagined the earth to be covered by an ocean of constant depth, subject to the force system (6.4.10), as well as the gravitational attraction of the earth. Equilibrium occurs when a water particle on the surface of the ocean has a constant potential energy due to both the moon and the earth, no matter where the particle is located.

If we let η be the height of the hypothetical ocean above mean sea level, then $V_E = g\eta$ is the potential energy per unit mass of water at the surface of the ocean due to the earth's gravitational field. The potential energy of the tide-generating force of the moon at the earth's surface is given by the third term in (6.4.9) with $x^2 + y^2 + z^2 = a^2$, where a is the radius of the earth. We can put this potential term in better form by going to spherical coordinates with $z = a \cos\theta$. Evidently for points on the earth's surface, $2z^2 - x^2 - y^2 = a^2(3 \cos^2\theta - 1)$, so that the tide-generating potential energy becomes

$$V'_M = -\frac{GM}{2R^3} a^2(3 \cos^2\theta - 1). \tag{6.4.11}$$

Hence the equilibrium shape of the ocean covering the earth is given by the requirement that

$$V'_M + V_E = -\frac{GM}{2R^3} a^2(3 \cos^2\theta - 1) + g\eta = C \qquad (C \equiv \text{const}), \tag{6.4.12}$$

and the height of the ocean above mean sea level becomes

$$\eta = \frac{GMa^2}{2gR^3}(3 \cos^2\theta - 1) + \frac{C}{g}. \tag{6.4.13}$$

In Prob. 6.4.2 it is found that $C = 0$. If we introduce the mass of the earth, $M_E = a^2 g/G$, we finally have that

$$\eta = \frac{1}{2} \frac{M}{M_E} \left(\frac{a}{R}\right)^3 a(3 \cos^2\theta - 1). \tag{6.4.14}$$

In Fig. 6.4.4 is shown a plot of (6.4.14) illustrating the two tidal "mountains" of the equilibrium theory. Their height due to the moon is about 0.6 m, and the height due to the sun is about one-half this value (Prob. 6.4.3).

The equilibrium theory ignores the dynamical aspects of the tides brought about by the rotation of the earth and the motion of the moon in its orbit. It predicts high water when the moon passes the observer's meridian, on one or

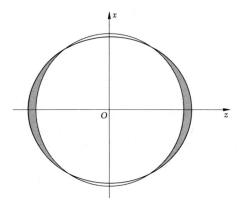

Fig. 6.4.4 The tidal mountains of the equilibrium theory.

the other sides of the earth. In contrast, high water is found to occur delayed many hours after the moon has passed this meridian. An explanation of the delay involves the notion of tidal waves. Nevertheless, the equilibrium theory explains the existence of two tides per day and how the joint effect of the tide-generating forces of the sun and moon combine to produce spring and neap tides. It also explains a diurnal inequality that arises as a result of the inclination of the moon's orbit with the celestial equator. When the moon passes the equator, the diurnal inequality due to the moon vanishes.

(d) The Dynamical Theory of Tides

About a century after Newton, Laplace proposed a dynamical theory of the tides. As a simplified model to illustrate his theory, let us consider that the earth has a canal of uniform depth encircling it at the equator and suppose that water in the canal is acted on by the tide-generating forces (6.4.10). We can further suppose that the earth does not rotate but that the moon revolves around it in 24 hr and 50 min, as it appears to do to an observer on earth. The diagram in Fig. 6.4.5 shows an arbitrary point P on the equator, with the longitude ϕ, the moon with an angular displacement θ from P traveling with an angular velocity ω. From the diagram we see that $\theta = \omega t - \phi$. Let us substitute this expression for θ in the tide-generating potential (6.4.11), which then reads

$$V'_M = -\frac{GM}{2R^3} a^2 [3 \cos^2(\omega t - \phi) - 1].$$ (6.4.15)

The tangential force on a unit mass of water in the canal at point P is therefore

$$F_\phi = -\frac{\partial V'_M}{a\, \partial \phi} = \frac{GMa}{R^3} 3 \cos(\omega t - \phi) \sin(\omega t - \phi)$$

$$= \frac{3}{2} \frac{GMa}{R^3} \sin(2\omega t - 2\phi).$$ (6.4.16)

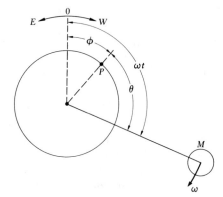

Fig. 6.4.5   Coordinates for the dynamical theory of tides.

This force contributes to the horizontal acceleration of water in the canal, $\partial^2\xi/\partial t^2$, and so must be added to the term $c^2(\partial^2\xi/\partial x^2)$ in the wave equation (6.4.6) due to pressure. With this driving force the wave equation becomes

$$\frac{\partial^2\xi}{\partial t^2} = c^2\,\frac{\partial^2\xi}{a^2\,\partial\phi^2} + k\,\sin(2\omega t - 2\phi), \tag{6.4.17}$$

where $k \equiv 3GMa/2R^3$ is the amplitude of the tide-generating force per unit mass. In (6.4.17) we have replaced dx by $a\,d\phi$, which measures incremental distance along the equatorial canal.

We expect that the equation for tidal waves in the canal with the perturbing force (6.4.16) has a particular solution of the form

$$\xi = \xi_m\,\sin(2\omega t - 2\phi) \tag{6.4.18}$$

to which may be added solutions of (6.4.6) representing free oscillations and waves of an arbitrary nature. On substituting (6.4.18) in (6.4.17), we find that the forced tidal wave has the form

$$\xi = \frac{ka^2}{4(c^2 - \omega^2 a^2)}\,\sin(2\omega t - 2\phi) \tag{6.4.19}$$

and, from (6.4.3), that the tidal wave has a vertical displacement

$$\eta_h = -h\,\frac{\partial\xi}{a\,\partial\phi} = \frac{kah}{2(c^2 - \omega^2 a^2)}\,\cos(2\omega t - 2\phi). \tag{6.4.20}$$

The equations just obtained show that the dynamical theory predicts, as expected, two tides for each apparent revolution of the moon. If $c^2 > \omega^2 a^2$, high water is in phase with the moon and the tides are *direct*. However, if $c^2 < \omega^2 a^2$, low water is in phase with the moon and the tides are *inverted*. If $c^2 \approx \omega^2 a^2$, resonance occurs and there should be tides of tremendous height. In Prob. 6.4.4 it is found that the critical depth for resonance is about 22 km, which is considerably deeper than the oceans. Hence the tides should be inverted near the equator. For a canal encircling the earth near a pole, the tides should be direct, with resonance occurring for a canal at an intermediate latitude.

Laplace showed theoretically for an extended ocean covering the earth, that north-south flow cancels tides at the intermediate latitude but that the equatorial tides should be inverted and those in polar regions, direct. In practice land masses considerably modify the tidal waves induced by the moon and the sun, as do Coriolis forces arising from the rotation of the earth. As a result, theory, coupled with the known motion of the moon and sun, is able to predict the frequency but not the amplitude and phase of the important harmonic components of tides at various points on the earth. The actual phase and amplitude of each component at any location must be found by harmonic analysis of the observed long-term behavior of the tide at that location. Once this has been done, it is then possible by harmonic synthesis to predict the local tides at all future times. Even so, variations in barometric pressure over the ocean and the action of wind can alter the actual tide pattern in a random fashion.†

Problems

6.4.1 Verify the expansion (6.4.9) using (*a*) the binomial theorem and (*b*) Taylor's theorem for a function of three variables. Which method is more efficient?

6.4.2 Show that $C = 0$ in (6.4.13) by requiring that the average of η over the surface of the earth vanish.

6.4.3 Find the amplitudes of high and low tides from mean sea level due to the moon by the equilibrium theory. Assume that $M_E/M = 81$ and $R/a = 60$. Compute the ratio of the solar tide to the lunar tide, given that $M_{sun}/M_E = 332{,}000$ and $R_{sun}/a = 23{,}200$.

6.4.4 Find the depth of an equatorial canal for which the tidal-wave velocity just keeps pace with the apparent motion of the moon around the earth. If the canal is only 2 miles deep, find the vertical and horizontal displacements of the lunar tide.

6.4.5 Find at what latitude lunar resonance should occur in a canal 2 miles deep encircling the earth.

† An interesting discussion of tidal processes is given by A. Defant, "Ebb and Flow," The University of Michigan Press, Ann Arbor, Mich., 1958.

6.5 Energy and Power Relations

The potential energy per unit area associated with a small-amplitude gravity wave, neglecting surface tension, is equal to the work required to displace the water from its level equilibrium state. At a point where the water surface is displaced vertically upward an amount η_h, an infinitesimal element of horizontal area $dx\, dz$ and height η_h has a mass $\rho_0 \eta_h\, dx\, dz$, and the work done in lifting its center of mass a distance $\frac{1}{2}\eta_h$ above the equilibrium level is therefore

$$d^2W = \tfrac{1}{2}\rho_0 g \eta_h{}^2\, dx\, dz. \tag{6.5.1}$$

The same amount of work is involved in displacing the surface when η_h is negative, instead of positive. Hence the potential energy per unit area is

$$V_1 = \frac{d^2W}{dx\, dz} = \tfrac{1}{2}\rho_0 g \eta_h{}^2. \tag{6.5.2}$$

For the gravity wave described by (6.2.23), $\eta_h = \eta_m \sin(\kappa x - \omega t)$, so that the potential energy density becomes

$$V_1 = \tfrac{1}{2}\rho_0 g \eta_m{}^2 \sin^2(\kappa x - \omega t), \tag{6.5.3}$$

which has the average value

$$\overline{V}_1 = \tfrac{1}{4}\rho_0 g \eta_m{}^2. \tag{6.5.4}$$

The kinetic energy density of the gravity wave (6.2.23) includes contributions from the vertical as well as the horizontal motion of the water in the region between the bottom and surface of the water,

$$K_1 = \tfrac{1}{2}\rho_0 \int_0^h (\dot{\xi}^2 + \dot{\eta}^2)\, dy. \tag{6.5.5}$$

If values of $\dot{\xi}$ and $\dot{\eta}$ are computed from (6.2.23) and ω is replaced by the value given by (6.2.16), K_1 takes the somewhat complicated form

$$K_1 = \tfrac{1}{4}\rho_0 g \eta_m{}^2 \left\{ 1 + \frac{2\kappa h}{\sinh 2\kappa h}\left[\sin^2(\kappa x - \omega t) - \cos^2(\kappa x - \omega t)\right]\right\}, \tag{6.5.6}$$

which, however, simplifies on averaging to become

$$\overline{K}_1 = \tfrac{1}{4}\rho_0 g \eta_m{}^2. \tag{6.5.7}$$

Hence, though the average potential and kinetic energy densities are equal, their instantaneous values are not equal, a relationship that exists for many traveling waves. The total average energy density,

$$\overline{E}_1 = \overline{K}_1 + \overline{V}_1 = \tfrac{1}{2}\rho_0 g \eta_m{}^2, \tag{6.5.8}$$

is equal to the work that would be required to give the surface its maximum displacement against the action of gravity. Since the equal average potential and kinetic energy densities just computed do not depend on the depth of the water, they continue to hold for waves on extremely deep water, as well as for waves on shallow water, i.e., for tidal waves.

The power density in gravity waves, i.e., the energy passing in one second through a transverse plane of unit width extending from bottom to surface of the water, is evidently given by the integral

$$P_1 = - \int_0^h p \frac{\partial \xi}{\partial t} \, dy, \tag{6.5.9}$$

where p is the time-dependent pressure in excess of the static-equilibrium value. The total pressure at a point (x,y), according to the Bernoulli equation (neglecting the $\frac{1}{2}v^2$ term), is given by

$$P = P_0 + \rho_0 g(h - y) + \rho_0 \frac{\partial \phi}{\partial t}, \tag{6.5.10}$$

whereas the static pressure at this depth is

$$P_{\text{static}} = P_0 + \rho_0 g(h - y). \tag{6.5.11}$$

Hence the excess pressure at any position and time, described by the velocity potential ϕ, is simply

$$p = P - P_{\text{static}} = \rho_0 \frac{\partial \phi}{\partial t}. \tag{6.5.12}$$

In terms of the velocity potential, the expression for the power density becomes

$$P_1 = -\rho_0 \int_0^h \frac{\partial \phi}{\partial t} \frac{\partial \phi}{\partial x} \, dy. \tag{6.5.13}$$

The velocity potential (6.2.21) for the surface wave (6.2.23) takes the form

$$\phi = \eta_m c \frac{\cosh \kappa y}{\sinh \kappa h} \cos(\kappa x - \omega t) \tag{6.5.14}$$

when A is replaced by its equal, $c\eta_m/\sinh \kappa h$. Using this expression for ϕ, we find that

$$P_1 = \tfrac{1}{2}\rho_0 g \eta_m^2 c \left(1 + \frac{2\kappa h}{\sinh 2\kappa h} \right) \sin^2(\kappa x - \omega t). \tag{6.5.15}$$

The average power, therefore, is given by

$$\overline{P}_1 = (\tfrac{1}{2}\rho_0 g \eta_m^2) \left[\tfrac{1}{2}c \left(1 + \frac{2\kappa h}{\sinh \kappa h} \right) \right]. \tag{6.5.16}$$

The first factor is the average total energy (6.5.8), and the second factor is the group velocity (6.2.27) of the wave train. We have found the same relation to hold for other dispersive waves.

Problems

6.5.1 Make a reasonable estimate of the average amplitude and wavelength of typical waves on the ocean and compute the power they carry. Express the result in horsepower per mile of wavefront.

6.5.2 Discuss the energy and power relations for pure surface-tension waves.

seven

*Elastic Waves in Solids

In Chap. 4 we discussed certain special cases of waves in solids, corresponding to the simple types of elastic deformation considered in Chap. 3. In order to treat more general cases, it is necessary to reformulate our description of stress and strain, expressing these quantities as *tensors of the second rank*. The mathematical techniques required here arise in many important branches of physics, e.g., the motion of rigid bodies, the electromagnetic properties of anisotropic materials, and relativity theory.

7.1 Tensors and Dyadics

The introductory discussion of elasticity and elastic waves in Chaps. 3 and 4 was restricted to simple structures for which symmetry of one sort or another generally allowed us to use only a single strain component in the analysis. We now develop a formal way of expressing all the components of strain as well as those of stress as a single entity, corresponding to the way a vector expresses a physical concept distinct from its three components. We find that stress and strain are in fact *second-rank tensors*, or *dyadics*, as such tensors are called when expressed in the vector notation of Gibbs. In the present section we start by reviewing some basic notions underlying scalars and vectors and then show how a dyadic can be defined in a way that is consistent with these notions. We complete the section with a brief account of some of the mathematical properties of dyadics.

Many quantities of interest in a physical theory can be expressed by single numbers that depend on position and time. We refer to descriptions of this sort as *scalar fields;* density and the velocity potential in a fluid are examples of such fields. Suppose that we let $\phi = \phi(x,y,z,t)$ stand for a particular scalar field, where x, y, z are the coordinates of a point referred to a reference (inertial) frame S and t is time. With respect to a different frame S', the same scalar field is now expressed by a different function $\phi' = \phi'(x',y',z',t)$. For the two functions to represent the same physical quantity, it is necessary that $\phi = \phi'$ whenever x, y, z and x', y', z' are the coordinates of the *same* point in space. We then say that ϕ is an *invariant* function of position and speak of the field as a *physical* scalar field.

Usually the definition of a particular scalar field ensures that it have this property of invariance. However, it is easy to exhibit a scalar field that is *not* a physical scalar field. For example, the x component of the velocity of a flowing liquid has a definite numerical value at each point in a reference frame S, but its value at any point depends on the orientation of the x axis of S. Hence $v_x(x,y,z,t)$, which is one component of a vector velocity field, does not itself constitute a *physical* scalar field, since it is not invariant with respect to a transformation of coordinates. The status of a scalar field arrived at by some means that does not clearly establish whether or not it is a physical scalar field can always be settled by testing its invariance with respect to a spatial rotation of the coordinate frame about a common origin.

A *vector field* is somewhat more complicated than a scalar field, since it involves both a magnitude and a direction that are assigned in some way to each point in space. If a mathematical representation of a vector field is to constitute a *physical* vector field, the vector at any point in space must be independent of the coordinate frame used in specifying the location of the point. Since the rectangular components of a vector take on values that depend on the orientation of the axes of the reference frame, it is convenient to devise a test for invariance that tells how the rectangular components transform under a rotation of the coordinate frame.

We can establish the test for invariance by examining the transformation

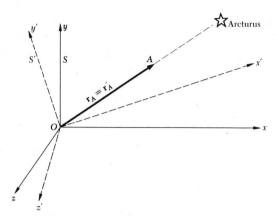

Fig. 7.1.1   A vector $\mathbf{r}_A$ referred to two reference frames having a common origin.

of a displacement vector OA between an origin O common to the two frames and some fixed point A. Let us designate this vector by $\mathbf{r}_A$, with a magnitude r_A and a direction fixed in space, say, toward the star Arcturus (see Fig. 7.1.1). With respect to S, the terminus of $\mathbf{r}_A$ has the coordinates x_A, y_A, z_A, which are also the components of the vector $\mathbf{r}_A$. With respect to a second frame S', the identical vector $\mathbf{r}'_A = \mathbf{r}_A$ has the components x'_A, y'_A, z'_A. In terms of unit vectors along the two sets of axes,

$$\mathbf{r}_A = \mathbf{i}x_A + \mathbf{j}y_A + \mathbf{k}z_A \tag{7.1.1}$$
$$\mathbf{r}'_A = \mathbf{i}'x'_A + \mathbf{j}'y'_A + \mathbf{k}'z'_A. \tag{7.1.2}$$

If γ_{11} is the cosine of the angle between the x' axis and the x axis, γ_{21} the cosine of the angle between the y' axis and the x axis, etc., it is clear that the unit vector $\mathbf{i}$ has the components γ_{11}, γ_{21}, γ_{31} along the x', y', and z' axes. Extending this result to the other two unit vectors in S, we can write that

$$\mathbf{i} = \gamma_{11}\mathbf{i}' + \gamma_{21}\mathbf{j}' + \gamma_{31}\mathbf{k}'$$
$$\mathbf{j} = \gamma_{12}\mathbf{i}' + \gamma_{22}\mathbf{j}' + \gamma_{32}\mathbf{k}' \tag{7.1.3}$$
$$\mathbf{k} = \gamma_{13}\mathbf{i}' + \gamma_{23}\mathbf{j}' + \gamma_{33}\mathbf{k}'.$$

If we now substitute (7.1.3) in (7.1.1) and compare the resulting expression with (7.1.2), we find that the set of equations

$$x'_A = \gamma_{11}x_A + \gamma_{12}y_A + \gamma_{13}z_A$$
$$y'_A = \gamma_{21}x_A + \gamma_{22}y_A + \gamma_{23}z_A \tag{7.1.4}$$
$$z'_A = \gamma_{31}x_A + \gamma_{32}y_A + \gamma_{33}z_A$$

expresses the transformation of the components of a displacement vector under a simple rotation of axes. We now assert that any set of three functions $V_x(x,y,z,t)$,

$V_y(x,y,z,t)$, and $V_z(x,y,z,t)$ that transforms under a rotation of axes according to the scheme (7.1.4) constitutes the components of a possible physical vector field. Instead of defining a physical vector field as one that associates a magnitude and direction with each point in space, we see that we can define it equally well as a field specified by three functions of position and time, one associated with each coordinate axis. The three functions, however, cannot be arbitrarily specified but must be such that they transform upon rotation of the coordinate axes as do the components of a displacement vector. This way of defining a physical vector field by its transformation properties is the one that can be most easily generalized to define tensor fields and, incidentally, is the approach used in the formal mathematical discussion of n-dimensional vector spaces.

We define a *second-rank tensor field* as one that associates a function of position with each of the *nine pairs* of coordinate axes, xx, xy, . . . , zz. The nine functions $T_{xx}(x,y,z,t)$, $T_{xy}(x,y,z,t)$, etc., cannot be arbitrarily chosen if they are to represent a *physical* tensor field, but they must transform in the same way as the *nine coordinate pairs*, xx, xy, etc. If we let the number 1 stand for the subscript x, 2 for the subscript y, and 3 for the subscript z, the transformation for a tensor field, corresponding to (7.1.4) for a vector field, takes the form

$$T'_{lm} = \sum_{s=1}^{3} \sum_{t=1}^{3} \gamma_{ls}\gamma_{mt}T_{st}. \tag{7.1.5}$$

We can write the tensor having the components T_{xx}, T_{xy}, . . . , as the *dyadic*

$$\mathsf{T} = \mathbf{ii}T_{xx} + \mathbf{ij}T_{xy} + \mathbf{ik}T_{xz} + \mathbf{ji}T_{yx} + \cdots , \tag{7.1.6}$$

where the nine *unit dyads*, $\mathbf{ii}$, $\mathbf{ij}$, etc., enable the tensor to be written as a single entity, just as the three unit vectors, $\mathbf{i}$, $\mathbf{j}$, $\mathbf{k}$, enable a vector to be written as a single entity $\mathbf{V} = \mathbf{i}V_x + \mathbf{j}V_y + \mathbf{k}V_z$. We shall often find it convenient to write the dyadic in the form of a 3×3 *matrix*†

$$\mathsf{T} = \begin{pmatrix} T_{xx} & T_{xy} & T_{xz} \\ T_{yx} & T_{yy} & T_{yz} \\ T_{zx} & T_{zy} & T_{zz} \end{pmatrix} \tag{7.1.7}$$

without the unit dyads appearing.

The most important mathematical operation with a dyadic consists in dotting a vector into it, either from the left side or from the right side. We define the operations on the unit dyads, $\mathbf{i} \cdot (\mathbf{ii}) = (\mathbf{i} \cdot \mathbf{i})\mathbf{i} = \mathbf{i}$, $\mathbf{i} \cdot (\mathbf{ji}) = (\mathbf{i} \cdot \mathbf{j})\mathbf{i} = 0$,

† Following P. M. Morse and H. Feshbach, "Methods of Theoretical Physics," McGraw-Hill Book Company, New York, 1953, we use the term *dyadic* as an unequivocal designation for a *second-rank* tensor, just as the term *vector* designates a *first-rank* tensor. Many authors, however, use the unmodified term *tensor* as implying second rank and reserve the term dyadic for the notation (7.1.6) that uses unit dyads explicitly, as opposed to the matrix notation (7.1.7).

etc., as obvious extensions of ordinary vector analysis. Hence we find that

$$\begin{aligned}
\mathbf{V} \cdot \mathbf{T} &= V_x(\mathbf{i}T_{xx} + \mathbf{j}T_{xy} + \mathbf{k}T_{xz}) \\
&+ V_y(\mathbf{i}T_{yx} + \mathbf{j}T_{yy} + \mathbf{k}T_{yz}) \\
&+ V_z(\mathbf{i}T_{zx} + \mathbf{j}T_{zy} + \mathbf{k}T_{zz}) \\
&= \mathbf{i}(V_x T_{xx} + V_y T_{yx} + V_z T_{zx}) + \cdots .
\end{aligned} \tag{7.1.8}$$

Similarly

$$\mathbf{T} \cdot \mathbf{V} = \mathbf{i}(V_x T_{xx} + V_y T_{xy} + V_z T_{xz}) + \cdots . \tag{7.1.9}$$

According to the above definitions, dotting a given vector into a dyadic from one or the other of its sides gives rise to one or the other of two new vectors, (7.1.8) and (7.1.9), generally differing in magnitude and direction from the given vector but with magnitudes linearly related to the magnitude of the given vector. The two new vectors are equal if $\mathbf{T}$ is a *symmetric* dyadic, i.e., if $T_{xy} = T_{yx}$, etc. Only in this event do $\mathbf{V}$ and $\mathbf{T}$ obey the commutative law of multiplication.

One can show that a symmetric dyadic $\mathbf{S}$ can always be reduced to *diagonal* or *normal form* by properly choosing the orientation of the coordinate axes, which are then called the *principal axes* of the dyadic. In normal form

$$\mathbf{S} = S_{xx}\mathbf{ii} + S_{yy}\mathbf{jj} + S_{zz}\mathbf{kk} = \begin{pmatrix} S_{xx} & 0 & 0 \\ 0 & S_{yy} & 0 \\ 0 & 0 & S_{zz} \end{pmatrix}. \tag{7.1.10}$$

The off-diagonal components, such as S_{xy}, all vanish.

A dyadic $\mathbf{A}$ is an *antisymmetric* dyadic when $A_{xy} = -A_{yx}$, $A_{yz} = -A_{zy}$, $A_{zx} = -A_{xz}$, and $A_{xx} = A_{yy} = A_{zz} = 0$. For such a dyadic

$$\begin{aligned}
\mathbf{A} \cdot \mathbf{V} &= -\mathbf{V} \cdot \mathbf{A} = \mathbf{i}(A_{xy}V_y - A_{zx}V_z) + \cdots \\
&= -\mathbf{A} \times \mathbf{V} = \mathbf{V} \times \mathbf{A},
\end{aligned} \tag{7.1.11}$$

where

$$\mathbf{A} \equiv \mathbf{i}A_{yz} + \mathbf{j}A_{zx} + \mathbf{k}A_{xy} \tag{7.1.12}$$

is the vector representation of the antisymmetric dyadic $\mathbf{A}$. It is easy to verify that the components of $\mathbf{A}$, as defined by (7.1.12), transform as components of a vector (Prob. 7.1.2). Thus the *cross*-product operation of ordinary vector algebra is equivalent to a *dot*-product operation in which one of the vectors is replaced by an antisymmetric dyadic.

We define the *transpose* of a dyadic to be a dyadic satisfying

$$\mathbf{T}^t \cdot \mathbf{V} = \mathbf{V} \cdot \mathbf{T}, \text{ or } \mathbf{V} \cdot \mathbf{T}^t = \mathbf{T} \cdot \mathbf{V}, \tag{7.1.13}$$

where $\mathbf{V}$ is an arbitrary vector. Evidently $T_{xx}^t = T_{xx}$, but $T_{xy}^t = T_{yx}$, etc. An arbitrary dyadic such as (7.1.6) can be separated into symmetric and antisymmetric parts by using the identity

$$\mathbf{T} \equiv \frac{\mathbf{T} + \mathbf{T}^t}{2} + \frac{\mathbf{T} - \mathbf{T}^t}{2}. \tag{7.1.14}$$

If we write $\mathbf{S}$ for the symmetric part and $\mathbf{A}$ for the antisymmetric part, it is evident that

$$\mathbf{S} = \begin{pmatrix} T_{xx} & \frac{1}{2}(T_{xy} + T_{yx}) & \frac{1}{2}(T_{xz} + T_{zx}) \\ \frac{1}{2}(T_{yx} + T_{xy}) & T_{yy} & \frac{1}{2}(T_{yz} + T_{zy}) \\ \frac{1}{2}(T_{zx} + T_{xz}) & \frac{1}{2}(T_{zy} + T_{yz}) & T_{zz} \end{pmatrix} \tag{7.1.15}$$

and

$$\mathbf{A} = \begin{pmatrix} 0 & \frac{1}{2}(T_{xy} - T_{yx}) & \frac{1}{2}(T_{xz} - T_{zx}) \\ \frac{1}{2}(T_{yx} - T_{xy}) & 0 & \frac{1}{2}(T_{yz} - T_{zy}) \\ \frac{1}{2}(T_{zx} - T_{xz}) & \frac{1}{2}(T_{zy} - T_{yz}) & 0 \end{pmatrix}. \tag{7.1.16}$$

The vector representing $\mathbf{A}$ can evidently be found by placing a vector-product symbol ($\times$) between the elements of the dyads in the original tensor $\mathbf{T}$ and making use of the fact that $\mathbf{i} \times \mathbf{i} = 0$, $\mathbf{i} \times \mathbf{j} = -\mathbf{j} \times \mathbf{i} = \mathbf{k}$, etc., and multiplying the resulting vector by $\frac{1}{2}$.

The dyadic

$$\mathbf{1} \equiv \mathbf{ii} + \mathbf{jj} + \mathbf{kk} = \begin{pmatrix} 1 & 0 & 0 \\ 0 & 1 & 0 \\ 0 & 0 & 1 \end{pmatrix} \tag{7.1.17}$$

is called the *idemfactor*, and plays the part of a unit dyadic, that is,

$$\mathbf{V} \cdot \mathbf{1} = \mathbf{1} \cdot \mathbf{V} = \mathbf{V}. \tag{7.1.18}$$

It is possible to define tensors of higher rank by associating functions of x, y, z, t with the 27 ($= 3^3$) coordinate triples xxx, xxy, . . . , zzz, the 81 ($= 3^4$) coordinate quadruples, etc. A tensor is said to be of rank r when it has 3^r elements. A scalar can be thought of as a tensor of zero rank; an ordinary vector, a tensor of first rank; a dyadic, a tensor of second rank; etc.

Very often a vector equation can be recast into a form involving dyadics with useful results. We can illustrate the procedure by an example drawn from mechanics. The angular momentum of a rigid body rotating with the angular velocity $\boldsymbol{\omega}$ is expressed by the integral

$$\mathbf{J} = \int \mathbf{r} \times (\boldsymbol{\omega} \times \mathbf{r}) \, dm, \tag{7.1.19}$$

where $\mathbf{r}$ is the position of the mass element dm. If we expand the triple vector product in the usual manner and then factor out the angular velocity $\boldsymbol{\omega}$,

$$\mathbf{J} = \int [\boldsymbol{\omega} r^2 - (\boldsymbol{\omega} \cdot \mathbf{r})\mathbf{r}] \, dm$$
$$= \boldsymbol{\omega} \cdot \int (r^2 \mathbf{1} - \mathbf{rr}) \, dm = \boldsymbol{\omega} \cdot \mathbf{I}. \tag{7.1.20}$$

The symmetric dyadic

$$I \equiv \int (r^2 \mathbf{1} - \mathbf{rr})\, dm \tag{7.1.21}$$

is the *moment-of-inertia dyadic*, having the nine components $I_{xx} = \int (y^2 + z^2)\, dm$, $I_{xy} = I_{yx} = -\int xy\, dm$, etc. The diagonal components, I_{xx}, I_{yy}, I_{zz}, are recognized as the *moments of inertia* about the three coordinate axes. The off-diagonal components, such as I_{xy}, are known as the *products of inertia*. The above example illustrates how a dyadic can arise naturally in an analysis that initially involves only scalar and vector quantities.

In ending this brief introduction to dyadics, let us recall again that the definition of a tensor field states how the field transforms under a rotation of coordinate axes. Such a definition assures us that we are dealing with a mathematical description which is independent of the orientation of the coordinate frame. Since the basic laws of physics presumably possess this independence, it follows that any valid quantitative physical law can be expressed by an equation that is some sort of a tensor equation—of zero rank if it is a scalar equation, of first rank if it is a vector equation, of second rank if it is a dyadic equation, etc. Furthermore the equation expressing the physical law must not contain any indication of the orientation of the axes of the reference frame. Hence when the equation is transformed by a rotation of axes, say from a frame S to a frame S', it must keep the same functional form in S' as in S.

Problems

7.1.1 (a) Express the unit vectors $\mathbf{i}', \mathbf{j}', \mathbf{k}'$ in terms of $\mathbf{i}, \mathbf{j}, \mathbf{k}$, and show that the transformation inverse to (7.1.4) is $x_j = \Sigma \gamma_{ij} x_i'$, where $x_1 = x$, $x_2 = y$, and $x_3 = z$. (b) Make use of the properties of the unit vectors (such as $\mathbf{i} \cdot \mathbf{i} = 1$, $\mathbf{i} \cdot \mathbf{j} = 0$, etc.) to establish that there are six relations of the type $\displaystyle\sum_{i=1}^{3} \gamma_{ij} \gamma_{ik} = \delta_{jk}$ connecting the nine directional cosines involved in the transformation between S and S' (δ_{jk} is the Kronecker delta). Interpret geometrically the fact that therefore only three of the γ_{ij}'s can be chosen independently. (c) Show that the determinant of γ_{ij} equals ± 1. Interpret the $\pm$ sign.

7.1.2 Prove that the three components $A_x \equiv A_{yz}$, $A_y \equiv A_{zx}$, $A_z \equiv A_{xy}$ of an antisymmetric dyadic transform as the components of a vector and hence enable $\mathbf{A}$ to be treated as a physical vector field when it is a function of x, y, z, and t.

7.1.3 Let $\mathbf{n}_1$ be a unit vector of variable direction with an origin at the coordinate origin. Let the position vector $\mathbf{r}$ from the origin be determined by the equation

$$\mathbf{r} = \mathbf{S} \cdot \mathbf{n}_1,$$

where S is a symmetric dyadic. When n_1 varies in direction, its terminus describes a unit sphere. Show that the terminus of r then describes an ellipsoid. *Hint:* Assume the coordinate system is oriented such that S is diagonal.

7.1.4 Investigate the geometrical significance of the equation

$$r \cdot S \cdot r = 1,$$

where $r = ix + jy + kz$ and S is a symmetric dyadic.

7.1.5 A vector has one scalar *invariant*—its magnitude or length—when its components are transformed by a rotation of axes. Show that a symmetric dyadic S is characterized by *three* scalar invariants: (*a*) its *trace*, $I_1 = S_{11} + S_{22} + S_{33}$; (*b*) the sum of its diagonal minors,

$$I_2 = \begin{vmatrix} S_{11} & S_{12} \\ S_{21} & S_{22} \end{vmatrix} + \begin{vmatrix} S_{22} & S_{23} \\ S_{32} & S_{33} \end{vmatrix} + \begin{vmatrix} S_{33} & S_{31} \\ S_{13} & S_{11} \end{vmatrix};$$

(*c*) its *determinant*, $I_3 = |S_{ij}|$.

7.1.6 Let S be any symmetric dyadic with the three invariants I_1, I_2, and I_3, as defined in Prob. 7.1.5. When referred to principal axes, S may be written in the form $S = iiS_1 + jjS_2 + kkS_3$. Show that the three principal components S_1, S_2, and S_3 satisfy the cubic equation

$$S^3 - I_1S^2 + I_2S - I_3 = 0.$$

7.1.7 Show that the nine directional cosines γ_{ij} cannot be considered the components of a (physical) dyadic by examining their transformation properties. Show that two successive rotations γ_{ij}, γ'_{ij} are in fact equivalent to the single rotation γ''_{ij}, whose components are found by the matrix multiplication of γ_{ij} and γ'_{ij}.

7.1.8 If V is a physical vector field, show that the divergence of V, $\nabla \cdot V$, is an invariant scalar field. *Hint:* By performing an axis rotation show that

$$\frac{\partial V_x}{\partial x} + \frac{\partial V_y}{\partial y} + \frac{\partial V_z}{\partial z} = \frac{\partial V'_x}{\partial x'} + \frac{\partial V'_y}{\partial y'} + \frac{\partial V'_z}{\partial z'}$$

when each side is calculated at the same point in space. It is also instructive to apply the test for invariance to the two products $V \cdot W$ and $V \times W$, where V and W are both physical vectors.

7.1.9 If A_1, A_2, and A_3 are three arbitrary vectors that are not coplanar, show that any arbitrary dyadic T can always be expressed as the sum of three dyads

$$T = A_1B_1 + A_2B_2 + A_3B_3$$

by suitably choosing the three vectors B_1, B_2, and B_3.

7.2 *Strain as a Dyadic*

We have already defined and made considerable use of the *components* of the strain dyadic in Chap. 3. We now wish to see how the dyadic notation discussed in the preceding section enables us to treat strain as a single mathematical entity. Let us start by assuming that we have an undistorted elastic medium at rest in an inertial frame. The position vector $\mathbf{r} = \mathbf{i}x + \mathbf{j}y + \mathbf{k}z$ serves to locate a point in the unstrained medium. When a (small) strain takes place, the medium initially at $\mathbf{r}$ suffers a displacement

$$\varrho = \mathbf{i}\xi(x,y,z) + \mathbf{j}\eta(x,y,z) + \mathbf{k}\zeta(x,y,z), \tag{7.2.1}$$

and the medium at a neighboring point $\mathbf{r} + d\mathbf{r}$ suffers the displacement

$$\varrho + d\varrho = \mathbf{i}\left(\xi + \frac{\partial\xi}{\partial x}\,dx + \frac{\partial\xi}{\partial y}\,dy + \frac{\partial\xi}{\partial z}\,dz\right) + \mathbf{j}(\cdot\;\cdot\;\cdot) + \mathbf{k}(\cdot\;\cdot\;\cdot). \tag{7.2.2}$$

By subtracting (7.2.1) from (7.2.2) we find that the relative displacement of the medium is given by the equation

$$d\varrho = d\mathbf{r} \cdot \nabla\varrho, \tag{7.2.3}$$

where $d\mathbf{r}$ has been factored out on the left side, leaving the dyadic $\nabla\varrho$ expressing the strain. The nine components of the *strain dyadic* $\nabla\varrho$ are most simply exhibited by writing the dyadic in matrix form,

$$\nabla\varrho = \begin{vmatrix} \dfrac{\partial\xi}{\partial x} & \dfrac{\partial\eta}{\partial x} & \dfrac{\partial\zeta}{\partial x} \\[2mm] \dfrac{\partial\xi}{\partial y} & \dfrac{\partial\eta}{\partial y} & \dfrac{\partial\zeta}{\partial y} \\[2mm] \dfrac{\partial\xi}{\partial z} & \dfrac{\partial\eta}{\partial z} & \dfrac{\partial\zeta}{\partial z} \end{vmatrix}. \tag{7.2.4}$$

We recognize the diagonal components as the three tension strains ϵ_{xx}, ϵ_{yy}, ϵ_{zz}, defined in Sec. 3.1, with a sum that is the dilatation (3.2.4)

$$\theta = \epsilon_{xx} + \epsilon_{yy} + \epsilon_{zz}, \tag{7.2.5}$$

which we know to be an invariant of the dyadic (Prob. 7.1.5).

The strain dyadic (7.2.4) in general is not symmetric, but it can always be decomposed into the sum of a symmetric and an antisymmetric part using the identity (7.1.14). Let us denote the symmetric part by $\mathbf{E}$ and the antisymmetric part by $\boldsymbol{\phi}$. In matrix form the two dyadics may be written

$$\mathbf{E} \equiv \begin{vmatrix} \dfrac{\partial\xi}{\partial x} & \dfrac{1}{2}\left(\dfrac{\partial\eta}{\partial x}+\dfrac{\partial\xi}{\partial y}\right) & \dfrac{1}{2}\left(\dfrac{\partial\zeta}{\partial x}+\dfrac{\partial\xi}{\partial z}\right) \\[3mm] \dfrac{1}{2}\left(\dfrac{\partial\xi}{\partial y}+\dfrac{\partial\eta}{\partial x}\right) & \dfrac{\partial\eta}{\partial y} & \dfrac{1}{2}\left(\dfrac{\partial\zeta}{\partial y}+\dfrac{\partial\eta}{\partial z}\right) \\[3mm] \dfrac{1}{2}\left(\dfrac{\partial\xi}{\partial z}+\dfrac{\partial\zeta}{\partial x}\right) & \dfrac{1}{2}\left(\dfrac{\partial\eta}{\partial z}+\dfrac{\partial\zeta}{\partial y}\right) & \dfrac{\partial\zeta}{\partial z} \end{vmatrix} \tag{7.2.6}$$

$$\phi \equiv \begin{pmatrix} 0 & \dfrac{1}{2}\left(\dfrac{\partial \eta}{\partial x} - \dfrac{\partial \xi}{\partial y}\right) & \dfrac{1}{2}\left(\dfrac{\partial \zeta}{\partial x} - \dfrac{\partial \xi}{\partial z}\right) \\[2mm] \dfrac{1}{2}\left(\dfrac{\partial \xi}{\partial y} - \dfrac{\partial \eta}{\partial x}\right) & 0 & \dfrac{1}{2}\left(\dfrac{\partial \zeta}{\partial y} - \dfrac{\partial \eta}{\partial z}\right) \\[2mm] \dfrac{1}{2}\left(\dfrac{\partial \xi}{\partial z} - \dfrac{\partial \zeta}{\partial x}\right) & \dfrac{1}{2}\left(\dfrac{\partial \eta}{\partial z} - \dfrac{\partial \zeta}{\partial y}\right) & 0 \end{pmatrix}. \tag{7.2.7}$$

Let us now regard $d\mathbf{r}$ as a small but finite vector of constant magnitude but of variable direction. Its origin is at a point P in the medium located by the vector $\mathbf{r}$. Its terminus therefore describes a small sphere surrounding P. As a result of a small elastic strain, the point P suffers a displacement ϱ, carrying it to some new position Q, as illustrated in Fig. 7.2.1. Points on the small sphere surrounding P now become a nonspherical surface surrounding Q, as described by the vector

$$d\mathbf{r} + d\varrho = d\mathbf{r} \cdot (1 + \mathbf{E} + \phi). \tag{7.2.8}$$

We can interpret (7.2.8) by first noting that $1 + \mathbf{E}$ is a symmetric dyadic, so that the terminus of the vector $d\mathbf{r} + d\varrho$, with an origin at Q, describes an ellipsoid (see Prob. 7.1.3). According to (7.1.11) and (7.1.12) relating to anti-symmetric dyadics, the remaining term $d\mathbf{r} \cdot \phi$ can be written

$$d\mathbf{r} \cdot \phi = \phi \times d\mathbf{r}, \tag{7.2.9}$$

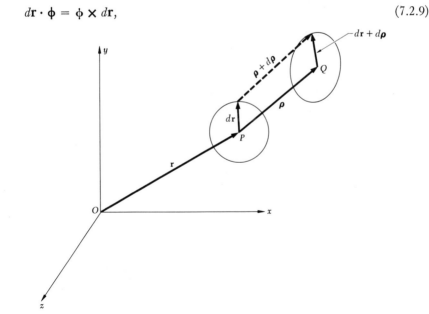

Fig. 7.2.1 A small strain.

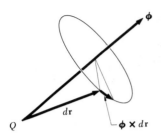

Fig. 7.2.2 The small bodily rotation represented by $\boldsymbol{\phi} \times d\mathbf{r}$ when $\boldsymbol{\phi}$ is small.

where the vector

$$\boldsymbol{\phi} \equiv \mathbf{i}\,\frac{1}{2}\left(\frac{\partial \zeta}{\partial y} - \frac{\partial \eta}{\partial z}\right) + \mathbf{j}\,\frac{1}{2}\left(\frac{\partial \xi}{\partial z} - \frac{\partial \zeta}{\partial x}\right) + \mathbf{k}\,\frac{1}{2}\left(\frac{\partial \eta}{\partial x} - \frac{\partial \xi}{\partial y}\right)$$

$$= \tfrac{1}{2} \nabla \times \boldsymbol{\varrho}, \tag{7.2.10}$$

crossed into $d\mathbf{r}$ from the left, is equivalent to the dyadic $\boldsymbol{\phi}$ dotted into $d\mathbf{r}$ from the right. Evidently (7.2.9) expresses a small (bodily) rotation of the medium about the point Q (see Fig. 7.2.2 and Prob. 3.3.7). No elastic distortion of the medium is connected with the *rotation dyadic* $\boldsymbol{\phi}$. The dyadic **E** is termed the *pure strain dyadic*, to distinguish it from the strain dyadic $\nabla \boldsymbol{\varrho}$, which includes rotation as well as distortion.

We can summarize our present findings by stating that the most general small strain is one that causes:

(1) A translation $\boldsymbol{\varrho}$ of each point in a medium.

(2) A small rotation $\boldsymbol{\phi}$ of the medium about each point.

(3) A small elastic distortion that changes a sphere surrounding each point in the unstrained medium into an ellipsoid.

In the case of a *homogeneous strain*, items 2 and 3 are independent of position in the medium. In such an event the principal axes of the pure strain dyadic do not change with position, and it is generally most convenient to choose coordinate axes x, y, z in the direction of the principal axes. The pure strain dyadic then takes the simple form

$$\mathbf{E} = \mathbf{ii}\epsilon_1 + \mathbf{jj}\epsilon_2 + \mathbf{kk}\epsilon_3, \tag{7.2.11}$$

where ϵ_1, ϵ_2, and ϵ_3 are the *principal extensions* of the medium. In the case of small strains we can usually neglect any small rotation of the medium, as given by (7.2.9), in designating the direction of the principal axes.

The three independent off-diagonal components of **E**, such as

$$\epsilon_{yz} = \frac{1}{2}\left(\frac{\partial \zeta}{\partial y} + \frac{\partial \eta}{\partial z}\right),$$

are one-half the shearing strains in the three coordinate planes, as designated by the subscripts of the components. In Sec. 3.3 we discussed shear distortion in a plane and defined the shearing strain in the yz plane by (3.3.3) to be $\gamma_{yz} = \gamma_{zy} = \partial\zeta/\partial y + \partial\eta/\partial z$. Hence we can now write that

$$\epsilon_{xy} = \epsilon_{yx} = \tfrac{1}{2}\gamma_{xy} = \tfrac{1}{2}\gamma_{yx} = \frac{1}{2}\left(\frac{\partial\eta}{\partial x} + \frac{\partial\xi}{\partial y}\right)$$

$$\epsilon_{yz} = \epsilon_{zy} = \tfrac{1}{2}\gamma_{yz} = \tfrac{1}{2}\gamma_{zy} = \frac{1}{2}\left(\frac{\partial\zeta}{\partial y} + \frac{\partial\eta}{\partial z}\right) \qquad (7.2.12)$$

$$\epsilon_{zx} = \epsilon_{xz} = \tfrac{1}{2}\gamma_{zx} = \tfrac{1}{2}\gamma_{xz} = \frac{1}{2}\left(\frac{\partial\xi}{\partial z} + \frac{\partial\zeta}{\partial x}\right).$$

Any set of six functions of position $\epsilon_{xx}(x,y,z)$, $\epsilon_{xy}(x,y,z)$, . . . , $\epsilon_{zz}(x,y,z)$ that transform properly, and hence are components of a possible physical symmetric dyadic, do not necessarily describe a possible state of strain of an elastic body. The strain components must also be consistent with their definition (7.2.6) in terms of space derivatives of the displacement components ξ, η, ζ. It is found that there exist six equations, called *equations of compatibility*, that the six strain components must satisfy identically.† For simplicity it is often preferable to formulate the basic differential equations for elastic waves in terms of the displacement ϱ rather than in terms of the strain dyadic **E**. We shall see how this can be done in Sec. 7.5.

Problems

7.2.1 Show that an arbitrary strain deformation **E** can be written as the sum of a pure shear and a hydrostatic compression, $\mathbf{E} = (\mathbf{E} - \tfrac{1}{3}\theta\mathbf{1}) + \tfrac{1}{3}\theta\mathbf{1}$, where θ is the dilatation.

7.2.2 Express the strain existing in a stretched rod as a dyadic. Do the same for a dilatation and for a shear distortion in the yz plane (refer to Secs. 3.1 to 3.3).

7.3 Stress as a Dyadic

In Secs. 3.1 and 3.3 we defined the various components of stress. For example, f_{xx} is a tension stress—a force per unit area—acting across a plane normal to the x axis and directed along the x axis; similarly, f_{xy} is a shearing stress—again a force per unit area—again acting across a plane normal to the x axis but now directed along the y axis, etc. For the rotational equilibrium of the medium, we

† See, for example, A. E. H. Love, "A Treatise on the Mathematical Theory of Elasticity," 4th ed., sec. 17, Dover Publications, Inc., New York, 1944.

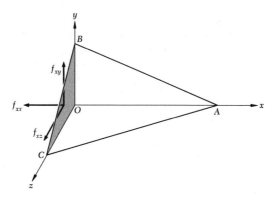

Fig. 7.3.1   Tetrahedral element in a stressed elastic medium. Only the stress components on the face *OBC* are shown.

showed that $f_{xy} = f_{yx}$. Hence we strongly suspect that the nine stress components, of which six are independent, constitute a symmetric dyadic. To show that this is indeed true, let us calculate the stress $\mathbf{f}_n$ on a plane surface whose normal, specified by the unit vector

$$\mathbf{n}_1 = \mathbf{i}\cos\alpha + \mathbf{j}\cos\beta + \mathbf{k}\cos\gamma = \mathbf{i}l + \mathbf{j}m + \mathbf{k}n, \qquad (7.3.1)$$

is inclined to the coordinate axes.

Figure 7.3.1 shows such a plane intercepting the coordinate axes at A, B, C, thus forming a small tetrahedron in which the stress components do not vary appreciably. (The tetrahedron can be made small by choosing the position of the origin O suitably.) As usual for a closed surface, the outward-drawn normals are taken as positive. If S is the area of the triangle ABC, then lS, mS, and nS are the areas of the triangles OBC, OAC, and OAB, respectively. Since the medium within the tetrahedron must be in static equilibrium, the force acting on the tetrahedron across the face ABC must equal the vector sum of the forces acting on the tetrahedron across the other three triangular faces. We thus have that

$$\mathbf{f}_n = \mathbf{i}(lf_{xx} + mf_{yx} + nf_{zx}) + \mathbf{j}(lf_{xy} + mf_{yy} + nf_{zy}) \\ + \mathbf{k}(lf_{xz} + mf_{yz} + nf_{zz}), \qquad (7.3.2)$$

where the area S has been canceled from the equation.

The result just found can be put in a more useful form by making the substitutions $l = \mathbf{n}_1 \cdot \mathbf{i}$, $m = \mathbf{n}_1 \cdot \mathbf{j}$, $n = \mathbf{n}_1 \cdot \mathbf{k}$. We then may write (7.3.2)

$$\mathbf{f}_n = \mathbf{n}_1 \cdot \mathbf{i}\mathbf{i}f_{xx} + \mathbf{n}_1 \cdot \mathbf{i}\mathbf{j}f_{xy} + \mathbf{n}_1 \cdot \mathbf{i}\mathbf{k}f_{xz} \\ + \mathbf{n}_1 \cdot \mathbf{j}\mathbf{i}f_{yx} + \mathbf{n}_1 \cdot \mathbf{j}\mathbf{j}f_{yy} + \mathbf{n}_1 \cdot \mathbf{j}\mathbf{k}f_{yz} \\ + \mathbf{n}_1 \cdot \mathbf{k}\mathbf{i}f_{zx} + \mathbf{n}_1 \cdot \mathbf{k}\mathbf{j}f_{zy} + \mathbf{n}_1 \cdot \mathbf{k}\mathbf{k}f_{zz}, \qquad (7.3.3)$$

where $\mathbf{n}_1$ occurs dotted from the left into each of the nine dyads. On factoring out $\mathbf{n}_1$, the stress acting across the plane ABC becomes

$$\mathbf{f}_n = \mathbf{n}_1 \cdot \mathbf{F}, \tag{7.3.4}$$

where

$$\mathbf{F} \equiv \begin{pmatrix} f_{xx} & f_{xy} & f_{xz} \\ f_{yx} & f_{yy} & f_{yz} \\ f_{zx} & f_{zy} & f_{zz} \end{pmatrix} \tag{7.3.5}$$

is the *stress dyadic* expressed as a matrix. We have thus succeeded in showing that the nine stress components constitute a symmetric dyadic and incidentally have developed a useful formula, (7.3.4), for finding the stress on any surface.

As a simple example of a stress dyadic, let us express the hydrostatic pressure P in a nonviscous fluid as a dyadic. It is evident that

$$P = -f_{xx} = -f_{yy} = -f_{zz}$$

relates pressure to the three tension stresses. Hence the stress dyadic for a hydrostatic pressure is

$$\mathbf{P} = -P\mathbf{1}, \tag{7.3.6}$$

where $\mathbf{1}$ is the idemfactor (7.1.17). The force per unit area on a surface whose orientation is specified by $\mathbf{n}_1$ is therefore

$$\mathbf{f}_{n_1} = \mathbf{n}_1 \cdot \mathbf{P} = -P\mathbf{n}_1 \cdot \mathbf{1} = -P\mathbf{n}_1. \tag{7.3.7}$$

This equation shows that the surface always experiences a force per unit area on it directed opposite to the (outward) normal of the surface. Hydrostatic pressure is thus seen to be represented fundamentally by a dyadic; it is not a simple scalar, as is often implied in elementary discussions.

Since stress is a symmetric dyadic, it can be referred to principal axes for which the three shearing stresses vanish. In such an event

$$\mathbf{F} = \mathbf{ii}f_1 + \mathbf{jj}f_2 + \mathbf{kk}f_3, \tag{7.3.8}$$

where f_1, f_2, and f_3 are the *principal tensions* along the three principal axes. For example, the elongation of a rod, as discussed in Sec. 3.1, has the stress dyadic

$$\mathbf{F} = \mathbf{ii}f_l, \tag{7.3.9}$$

where $f_1 = f_l$ is the applied tension (3.1.1) and $f_2 = f_3 = 0$.

Consider next an elastic medium in which the stress dyadic varies from point to point for some reason. In such a situation there is generally a net force acting on elements of the medium, since the forces on opposite faces of a volume element no longer balance each other. Let us endeavor to compute the body force per unit volume that arises in this way.

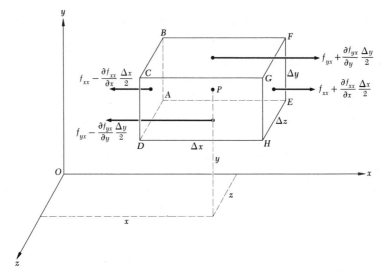

Fig. 7.3.2   Element of a stressed elastic medium when a body force is present.

First consider the net force in the x direction on the volume element $\Delta x \, \Delta y \, \Delta z$ centered at the point P having the coordinates x, y, z, using Fig. 7.3.2 to guide our thinking. At the face $EFGH$ the force due to the tension in the x direction is

$$\left(f_{xx} + \frac{\partial f_{xx}}{\partial x} \frac{\Delta x}{2} \right) \Delta y \, \Delta z,$$

whereas at the face $ABCD$ the force due to the same tension is

$$- \left(f_{xx} - \frac{\partial f_{xx}}{\partial x} \frac{\Delta x}{2} \right) \Delta y \, \Delta z.$$

Hence the net force in the x direction due to the variation in f_{xx} is

$$\frac{\partial f_{xx}}{\partial x} \Delta x \, \Delta y \, \Delta z. \tag{7.3.10}$$

Next consider the shearing stress f_{yx}. On the face $BCGF$ the force in the x direction is

$$\left(f_{yx} + \frac{\partial f_{yx}}{\partial y} \frac{\Delta y}{2} \right) \Delta z \, \Delta x,$$

whereas the force on the opposite face $ADHE$ is

$$- \left(f_{yx} - \frac{\partial f_{yx}}{\partial y} \frac{\Delta y}{2} \right) \Delta z \, \Delta x.$$

Hence the net force in the x direction due to the variation in f_{yx} is

$$\frac{\partial f_{yx}}{\partial y} \Delta x \, \Delta y \, \Delta z. \tag{7.3.11}$$

Similarly, that due to the variation in f_{zx} is

$$\frac{\partial f_{zx}}{\partial z} \Delta x \, \Delta y \, \Delta z. \tag{7.3.12}$$

On adding the three contributions (7.3.10) to (7.3.12), we find for the total net force in the x direction

$$\Delta F_x = \left(\frac{\partial f_{xx}}{\partial x} + \frac{\partial f_{yx}}{\partial y} + \frac{\partial f_{zx}}{\partial z} \right) \Delta x \, \Delta y \, \Delta z, \tag{7.3.13}$$

with similar expressions for the components in the y and z directions. We thus find that there is a net elastic body force per unit volume in the medium, as given by

$$\mathbf{F}_1 = \mathbf{i} \left(\frac{\partial f_{xx}}{\partial x} + \frac{\partial f_{yx}}{\partial y} + \frac{\partial f_{zx}}{\partial z} \right) + \mathbf{j} \left(\frac{\partial f_{xy}}{\partial x} + \frac{\partial f_{yy}}{\partial y} + \frac{\partial f_{zy}}{\partial z} \right)$$

$$+ \mathbf{k} \left(\frac{\partial f_{xz}}{\partial x} + \frac{\partial f_{yz}}{\partial y} + \frac{\partial f_{zz}}{\partial z} \right)$$

$$= \boldsymbol{\nabla} \cdot \mathbf{F}, \tag{7.3.14}$$

where $\mathbf{F}$ is the stress dyadic (7.3.5). The quantity $\boldsymbol{\nabla} \cdot \mathbf{F}$ is the divergence of the dyadic $\mathbf{F}$; it is evidently a vector.

If an elastic member is in equilibrium, ordinarily the only body force is that due to the gravitational attraction of the earth. We then have that $\boldsymbol{\nabla} \cdot \mathbf{F} + \rho_0 \mathbf{g} = 0$, where $\mathbf{g}$ is the (vector) acceleration of gravity. The total body force on the elastic member in equilibrium is balanced by external forces applied to the surface of the member. In the case of a nonviscous fluid having the stress dyadic (7.3.6), the force per unit volume becomes

$$\mathbf{F}_1 = \boldsymbol{\nabla} \cdot \mathbf{P} = -\boldsymbol{\nabla} \cdot (P\mathbf{1}) = -\boldsymbol{\nabla} P, \tag{7.3.15}$$

which agrees with the calculation (5.1.4).

Problems

7.3.1   Write the pressure at any depth below the surface of a liquid as a dyadic. Find the force $d\mathbf{F}$ on an element of area $d\mathbf{S}$ of a submerged body. Integrate this force over the surface of the body to establish Archimedes' principle. *Hint:* Use Gauss' theorem for a dyadic

$$\oint d\mathbf{S} \cdot \mathbf{T} = \int \boldsymbol{\nabla} \cdot \mathbf{T} \, dv \tag{7.3.16}$$

as an aid in evaluating the integrals.

7.3.2   Solve Probs. 3.3.1 and 3.3.2 by using (7.1.5) to rotate the y and z coordinate axes $45°$ about the x axis. *Hint:* Initially

$$\mathbf{F} = f_{yz}(\mathbf{jk} + \mathbf{kj}) \quad \text{and} \quad \mathbf{E} = \tfrac{1}{2}\gamma_{yz}(\mathbf{jk} + \mathbf{kj})$$

★**7.3.3**   Use (7.3.4) and Gauss' theorem for a dyadic (7.3.16) to establish (7.3.14).

7.3.4   Define the *mean pressure* in a stressed elastic medium to be $P = -\tfrac{1}{3}(f_{xx} + f_{yy} + f_{zz})$, an invariant of the stress dyadic. Show that any stress dyadic $\mathbf{F}$ can then be written as the sum of a pure shear dyadic and a dyadic representing mean pressure.

7.4   Hooke's Law

In Chap. 3 we applied Hooke's law to individual components of the stress and strain dyadics and defined a number of different but related elastic constants for an isotropic medium. We now wish to exhibit Hooke's law as a dyadic equation in which each stress component is a linear function of the nine strain components. Such an equation may be written in component form

$$f_{ij} = \sum_{k=1}^{3} \sum_{l=1}^{3} c_{ijkl}\epsilon_{kl}, \tag{7.4.1}$$

where c_{ijkl} is one of the 81 components of a fourth-rank tensor (or *tetradic*). Because of spatial symmetries of one sort or another, the number of independent elastic moduli is not 81 but reduces to 21 for the most anisotropic crystal and to only 2 for an isotropic medium (see Probs. 7.4.1 and 7.4.2). Let us now see how (7.4.1) can be written for an isotropic medium for which many of the c's are either zero or equal to one or the other of two independent moduli.

Instead of applying the restrictions of symmetry to (7.4.1) directly, it is easier to start afresh, making use of the fact that the principal axes for the stress and strain dyadics necessarily coincide in an isotropic medium. With respect to principal axes, the two dyadics have only diagonal components, so that we can write Hooke's law in the simple form

$$\begin{aligned}
f_{xx} &= a\epsilon_{xx} + b\epsilon_{yy} + b\epsilon_{zz} \\
f_{yy} &= b\epsilon_{xx} + a\epsilon_{yy} + b\epsilon_{zz} \\
f_{zz} &= b\epsilon_{xx} + b\epsilon_{yy} + a\epsilon_{zz}.
\end{aligned} \tag{7.4.2}$$

where a and b are elastic moduli. To make the two elastic moduli agree with those generally adopted, known as *Lamé coefficients*, (7.4.2) can be altered to read

$$\begin{aligned}
f_{xx} &= \lambda(\epsilon_{xx} + \epsilon_{yy} + \epsilon_{zz}) + 2\mu\epsilon_{xx} \\
f_{yy} &= \lambda(\epsilon_{xx} + \epsilon_{yy} + \epsilon_{zz}) + 2\mu\epsilon_{yy} \\
f_{zz} &= \lambda(\epsilon_{xx} + \epsilon_{yy} + \epsilon_{zz}) + 2\mu\epsilon_{zz}.
\end{aligned} \tag{7.4.3}$$

We shall presently show that the Lamé coefficient μ is identical with the shear modulus defined by (3.3.5) and that the other Lamé coefficient is given by

$$\lambda = B - \tfrac{2}{3}\mu, \tag{7.4.4}$$

where B is the bulk modulus defined by (3.2.6). As a single equation among dyadics, the component equations (7.4.3) take the form

$$\mathbf{F} = \lambda(\boldsymbol{\nabla} \cdot \boldsymbol{\varrho})\mathbf{1} + 2\mu\mathbf{E}, \tag{7.4.5}$$

which replaces (7.4.1) for isotropic media but does not reveal that the elastic moduli are basically components of a tetradic. Although we obtained (7.4.5) with both $\mathbf{F}$ and $\mathbf{E}$ referred to principal axes, the equation continues to hold for arbitrary axes, since it expresses a relationship among symmetric dyadics.

Let us now establish that μ is indeed the shear modulus previously defined and that λ is related to B and μ by (7.4.4). If we examine any of the off-diagonal terms of (7.4.5) when principal axes are not used, we find, for example, that

$$f_{yz} = 2\mu\epsilon_{yz} = \mu\gamma_{yz}, \tag{7.4.6}$$

which agrees with the earlier definition of the shear modulus (3.3.5). If we then set $\mathbf{F}$ equal to the stress dyadic for hydrostatic pressure (7.3.6), we find that each of the diagonal components becomes

$$P = -(\lambda + \tfrac{2}{3}\mu)\boldsymbol{\nabla} \cdot \boldsymbol{\varrho} = -(\lambda + \tfrac{2}{3}\mu)\theta. \tag{7.4.7}$$

In view of the definition of the bulk modulus B by (3.2.6), we arrive at (7.4.4).

Hooke's law for an isotropic medium can be easily written in an inverted form expressing strain as a function of stress. All we need to do is to write (3.1.13), namely,

$$\epsilon_{xx} = +\frac{1}{Y}f_{xx} - \frac{\sigma}{Y}f_{yy} - \frac{\sigma}{Y}f_{zz}$$

$$\epsilon_{yy} = -\frac{\sigma}{Y}f_{xx} + \frac{1}{Y}f_{yy} - \frac{\sigma}{Y}f_{zz} \tag{7.4.8}$$

$$\epsilon_{zz} = -\frac{\sigma}{Y}f_{xx} - \frac{\sigma}{Y}f_{yy} + \frac{1}{Y}f_{zz}$$

in dyadic form. We recall that Y is Young's modulus (for a rod) and σ is Poisson's ratio. The absence of off-diagonal components in (7.4.8) shows that the stress-strain system is referred to principal axes. As a single dyadic equation, (7.4.8) becomes

$$\mathbf{E} = \frac{1+\sigma}{Y}\mathbf{F} + \frac{3\sigma}{Y}\bar{P}\mathbf{1}, \tag{7.4.9}$$

where

$$P \equiv -\tfrac{1}{3}(f_{xx} + f_{yy} + f_{zz}) \tag{7.4.10}$$

is the *mean pressure* in the medium. When the stress-strain system is not referred to principal axes, (7.4.9) continues to hold, with off-diagonal components such as

$$\epsilon_{yz} = \frac{1 + \sigma}{Y} f_{yz}. \tag{7.4.11}$$

Comparing with (7.4.6), we find the relation among the elastic constants

$$Y = 2\mu(1 + \sigma), \tag{7.4.12}$$

which was earlier established as (3.3.6).

If we now form the sum of the diagonal elements of (7.4.9), i.e., we take the invariant *trace* of the equation (Prob. 7.1.5), we find that

$$\theta = \epsilon_{xx} + \epsilon_{yy} + \epsilon_{zz} = -\frac{3(1 + \sigma)}{Y} P + \frac{9\sigma}{Y} P. \tag{7.4.13}$$

On rearrangement the equation becomes

$$P = -\frac{Y}{3(1 - 2\sigma)} \theta = -B\theta. \tag{7.4.14}$$

Hence we have established another relation among the elastic constants,

$$Y = 3B(1 - 2\sigma), \tag{7.4.15}$$

which appeared as (3.2.8). The pressure involved here is the mean pressure (7.4.10), which differs from the hydrostatic pressure in a fluid, for which

$$f_{xx} = f_{yy} = f_{zz} = -P.$$

Problems

7.4.1 Show (*a*) that there are a maximum of 21 independent elastic constants, using the fact that both stress and strain are symmetric dyadics, and (*b*) that there exists a symmetry among the constants arising from the *reciprocity relations*

$$\frac{\partial f_{ij}}{\partial \epsilon_{kl}} = \frac{\partial f_{kl}}{\partial \epsilon_{ij}},$$

where *ij* and *kl* stand for pairs of values of *x*, *y*, *z*.

7.4.2 Give arguments establishing that the number of elastic constants reduces to two for an isotropic medium and therefore that Hooke's law can be written in the form of (7.4.2).

7.4.3 Express each side of Hooke's law (7.4.9) as the sum of a pure shear term and a hydrostatic term, using the results of Probs. 7.2.1 and 7.3.4. Hooke's law can now be separated in an invariant manner into two equations, one involving shear, the other hydrostatic compression. Carry out this separation and show that it leads to the two relations among the elastic constants (7.4.12) and (7.4.15).

7.5 Waves in an Isotropic Medium

We are now prepared to derive a wave equation for elastic waves in an extended homogeneous isotropic medium in which the perturbing influence of surfaces can be ignored. Seismic waves in the earth constitute an important example of such waves. In our idealized analysis we ignore the static body force of gravity and suppose that the medium has a uniform density ρ_0.

An elastic wave in the medium causes a spatial and temporal variation in the stress dyadic $\mathbf{F}$, which in turn produces a time-varying body force $\mathbf{F}_1$ per unit volume, as given by (7.3.14). The elastic body force is then responsible for the acceleration $\partial^2 \varrho / \partial t^2$ of the mass ρ_0 in a unit volume. Newton's second law therefore takes the form

$$\mathbf{F}_1 = \mathbf{\nabla} \cdot \mathbf{F} = \rho_0 \frac{\partial^2 \varrho}{\partial t^2}. \tag{7.5.1}$$

To obtain a wave equation for the displacement ϱ, we need to express $\mathbf{F}$ in terms of the spatial derivatives of ϱ, using Hooke's law (7.4.5),

$$\mathbf{F} = \lambda(\mathbf{\nabla} \cdot \varrho)\mathbf{1} + 2\mu\mathbf{E}. \tag{7.5.2}$$

The divergence of the first term in (7.5.2) is

$$\mathbf{\nabla} \cdot [(\mathbf{\nabla} \cdot \varrho)\mathbf{1}] = \mathbf{\nabla}\mathbf{\nabla} \cdot \varrho, \tag{7.5.3}$$

and the divergence of the second is (Prob. 7.5.1)

$$\mathbf{\nabla} \cdot \mathbf{E} = \mathbf{\nabla} \cdot \tfrac{1}{2}[\mathbf{\nabla}\varrho + (\mathbf{\nabla}\varrho)^t] = \tfrac{1}{2}(\mathbf{\nabla} \cdot \mathbf{\nabla}\varrho + \mathbf{\nabla}\mathbf{\nabla} \cdot \varrho). \tag{7.5.4}$$

Hence (7.5.1) becomes

$$(\lambda + \mu)\mathbf{\nabla}\mathbf{\nabla} \cdot \varrho + \mu\mathbf{\nabla} \cdot \mathbf{\nabla}\varrho = \rho_0 \frac{\partial^2 \varrho}{\partial t^2}, \tag{7.5.5}$$

which is often put into the alternative form (7.5.18) of Prob. 7.5.2.

We recognize (7.5.5) as some form of a vector wave equation. To make further progress let us make use of *Helmholtz' theorem* that any vector field—subject to certain mathematical restrictions—can be expressed as the sum of an *irrotational* field having a vanishing curl and a *solenoidal* field having a vanishing divergence (see Appendix A). This theorem suggests that we separate the discussion of (7.5.5) into two cases: irrotational waves, for which $\nabla \times \varrho = 0$, and solenoid waves, for which $\nabla \cdot \varrho = 0$.

(a) *Irrotational Waves*

When $\nabla \times \varrho$ vanishes, $\nabla \times (\nabla \times \varrho) = \nabla\nabla \cdot \varrho - \nabla \cdot \nabla\varrho = 0$, so that

$$\nabla\nabla \cdot \varrho = \nabla \cdot \nabla\varrho. \tag{7.5.6}$$

The wave equation (7.5.5) then becomes

$$\nabla \cdot \nabla\varrho = \frac{1}{c_l{}^2}\frac{\partial^2\varrho}{\partial t^2}, \tag{7.5.7}$$

where

$$c_l{}^2 \equiv \frac{\lambda + 2\mu}{\rho_0}$$

$$= \frac{Y(1 - \sigma)}{(1 + \sigma)(1 - 2\sigma)\rho_0} = \frac{B + \frac{4}{3}\mu}{\rho_0}. \tag{7.5.8}$$

We have expressed the elastic constant $\lambda + 2\mu$ involved in the wave velocity c_l in terms of more familiar constants (see Prob. 7.5.3). The elastic constant is that for *pure linear strain* (see Prob. 3.1.3).

We recognize (7.5.7) as an ordinary wave equation in three dimensions for the vector displacement ϱ. When it applies to a plane wave traveling in the x direction, it is equivalent to the three scalar wave equations

$$\frac{\partial^2\xi}{\partial x^2} = \frac{1}{c_l{}^2}\frac{\partial^2\xi}{\partial t^2}$$

$$\frac{\partial^2\eta}{\partial x^2} = \frac{1}{c_l{}^2}\frac{\partial^2\eta}{\partial t^2} \tag{7.5.9}$$

$$\frac{\partial^2\zeta}{\partial x^2} = \frac{1}{c_l{}^2}\frac{\partial^2\zeta}{\partial t^2}.$$

We have discussed the solution of wave equations such as (7.5.9) in considerable detail in Chap. 1 and have applied this knowledge to three-dimensional acoustic waves in fluids in Chap. 5. Here, for a *plane wave* of arbitrary shape traveling in the positive x direction, the solutions have the form

$$\xi(x,t) = f_1(x - c_l t)$$
$$\eta(x,t) = f_2(x - c_l t) \tag{7.5.10}$$
$$\zeta(x,t) = f_3(x - c_l t),$$

where the three wave functions must satisfy the vanishing-curl condition

$$\nabla \times \varrho = \mathbf{i}\left(\frac{\partial\zeta}{\partial y} - \frac{\partial\eta}{\partial z}\right) + \mathbf{j}\left(\frac{\partial\xi}{\partial z} - \frac{\partial\zeta}{\partial x}\right) + \mathbf{k}\left(\frac{\partial\eta}{\partial x} - \frac{\partial\xi}{\partial y}\right) = 0. \tag{7.5.11}$$

Since ξ is not a function of y or z, it is necessary that

$$\frac{\partial \eta}{\partial x} = 0 \qquad \frac{\partial \zeta}{\partial x} = 0, \tag{7.5.12}$$

showing that the y and z components of the wave (7.5.10), that is, f_2 and f_3, vanish (or are at most constant displacements that do not concern us). Hence the only solution of (7.5.7) that represents a *plane wave* traveling in the x direction is one in which the wave displacement is in the direction of wave travel. There can be no transverse motion of the medium that travels with the wave velocity c_l. Such a wave is usually called a *longitudinal* wave, hence the subscript l on the wave velocity. The name *compressional* wave is also used. When Poisson's ratio has the typical value of 0.3, the longitudinal wave velocity in an extended medium is about 16 percent greater than that of longitudinal waves on a slender rod of the same material. For a fluid, for which the rigidity modulus μ is necessarily zero, the wave velocity (7.5.8) is identical with (5.1.9) found from the scalar wave equation for pressure.

(b) *Solenoidal Waves*

When $\nabla \cdot \varrho$ vanishes, the wave equation (7.5.5) becomes immediately

$$\nabla \cdot \nabla \varrho = \frac{1}{c_t^2} \frac{\partial^2 \varrho}{\partial t^2}, \tag{7.5.13}$$

where

$$c_t^2 \equiv \frac{\mu}{\rho_0} = \frac{Y}{2(1 + \sigma)\rho_0}. \tag{7.5.14}$$

The wave velocity c_t is that found in Sec. 4.5 for torsional (shear) waves on a rod or tube of circular section. Except for a different wave velocity, the present wave equation is the same as that for longitudinal waves.

Let us again look at the equations for a plane wave traveling in the x direction. The wave equation (7.5.13) can be written as three scalar wave equations similar to (7.5.9), with solutions similar to (7.5.10). The only change consists in replacing c_l by c_t. Now, however, instead of the vanishing-curl condition, the divergence $\nabla \cdot \varrho$ must vanish, that is,

$$\nabla \cdot \varrho = \frac{\partial \xi}{\partial x} + \frac{\partial \eta}{\partial y} + \frac{\partial \zeta}{\partial z} = 0. \tag{7.5.15}$$

Since η and ζ are functions of x and t, not of y and z, we find that

$$\frac{\partial \xi}{\partial x} = 0. \tag{7.5.16}$$

Hence there can be no wave displacement in the direction of wave travel: the wave is entirely transverse, with only y and z components. Such a wave in general is called a *transverse* wave, and it may also be called a *shear* or *dilatationless* wave in the present instance.

Irrotational, or longitudinal, waves involve a vibration in only one direction, and are said to possess one degree of freedom. Solenoidal, or transverse, waves can vibrate independently in two directions (or *polarizations*) and therefore possess two degrees of freedom. Longitudinal and transverse waves have different velocities in an elastic medium, as given by (7.5.8) and (7.5.14), with the ratio

$$\frac{c_l}{c_t} = \left[\frac{2(1 - \sigma)}{1 - 2\sigma} \right]^{1/2}. \tag{7.5.17}$$

Thus, when $\sigma \approx 0.3$, $c_l/c_t \approx 1.7$. Although the two kinds of waves propagate independently in a homogeneous medium, at an interface between two media there is usually a partial conversion of one kind of wave into the other. What happens in a particular case is dictated by the boundary conditions that must be satisfied at the interface: continuity of the displacement vector ϱ and of the stress dyadic **F**.

An extended solid medium bounded by a surface, e.g., the earth, can support elastic surface waves somewhat similar to the surface waves on a deep body of water. The wave displacement is restricted to the material near the surface, falling off exponentially with depth. It is found that a surface wave consists of a mixture of the solenoidal and irrotational waves that can exist in the interior of the solid. The wave velocity of a surface wave is somewhat less than the transverse wave velocity c_t, the amount depending on Poisson's ratio.[†]

Seismic waves arising from local movements in the earth's crust involve surface waves, as well as waves passing through the interior of the earth. Since the energy involved in surface waves spreads out in two dimensions, instead of three, surface waves contribute greatly to the destructiveness of earthquakes. By studying the time of arrival of various components of seismic waves from distant earthquakes, it is possible to gain information regarding the physical properties of the material making up the interior of the earth. For example, there appears to be a sharp discontinuity in properties at a depth of about 2,900 km. Below this depth there is no evidence for the propagation of a transverse wave, which is consistent with the hypothesis that the earth has a liquid core, probably of compressed metallic iron.[‡]

[†] For a discussion of the theory of surface waves, see L. D. Landau and E. M. Lifshitz, "Theory of Elasticity," sec. 24, Addison-Wesley Publishing Company, Inc., Reading, Mass., 1959.

[‡] For further information regarding this application of wave theory, see K. E. Bullen, "An Introduction to the Theory of Seismology," Cambridge University Press, New York, 1963.

Problems

7.5.1   Verify the vector operations used in establishing (7.5.3) and (7.5.4).

7.5.2   Show that the wave equation (7.5.5) can be put in the alternative form

$$(\lambda + 2\mu)\nabla\nabla \cdot \varrho - \mu\nabla \times (\nabla \times \varrho) = \rho_0 \frac{\partial^2 \varrho}{\partial t^2}, \tag{7.5.18}$$

which exhibits directly the parts of the wave equation that vanish for solenoidal and for irrotational waves.

7.5.3   Show that

$$\lambda + 2\mu = \frac{Y(1 - \sigma)}{(1 + \sigma)(1 - 2\sigma)} = B + \tfrac{4}{3}\mu. \tag{7.5.19}$$

Show that when an external body force per unit volume $\mathbf{F}_{1e}$ is present, the equation for the elastic displacement (7.5.5) becomes

$$(\lambda + \mu)\nabla\nabla \cdot \varrho + \mu\nabla \cdot \nabla\varrho + \mathbf{F}_{1e} = \rho_0 \frac{\partial^2 \varrho}{\partial t^2}. \tag{7.5.20}$$

(This equation becomes *an equation for elastic equilibrium* when ϱ is not a function of time. Its solution $\varrho(x,y,z)$ must then satisfy specified conditions at the boundary of a body. The stress and strain at each point of the body can be calculated knowing ϱ. We thus have formulated the problem of elastic equilibrium in a fundamental way.)

7.6   *Energy Relations*

External work must be expended to increase the elastic distortion of a medium. When the stress-strain relation is *single-valued*, the work done in producing the distortion can be recovered, at least in principle, by reversing the loading procedure and allowing the medium to do work on its surroundings as it returns to its undistorted state. We can then define an elastic potential energy equal to the work done in (reversibly) distorting the medium. This potential energy can be thought of as localized in the medium and is described by a scalar function of position V_1, the potential energy per unit volume.

Let us see what form V_1 takes for a homogeneous isotropic medium that obeys Hooke's law. The stress at any point is given by the stress dyadic $\mathbf{F}$, and the strain by the strain dyadic $\mathbf{E}$, the two being related by Hooke's law (7.4.5) or (7.4.9). For convenience we can assume that the medium under consideration is in a state of uniform stress.

Perhaps the simplest way of finding the expression for the potential energy density V_1 consists in imagining that a rectangular block of material of dimen-

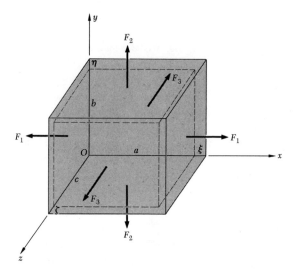

Fig. 7.6.1   A rectangular block stressed by tensions applied to its faces.

sions a, b, c is stressed by uniform tensions applied to its three pairs of faces. The edges of the block are then parallel to principal axes, with respect to which stress and strain shear components vanish. The situation is sketched in Fig. 7.6.1.

Suppose that at some stage in the process of applying tensions to the faces of the block, the three principal tensions have the values f_1', f_2', f_3'. The total forces, exerted on each of the three pairs of faces of the block, are then

$$F_1' = bcf_1' \qquad F_2' = caf_2' \qquad F_3' = abf_3'. \tag{7.6.1}$$

If ϵ_1', ϵ_2', and ϵ_3' denote the corresponding principal extensions, the block has been elongated in the three coordinate directions by the amounts

$$\xi' = \left(\frac{\partial \xi}{\partial x}\right)' a = a\epsilon_1'$$

$$\eta' = \left(\frac{\partial \eta}{\partial y}\right)' b = b\epsilon_2' \tag{7.6.2}$$

$$\zeta' = \left(\frac{\partial \zeta}{\partial z}\right)' c = c\epsilon_3'.$$

The external work required to increase these elongations by additional infinitesimal amounts $d\xi'$, $d\eta'$, $d\zeta'$ is clearly

$$dW = F_1' \, d\xi' + F_2' \, d\eta' + F_3' \, d\zeta'$$
$$= abc(f_1' \, d\epsilon_1' + f_2' \, d\epsilon_2' + f_3' \, d\epsilon_3'), \tag{7.6.3}$$

where substitutions have been made from (7.6.1) and (7.6.2). The total work done in straining the block can be calculated from the integral

$$W = abc \int \sum_{i=1}^{3} f'_i \, d\epsilon'_i. \tag{7.6.4}$$

The integral (7.6.4) is a *line integral*, whose value depends on the initial and final stress-strain state, and not on intermediate states, provided the stress-strain relation is single-valued, which is certainly true if the medium obeys Hooke's law. The integration is easily carried out by a well-known artifice when the stress-strain relation is linear. Let f_1, f_2, f_3 be the final values of the principal tensions, corresponding to the principal extensions $\epsilon_1, \epsilon_2, \epsilon_3$. We now imagine a process in which each stress component, and therefore each strain component, increases from zero to its final value in proportion to some parameter α, which increases from zero to unity. Accordingly, we can write that

$$\begin{aligned} f'_1 &= \alpha f_1 & d\epsilon'_1 &= \epsilon_1 \, d\alpha \\ f'_2 &= \alpha f_2 & d\epsilon'_2 &= \epsilon_2 \, d\alpha \\ f'_3 &= \alpha f_3 & d\epsilon'_3 &= \epsilon_3 \, d\alpha. \end{aligned} \tag{7.6.5}$$

The integral (7.6.4) then becomes

$$W = abc(f_1\epsilon_1 + f_2\epsilon_2 + f_3\epsilon_3) \int_0^1 \alpha \, d\alpha. \tag{7.6.6}$$

The potential energy density, which equals to the work done per unit volume in distorting the medium, is therefore given by

$$V_1 = \frac{W}{abc} = \frac{1}{2} \sum_{i=1}^{3} f_i\epsilon_i \tag{7.6.7}$$

since $\int_0^1 \alpha \, d\alpha = \frac{1}{2}$. This equation, and the various other forms into which it can be put, constitutes an *equation of state* for an elastic solid. Although (7.6.7) contains no explicit reference to Hooke's law, it is a valid equation only for media having a linear stress-strain relation.

We can write (7.6.7) in a particularly elegant form if we define the *double dot* (or *scalar*) *product* of two dyadics A and B by the equation

$$\mathsf{A} : \mathsf{B} \equiv \sum_{i=1}^{3} \sum_{j=1}^{3} a_{ij}b_{ij} = a_{xx}b_{xx} + a_{xy}b_{xy} + \cdots + a_{zz}b_{zz}. \tag{7.6.8}$$

Each component of the dyadic A is multiplied by the corresponding component of the dyadic B, and the sum of the nine terms taken. The definition is directly analogous to the definition of the dot product of two vectors,

$$\mathbf{A} \cdot \mathbf{B} = A_x B_x + A_y B_y + A_z B_z.$$

The double dot product (7.6.8) is easily shown to be a scalar invariant; i.e., its value does not depend on the orientation of the coordinate axes (see Prob. 7.6.1).

We therefore can write (7.6.7) in the form

$$V_1 = \tfrac{1}{2}\mathbf{F}\!:\!\mathbf{E}. \tag{7.6.9}$$

Since the double dot product is a scalar invariant, (7.6.9) continues to hold when nonprincipal axes are used, whereupon $\mathbf{F}$ and $\mathbf{E}$ have off-diagonal components. In this case, the potential energy density in expanded form reads

$$V_1 = \tfrac{1}{2}[f_{xx}\epsilon_{xx} + f_{yy}\epsilon_{yy} + f_{zz}\epsilon_{zz} + 2(f_{yz}\epsilon_{yz} + f_{zx}\epsilon_{zx} + f_{zy}\epsilon_{xy})]. \tag{7.6.10}$$

If now we replace $\mathbf{F}$ by its value (7.4.5),

$$\mathbf{F} = \lambda(\boldsymbol{\nabla}\cdot\boldsymbol{\varrho})\mathbf{1} + 2\mu\mathbf{E}, \tag{7.6.11}$$

we find that

$$\begin{aligned}
V_1 &= \tfrac{1}{2}\lambda(\boldsymbol{\nabla}\cdot\boldsymbol{\varrho})\mathbf{E}\!:\!\mathbf{1} + \mu\mathbf{E}\!:\!\mathbf{E} \\
&= \tfrac{1}{2}\lambda(\epsilon_{xx} + \epsilon_{yy} + \epsilon_{zz})^2 \\
&\quad + \mu[\epsilon_{xx}{}^2 + \epsilon_{yy}{}^2 + \epsilon_{zz}{}^2 + 2(\epsilon_{yz}{}^2 + \epsilon_{zx}{}^2 + \epsilon_{xy}{}^2)],
\end{aligned} \tag{7.6.12}$$

where we have made use of the fact that

$$\mathbf{E}\!:\!\mathbf{1} = \epsilon_{xx} + \epsilon_{yy} + \epsilon_{zz} = \boldsymbol{\nabla}\cdot\boldsymbol{\varrho}.$$

Evidently the double dot product of the idemfactor and a dyadic gives the trace of the dyadic, and incidentally confirms the fact that the trace is a scalar invariant (see Prob. 7.1.5).

Alternatively, we can replace $\mathbf{E}$ by its value (7.4.9),

$$\mathbf{E} = \frac{1+\sigma}{Y}\mathbf{F} + \frac{3\sigma}{Y}\bar{P}\mathbf{1}, \tag{7.6.13}$$

where $\bar{P}$ is given by (7.4.10). In this case we find that

$$\begin{aligned}
V_1 = \frac{1+\sigma}{2Y}&[f_{xx}{}^2 + f_{yy}{}^2 + f_{zz}{}^2 + 2(f_{yz}{}^2 + f_{zz}{}^2 + f_{xy}{}^2)] \\
&\qquad - \frac{\sigma}{2Y}(f_{xx} + f_{yy} + f_{zz})^2,
\end{aligned} \tag{7.6.14}$$

since

$$\mathbf{F}\!:\!\mathbf{1} = f_{xx} + f_{yy} + f_{zz} = -3\bar{P}.$$

It is left as an exercise (Prob. 7.6.3) to show that the expressions found in Chap. 3 for potential energy densities associated with various simple stress-strain systems are contained in these general expressions (7.6.9), (7.6.12), and (7.6.14).

To keep the discussion of elasticity as simple as possible, we have so far made no distinction between the almost equal adiabatic and isothermal elastic constants of solids. A quantitative discussion of this distinction requires a considerable excursion into the field of thermodynamics and cannot be undertaken here. The elastic potential energy of a solid just computed is known in thermodynamics as a (Helmholtz) *free energy* when it applies to *isothermal* conditions, with isothermal elastic constants, and as an *internal energy* when it applies to *adiabatic* conditions, with adiabatic elastic constants. If the rate of distortion of an elastic body is too fast for isothermal conditions but too slow for adiabatic conditions to prevail, no elastic equation of state exists. The value of the line integral in (7.6.4) then depends on the path followed as the distortion progresses. For problems involving elastic waves we can normally assume that conditions are adiabatic, whereas for problems in static elasticity, the conditions are isothermal. Since accurate values of the elastic constants are nearly always measured using standing- or traveling-wave techniques, the elastic constants so obtained are adiabatic values. Isothermal values can then be calculated, if needed, using certain formulas from thermodynamics.†

Problems

7.6.1 Show that the double dot product of two dyadics, as defined by (7.6.8), gives a scalar invariant of the two dyadics. *Hint:* Rotate the coordinate frame and use results quoted in Prob. 7.1.1.

7.6.2 Show that when V_1 can be expressed as a function of the stress components f_{ij}, the corresponding strain components are given by $\epsilon_{ij} = \partial V_1/\partial f_{ij}$. Similarly, if V_1 is expressed as a function of the strain components ϵ_{ij}, show that $f_{ij} = \partial V_1/\partial \epsilon_{ij}$. Establish the reciprocity relations $\partial f_{ij}/\partial \epsilon_{kl} = \partial f_{kl}/\partial \epsilon_{ij}$ used in Prob. 7.4.1.

7.6.3 Show that the potential energy densities (3.1.16), (3.2.10), and (3.3.8) obtained earlier are special cases of the more general equations of this section.

7.6.4 Discuss the energy content of sinusoidal plane waves traveling in the positive x direction in a solid isotropic elastic medium. Treat separately irrotational waves and solenoidal waves.

7.6.5 Show that

$$\mathbf{P}_1 = \mathbf{F} \cdot \frac{\partial \boldsymbol{\varrho}}{\partial t} \tag{7.6.15}$$

† See, for example, Landau and Lifshitz, *op. cit.*, sec. 6.

gives the instantaneous power flow in an elastic wave in a solid, where **F** is the stress dyadic and $\partial\boldsymbol{\varrho}/\partial t$ is the displacement velocity. Apply this result to the waves discussed in the previous problem and show for each sort of wave that the average power flow equals the average total energy density times the wave velocity.

*7.7 Momentum Transport by a Shear Wave

In Sec. 1.11, and again in Secs. 4.1d and 5.3a, we examined the mechanism by which particular elastic waves transport linear momentum. A close connection was found between the transport of momentum and the transport of energy by a traveling wave. It would appear that in a dispersionless medium these two aspects of wave motion are always related by an equation of the form

$$\bar{P} = c\bar{E} = c^2\bar{g}, \tag{7.7.1}$$

where $\bar{P}$ = average rate of energy flow
$\bar{E}$ = average energy density
$\bar{g}$ = average linear momentum density
c = wave (phase) velocity.

Here we wish to examine how a shear (solenoidal) wave in an extended elastic medium, described for instance by the solution $\eta(x - ct)$ of the wave equation (7.5.13), transports momentum as implied by (7.7.1). The mechanism is not immediately obvious, since the motion of the medium is basically transverse, apparently with no force component in the direction of wave travel.

Let us start our analysis with a simplified derivation of the transverse wave equation

$$\mu\frac{\partial^2\eta}{\partial x^2} = \rho_0\frac{\partial^2\eta}{\partial t^2}, \tag{7.7.2}$$

which is the y component of (7.5.13). When the only displacement component is η, which is a function of only x (and t), the pure strain dyadic (7.2.6) takes the form

$$\mathbf{E} = \frac{1}{2}\frac{\partial\eta}{\partial x}(\mathbf{ij} + \mathbf{ji}) = \tfrac{1}{2}\gamma(\mathbf{ij} + \mathbf{ji}), \tag{7.7.3}$$

where $\gamma \equiv \partial\eta/\partial x$ is the shearing strain, discussed in Secs. 3.3 and 7.2. Since the dilatation (7.5.15) vanishes, Hooke's law (7.4.5) gives the force dyadic

$$\mathbf{F} = \mu\frac{\partial\eta}{\partial x}(\mathbf{ij} + \mathbf{ji}). \tag{7.7.4}$$

The force per unit volume acting on the medium, according to (7.3.14), is then

$$\mathbf{F}_1 = \boldsymbol{\nabla}\cdot\mathbf{F} = \mathbf{j}\mu\frac{\partial^2\eta}{\partial x^2} = \rho_0\frac{\partial^2\boldsymbol{\varrho}}{\partial t^2}, \tag{7.7.5}$$

where we have equated the force density to the mass of a unit volume times its acceleration. Hence we arrive at the wave equation (7.7.2), with no indication of momentum transport in the direction of wave travel. What has gone wrong?

A hint of how our theoretical model is inadequate can be found in Sec. 1.11, where the momentum transport by a transverse wave on a string was considered. There we found it necessary to take second-order terms into account in deriving the wave equation. We then discovered that the small stretch in the longitudinal direction accompanying a transverse wave, which arises geometrically and is needed to account for the storage of potential energy in a wave, was also responsible for momentum transport, related to energy transport by (7.7.1).

We can improve our present model for shear-wave propagation if we note that the pure strain dyadic (7.7.3) ignores the small rotation of the medium and, therefore, the small rotation of the principal axes of the stress-strain system accompanying a transverse wave. In Prob. 7.7.1 it is established that a (small) simple shear specified by $\gamma = \partial\eta/\partial x$ is accompanied by a rotation $\gamma/4$ of the principal axes beyond their position at 45° holding for the pure shear dyadic (7.7.3). This rotation is just *half* the rotation of the medium

$$\phi_z = \frac{1}{2}\left(\frac{\partial\eta}{\partial x} - \frac{\partial\xi}{\partial y}\right) = \tfrac{1}{2}\gamma \tag{7.7.6}$$

given by (7.2.10). With respect to the principal axes, denoted by primes, the shearing strain γ takes the form

$$\mathbf{E}' = \tfrac{1}{2}\gamma(\mathbf{i}'\mathbf{i}' - \mathbf{j}'\mathbf{j}') \tag{7.7.7}$$

(see Prob. 3.3.2).† If now we refer $\mathbf{E}'$ to the original x and y axes by a rotation about the z axis of $-(\pi + \gamma)/4$ rad (see Prob. 7.7.2), we find that

$$\mathbf{E} = \tfrac{1}{2}\gamma(\mathbf{ij} + \mathbf{ji}) - \tfrac{1}{4}\gamma^2(\mathbf{ii} - \mathbf{jj}) + \cdots, \tag{7.7.8}$$

where higher-order terms in γ have been omitted. Hence the stress dyadic is

$$\mathbf{F} = 2\mu\mathbf{E} = \mu\gamma(\mathbf{ij} + \mathbf{ji}) - \tfrac{1}{2}\mu\gamma^2(\mathbf{ii} - \mathbf{jj}). \tag{7.7.9}$$

The additional terms, beyond the one in γ^2, can be safely thrown away, just as the higher-order terms in Sec. 1.11 were not needed in treating waves of small amplitude on a string. The force density now becomes

$$\mathbf{F}_1 = \nabla \cdot \mathbf{F} = \mathbf{j}\mu \frac{\partial^2\eta}{\partial x^2} - \mathbf{i}\mu \frac{\partial\eta}{\partial x}\frac{\partial^2\eta}{\partial x^2} = \rho_0 \frac{\partial^2\boldsymbol{\varrho}}{\partial t^2}, \tag{7.7.10}$$

where we have again equated it to the mass density times the displacement acceleration. We have thus again found the wave equation (7.7.2) that describes

---

† Problem 7.7.1 shows that the principal strains have the magnitude $\tfrac{1}{2}\gamma + \tfrac{1}{8}\gamma^2 + \cdots$. The higher-order terms can be ignored, since they lead to terms in the final result that are negligible.

the dominant process but have also found an associated wave equation

$$-\mu \frac{\partial \eta}{\partial x} \frac{\partial^2 \eta}{\partial x^2} = \rho_0 \frac{\partial^2 \xi}{\partial t^2} \tag{7.7.11}$$

governing a very small (second-order) motion in the longitudinal direction. We emphasize that (7.7.11) has arisen physically from the rotation of the stress-strain system that always accompanies transverse waves. We recognize (7.7.11) to be basically the same as (1.11.7) found earlier for transverse waves on a string. Hence the subsequent analysis and discussion of Sec. 1.11 pertains with slight change to the present case of shear waves. Here again we have found that a theoretical analysis at the simplest level fails to account for momentum transport, and it is only when greater care is taken that we discover what is taking place physically.

Problems

7.7.1 An elastic medium suffers a simple shear in the y direction such that a point originally at (x,y) moves to the position $(x, y + \gamma x)$, where $\gamma \equiv \partial \eta / \partial x \ll 1$ is the shearing strain. Examine the transformation of the circle $x^2 + y^2 = a^2$ into an ellipse to find the magnitudes and orientation of the principal axes due to a simple shear. *Answer:* To a first order in γ the angles between the principal axes and the x axis are $(\pi + \gamma)/4$, $(3\pi + \gamma)/4$; their magnitudes are $a[1 \pm (\gamma/2 + \gamma^2/8 + \cdots)]$.

7.7.2 Use the transformation properties of a dyadic described in Sec. 7.1 to obtain the strain dyadic (7.7.8) from the strain dyadic (7.7.7). *Hint:* $\cos(\pi + \gamma)/4 \approx (1 - \gamma/4)/\sqrt{2}$; $\sin(\pi + \gamma)/4 \approx (1 + \gamma/4)/\sqrt{2}$.

eight

★Electromagnetic Waves

The wave processes associated with electromagnetism are more difficult to visualize than those discussed so far. Physicists in the last century sought an "ether" as a mechanical medium that could be elastically distorted and hence support a wave. Today, we must train ourselves to visualize a time-dependent *vector electric field* existing in a three-dimensional region that is otherwise empty and, simultaneously, to visualize a separate, but closely interrelated, time-dependent *vector magnetic field*. These two vector fields exist in vacuum; no material medium is necessarily stressed or strained in the process of transmitting a wave. Indeed, the situation is further complicated by the presence of matter (in the form of polarizable dielectrics, magnetizable materials, and imperfect conductors).

Although electromagnetic phenomena are most fundamentally described in terms of electric and magnetic fields, certain cases are more conveniently described in terms of *voltages* and *currents*. This latter *circuit* point of view is generally more familiar and less abstract than the *field* description. Accordingly, we first present the circuit theory of waves on a two-conductor transmission line. We then discuss Maxwell's field equations and apply them to a number of important cases of wave propagation. The theory of electromagnetism stands as one of the great cornerstones of physics. Since we cannot hope to do justice to its richness within the scope of this book, we attempt only a very sketchy treatment of the basic physics needed to discuss electromagnetic waves.

8.1 Two-conductor Transmission Line

Consider a system of two parallel conductors, such as the examples of Fig. 8.1.1. The conductors extend indefinitely in the z direction with a cross section that is independent of z. The special cases of Fig. 8.1.1 are of practical importance and have a geometry simple enough to permit explicit calculation of the electromagnetic fields. Our analysis applies, however, to systems whose cross sections are of the general forms shown in Fig. 8.1.2. For simplicity we assume that the conductors have no resistance and that the surrounding medium is vacuum (see Probs. 8.1.3 and 8.1.4).

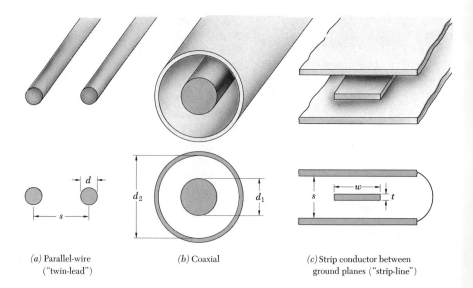

(*a*) Parallel-wire   (*b*) Coaxial   (*c*) Strip conductor between
("twin-lead")   ground planes ("strip-line")

Fig. 8.1.1   Common two-conductor transmission lines.

(*a*) Open   (*b*) Enclosed

Fig. 8.1.2   Cross sections of general classes of two-conductor transmission lines.

(*a*) *Circuit Equations*

An element of length Δz is shown schematically in Fig. 8.1.3. Suppose that a loop current i flows symmetrically across the left-hand boundary as shown; similarly, a loop current $i + (\partial i/\partial z)\,\Delta z$ flows across the right-hand boundary. Hence a net current $-(\partial i/\partial z)\,\Delta z$ is flowing into the upper conductor, and an equal current of opposite sign is flowing into the lower. The potential difference (*voltage*) between conductors is v at the left, $v + (\partial v/\partial z)\,\Delta z$ at the right. A net increase in potential difference of $(\partial v/\partial z)\,\Delta z$ occurs across the element. We wish to use the circuit relations of elementary electromagnetism to interrelate the current and voltage. These relations involve the familiar coefficients of *capacitance* and *self-inductance*, which are functions only of the geometry of the conductors.†

The two-conductor line can be described by its *capacitance per unit length* C_1 and its *inductance per unit length* L_1. We need not know explicitly how these coefficients depend upon the dimensions of the conductors, although the relation can be worked out easily in simple cases (Probs. 8.1.1 and 8.1.2). The capacitance C of a pair of conductors relates the potential difference v to equal but opposite charges $\pm q$ on the conductors by the definition $C \equiv q/v$. In a time Δt, the net current into the upper conductor of the element Δz delivers the incre-

---

† The capacitance and inductance also depend upon the electromagnetic properties of the surrounding medium. Here we assume that the medium is vacuum (or air, to good approximation), so that the relative permittivity and permeability are unity.

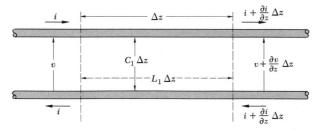

Fig. 8.1.3   Element of a parallel-wire line.

ment of charge $-(\partial i/\partial z) \, \Delta z \, \Delta t$; the lower conductor receives an equal and opposite increment. Since the capacitance is $C_1 \, \Delta z$, the mean potential difference v of the element is changed by the amount

$$\Delta v = \frac{-(\partial i/\partial z) \, \Delta z \, \Delta t}{C_1 \, \Delta z}.$$

Accordingly, the time rate of change of potential difference is

$$\frac{\partial v}{\partial t} = -\frac{1}{C_1} \frac{\partial i}{\partial z}. \tag{8.1.1}$$

This equation must hold at all times at all positions along the line.

The self-inductance L of a circuit loop relates the emf $\mathcal{E}$ induced in the loop to the time rate of change of current by the definition $L \equiv -\mathcal{E}/(\partial i/\partial t)$. This induced emf provides the assumed increase in potential difference across the element Δz of the transmission line,

$$\frac{\partial v}{\partial z} \, \Delta z = -L_1 \, \Delta z \frac{\partial i}{\partial t}.$$

Accordingly, the time rate of change of current is

$$\frac{\partial i}{\partial t} = -\frac{1}{L_1} \frac{\partial v}{\partial z}. \tag{8.1.2}$$

Along with (8.1.1), this equation must hold at all times at all positions.

(b)   Wave Equation

By taking the time derivative of one equation and the space derivative of the other, we readily obtain from (8.1.1) and (8.1.2) the familiar one-dimensional wave equations

$$\frac{\partial^2 v}{\partial z^2} = L_1 C_1 \frac{\partial^2 v}{\partial t^2} \tag{8.1.3}$$

$$\frac{\partial^2 i}{\partial z^2} = L_1 C_1 \frac{\partial^2 i}{\partial t^2}. \tag{8.1.4}$$

These equations show that a wave of potential difference, and simultaneously of current, may travel in either direction along the transmission line at the velocity

$$c = \frac{1}{(L_1 C_1)^{1/2}}. \tag{8.1.5}$$

Since both the capacitance and the inductance are functions of the geometry of the conductors, they are not independent of each other. There exists a general theorem to the effect that for loss-free lines the velocity (8.1.5) is precisely that of a plane wave in an unbounded medium whose properties are the same as those of the medium surrounding the conductors.† Thus for vacuum and for perfect conductors, (8.1.5) is the free-space velocity of light ($\approx 3 \times 10^8$ m/sec) independent of the size and shape of conductors (Fig. 8.1.2). This theorem may be readily verified for simple geometries (Probs. 8.1.1 and 8.1.2). It fails when the resistance of the conductors is significant, in which case the wave is damped as it travels along the line. It also fails when the conductors are not uniform in the z direction.

The type of wave considered here is known as *transverse electromagnetic* (TEM) since it can be shown that the electric and magnetic fields constituting the wave have no component in the z direction. The results are correct even for such high frequencies that the wavelength is comparable with the dimensions of the transmission-line cross section. At such high frequencies, however, other classes of electromagnetic waves may also exist known as *transverse electric* (TE) and *transverse magnetic* (TM), which are analogous to the waves that can propagate in hollow-pipe waveguides (Sec. 8.7).

(c) Characteristic Impedance

Consider a semi-infinite transmission line. At the input or generator end at $z = 0$, connect a sinusoidal emf of magnitude given by the real part of‡

$$v(0,t) = v_0 e^{j\omega t}. \tag{8.1.6}$$

This source sends voltage and current waves down the line, which never return because of its infinite length. The solution of the wave equation (8.1.3) that matches the boundary condition (8.1.6) is clearly

$$v(z,t) = v_0 e^{j(\omega t - \kappa z)}, \tag{8.1.7}$$

where $\kappa = \omega/c = 2\pi/\lambda$ is the wave number. The related current can be found from either (8.1.1) or (8.1.2), with the result (Prob. 8.1.5)

$$i(z,t) = \left(\frac{C_1}{L_1}\right)^{1/2} v_0 e^{j(\omega t - \kappa z)}. \tag{8.1.8}$$

Evidently the current wave is in phase with the voltage wave. The impedance

---

† See footnote, page 285. For a more complete discussion, see for instance R. E. Collin, "Field Theory of Guided Waves," chap. 4, McGraw-Hill Book Company, New York, 1960.

‡ In this chapter, as in Sec. 4.2, we follow convention in using the symbol j for $\sqrt{-1}$. We also choose the time factor $e^{+j\omega t}$. All results may easily be converted to the alternative time factor $e^{-i\omega t}$ by the substitution $j = -i$.

looking into the line at $z = 0$,

$$Z_0 \equiv \frac{v(0,t)}{i(0,t)} = \left(\frac{L_1}{C_1}\right)^{1/2},$$ (8.1.9)

is known as the *characteristic impedance* of the transmission line. For a lossless line, Z_0 is a pure resistance independent of frequency.

We can cut the line at $z = l$ and replace the line from this point to infinity by a lumped resistance $R = Z_0$ connected between the conductors. The wave between $z = 0$ and l is not affected; at l, the wave is absorbed by the *terminating* resistance R instead of continuing along the line. The line transfers power from a generator at the input end to a distant load, with no loss en route. The load is said to be *matched* to the line when $R = Z_0$, so that there is no effective discontinuity causing a reflected wave to travel back toward the generator.

(d) Reflection from Terminal Impedance

If an arbitrary terminal impedance $\check{Z}_l \neq Z_0$ is connected at the far end of the line, reflection occurs. Waves of voltage and current then travel in both directions, represented by

$$v(z,t) = \check{v}_+ e^{j(\omega t - \kappa z)} + \check{v}_- e^{j(\omega t + \kappa z)}$$ (8.1.10)
$$i(z,t) = \check{\imath}_+ e^{j(\omega t - \kappa z)} + \check{\imath}_- e^{j(\omega t + \kappa z)}.$$ (8.1.11)

The (complex) terminating impedance $\check{Z}_l$ constrains the ratio of voltage to current at $z = l$,

$$\frac{v(l,t)}{i(l,t)} = \frac{\check{v}_+ e^{-j\kappa l} + \check{v}_- e^{j\kappa l}}{\check{\imath}_+ e^{-j\kappa l} + \check{\imath}_- e^{j\kappa l}} = \check{Z}_l.$$ (8.1.12)

However, from (8.1.1) or (8.1.2), the respective voltage and current amplitudes are related by the characteristic impedance

$$Z_0 = \frac{\check{v}_+}{\check{\imath}_+} = -\frac{\check{v}_-}{\check{\imath}_-}.$$ (8.1.13)

The input impedance, at $z = 0$,

$$\check{Z}_g = \frac{v(0,t)}{i(0,t)} = \frac{\check{v}_+ + \check{v}_-}{\check{\imath}_+ + \check{\imath}_-},$$ (8.1.14)

thus becomes, upon eliminating v_+, v_-, i_+, i_- (Prob. 8.1.6),

$$\check{Z}_g = Z_0 \frac{\check{Z}_l + jZ_0 \tan\kappa l}{Z_0 + j\check{Z}_l \tan\kappa l}.$$ (8.1.15)

This result is identical with (4.2.16). Indeed, once we have established the characteristic impedance (8.1.9) for the electromagnetic transmission line, all the results of Sec. 4.2 that are cast in terms of impedances can be carried over without change to the electromagnetic case. For instance, from (4.2.15), and its related footnote, we immediately have the voltage amplitude reflection coefficient

$$\check{R} \equiv \frac{\check{v}_-e^{j\kappa l}}{\check{v}_+e^{-j\kappa l}} = \frac{\check{Z}_l - Z_0}{\check{Z}_l + Z_0}. \tag{8.1.16}$$

We need not repeat the discussion there of impedance matching with quarter-wave sections and other special cases (see Prob. 8.1.7). When the load impedance is infinite or zero, there exist positions on the line, repeated every half-wavelength, where a voltage *node* and a current *loop* occur. Halfway between these positions, the extrema are interchanged. Only when the line is terminated by a resistance equal to its characteristic impedance do waves travel from generator to load without reflection. For any other termination, reflected waves give rise to a standing-wave pattern with positions of maximum and minimum (but not zero) voltage.

(e) *Impedance Measurement*

An unknown complex load impedance $\check{Z}_l$ can be calculated from measurements of (1) the *standing-wave ratio* of the voltage amplitudes

$$\text{VSWR} \equiv \frac{|v_{\max}|}{|v_{\min}|}, \tag{8.1.17}$$

and (2) the (finite) distance $\Delta z_{\max}$ from the load end to a position where $|v_{\max}|$ occurs. If the magnitude and phase of the complex reflection coefficient can be found,

$$\check{R} = |\check{R}|e^{j\phi}, \tag{8.1.18}$$

then by (8.1.16) one can compute $\check{Z}_l$ given Z_0. It remains to show how $|\check{R}|$ and ϕ can be found. From (8.1.16) we may write the amplitudes of the counter-traveling waves in terms of a reference amplitude $\check{v}_1$ such that

$$\check{v}_+ = \check{v}_1 e^{j\kappa l} \tag{8.1.19}$$
$$\check{v}_- = \check{v}_1|\check{R}|e^{-j(\kappa l-\phi)}. \tag{8.1.20}$$

Substituting into (8.1.10) for a position Δz from the load end, we obtain

$$v(l-\Delta z, t) = \check{v}_1(e^{j\kappa\Delta z} + |\check{R}|e^{-j(\kappa\Delta z-\phi)})e^{j\omega t}. \tag{8.1.21}$$

By choosing a time origin so as to make $\check{v}_1$ real, one can easily see that the actual wave, i.e., the real part of (8.1.21), is

$$v(l-\Delta z, t) = v_1[\cos(\omega t + \kappa\,\Delta z) + |\check{R}|\cos(\omega t - \kappa\,\Delta z + \phi)]. \tag{8.1.22}$$

By suitable choice of t and Δz, the cosine factors both reach their extreme values ± 1. Hence,

$$\text{VSWR} = \frac{1 + |\breve{R}|}{1 - |\breve{R}|},$$ (8.1.23)

from which

$$|\breve{R}| = \frac{\text{VSWR} - 1}{\text{VSWR} + 1}.$$ (8.1.24)

The position where $|v_{\max}|$ occurs is given by

$$(\omega t + \kappa \Delta z) - (\omega t - \kappa \Delta z + \phi) = \pm n(2\pi),$$ (8.1.25)

where n is an integer chosen to make ϕ lie in the convenient range $-\pi$ to $+\pi$. Hence

$$\phi = 2\kappa \Delta z_{\max} \mp n(2\pi) \qquad -\pi < \phi \leq \pi.$$ (8.1.26)

We have thus shown how measurements of the voltage standing-wave ratio and of the position of $|v_{\max}|$ give values of $|R|$ and ϕ, respectively. The unknown load impedance can be computed, in turn, from (8.1.16) using the formulas of Prob. 8.1.8. In practice it is more convenient to make use of the *Smith transmission-line calculator*, a special slide rule for doing this calculation in one setting, as explained in Appendix B. A parallel-wire transmission line (Fig. 8.1.1*a*), equipped to observe the standing-wave pattern, is a practical apparatus for measuring an unknown impedance at high frequencies (~ 300 MHz) as well as for measuring the frequency of a source by observation of the wavelength. In this context it is known as a *Lecher wire* line. A coaxial line (Fig. 8.1.1*b*) or a waveguide (Sec. 8.7) equipped with a longitudinal slot in the outer conductor to permit probing the internal standing-wave field is called a *slotted line*.

Problems

8.1.1   For the coaxial line of Fig. 8.1.1*b*, calculate the capacitance per unit length C_1 and the inductance per unit length L_1. Verify that the speed of propagation is $c \approx 3 \times 10^8$ m/sec. Show that the characteristic impedance is

$$Z_0 = \frac{1}{2\pi} \left(\frac{\mu_0}{\epsilon_0}\right)^{1/2} \ln \frac{d_2}{d_1} \approx 138 \log \frac{d_2}{d_1} \text{ ohms.}$$

8.1.2   (*a*) Repeat Prob. 8.1.1 for the parallel-wire line of Figure 8.1.1*a* in the limit $d \ll s$ to show that

$$Z_0 = \frac{1}{\pi} \left(\frac{\mu_0}{\epsilon_0}\right)^{1/2} \ln \frac{2s}{d} \approx 276 \log \frac{2s}{d} \text{ ohms.}$$

★(b) Show that without the simplifying limit

$$Z_0 = \frac{1}{\pi}\left(\frac{\mu_0}{\epsilon_0}\right)^{1/2} \cosh^{-1}\frac{s}{d}.$$

8.1.3   Repeat Prob. 8.1.1. for the case where the space between the conductors is filled with a nonconducting medium of relative permittivity (dielectric constant) κ_e and relative permeability κ_m. Show that the speed of propagation is then multiplied by the factor $(\kappa_e\kappa_m)^{-1/2}$ and the characteristic impedance by the factor $(\kappa_m/\kappa_e)^{1/2}$.

★**8.1.4**   Consider a general two-conductor transmission line for which the conductors have a (round-trip) series resistance per unit length R_1 and the medium between conductors has a leakage conductance per unit length G_1. Show that the voltage and current waves then obey the *telegrapher's equation*

$$\frac{\partial^2 v}{\partial z^2} = L_1 C_1 \frac{\partial^2 v}{\partial t^2} + (R_1 C_1 + G_1 L_1)\frac{\partial v}{\partial t} + R_1 G_1 v.$$

What can you discover about the solutions of this equation? Show that the characteristic impedance for monochromatic waves is

$$\check{Z}_0 = \left(\frac{R_1 + j\omega L_1}{G_1 + j\omega C_1}\right)^{1/2}.$$

8.1.5   Establish from (8.1.1) or (8.1.2) that the voltage and current of a sinusoidal traveling wave are in phase and related in magnitude by the characteristic impedance $Z_0 = (L_1/C_1)^{1/2}$. Hence verify (8.1.8).

8.1.6   Eliminate v_+, v_-, i_+, i_- among (8.1.12) to (8.1.14) to obtain (8.1.15).

8.1.7   Show from (8.1.15) that a short length of transmission line ($l \ll \lambda/2\pi$) is equivalent to a lumped capacitor of value $C_1 l$ when open-circuited and to a lumped inductor of value $L_1 l$ when short-circuited.

8.1.8   Show that the resistive and reactive parts of an unknown load impedance $\check{Z}_l \equiv R_l + jX_l$ are given by

$$R_l = Z_0 \frac{1 - |\check{R}|^2}{1 - 2|\check{R}|\cos\phi + |\check{R}|^2}$$

$$X_l = Z_0 \frac{2|\check{R}|\sin\phi}{1 - 2|\check{R}|\cos\phi + |\check{R}|^2},$$

where $|\check{R}|$ and ϕ specify the complex reflection coefficient $\check{R}$ and Z_0 is the characteristic impedance. *Note:* See Prob. 1.4.3.

8.1.9 Practical coaxial lines used for the distribution of high-frequency signals often consist of a thin copper wire in a polyethylene sleeve on which a copper braid is woven (usually there is also a protective plastic jacket over the braid). Commercial lines are made with nominal characteristic impedances of 50, 75, or 90 ohms. A common 50-ohm variety has a center conductor of diameter 0.035 in. The dielectric constant of polyethylene is 2.3. What is the nominal (inside) diameter of the copper braid? What are the capacitance and inductance per foot? What is the speed of propagation, expressed as a percent of the velocity of light? *Answer:* 0.120 in.; 30 pF/ft; 0.074 μH/ft; 66 percent.

8.1.10 A transmission line, short-circuited at its far end, is connected to a dc battery of emf $\mathcal{E}$ through a switch and a resistance of magnitude Z_0. The line is of length l and has propagation speed c. The key is now closed at $t = 0$. Make a plot of the voltage waveform at the input end of the line. Can you see any practical application for this arrangement? Repeat, assuming the far end is open-circuited and the line initially uncharged.

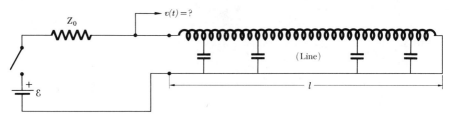

Prob 8.1.10

8.1.11 Show that the average power transmitted to a load impedance $\check{Z}_l$ is given by

$$P = \frac{1}{2Z_0} (|\check{v}_+|^2 - |\check{v}_-|^2)$$

$$= \frac{1}{2Z_0} |\check{v}_{\mathrm{max}}||\check{v}_{\mathrm{min}}|$$

$$= \frac{1}{2Z_0} \frac{|\check{v}_{\mathrm{max}}|^2}{\mathrm{VSWR}} = \frac{1}{2Z_0} |\check{v}_{\mathrm{min}}|^2 \mathrm{VSWR},$$

where $|\check{v}_{\mathrm{max}}|$ and $|\check{v}_{\mathrm{min}}|$ are the amplitudes of the voltage at maxima and minima of the standing-wave pattern.

★8.1.12 An artificial line can be made from a number of lumped inductors and capacitors by arranging them alternately as shown in the diagram. The circuit can be analyzed in terms of an elementary *T section*, also shown. Show that in the low-frequency limit the characteristic impedance of the artificial line of arbitrary length is

$$Z_0 = \left(\frac{L}{C}\right)^{1/2}$$

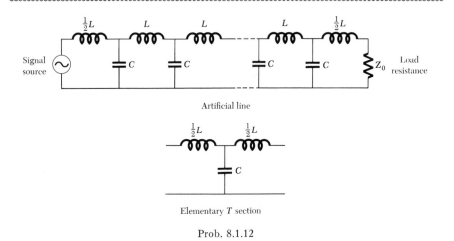

Artificial line

Elementary T section

Prob. 8.1.12

and that the speed of propagation, in sections per second, is

$$c' = \frac{1}{(LC)^{1/2}}.$$

Further show that the line does not transmit waves at frequencies greater than the *cutoff frequency*

$$\omega_c = \frac{2}{(LC)^{1/2}}.$$

How do Z_0 and c' vary with frequency near ω_c? What if each inductor has a small amount of stray capacitance in parallel with it? What if adjacent inductors are coupled by mutual inductance? What practical applications might such circuits have?

8.2 *Maxwell's Equations*

Electromagnetic phenomena are produced and detected by the displacement of a fundamental constituent of matter called *electric charge*. Instead of describing the interactions between charges directly (as is done, for instance, by Coulomb's force law), it is more convenient to break the problem into two parts: (1) the *source* charge (or current, etc.) generates a (vector) *field*, which pervades all space and (2) the *detector* charge reacts to the local value of the field by experiencing a force. The description of electromagnetic interactions in terms of fields simplifies the treatment not only of complicated situations with many sources but also of problems where it is not necessary to specify the sources explicitly. In particular, a discussion of electromagnetic waves would be very awkward without invoking the field concept. Moreover, the empirical existence of electromagnetic waves, carrying energy and momentum, is important evidence for the physical reality of the fields themselves.

Electric fields are generated by electric charges. Magnetic fields are generated by charges in motion, i.e., by electric currents. In addition, the two fields are cross-coupled in that each can be generated (or *induced*) by time variation of the other. The respective fields, **E** and **B**, can be expressed in terms of integrals over the spatial region containing their sources, but this description by *integral* equations is mathematically and physically awkward. A far more useful formulation is the set of four *differential* equations, known as *Maxwell's equations*, that relate the spatial derivatives (divergence and curl) of the fields to their sources. Helmholtz's theorem of vector analysis assures us that a vector field is uniquely specified when its divergence and curl are given.† For the fields in vacuum, Maxwell's equations are‡

$$\nabla \cdot \mathbf{E} = \frac{\rho}{\epsilon_0} \tag{8.2.1}$$

$$\nabla \times \mathbf{E} = -\frac{\partial \mathbf{B}}{\partial t} \tag{8.2.2}$$

$$\nabla \cdot \mathbf{B} = 0 \tag{8.2.3}$$

$$\nabla \times \mathbf{B} = \mu_0 \mathbf{J} + \epsilon_0 \mu_0 \frac{\partial \mathbf{E}}{\partial t}, \tag{8.2.4}$$

where the (scalar) charge density ρ (coulombs per cubic meter) and the (vector) current density **J** (amperes per square meter) are the ultimate sources of the fields, while the time-derivative terms couple the two fields. Maxwell's equations, like Newton's equation of motion $\mathbf{F} = m\mathbf{a}$, are valid only in an *inertial* reference frame, i.e., one defined operationally by Newton's first law such that a force-free particle travels in a straight line at constant speed.

Equations (8.2.1) and (8.2.3) express *Gauss'* law for electric and magnetic fields, respectively; the vanishing of the divergence of **B** follows from the fact

---

† The vector differential operators are discussed briefly in Appendix A. For a discussion of Helmholtz's theorem in the context of electromagnetism, see R. Plonsey and R. E. Collin, "Principles and Applications of Electromagnetic Fields," sec. 1.18, McGraw-Hill Book Company, New York, 1961; and W. Panofsky and M. Phillips, "Classical Electricity and Magnetism," 2d ed., sec. 1-1, Addison-Wesley Publishing Company, Inc., Reading, Mass., 1962.

‡ The constants $\mu_0 \equiv 4\pi \times 10^{-7}$ henry/meter and $\epsilon_0 \approx 10^{-9}/36\pi$ farad/meter are the dimensional parameters of rationalized mksa units. The coefficients have a somewhat different form in other systems of units, e.g., gaussian. For a discussion of the relation of these equations (8.2.1) to (8.2.4) to more elementary formulations of the electromagnetic laws, see for instance D. Halliday and R. Resnick, "Physics," 2d ed., supplementary topic V, John Wiley & Sons, Inc., New York, 1966; E. M. Purcell, "Electricity and Magnetism," Berkeley Physics Course, vol. 2, pp. 53–73, 194, 242–243, McGraw-Hill Book Company, New York, 1965.

that magnetic monopoles do not occur in nature. Equation (8.2.2) expresses *Faraday's* law of induction. Equation (8.2.4) expresses *Ampère's* law, including the contribution of Maxwell's displacement current $\epsilon_0\, \partial E/\partial t$. The physical sources ρ and J occur in the divergence equation for E but in the curl equation for B, while the cross-coupling (time-derivative) terms occur symmetrically in the two curl equations. Thus (8.2.1) and (8.2.4) state how the fields depend on their local sources, while (8.2.2) and (8.2.3) express formal constraints on the fields. The four differential equations pertain to points in the spatial region of interest. Nevertheless, a nontrivial field configuration can occur in a finite, bounded region that contains *no* source charges or currents if there are sources outside the region. The external sources need not be specified explicitly since their effects can be represented by suitable *boundary conditions* on the fields at the surface of the region of interest.

In empty space with no sources (ρ, $J = 0$), the Maxwell curl equations (8.2.2) and (8.2.4) are simultaneous equations in the two fields E and B. Since the partial differential operators $\nabla = i\, \partial/\partial x + j\, \partial/\partial y + k\, \partial/\partial z$ and $\partial/\partial t$ are independent of each other, i.e., can commute, we can readily take the curl of one equation and the time derivative of the other to obtain, for instance,

$$\nabla \times (\nabla \times E) = -\nabla \times \frac{\partial B}{\partial t} = -\epsilon_0 \mu_0 \frac{\partial^2 E}{\partial t^2}. \tag{8.2.5}$$

Applying the important vector identity

$$A \times (B \times C) = B(A \cdot C) - C(A \cdot B) \tag{8.2.6}$$

to the double curl, we have

$$\nabla \times (\nabla \times E) = \nabla(\nabla \cdot E) - \nabla \cdot \nabla E. \tag{8.2.7}$$

Finally invoking (8.2.1) (with $\rho = 0$), we obtain the three-dimensional wave equation

$$\nabla^2 E = \epsilon_0 \mu_0 \frac{\partial^2 E}{\partial t^2}, \tag{8.2.8}$$

where $\nabla^2 = \nabla \cdot \nabla = \partial^2/\partial x^2 + \partial^2/\partial y^2 + \partial^2/\partial z^2$ is the laplacian operator (5.1.10). By a similar argument,

$$\nabla^2 B = \epsilon_0 \mu_0 \frac{\partial^2 B}{\partial t^2}. \tag{8.2.9}$$

Thus it is possible for waves of electric and magnetic field to exist in free space. Since E and B are coupled by Maxwell's curl equations, the wave equations (8.2.8) and (8.2.9) are not independent; either can serve as the basic equation for *electromagnetic waves*.

Equations (8.2.8) and (8.2.9) show that electromagnetic waves propagate at the speed

$$c = \frac{1}{(\epsilon_0\mu_0)^{1/2}} = 2.998 \times 10^8 \text{ m/sec}, \tag{8.2.10}$$

the familiar velocity of light, a fundamental constant of nature. In the late nineteenth century, laboratory measurements at low frequencies of ϵ_0 and μ_0 (or their equivalent in other systems of units) gave a numerical value for (8.2.10) remarkably close to the speed measured for visible light.† This evidence helped confirm the interpretation that visible light is electromagnetic radiation. From our modern perspective, the agreement between low-frequency and visible-light measurements is not a coincidence but an identity.

The wave equations (8.2.8) and (8.2.9) involve a three-dimensional scalar differential operator (the laplacian ∇^2) operating on a three-dimensional vector function ($\mathbf{E}$ or $\mathbf{B}$) of position and time. To make any practical use of these formal equations, we must choose a coordinate system (set of three basis vectors) in terms of which the vector ($\mathbf{E}$, say) can be expanded as a sum of orthogonal components. The single vector equation (8.2.8) can then be written out as a system of three scalar equations. We must also choose a coordinate system (set of three independent variables) in terms of which to express the partial derivatives ∇^2 and the components of $\mathbf{E}$. These two choices need not be the same. For the first choice, there is strong reason to represent the vector field in *rectangular* (cartesian) components since the three scalar equations are then independent. If the vector is resolved into cylindrical, spherical, or other components, the respective unit vectors are not fixed in direction and hence have spatial derivatives that couple the three equations together. On the other hand, it is frequently helpful to use cylindrical or spherical coordinates, etc., for the laplacian derivatives and the arguments of the (cartesian) field components. Examples are given in Secs. 8.7*d* and 8.9.

Problems

8.2.1 Use Gauss' and Stokes' theorems (Appendix A) to convert Maxwell's differential equations for vacuum, (8.2.1) to (8.2.4), to their integral form

$$\oint_S \mathbf{E} \cdot d\mathbf{S} = \frac{q}{\epsilon_0} \tag{8.2.11}$$

$$\oint_L \mathbf{E} \cdot d\mathbf{l} = -\frac{d\Phi_m}{dt} \tag{8.2.12}$$

$$\oint_S \mathbf{B} \cdot d\mathbf{S} = 0 \tag{8.2.13}$$

$$\oint_L \mathbf{B} \cdot d\mathbf{l} = \mu_0 I + \mu_0 \frac{d\Phi_e}{dt}, \tag{8.2.14}$$

† See Sec. 5.4 and Prob. 8.2.4.

where the closed surface S contains the net charge q and the closed line (loop) L is linked by the net current I, the magnetic flux $\Phi_m = \int \mathbf{B} \cdot d\mathbf{S}$, and the electric flux $\Phi_e = \epsilon_0 \int \mathbf{E} \cdot d\mathbf{S}$. *Note:* The corresponding equations for a general electromagnetic medium are developed in Prob. 8.6.1.

8.2.2 When matter is present, the phenomenon of *polarization* (electrical displacement of charge in a molecule or alignment of polar molecules) can produce unneutralized (bound) charge that properly contributes to ρ in (8.2.1). Similarly the *magnetization* of magnetic materials, as well as time-varying polarization, can produce effective currents that contribute to $\mathbf{J}$ in (8.2.4). These dependent source charges and currents, as opposed to the independent or "causal" free charges and currents, can be taken into account implicitly by introducing two new fields, the *electric displacement* $\mathbf{D}$ and the *magnetic intensity* $\mathbf{H}$.† For linear isotropic media,

$$\mathbf{D} \equiv \kappa_e \epsilon_0 \mathbf{E} \tag{8.2.15}$$

$$\mathbf{H} \equiv \frac{\mathbf{B}}{\kappa_m \mu_0}, \tag{8.2.16}$$

where κ_e is the *relative permittivity* (or *dielectric constant*) and κ_m is the *relative permeability* of the medium. In this more general situation, Maxwell's equations are

$$\nabla \cdot \mathbf{D} = \rho_{\text{free}} \tag{8.2.17}$$

$$\nabla \times \mathbf{E} = -\frac{\partial \mathbf{B}}{\partial t} \tag{8.2.18}$$

$$\nabla \cdot \mathbf{B} = 0 \tag{8.2.19}$$

$$\nabla \times \mathbf{H} = \mathbf{J}_{\text{free}} + \frac{\partial \mathbf{D}}{\partial t}. \tag{8.2.20}$$

Show that in a homogeneous material medium without free charges or currents, the fields obey the simple wave equation with a velocity of propagation

$$c' = \frac{1}{(\kappa_e \epsilon_0 \kappa_m \mu_0)^{1/2}} = \frac{c}{(\kappa_e \kappa_m)^{1/2}} \tag{8.2.21}$$

and that consequently the refractive index of the medium is given by

$$n = (\kappa_e \kappa_m)^{1/2}. \tag{8.2.22}$$

8.2.3 A coaxial transmission line (Fig. 8.1.1*b*) delivers the power P to a matched terminating resistance Z_0. Describe quantitatively the electric and magnetic fields associated with the

---

† For insight into the physical distinction between the fundamental fields $\mathbf{E}$, $\mathbf{B}$ and the auxiliary fields $\mathbf{D}$, $\mathbf{H}$, the reader is referred to an electromagnetism text such as Purcell, *op. cit.*, chaps. 9 and 10. For historical reasons the $\mathbf{H}$ field is sometimes called the *magnetic field*, in which case the $\mathbf{B}$ field is called the *magnetic induction* or *flux density*. Usage of the symbols $\mathbf{B}$ and $\mathbf{H}$ for the respective fields is more standardized in the literature than is that of the verbal names. The $\mathbf{D}$ field is also known as the *electric flux density*.

traveling wave. How would you design the terminating resistor so that the field configuration is not perturbed near it?

8.2.4   The text following (8.2.10) refers to low-frequency (or dc) laboratory measurements of ϵ_0 and μ_0. How could you determine these constants? What logical chain of definitions and calibrations would be needed?

★8.2.5   It is often convenient to discuss electromagnetic problems in terms of *potentials* rather than *fields*. For instance, elementary treatments show that the electro*static* field $\mathbf{E}(\mathbf{r})$ is conservative and can be derived from a scalar potential function $\phi(\mathbf{r})$, which is related to $\mathbf{E}$ by

$$\phi \equiv -\int_{\mathbf{r}_0}^{\mathbf{r}} \mathbf{E} \cdot d\mathbf{l}$$
$$\mathbf{E} = -\nabla\phi.$$

Mathematically, the conservative nature of the static field $\mathbf{E}$ is expressed by the vanishing of its curl. Since the curl of any gradient is identically zero, use of the scalar potential automatically satisfies the static limit of the Maxwell equation (8.2.2); the other constraint on ϕ is Gauss' law (8.2.1), which becomes *Poisson's equation*

$$\nabla^2\phi = -\frac{\rho}{\epsilon_0}.$$

(*a*) Show that (8.2.3) is satisfied automatically if we introduce the *magnetic vector potential* $\mathbf{A}$, related to the magnetic field by

$$\mathbf{B} \equiv \nabla \times \mathbf{A}.$$

(*b*) Show that in the general (nonstatic) case, the electric field is given in terms of the scalar and vector potentials by

$$\mathbf{E} = -\nabla\phi - \frac{\partial\mathbf{A}}{\partial t}.$$

(*c*) Complete the prescription of $\mathbf{A}$ by defining its divergence by the *Lorentz condition*

$$\nabla \cdot \mathbf{A} = -\frac{1}{c^2}\frac{\partial\phi}{\partial t},$$

and show that the two potentials obey the symmetrical inhomogeneous wave equations

$$\nabla^2\phi - \frac{1}{c^2}\frac{\partial^2\phi}{\partial t^2} = -\frac{\rho}{\epsilon_0}$$
$$\nabla^2\mathbf{A} - \frac{1}{c^2}\frac{\partial^2\mathbf{A}}{\partial t^2} = -\mu_0\mathbf{J}.$$

These equations connect the potentials associated with radiation fields with their sources ρ and $\mathbf{J}$.

8.2.6 Since electric charge is conserved, the charge density ρ and the current density $\mathbf{J}$ must be linked by an *equation of continuity*

$$\nabla \cdot \mathbf{J} = -\frac{\partial \rho}{\partial t},$$

analogous to (6.1.11). Show that Maxwell's equations (8.2.1) and (8.2.4) are indeed consistent with conservation of charge. [The violation of the equation of continuity by the experimental laws of Gauss and Ampère motivated Maxwell to postulate the *displacement-current* term $\partial \mathbf{D}/\partial t$ in (8.2.20) and thus to deduce the existence of electromagnetic waves.]

8.3 Plane Waves

Consider a sinusoidal plane wave traveling in the $+z$ direction in vacuum. We may represent it by (real-part convention)

$$\mathbf{E} = \mathbf{E}_0 e^{j(\omega t - \kappa z)}. \tag{8.3.1}$$

The amplitude $\mathbf{E}_0$ is a constant vector whose direction and magnitude are independent of x and y by our assumption of an unbounded plane wave. Clearly (8.3.1) is a solution of the wave equation (8.2.8), with $\omega/\kappa = c = (\epsilon_0 \mu_0)^{-1/2}$. To obtain additional information from Maxwell's equations, we write the wave amplitude in rectangular coordinates,

$$\mathbf{E}_0 = \mathbf{i} E_{0x} + \mathbf{j} E_{0y} + \mathbf{k} E_{0z} \tag{8.3.2}$$

and recognize that associated with the $\mathbf{E}$-field wave there is a $\mathbf{B}$-field traveling wave of the form

$$\mathbf{B} = \mathbf{B}_0 e^{j(\omega t - \kappa z)}, \tag{8.3.3}$$

where

$$\mathbf{B}_0 = \mathbf{i} B_{0x} + \mathbf{j} B_{0y} + \mathbf{k} B_{0z}. \tag{8.3.4}$$

The two divergence equations (8.2.1) and (8.2.3) enable us to establish that

$$E_{0z} = 0 \tag{8.3.5}$$
$$B_{0z} = 0; \tag{8.3.6}$$

that is, the (vector) electric and magnetic fields of a plane wave necessarily lie in the plane perpendicular to the direction of travel.† Hence electromagnetic waves are called *transverse* waves and can be *polarized*. The curl equations (8.2.2) and

---

† Nonplane electromagnetic waves may have field components parallel to the direction of propagation; see Secs. 8.7 (waveguides) and 8.9 (spherical waves).

(8.2.4) enable us to relate the electric and magnetic field components (Prob. 8.3.1),

$$E_{0x} = cB_{0y} \tag{8.3.7}$$
$$E_{0y} = -cB_{0x}. \tag{8.3.8}$$

Since these latter equations are independent, we may say that the y component of $\mathbf{B}$ "belongs" to the x component of $\mathbf{E}$, and vice versa. If we align the x axis with the $\mathbf{E}$ field of a plane wave, then the $\mathbf{B}$ field necessarily lies in the y direction. Hence, without loss of generality, we may take E_{0x} and B_{0y} as the only nonzero components of a *linearly polarized* plane electromagnetic wave (see Probs. 8.3.5 and 8.3.6). Note that the directions of the electric field, the magnetic field, and the propagation, respectively, form a right-handed orthogonal set.

In discussing mechanical waves (see especially Sec. 4.2), we found it useful to introduce the concept of *impedance* as the ratio of the force or pressure associated with the wave to the instantaneous velocity of the medium. Similarly, impedance is a familiar concept in electric-circuit theory, where the ratio is between voltage and current (see Sec. 8.1). In the case of electromagnetic waves, the analogous variables are the electric field (closely related to voltage) and the magnetic field (closely related to current). Equations (8.3.7) and (8.3.8) show that, for a plane wave in vacuum, the magnitudes of the vectors $\mathbf{E}$ and $\mathbf{B}$ are in the ratio of the velocity of light

$$\frac{|\mathbf{E}|}{|\mathbf{B}|} = c. \tag{8.3.9}$$

It turns out that in the more general case embracing material media, the auxiliary magnetic field $\mathbf{H}$ (see Prob. 8.2.2) should be used in place of the fundamental field $\mathbf{B}$. In vacuum, the $\mathbf{H}$ field differs from the $\mathbf{B}$ field only by the dimensional scale factor μ_0, $\mathbf{H} = \mathbf{B}/\mu_0$.† Thus the *wave impedance of free space* is

$$Z_0 = \frac{|\mathbf{E}|}{|\mathbf{B}|/\mu_0} = \mu_0 c = \left(\frac{\mu_0}{\epsilon_0}\right)^{1/2} \approx 377 \text{ ohms.} \tag{8.3.10}$$

Use of the wave impedance is illustrated in Sec. 8.6.

Problems

8.3.1 Substitute the assumed form of a plane wave, (8.3.1) and (8.3.3), in Maxwell's equations (8.2.1) to (8.2.4) to verify the results stated in (8.3.5) to (8.3.8).

† The form of the relations stated in (8.3.9) and (8.3.10) depends upon the system of electrical units used. For instance, in gaussian units, $\mathbf{B}$ and $\mathbf{H}$ are identical in vacuum, and furthermore the electric and magnetic amplitudes of a wave are numerically equal; i.e., the wave impedance is unity. The Lorentz force law (8.3.11) then contains an explicit factor of c in the denominator of the $\mathbf{v} \times \mathbf{B}$ term.

8.3.2 A charged particle in an electromagnetic field experiences the Lorentz force

$$\mathbf{F} = q(\mathbf{E} + \mathbf{v} \times \mathbf{B}), \tag{8.3.11}$$

where q is the charge and $\mathbf{v}$ the (vector) velocity of the particle. Show that an electromagnetic wave in free space acts on a charged particle primarily through its electric field, the magnetic interaction being smaller by at least the ratio $|\mathbf{v}|/c$.

8.3.3 The wave impedance of free space (8.3.10), in ohms, has the same numerical magnitude as the angular frequency $\omega = 2\pi f$ of the 60-cycle power-line frequency: both are 377. Is the similarity significant or a pure coincidence? Compare Washington's birth year and the square root of 3.

8.3.4 Extend Prob. 8.2.2 to show that in a material medium of (relative) permittivity κ_e and permeability κ_m the wave impedance for a plane wave is

$$Z_0' = \frac{|\mathbf{E}|}{|\mathbf{H}|} = \left(\frac{\kappa_m \mu_0}{\kappa_e \epsilon_0}\right)^{1/2} \approx 377 \left(\frac{\kappa_m}{\kappa_e}\right)^{1/2} \text{ ohms.} \tag{8.3.12}$$

8.3.5 Show that

$$\mathbf{E} = (\mathbf{i} + j\mathbf{j})E_1 e^{j(\omega t - \kappa z)}$$

$$\mathbf{B} = (-\mathbf{i}j + \mathbf{j}) \frac{E_1}{c} e^{j(\omega t - \kappa z)}$$

represent a *circularly polarized* plane wave. (Note that $j = \sqrt{-1}$, while $\mathbf{i}, \mathbf{j}$ are the cartesian unit vectors in the x and y directions!) If you watch the time variation of the electric field at a fixed position, will the direction of the field rotate in the right- or left-handed sense with respect to the direction of travel $(+z)$? If you could take a snapshot of the electric field over space, in which sense would the direction rotate? Repeat these questions for the magnetic field. How would you represent a circularly polarized wave of the opposite handedness? *Answer:* Left-handed; right-handed; magnetic same as electric; reverse sign of j in coefficients.

8.3.6 Postulate wave fields of the form

$$\mathbf{E} = \mathbf{i}f(z - ct) + \mathbf{j}g(z - ct) + \mathbf{k}h(z - ct)$$

$$\mathbf{B} = \mathbf{i}q(z - ct) + \mathbf{j}r(z - ct) + \mathbf{k}s(z - ct),$$

where f, g, h, q, r, s are arbitrary (nonsinusoidal) functions, independent of x and y. Show that such waves are a solution of the wave equations (8.2.8) and (8.2.9) and that Maxwell's equations (8.2.1) to (8.2.4) require

$$h = s = 0$$
$$f = cr$$
$$g = -cq,$$

that is, that only two of the six functions are really arbitrary.

8.3.7   (*a*) Recall that for functions with time dependence $e^{j\omega t}$, the differential operator $\partial/\partial t$ reduces to simple multiplication by the algebraic factor $j\omega$. Now show that for functions with space dependence $\exp(-j\boldsymbol{\kappa} \cdot \mathbf{r})$, the differential operation $\boldsymbol{\nabla}$ reduces to the algebraic vector $-j\boldsymbol{\kappa}$, that is,

$$\boldsymbol{\nabla} f = -j\boldsymbol{\kappa} f$$
$$\boldsymbol{\nabla} \cdot \mathbf{A} = -j\boldsymbol{\kappa} \cdot \mathbf{A}$$
$$\boldsymbol{\nabla} \times \mathbf{A} = -j\boldsymbol{\kappa} \times \mathbf{A}$$

where f and $\mathbf{A}$ represent any scalar and vector, respectively, whose space dependence is of the form $\exp(-j\boldsymbol{\kappa} \cdot \mathbf{r})$. (*b*) Show that for a plane wave traveling in the direction specified by the vector wave number $\boldsymbol{\kappa}$,

$$\mathbf{B} = \frac{\boldsymbol{\kappa} \times \mathbf{E}}{\omega} = \frac{1}{c}\hat{\boldsymbol{\kappa}} \times \mathbf{E} \tag{8.3.13}$$

where $\hat{\boldsymbol{\kappa}}$ is a unit vector in the direction of $\boldsymbol{\kappa}$.

★8.3.8   Consider an inhomogeneous dielectric medium, i.e., one for which the dielectric constant is a function of position, $\kappa_e = \kappa_e(x,y,z)$. Show that the fields obey the wave equations

$$\nabla^2 \mathbf{E} - \frac{\kappa_e}{c^2}\frac{\partial^2 \mathbf{E}}{\partial t^2} = -\boldsymbol{\nabla}\left(\frac{\boldsymbol{\nabla}\kappa_e}{\kappa_e} \cdot \mathbf{E}\right)$$
$$\nabla^2 \mathbf{B} - \frac{\kappa_e}{c^2}\frac{\partial^2 \mathbf{B}}{\partial t^2} = -\frac{\boldsymbol{\nabla}\kappa_e}{\kappa_e} \times (\boldsymbol{\nabla} \times \mathbf{B}),$$

where, in general, the terms on the right-hand sides couple the cartesian components of the fields. Now introduce the special case that the permittivity changes only in the direction of propagation (the z direction, say) and show that for monochromatic plane waves the equations become

$$\frac{d^2\mathbf{E}}{dz^2} + \frac{\omega^2}{c^2}\kappa_e(z)\mathbf{E} = 0$$
$$\frac{d^2\mathbf{B}}{dz^2} + \frac{\omega^2}{c^2}\kappa_e(z)\mathbf{B} = \frac{1}{\kappa_e(z)}\frac{d\kappa_e}{dz}\frac{d\mathbf{B}}{dz}.$$

Approximate solution of this type of equation is discussed in Sec. 9.1.

8.4   *Electromagnetic Energy and Momentum*

We know that the power carried by a mechanical wave is the product of the force and velocity amplitudes (while the impedance is the ratio of these two amplitudes). To establish an energy theorem for electromagnetic fields, therefore, we are motivated to look for relations involving the product of the analogous variables, the fields **E** and **B**.† Moreover, since we have seen that the fields **E** and **B** are perpendicular to each other for a plane wave, we suspect that the vector (cross) product is more interesting than the scalar (dot) product. How then can we manipulate Maxwell's equations to obtain a relation involving the quantity **E** × **B**, or at least its spatial derivatives? The identity

$$\nabla \cdot (\mathbf{E} \times \mathbf{B}) = \mathbf{B} \cdot \nabla \times \mathbf{E} - \mathbf{E} \cdot \nabla \times \mathbf{B} \tag{8.4.1}$$

serves the purpose since the right-hand side can be readily evaluated by dotting **B** and **E** into the Maxwell curl equations (8.2.2) and (8.2.4), respectively, with the result

$$\frac{1}{\mu_0} \nabla \cdot (\mathbf{E} \times \mathbf{B}) + \left(\epsilon_0 \mathbf{E} \cdot \frac{\partial \mathbf{E}}{\partial t} + \frac{1}{\mu_0} \mathbf{B} \cdot \frac{\partial \mathbf{B}}{\partial t} \right) + \mathbf{E} \cdot \mathbf{J} = 0. \tag{8.4.2}$$

We now seek the physical interpretation of the three terms into which we have chosen to divide (8.4.2).

The differential equation (8.4.2) interrelates the fields and the current density at a point in space. We have assumed a space filled with vacuum except for the source charge density ρ and current density **J** shown explicitly in Maxwell's equations (8.2.1) and (8.2.4).‡ Thus the only interaction between fields and matter is the Lorentz force density on the sources,

$$\mathbf{F}_1 = \rho(\mathbf{E} + \mathbf{v} \times \mathbf{B}) = \rho \mathbf{E} + \mathbf{J} \times \mathbf{B}, \tag{8.4.3}$$

where we recognize that at the microscopic level the current density **J** is simply the local charge density ρ in motion at the velocity v. Consequently, *the rate (per unit volume) at which the electromagnetic field does work on the charge*, i.e., the rate of energy exchange between field and matter, is

$$\frac{\partial}{\partial t} (W_1)_{\text{matter}} \equiv \mathbf{F}_1 \cdot \mathbf{v} = \rho \mathbf{E} \cdot \mathbf{v} = \mathbf{E} \cdot \mathbf{J}, \tag{8.4.4}$$

which of course is independent of **B**. We recognize (8.4.4) as the third term of (8.4.2). Clearly, a volume integral of **E** · **J** gives the rate of work done on all the charge (matter) in a finite region of space.

In order to understand the remaining two terms in (8.4.2), we must integrate them over a fixed finite volume V. The second term then becomes

$$\int_V \left(\epsilon_0 \mathbf{E} \cdot \frac{\partial \mathbf{E}}{\partial t} + \frac{1}{\mu_0} \mathbf{B} \cdot \frac{\partial \mathbf{B}}{\partial t} \right) dv = \frac{d}{dt} \int_V \left(\tfrac{1}{2}\epsilon_0 E^2 + \frac{B^2}{2\mu_0} \right) dv. \tag{8.4.5}$$

---

† Strictly, the magnetic **H** field, rather than the **B** field, is the proper analog of velocity (see Probs. 8.2.2 and 8.4.1).

‡ The reader is asked to develop the more general case of a polarizable, magnetizable material medium in Probs. 8.2.2 and 8.4.1.

In elementary treatments of electromagnetism it is shown that the potential energy of an electrostatic configuration, e.g., a charged capacitor, can be expressed as the volume integral $\int \frac{1}{2}\epsilon_0 E^2 \, dv$. Similarly, the energy of quasi-static magnetic configurations can be expressed as the integral $\int \frac{1}{2}B^2/\mu_0 \, dv$. Thus we infer that in general (including time-dependent processes such as waves) we can associate the energy density

$$(W_1)_{\text{field}} \equiv \tfrac{1}{2}\epsilon_0 E^2 + \frac{1}{2\mu_0} B^2 \tag{8.4.6}$$

with the electromagnetic field, so that (8.4.5) represents the *time rate of change of energy stored in the electric and magnetic fields* within the volume V.

Finally, since the first term of (8.4.2) has the form of a divergence, its volume integral can be transformed into a surface integral by invoking Gauss' theorem (Appendix A),

$$\frac{1}{\mu_0} \int_V \boldsymbol{\nabla} \cdot (\mathbf{E} \times \mathbf{B}) \, dv = \frac{1}{\mu_0} \oint_S (\mathbf{E} \times \mathbf{B}) \cdot d\mathbf{S}. \tag{8.4.7}$$

Thus this term, which arose from our original motivation by analogy, is the *flux* of the *Poynting vector*†

$$\mathbf{s} \equiv \frac{1}{\mu_0} \mathbf{E} \times \mathbf{B} \tag{8.4.8}$$

through the closed surface S bounding the volume V. By invoking the principle of conservation of energy, we are led to infer that the Poynting vector (8.4.8) does indeed represent *the rate of flow of electromagnetic energy carried across the bounding surface*, as we originally hypothesized.

To summarize, we interpret the volume integral of (8.4.2),

$$\frac{1}{\mu_0} \oint (\mathbf{E} \times \mathbf{B}) \cdot d\mathbf{S} + \frac{d}{dt} \int_V \left(\tfrac{1}{2}\epsilon_0 E^2 + \frac{B^2}{2\mu_0} \right) dv + \int_V \mathbf{E} \cdot \mathbf{J} \, dv = 0, \tag{8.4.9}$$

as a statement of conservation of energy, in which the three terms are, respectively:

(1) The outward transport of electromagnetic energy across the surface S enclosing the volume V.

(2) The time rate of increase of electromagnetic energy stored within the volume.

† The Poynting vector is usually written $\mathbf{E} \times \mathbf{H}$, where $\mathbf{H} \equiv \mathbf{B}/\mu_0$ in vacuum (see Prob. 8.4.1).

(*3*)   The rate at which the electromagnetic field does work on charged particles within the volume.

This expression of energy conservation is known as *Poynting's theorem*. It applies equally well to static and to wave fields. With the notation W_1 for energy density and $W = \int W_1 \, dv$ for total energy, we may write (8.4.2) and (8.4.9) in the more compact forms

$$\mathbf{\nabla} \cdot \mathbf{s} + \frac{\partial}{\partial t} [(W_1)_{\text{field}} + (W_1)_{\text{matter}}] = 0 \tag{8.4.10}$$

$$\oint \mathbf{s} \cdot d\mathbf{S} + \frac{d}{dt} (W_{\text{field}} + W_{\text{matter}}) = 0. \tag{8.4.11}$$

The assignment of energy to particular locations of a distributed system raises difficulties. We have seen in Sec. 1.8, for a stretched string, that such an assignment could be made only by specifying the elastic properties of the string, a refinement that had not been required in obtaining the wave equation itself. The localization of energy is particularly subtle in the electromagnetic case since it is often not unique. The interpretation of Poynting's theorem (8.4.9) is clear since the algebraic form of the surface and volume integrals gives an unambiguous physical distinction between the two terms that involve field quantities only, even though it is not possible to prove the meaning of each of these terms independently with the same rigor available for the $\mathbf{E} \cdot \mathbf{J}$ term. It is tempting to think of the integrand $(W_1)_{\text{field}} \, dv = (\frac{1}{2}\epsilon_0 E^2 + \frac{1}{2} B^2/\mu_0) \, dv$ as the energy stored locally in a particular volume element dv and of the integrand $\mathbf{s} \cdot d\mathbf{S}$ as the energy flow across a particular element of surface $d\mathbf{S}$, but this interpretation may not be unique. In particular, the differential version (8.4.10) shows that $\mathbf{\nabla} \cdot \mathbf{s}$, rather than $\mathbf{s}$ itself, has physical significance at a point (see Prob. 8.4.5).

If a wave transports energy, it should also transport momentum. Consider an electromagnetic wave traveling through vacuum and then impinging on a region containing charges and currents. The force density exerted on the matter is given by the Lorentz relation (8.4.3),

$$\mathbf{F}_1 = \rho \mathbf{E} + \mathbf{J} \times \mathbf{B}. \tag{8.4.12}$$

This force can be expressed entirely in terms of field quantities by using the Maxwell equations (8.2.1) and (8.2.4) to eliminate ρ and $\mathbf{J}$. The result,

$$\mathbf{F}_1 = \epsilon_0(\mathbf{\nabla} \cdot \mathbf{E})\mathbf{E} + \frac{1}{\mu_0} (\mathbf{\nabla} \times \mathbf{B}) \times \mathbf{B} - \epsilon_0 \frac{\partial \mathbf{E}}{\partial t} \times \mathbf{B}, \tag{8.4.13}$$

is cumbersome and does not provide much insight. The product $\mathbf{E} \times \mathbf{B}$, which we know to be significant in energy flow, appears in the final term on the right-hand side but with an asymmetrical time derivative. Using the expansion identity

$$\frac{\partial}{\partial t} (\mathbf{E} \times \mathbf{B}) = \frac{\partial \mathbf{E}}{\partial t} \times \mathbf{B} + \mathbf{E} \times \frac{\partial \mathbf{B}}{\partial t} \tag{8.4.14}$$

and the Maxwell equation (8.2.2), we can recast (8.4.13) in the symmetrical form

$$\mathbf{F}_1 + \epsilon_0 \frac{\partial}{\partial t} (\mathbf{E} \times \mathbf{B}) = \epsilon_0 (\boldsymbol{\nabla} \cdot \mathbf{E})\mathbf{E} - \epsilon_0 \mathbf{E} \times (\boldsymbol{\nabla} \times \mathbf{E})$$

$$+ \frac{1}{\mu_0} (\boldsymbol{\nabla} \cdot \mathbf{B})\mathbf{B} - \frac{1}{\mu_0} \mathbf{B} \times (\boldsymbol{\nabla} \times \mathbf{B}), \tag{8.4.15}$$

where we have inserted the term containing $\boldsymbol{\nabla} \cdot \mathbf{B} = 0$ to achieve a symmetry on the right-hand side of the equation.

Let us integrate (8.4.15) over a fixed finite volume V. The net force $\mathbf{F}$ on the matter contained in the volume is related to the fields by

$$\mathbf{F} + \frac{d}{dt} \int_V \epsilon_0 \mathbf{E} \times \mathbf{B} \, dv = \int_V \left[\epsilon_0 (\boldsymbol{\nabla} \cdot \mathbf{E})\mathbf{E} - \epsilon_0 \mathbf{E} \times (\boldsymbol{\nabla} \times \mathbf{E}) \right.$$

$$\left. + \frac{1}{\mu_0} (\boldsymbol{\nabla} \cdot \mathbf{B})\mathbf{B} - \frac{1}{\mu_0} \mathbf{B} \times (\boldsymbol{\nabla} \times \mathbf{B}) \right] dv. \tag{8.4.16}$$

By Newton's second law, the force $\mathbf{F}$ is just the rate of change of momentum of the matter, $(d\mathbf{p}/dt)_{\text{matter}}$. Since the other term on the left also has the form of a time derivative, we associate with the electromagnetic field a *linear momentum*

$$\mathbf{p}_{\text{field}} \equiv \int_V \epsilon_0 \mathbf{E} \times \mathbf{B} \, dv = \int \frac{\mathbf{s}}{c^2} \, dv. \tag{8.4.17}$$

Thus (8.4.16) can be written in the form

$$\frac{d}{dt} (\mathbf{p}_{\text{matter}} + \mathbf{p}_{\text{field}}) = \int_V [\cdot \ \cdot \ \cdot] \, dv \tag{8.4.18}$$

where the right-hand side appears as a force acting to accelerate (increase the momentum of) both matter and field. This right-hand side can be converted (Prob. 8.4.9) into a surface integral; i.e., the field may be regarded as exerting forces on the bounding surface of the volume V. We infer that an electromagnetic field carries the *momentum density*

$$(\mathbf{p}_1)_{\text{field}} = \frac{\mathbf{s}}{c^2} \tag{8.4.19}$$

where $\mathbf{s}$ is the Poynting vector (8.4.8), and that this component of momentum must be included in applications of the conservation of momentum.

To understand the significance of the relations (8.4.15) and (8.4.16), consider a linearly polarized plane wave impinging on a region containing matter in the form of charges and currents. For simplicity, assume that the densities ρ and **J** change only in the direction in which the wave is traveling (the z direction, say). The time average of the force density $\mathbf{F}_1$ is, by (8.4.15) and Prob. 8.4.2,

$$\mathbf{F}_1 = -\frac{d}{dz} \left(\frac{1}{2} \epsilon_0 \overline{E_x^2} + \frac{1}{2\mu_0} \overline{B_y^2} \right) \mathbf{k} = -\frac{d}{dz} (\overline{W}_1)_{\text{field}} \mathbf{k}. \tag{8.4.20}$$

As the wave progresses, it does work on the charged matter and gradually damps out. The magnitude of the net force per unit area on the surface of the absorbing region, i.e., the *radiation pressure* $\mathfrak{p}$, is then

$$\mathfrak{p} = \int_{z_0}^{\infty} \overline{F}_1 \, dz = -\int_{z_0}^{\infty} \frac{d\overline{W}_1}{dz} \, dz = \int_{\infty}^{z_0} d\overline{W}_1 = [\overline{W}_1(z_0)]_{\text{field}}, \tag{8.4.21}$$

where z_0 is the coordinate of the front surface of the absorbing region. We conclude that the radiation pressure of a wave incident normally on a perfectly absorbing surface is equal to the energy density of the wave (see Prob. 8.4.6). If the surface were completely reflecting, the rate of change of electromagnetic momentum would be doubled, and so would the pressure. But since the energy density would also be doubled, the relation $\mathfrak{p} = (\overline{W}_1)_{\text{field}}$ continues to hold.

Problems

8.4.1 Develop Poynting's theorem for the general material medium of relative permittivity κ_e and permeability κ_m introduced in Prob. 8.2.2; i.e., substitute the Maxwell curl equations (8.2.18) and (8.2.20) in the expansion of $\nabla \cdot (\mathbf{E} \times \mathbf{H})$ to obtain

$$\oint_S (\mathbf{E} \times \mathbf{H}) \cdot d\mathbf{S} + \int_V \left(\mathbf{E} \cdot \frac{\partial \mathbf{D}}{\partial t} + \mathbf{H} \cdot \frac{\partial \mathbf{B}}{\partial t} \right) dv + \int_V \mathbf{E} \cdot \mathbf{J} \, dv = 0, \tag{8.4.22}$$

from which it follows that the Poynting vector is

$$\mathbf{s} = \mathbf{E} \times \mathbf{H} \tag{8.4.23}$$

and the energy density is

$$(W_1)_{\text{field}} = \tfrac{1}{2} \mathbf{E} \cdot \mathbf{D} + \tfrac{1}{2} \mathbf{H} \cdot \mathbf{B}$$

$$= \tfrac{1}{2} \kappa_e \epsilon_0 E^2 + \frac{1}{2\kappa_m \mu_0} B^2. \tag{8.4.24}$$

What restrictions on κ_e and κ_m are necessary to obtain (8.4.24)?

8.4.2 (*a*) Evaluate the energy density W_1 and the Poynting vector **s** for the simple plane wave of Sec. 8.3 to show that the electric and magnetic energy densities are equal and that

$$|\mathbf{s}| = c(W_1)_{\text{field}}.$$

(b) Show that the time-average Poynting vector is

$$\bar{\mathbf{s}} = \frac{E_0{}^2}{2Z_0} \hat{\mathbf{\kappa}} = \tfrac{1}{2}H_0{}^2Z_0\hat{\mathbf{\kappa}}, \tag{8.4.25}$$

where E_0 and $H_0 = B_0/\mu_0$ are the (peak) field amplitudes, Z_0 is the wave impedance given by (8.3.10), and $\hat{\mathbf{\kappa}}$ is a unit vector in the direction of propagation. (c) Evaluate the field-dependent terms in (8.4.15) for a plane wave to show that the time-average force density is

$$\overline{F}_1 = -\mathbf{\nabla}(\overline{W}_1)_{\text{field}},$$

which is a simple generalization of the result quoted in (8.4.20).

8.4.3   Use the results of Prob. 8.2.3 to compute the Poynting vector for a coaxial transmission line. Integrate it over the annular area between conductors and show that the power carried down the line by the wave is

$$P = i^2Z_0 = \frac{v^2}{Z_0},$$

where i and v are the instantaneous current and voltage and Z_0 is the characteristic impedance (8.1.9), that is, just the result one would expect from elementary circuit analysis.

8.4.4   A long straight wire of radius a carries a current I and has resistance R_1 per unit length. Compute $\mathbf{E}$ and $\mathbf{B}$ at its surface and show that the rate of energy flow into the wire via the Poynting flux is I^2R_1 per unit length.

8.4.5   A small permanent magnet of dipole moment $\mathbf{m}$, supported by an insulating thread, is given an electrostatic charge q. Calculate the Poynting vector (at distances large compared to the size of the magnet). Also calculate its divergence. How does this problem differ from Probs. 8.4.3 and 8.4.4?

8.4.6   A plane electromagnetic wave, with momentum density $\mathbf{p}_1$, is incident on a plane absorbing surface at an angle θ with respect to the normal. (a) Show that the normal force per unit area, i.e., the pressure, is

$$\mathfrak{p} = \overline{W}_1 \cos^2\theta.$$

(b) Show that if waves are incident on the surface at all angles, the pressure is

$$\mathfrak{p} = \tfrac{1}{3}\overline{W}_1.$$

(c) If the surface has a power reflection coefficient $R(\theta)$, how are these results affected?

8.4.7   Electromagnetic radiation from the sun reaches the earth at the rate of about 0.14 W/cm^2. Find the energy density and the rms values of E and B. Compute the pressure of the sun's radiation on a perfect absorber. Compute the total force on the earth, assuming the earth to be a blackbody, and compare with the gravitational attraction. Under what circumstances would radiation pressure be significant for an object such as a space vehicle? (In practice, a space vehicle is also subject to the corpuscular radiation of the solar wind.)

8.4.8 If oscillatory fields are represented by $\mathbf{E} = \check{\mathbf{E}}_0 e^{j\omega t}$ and $\mathbf{H} = \check{\mathbf{H}}_0 e^{j\omega t}$, using the real-part convention, show that the (real) Poynting vector is given by

$$\mathbf{s} = \tfrac{1}{2}(\mathbf{E} \times \mathbf{H}^* + \mathbf{E}^* \times \mathbf{H}) = \mathrm{Re}(\check{\mathbf{E}} \times \check{\mathbf{H}}^*) = \mathrm{Re}(\check{\mathbf{E}}^* \times \check{\mathbf{H}}),$$

where the asterisk denotes complex conjugate. Also, show that the time-average Poynting vector is

$$\bar{\mathbf{s}} = \tfrac{1}{4}(\check{\mathbf{E}}_0 \times \check{\mathbf{H}}_0^* + \check{\mathbf{E}}_0^* \times \check{\mathbf{H}}_0) = \tfrac{1}{2}\,\mathrm{Re}(\check{\mathbf{E}}_0 \times \check{\mathbf{H}}_0^*).$$

★8.4.9 Show that the right-hand side of (8.4.16) can be written in the form

$$\int_V \boldsymbol{\nabla} \cdot \mathbf{T}\, dv,$$

where $\mathbf{T}$ is the *Maxwell stress tensor*

$$\mathbf{T} \equiv \epsilon_0 \mathbf{E}\mathbf{E} + \frac{1}{\mu_0}\mathbf{B}\mathbf{B} - \mathbf{1}\left(\tfrac{1}{2}\epsilon_0 E^2 + \frac{B^2}{2\mu_0}\right).$$

Then apply Gauss' theorem to convert the integral to the surface-integral form

$$\int_S \mathbf{T} \cdot d\mathbf{S}.$$

8.5 Waves in a Conducting Medium

Our discussion so far has emphasized waves propagating in vacuum, although in Probs. 8.2.2, 8.3.4, and 8.4.1 the reader is asked to extend the treatment to waves in a nonconducting material medium. We now consolidate these results and, in particular, include the case where the medium is a conductor.

A material medium can be described by three phenomenological constants:

(1) Conductivity g.

(2) Relative permittivity (dielectric constant) κ_e.

(3) Relative permeability κ_m.

These constants are the coefficients of proportionality in Ohm's law

$$\mathbf{J} = g\mathbf{E} \tag{8.5.1}$$

and the *constitutive relations*

$$\mathbf{D} = \kappa_e \epsilon_0 \mathbf{E} \tag{8.5.2}$$
$$\mathbf{B} = \kappa_m \mu_0 \mathbf{H}, \tag{8.5.3}$$

where the auxiliary fields **D** and **H** have been introduced in Prob. 8.2.2.† This is a very general description of an electromagnetic medium, but it is still subject to some important restrictions:‡

(1) The medium is *linear;* that is, g, κ_e, and κ_m are constants independent of the magnitude of the fields appearing in their respective equations.

(2) The medium is *homogeneous;* that is, g, κ_e, κ_m are independent of position.

(3) The median is *isotropic;* that is, g, κ_e, κ_m, are scalar coefficients, not tensors.

Substituting (8.5.1) to (8.5.3) in the general Maxwell's equations (8.2.17) to (8.2.20) and assuming the absence of net free charge, we have

$$\nabla \cdot \mathbf{E} = 0 \tag{8.5.4}$$

$$\nabla \times \mathbf{E} = - \frac{\partial \mathbf{B}}{\partial t} \tag{8.5.5}$$

$$\nabla \cdot \mathbf{B} = 0 \tag{8.5.6}$$

$$\nabla \times \mathbf{B} = \mu_0 \kappa_m g \mathbf{E} + \frac{\kappa_e \kappa_m}{c^2} \frac{\partial \mathbf{E}}{\partial t}. \tag{8.5.7}$$

Only (8.5.7) is affected by the characteristics of the medium. Following the same procedure used in Sec. 8.2 (Prob. 8.5.1), we now obtain the modified wave equation

$$\nabla^2 \mathbf{E} = \mu_0 \kappa_m g \frac{\partial \mathbf{E}}{\partial t} + \frac{\kappa_e \kappa_m}{c^2} \frac{\partial^2 \mathbf{E}}{\partial t^2}. \tag{8.5.8}$$

A similar equation holds for **B**. The assumption of a conducting medium has thus added a new term containing the *first*-order derivative $\partial \mathbf{E}/\partial t$. The added term causes the wave to travel more slowly and to die out exponentially as it travels in the conductor.

Assume a sinusoidal plane wave traveling in the $+z$ direction and linearly polarized in the x direction. It is thus of the form§

$$\mathbf{E} = \mathbf{i} E_0 \exp j(\omega t - \check{k}z). \tag{8.5.9}$$

---

† If the reader is unfamiliar with the auxiliary fields **D** and **H**, the important results of this section (for $g \neq 0$ but with $\kappa_e = \kappa_m = 1$) can be obtained using only (8.5.1) and the vacuum form of Maxwell's equations (8.2.1) to (8.2.4).

‡ Inhomogeneity is considered in Prob. 8.3.8 and anisotropy in Prob. 8.8.6. Dispersion results if the coefficients are functions of frequency (see Secs. 4.7 and 12.5).

§ The wave number $\check{k} = \kappa_r + j\kappa_i$ should not be confused with the constitutive constants κ_e and κ_m!

Substitution of (8.5.9) in (8.5.8) shows that the wave number $\check{\kappa}$ is *complex* (indicated by the cup over the symbol) and is given by the relation (Prob. 8.5.2)

$$\check{\kappa} = \frac{\omega}{c'}\left(1 - j\frac{g}{\omega\epsilon_0\kappa_e}\right)^{1/2},\tag{8.5.10}$$

where

$$c' = \frac{c}{(\kappa_e\kappa_m)^{1/2}}\tag{8.5.11}$$

is the wave speed (8.2.21) for a nonconducting material medium. The meaning of a complex wave number is readily apparent upon putting $\check{\kappa} \equiv \kappa_r + j\kappa_i$ and substituting in (8.5.9). The real part κ_r is the familiar wave number $2\pi/\lambda$, controlling the phase of the oscillatory wave field. The imaginary part κ_i contributes a *real* exponential factor $e^{\kappa_i z}$ signifying that the wave grows or decays in amplitude as it progresses, depending on the sign of κ_i. The dimensionless ratio $g/\omega\epsilon_0\kappa_e$ is often referred to as the *loss tangent* of an electromagnetic medium.

Let us consider a highly conducting medium for which

$$g \gg \omega\epsilon_0\kappa_e,\tag{8.5.12}$$

a limit that is satisfied by metals up to visible-light frequencies and by many semiconductors at radio frequencies. Then (8.5.10) simplifies to

$$\check{\kappa} \to \frac{\omega}{c'}\left(-j\frac{g}{\omega\epsilon_0\kappa_e}\right)^{1/2} = (1 - j)\left(\frac{\omega\mu_0\kappa_m g}{2}\right)^{1/2}.\tag{8.5.13}$$

Defining the *skin depth* as

$$\delta \equiv \left(\frac{2}{\omega\mu_0\kappa_m g}\right)^{1/2},\tag{8.5.14}$$

we may write (8.5.9) in the form

$$\mathbf{E} = \mathbf{i}E_0 e^{-z/\delta}e^{j(\omega t - z/\delta)},\tag{8.5.15}$$

which shows explicitly that the wave amplitude attenuates by a factor of e in each skin depth. The inability of high-frequency waves to penetrate very far into a conductor is known as *skin effect*. The form (8.5.15) also shows that the effective wavelength $\lambda = 2\pi\delta$ is reduced from the vacuum value $\lambda_0 = 2\pi c/\omega$ by the large factor

$$\frac{\lambda_0}{\lambda} = \frac{c}{\omega\delta} = \left(\frac{\kappa_e\kappa_m}{2}\frac{g}{\omega\epsilon_0\kappa_e}\right)^{1/2} \gg 1.\tag{8.5.16}$$

Clearly, the wave speed is less than the vacuum value c by the same factor (8.5.16).

The magnetic component of the wave (8.5.9) is found from Faraday's law (8.5.5) (Prob. 8.5.4),

$$\mathbf{B} = \mathbf{j} \frac{\check{\kappa}}{\omega} E_0 \exp j(\omega t - \check{\kappa} z). \tag{8.5.17}$$

Since the complex wave number $\check{\kappa}$ appears as an amplitude factor, the $\mathbf{E}$ and $\mathbf{B}$ fields are not in phase. In a good conductor, to which the limit (8.5.13) applies, the magnetic field lags the electric by $45°$ in time. The amplitudes of the two field components are in the (complex) ratio

$$\frac{E}{B} = \frac{\omega}{\check{\kappa}} \rightarrow \begin{cases} c' = \dfrac{c}{(\kappa_e \kappa_m)^{1/2}} & \text{nonconductor} \\[2ex] \left(\dfrac{\omega \epsilon_0}{\kappa_m g}\right)^{1/2} c e^{j\pi/4} & \text{good conductor.} \end{cases} \tag{8.5.18}$$

The wave impedance is, generalizing (8.3.12),

$$\check{Z}_0' \equiv \frac{E}{H} = \frac{\omega \mu_0 \kappa_m}{\check{\kappa}} \rightarrow \begin{cases} \left(\dfrac{\kappa_m \mu_0}{\kappa_e \epsilon_0}\right)^{1/2} & \text{nonconductor} \\[2ex] \left(\dfrac{\omega \mu_0 \kappa_m}{g}\right)^{1/2} e^{j\pi/4} & \text{good conductor.} \end{cases} \tag{8.5.19}$$

Thus the magnitude of $\check{Z}_0'$ for a conductor is much smaller than for a comparable nonconductor, the ratio being

$$\frac{|\check{Z}_0'|_{\text{conductor}}}{(Z_0')_{\text{nonconductor}}} = \left(\frac{\omega \epsilon_0 \kappa_e}{g_{\text{conductor}}}\right)^{1/2} \ll 1. \tag{8.5.20}$$

The ratio of average electric to magnetic energy densities is, from (8.4.24),

$$\frac{\kappa_e \epsilon_0 \overline{E^2}}{\dfrac{1}{\kappa_m \mu_0} \overline{B^2}} = \left|\frac{\omega}{c' \check{\kappa}}\right|^2 \rightarrow \begin{cases} 1 & \text{nonconductor} \\[2ex] \dfrac{\omega \epsilon_0 \kappa_e}{g} & \text{good conductor.} \end{cases} \tag{8.5.21}$$

Thus in a conductor a large fraction of the energy is carried by the magnetic field. The time-averaged magnitude of the Poynting vector (8.4.23) is

$$\bar{s} = \overline{|\mathbf{E} \times \mathbf{H}|} \rightarrow \frac{\text{Re}(\check{\kappa})}{\omega \mu_0 \kappa_m} \frac{E_0^2}{2} e^{-2z/\delta} = \frac{E_0^2}{2\omega \mu_0 \kappa_m \delta} e^{-2z/\delta} \qquad \text{good conductor.} \tag{8.5.22}$$

Problems

8.5.1 (*a*) Carry out the manipulations of Maxwell's equations (8.5.4) to (8.5.7) to obtain the wave equation (8.5.8). (*b*) Show that an identical equation holds for $\mathbf{B}$.

8.5.2 Substitute the plane-wave function (8.5.9) in the wave equation (8.5.8) to verify (8.5.10). Show that the exact value of the complex square root gives

$$\frac{\check{\kappa}c'}{\omega} = \left\{\frac{1}{2} + \frac{1}{2}\left[1 + \left(\frac{g}{\omega\epsilon_0\kappa_e}\right)^2\right]^{1/2}\right\}^{1/2} - j\left\{-\frac{1}{2} + \frac{1}{2}\left[1 + \left(\frac{g}{\omega\epsilon_0\kappa_e}\right)^2\right]^{1/2}\right\}^{1/2}, \quad (8.5.23)$$

of which (8.5.13) is a first approximation. Justify the choice of signs for the square-root operations in (8.5.23).

8.5.3 (a) Show that the skin depth δ can be put in the form

$$\delta = \left(\frac{\lambda_0}{\pi Z_0 \kappa_m g}\right)^{1/2},$$

where Z_0 is the free-space wave impedance (8.3.10) and $\lambda_0 = 2\pi c/\omega$ is the vacuum wavelength. (b) Evaluate δ for copper, for waves having wavelengths in vacuum of 5,000 km (60-Hz power line); 100 m ($\sim$AM broadcast band); 1 m ($\sim$television and FM broadcast); 3 cm ($\sim$radar); 500 nm ($\sim$visible light). How does the size of the skin depth affect the technology of these various applications? (c) The electrical conductivity of sea water is about 4 mhos/m. How would you communicate by radio with a submarine 100 m below the surface?

8.5.4 Verify in detail the results stated in (8.5.17) to (8.5.22).

8.5.5 Show that for a medium having a very slight conductivity $g \ll \omega\epsilon_0\kappa_e$, the skin depth or attenuation distance is

$$\delta \equiv -\frac{1}{\kappa_i} \longrightarrow \frac{2}{Z_0'g},$$

where Z_0' is the wave impedance (8.3.12) of the medium.

★8.5.6 (a) For waves varying sinusoidally with time as $e^{j\omega t}$, show that the conductivity can be eliminated from (8.5.7) and (8.5.8) by substituting for the relative permittivity the complex quantity

$$\check{\kappa}_e \equiv \kappa_e - j\frac{g}{\omega\epsilon_0}.$$

Then all electromagnetic properties of the medium are contained in only *two* constants, $\check{\kappa}_e$ and κ_m. (b) When currents flow nonuniformly in space, it is possible that a net charge density ρ_{free} builds up at certain locations. Show that the complex permittivity formalism of part (a) not only eliminates the $\mathbf{J}_{free}$ term in Maxwell's equation (8.2.20) but also eliminates the ρ_{free} term in (8.2.17). (c) As an alternative to the formalism of part (a), show that the relative permittivity can be disregarded, i.e., set equal to unity, by introducing the complex conductivity

$$\check{g} \equiv g + j\omega\epsilon_0(\kappa_e - 1).$$

In this case, the properties of the medium are specified by the two constants $\check{g}$ and κ_m.

8.6 Reflection and Refraction at a Plane Interface

It is a general property of traveling waves that they are partially *reflected* and partially *transmitted* at an abrupt discontinuity in the wave-supporting structure or medium (see Secs. 1.9 and 5.6 and Probs. 2.4.3, 4.1.2, and 4.2.4). Let us examine this effect for the special case of plane electromagnetic waves incident on a plane interface separating two distinct media.

(a) Boundary Conditions

Wave solutions for two uniform media in contact are joined together at the interface by invoking boundary conditions on the wave variables. The boundary conditions for mechanical waves are reasonably intuitive, i.e., continuity of particle displacement and of forces (Newton's third law) across the boundary. The corresponding conditions for electromagnetic waves are more formal, taking the form of constraints on the field components. We deduce the boundary conditions from the integral versions of Maxwell's equations obtained in Prob. 8.6.1 for the general electromagnetic medium described by the three parameters g, κ_e, κ_m of (8.5.1) to (8.5.3), namely,

$$\oint_S \kappa_e \epsilon_0 \mathbf{E} \cdot d\mathbf{S} = q_{\text{free}} \tag{8.6.1}$$

$$\oint_L \mathbf{E} \cdot d\mathbf{l} = -\frac{d\Phi_m}{dt} \tag{8.6.2}$$

$$\oint_S \mathbf{B} \cdot d\mathbf{S} = 0 \tag{8.6.3}$$

$$\oint_L \frac{1}{\kappa_m \mu_0} \mathbf{B} \cdot d\mathbf{l} = I_{\text{free}} + \frac{d\Phi_e}{dt}, \tag{8.6.4}$$

where S and L signify a closed surface and a closed line, respectively, and where Φ_m and Φ_e are the fluxes linking the loop L.

We apply the integral equations (8.6.1) to (8.6.4) to the simple geometrical constructions of Fig. 8.6.1. For simplicity we assume that no free charge or cur-

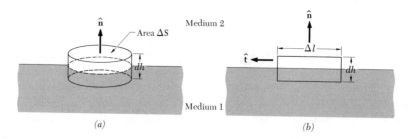

Fig. 8.6.1 (*a*) Gaussian surface and (*b*) amperian loop, used for establishing electromagnetic boundary conditions.

rent exists on the boundary surface (see Prob. 8.6.2). Figure 8.6.1a shows a "pillbox" (right circular cylinder) whose faces lie parallel to, and on opposite sides of, the surface separating the two media. The orientation of the interface is specified by the unit normal $\hat{n}$, with its sense from medium 1 to medium 2. Using this *gaussian surface* with (8.6.1) and (8.6.3) and shrinking the height dh to an infinitesimal, we establish that the *normal components* of the respective fields are related by

$$\kappa_{e2}\mathbf{E}_2 \cdot \hat{n} - \kappa_{e1}\mathbf{E}_1 \cdot \hat{n} \equiv \kappa_{e2}E_{n2} - \kappa_{e1}E_{n1} = 0 \tag{8.6.5}$$

$$\mathbf{B}_2 \cdot \hat{n} - \mathbf{B}_1 \cdot \hat{n} \equiv B_{n2} - B_{n1} = 0. \tag{8.6.6}$$

Thus the normal component of $\mathbf{B}$ is continuous across the interface, while the normal components of $\mathbf{E}$ are in the inverse ratio of the permittivities.

Similarly, Fig. 8.6.1b shows a rectangular loop intersecting the boundary surface. The unit vector $\hat{t}$ is parallel to the sides of length Δl and to the surface. Using this *amperian loop* with (8.6.2) and (8.6.4) and shrinking the dimension dh to an infinitesimal, we establish that the *tangential components* are related by

$$\mathbf{E}_2 \cdot \hat{t} - \mathbf{E}_1 \cdot \hat{t} \equiv E_{t2} - E_{t1} = 0 \tag{8.6.7}$$

$$\frac{1}{\kappa_{m2}}\mathbf{B}_2 \cdot \hat{t} - \frac{1}{\kappa_{m1}}\mathbf{B}_1 \cdot \hat{t} \equiv \frac{B_{t2}}{\kappa_{m2}} - \frac{B_{t1}}{\kappa_{m1}} = 0. \tag{8.6.8}$$

It is always possible to orient $\hat{t}$ in the tangent plane so that the field lies in the $\hat{n}\hat{t}$ plane; the field components in (8.6.7) and (8.6.8) then represent the entire tangential components. Thus the tangential component of $\mathbf{E}$ is continuous across the interface, while the tangential components of $\mathbf{B}$ are in the direct ratio of the permeabilities.

Except for the neglect of free charge and current on the interface (Prob. 8.6.2) and the assumption of isotropy, the boundary conditions (8.6.5) to (8.6.8) are completely general, applying to static as well as to wave fields. They represent a kind of "Snell's law" for the deflection of a line of force of $\mathbf{E}$ or $\mathbf{B}$ passing across the boundary (Prob. 8.6.3).

(b)   Normal Incidence on a Conductor

Consider the special case of a wave in vacuum ($\approx$ air) impinging on a good conductor ($g \gg \omega\epsilon_0\kappa_e$) at normal incidence. As suggested in Fig. 8.6.2, the incident wave of amplitude E_1, traveling in the $+z$ direction, sets up the reflected wave of amplitude $\breve{E}_1'$ and the (damped) transmitted wave of initial amplitude $\breve{E}_2$. When the conducting region is sufficiently thick, we may neglect a reflected wave approaching the interface from the right.

The electric fields of the three linearly polarized, monochromatic, plane

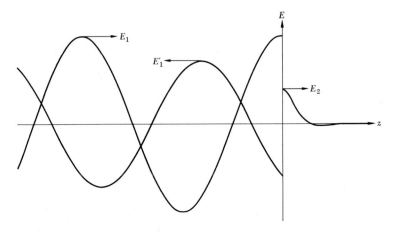

Fig. 8.6.2   Reflection of a plane wave at the surface of a conductor (normal incidence).

waves have the form, from (8.3.1) and (8.5.15):

Incident:

$$E_1 e^{j(\omega t - \kappa_0 z)} \tag{8.6.9}$$

Reflected:

$$\breve{E}_1' e^{j(\omega t + \kappa_0 z)} \tag{8.6.10}$$

Transmitted:

$$\breve{E}_2 e^{-z/\delta} e^{j(\omega t - z/\delta)}, \tag{8.6.11}$$

where $\kappa_0 = \omega/c$ is the vacuum wave number and δ is the skin-depth parameter (8.5.14) of the conductor. The plane-wave fields are transverse; hence only tangential components exist at the boundary plane ($z = 0$). By symmetry, the electric field vectors all lie in the same plane, so that no generality is lost in writing (8.6.9) to (8.6.11) as scalar quantities. The cups on the symbols $\breve{E}_1'$ and $\breve{E}_2$ signify that the amplitudes are complex quantities, i.e., that the reflected and transmitted waves may not be in phase with the incident wave of prescribed amplitude and phase.

Similarly, from (8.3.3), (8.3.9), (8.5.13), and (8.5.17), the magnetic fields of the three waves have the form:

Incident:

$$B_1 e^{j(\omega t - \kappa_0 z)} = \frac{E_1}{c} e^{j(\omega t - \kappa_0 z)} \tag{8.6.12}$$

Reflected:

$$\breve{B}_1' e^{j(\omega t + \kappa_0 z)} = -\frac{\breve{E}_1'}{c} e^{j(\omega t + \kappa_0 z)} \tag{8.6.13}$$

Transmitted:

$$\breve{B}_2 e^{-z/\delta} e^{j(\omega t - z/\delta)} \approx (1 - j)\frac{\breve{E}_2}{\omega\delta} e^{-z/\delta} e^{j(\omega t - z/\delta)}. \tag{8.6.14}$$

Again, the magnetic vectors lie in the same plane, perpendicular to that of the electric vectors, and are tangential to the boundary.

The boundary conditions (8.6.5) and (8.6.6) are satisfied trivially at $z = 0$, since there are no normal field components. Conditions (8.6.7) and (8.6.8) provide the simultaneous equations

$$E_1 + \breve{E}_1' = \breve{E}_2 \tag{8.6.15}$$

$$E_1 - \breve{E}_1' = (1 - j)\frac{c}{\omega\delta\kappa_m}\breve{E}_2 \tag{8.6.16}$$

for the tangential components, where κ_m is the relative permeability of the conductor. These yield the (complex) reflection and transmission coefficients for the electric field amplitudes (see Prob. 8.6.4)

$$\breve{R}_E \equiv \frac{\breve{E}_1'}{E_1} = \frac{-(1-j)(c/\omega\delta\kappa_m) + 1}{(1-j)(c/\omega\delta\kappa_m) + 1} \approx -\left[1 - (1+j)\frac{\omega\delta\kappa_m}{c}\right] \tag{8.6.17}$$

$$\breve{T}_E \equiv \frac{\breve{E}_2}{E_1} = \frac{2}{(1-j)(c/\omega\delta\kappa_m) + 1} \approx (1+j)\frac{\omega\delta\kappa_m}{c}. \tag{8.6.18}$$

Since the incident and reflected waves are in the same medium, the power reflection coefficient is simply

$$R_p = |\breve{R}_E|^2 = \frac{1 + (1 - \omega\delta\kappa_m/c)^2}{1 + (1 + \omega\delta\kappa_m/c)^2} \approx 1 - \frac{2\omega\delta\kappa_m}{c}. \tag{8.6.19}$$

Conservation of energy requires that the power transmission coefficient, i.e., the fraction of the incident power dissipated in the conductor, be (see Prob. 8.6.6)

$$T_p = 1 - R_p = \frac{4\omega\delta\kappa_m/c}{1 + (1 + \omega\delta\kappa_m/c)^2} \approx \frac{2\omega\delta\kappa_m}{c}. \tag{8.6.20}$$

For a good conductor ($g \to \infty$, $\delta \to 0$), the wave is almost perfectly reflected (see Prob. 8.6.5).

(c)   Oblique Incidence on a Nonconductor

We consider here the case of a sinusoidal plane wave incident obliquely on a plane interface separating two uniform nonconducting media. An extensive discussion in Sec. 5.6 has established three important theorems for this case that follow from geometrical considerations alone and are independent of the particular type of wave (acoustic, electromagnetic, etc.).

(1) The vector wave numbers of the reflected and transmitted (refracted) waves lie in the *plane of incidence*, i.e., the plane defined by the wave number κ_1 of the incident wave and the normal to the interface. This is taken to be the xz plane in Fig. 8.6.3.

(2) The angles of incidence and reflection are equal (both θ_1 in Fig. 8.6.3).

(3) The angle of refraction θ_2 is related to the angle of incidence by *Snell's law*

$$\frac{1}{c_1}\sin\theta_1 = \frac{1}{c_2}\sin\theta_2, \tag{8.6.21}$$

where $c_1 = c/(\kappa_{e1}\kappa_{m1})^{1/2}$ and $c_2 = c/(\kappa_{e2}\kappa_{m2})^{1/2}$ are the respective wave speeds in the two media.

These theorems ensure that the exponential space-time factors $\exp j(\omega t - \kappa \cdot \mathbf{r})$ for the three waves ($\kappa \rightarrow \kappa_1, \kappa_1', \kappa_2$, respectively) are identical at all points in the interface.

We are now ready to invoke the electromagnetic boundary conditions (8.6.5) to (8.6.8). It is helpful to consider separately the two cases:

Case I.   The incident electric field $\mathbf{E}_1$ is perpendicular to the plane of incidence, i.e., parallel to the interface.

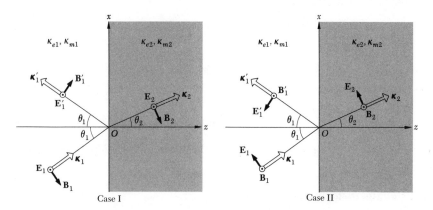

Case I   Case II

Fig. 8.6.3   Reflection and refraction of plane waves at a plane interface. The symbol $\odot$ indicates that the vector is directed out of the paper, in the $+y$ direction.

Case II. The field $\mathbf{E}_1$ lies in the plane of incidence; the incident magnetic field $\mathbf{B}_1$ is parallel to the interface.

The general case consists of a superposition of these two. An appeal to symmetry establishes that in case I all three $\mathbf{E}$ fields have only y components while in case II all three $\mathbf{B}$ fields have only y components. Figure 8.6.3 defines our conventions as to the positive senses of the field vectors.

CASE I ($\mathbf{E} \perp$ plane of incidence)
From Fig. 8.6.3 and Eq. (8.3.13), the wave fields have the form:

Incident:

$$\mathbf{E}_1 = \mathbf{j}E_1 \exp j(\omega t - \boldsymbol{\kappa}_1 \cdot \mathbf{r})$$

$$\mathbf{B}_1 = (-\cos\theta_1 \mathbf{i} + \sin\theta_1 \mathbf{k}) \frac{E_1}{c_1} \exp j(\omega t - \boldsymbol{\kappa}_1 \cdot \mathbf{r}) \qquad (8.6.22)$$

Reflected:

$$\mathbf{E}_1' = \mathbf{j}E_1' \exp j(\omega t - \boldsymbol{\kappa}_1' \cdot \mathbf{r})$$

$$\mathbf{B}_1' = (\cos\theta_1 \mathbf{i} + \sin\theta_1 \mathbf{k}) \frac{E_1'}{c_1} \exp j(\omega t - \boldsymbol{\kappa}_1' \cdot \mathbf{r}) \qquad (8.6.23)$$

Transmitted (refracted):

$$\mathbf{E}_2 = \mathbf{j}E_2 \exp j(\omega t - \boldsymbol{\kappa}_2 \cdot \mathbf{r})$$

$$\mathbf{B}_2 = (-\cos\theta_2 \mathbf{i} + \sin\theta_2 \mathbf{k}) \frac{E_2}{c_2} \exp j(\omega t - \boldsymbol{\kappa}_2 \cdot \mathbf{r}). \qquad (8.6.24)$$

The boundary condition (8.6.5) is satisfied trivially since there are no normal components of the $\mathbf{E}$ fields at the interface. The tangential $\mathbf{E}$-field condition (8.6.7) demands that

$$E_1 + E_1' = E_2. \qquad (8.6.25)$$

The normal and tangential $\mathbf{B}$-field conditions, (8.6.6) and (8.6.8), lead respectively to

$$\frac{1}{c_1} \sin\theta_1 (E_1 + E_1') = \frac{1}{c_2} \sin\theta_2 E_2 \qquad (8.6.26)$$

$$\frac{1}{K_{m1}c_1} \cos\theta_1 (E_1 - E_1') = \frac{1}{K_{m2}c_2} \cos\theta_2 E_2. \qquad (8.6.27)$$

By Snell's law (8.6.21), (8.6.26) is redundant with (8.6.25). Thus (8.6.25) and

(8.6.27) determine the coefficients of reflection and transmission for the electric field,

$$R_{E\perp} \equiv \frac{E_1'}{E_1} = \frac{-\kappa_{m1}\tan\theta_1 + \kappa_{m2}\tan\theta_2}{\kappa_{m1}\tan\theta_1 + \kappa_{m2}\tan\theta_2} \xrightarrow[\kappa_{m1}=\kappa_{m2}]{} -\frac{\sin(\theta_1 - \theta_2)}{\sin(\theta_1 + \theta_2)} \qquad (8.6.28)$$

$$T_{E\perp} \equiv \frac{E_2}{E_1} = \frac{2\kappa_{m2}\tan\theta_2}{\kappa_{m1}\tan\theta_1 + \kappa_{m2}\tan\theta_2} \xrightarrow[\kappa_{m1}=\kappa_{m2}]{} \frac{2\cos\theta_1\sin\theta_2}{\sin(\theta_1 + \theta_2)}. \qquad (8.6.29)$$

The simplified forms shown for magnetically equivalent materials apply, a fortiori, to the common special case of nonmagnetic materials. These formulas may be cast in several other forms by using Snell's law or trigonometric identities (see Prob. 8.6.13). Since the angles θ_1 and θ_2 are not independent, being related by Snell's law, the coefficients (8.6.28) and (8.6.29) should be regarded as functions of only one angular variable, the other appearing as a parameter to simplify the algebraic form.

CASE II   (**E** ∥ plane of incidence)

Again from Fig. 8.6.3 and Eq. (8.3.13), the wave fields now have the form:

Incident:

$$\mathbf{E}_1 = (\cos\theta_1\mathbf{i} - \sin\theta_1\mathbf{k})E_1 \exp j(\omega t - \mathbf{\kappa}_1 \cdot \mathbf{r})$$

$$\mathbf{B}_1 = \mathbf{j}\,\frac{E_1}{c_1} \exp j(\omega t - \mathbf{\kappa}_1 \cdot \mathbf{r}) \qquad (8.6.30)$$

Reflected:

$$\mathbf{E}_1' = (-\cos\theta_1\mathbf{i} - \sin\theta_1\mathbf{k})E_1' \exp j(\omega t - \mathbf{\kappa}_1' \cdot \mathbf{r})$$

$$\mathbf{B}_1' = \mathbf{j}\,\frac{E_1'}{c_1} \exp j(\omega t - \mathbf{\kappa}_1' \cdot \mathbf{r}) \qquad (8.6.31)$$

Transmitted (refracted):

$$\mathbf{E}_2 = (\cos\theta_2\mathbf{i} - \sin\theta_2\mathbf{k})E_2 \exp j(\omega t - \mathbf{\kappa}_2 \cdot \mathbf{r})$$

$$\mathbf{B}_2 = \mathbf{j}\,\frac{E_2}{c_2} \exp j(\omega t - \mathbf{\kappa}_2 \cdot \mathbf{r}). \qquad (8.6.32)$$

In this case, the normal-**B** boundary condition (8.6.6) is satisfied trivially. The remaining conditions (8.6.5) to (8.6.8) demand

$$\kappa_{e1}\sin\theta_1(E_1 + E_1') = \kappa_{e2}\sin\theta_2 E_2 \qquad (8.6.33)$$

$$\cos\theta_1(E_1 - E_1') = \cos\theta_2 E_2 \qquad (8.6.34)$$

$$\frac{1}{\kappa_{m1}c_1}(E_1 + E_1') = \frac{1}{\kappa_{m2}c_2}E_2 \qquad (8.6.35)$$

with (8.6.33) being redundant with (8.6.35). The reflection and transmission coefficients become

$$R_{E\parallel} = \frac{-\kappa_{e1}\tan\theta_1 + \kappa_{e2}\tan\theta_2}{\kappa_{e1}\tan\theta_1 + \kappa_{e2}\tan\theta_2} \xrightarrow[\kappa_{m1}=\kappa_{m2}]{} \frac{\tan(\theta_1 - \theta_2)}{\tan(\theta_1 + \theta_2)} \tag{8.6.36}$$

$$T_{E\parallel} = \frac{2\kappa_{e1}\sin\theta_1/\cos\theta_2}{\kappa_{e1}\tan\theta_1 + \kappa_{e2}\tan\theta_2} \xrightarrow[\kappa_{m1}=\kappa_{m2}]{} \frac{2\cos\theta_1\sin\theta_2}{\sin(\theta_1 + \theta_2)\cos(\theta_1 - \theta_2)}. \tag{8.6.37}$$

Simplified forms again result for magnetically equivalent materials.

Formulas equivalent to (8.6.28) and (8.6.29) and (8.6.36) and (8.6.37) can be readily worked out for the magnetic fields (Prob. 8.6.8). The *power* reflection and transmission coefficients can also be computed by evaluating the Poynting vector (8.4.23). Using the results of Prob. 8.6.8, the power coefficients may be written

$$R_p = R_E{}^2 \tag{8.6.38}$$

$$T_p = \frac{\cos\theta_2}{\cos\theta_1}\frac{Z_{01}}{Z_{02}} T_E{}^2, \tag{8.6.39}$$

where the Z_0's are the wave impedances (8.3.12) of the two media. A convenient check on computations and algebraic manipulations is to confirm the necessary condition

$$R_p + T_p = 1. \tag{8.6.40}$$

The formulas giving reflection and transmission coefficients for electromagnetic waves at a plane interface, subdivided into the two polarization cases, are known as the *Fresnel formulas*.

Figure 8.6.4 illustrates numerical results for an air-glass interface. Two noteworthy features are the disappearance of the case II reflected wave at the *Brewster angle* and of the transmitted wave under conditions of *total internal reflection*. The Brewster condition is readily derived in the final form of (8.6.36) (for $\kappa_{m1} = \kappa_{m2}$), which clearly vanishes for $\theta_1 + \theta_2 = \pi/2$. With the elimination of θ_2 using Snell's law, the Brewster angle of incidence is given by (see Prob. 8.6.10)

$$\tan(\theta_1)_{\text{Brewster}} = \frac{c_1}{c_2} = \left(\frac{\kappa_{e2}}{\kappa_{e1}}\right)^{1/2} \quad (\kappa_{m1} = \kappa_{m2}), \tag{8.6.41}$$

where the velocity ratio c_1/c_2 is often termed the relative refractive index $n = n_2/n_1$ of the two media. Total internal reflection occurs when the kinematical constraints implied by Snell's law (8.6.21) cannot be met for the refracted wave, as shown formally by the fact that $\sin\theta_2$ cannot exceed unity, i.e., for angles of incidence greater than the critical angle given by (see Prob. 8.6.11)

$$\sin(\theta_1)_{\text{critical}} = \frac{c_1}{c_2}. \tag{8.6.42}$$

Nevertheless nonzero fields exist in the second medium, dying off exponentially away from the interface. These fields constitute an *evanescent* wave, which carries no power in the steady state.

The theory of this section can be extended to cover multiple reflections set up at two or more parallel interfaces.† The full generality of several interfaces, oblique incidence, arbitrary polarization, and conducting and magnetic materials is straightforward but cumbersome. Useful formulas are available only for

† For a more complete treatment, see J. A. Stratton, "Electromagnetic Theory," chap. 9, McGraw-Hill Book Company, New York, 1941.

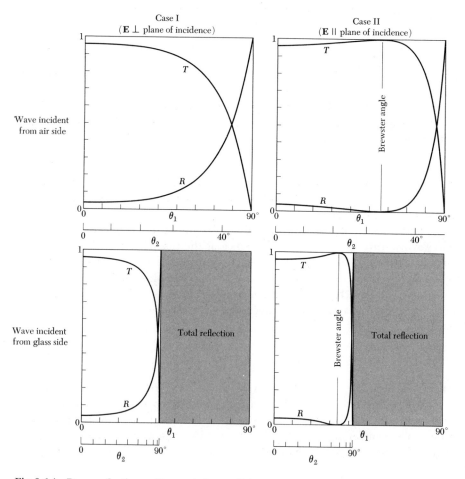

Fig. 8.6.4   Power reflection and transmission coefficients at an air-glass interface; $(\kappa_e)_{glass}/(\kappa_e)_{air}$ = 2.25.

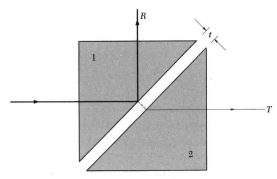

Fig. 8.6.5   Frustrated internal reflection between two dielectric prisms.

special cases. A familiar example is the use of a quarter-wave layer of material of intermediate refractive index to reduce reflection from the surface of a lens (Prob. 8.6.12). Another interesting example involves the evanescent wave associated with total reflection. In Fig. 8.6.5, the wave incident from the left is "totally" reflected from the hypotenuse of the first prism. However, when a second prism is brought up to a small separation t, the evanescent wave excites a power-carrying transmitted wave in the second prism (and the reflected wave is correspondingly reduced). The transmission coefficient thus varies exponentially with the separation t. This coupling through a "forbidden" region is the electromagnetic analog of a quantum-mechanical phenomenon known as the *tunnel effect*.

Problems

8.6.1 Use the general form of Maxwell's equations (8.2.17) to (8.2.20), together with Gauss' and Stokes' theorems, to obtain the corresponding integral equations for a material medium,

$$\oint_S \mathbf{D} \cdot d\mathbf{S} = q_{\text{free}} \tag{8.6.43}$$

$$\oint_L \mathbf{E} \cdot d\mathbf{l} = -\frac{d\Phi_m}{dt} \tag{8.6.44}$$

$$\oint_S \mathbf{B} \cdot d\mathbf{S} = 0 \tag{8.6.45}$$

$$\oint_L \mathbf{H} \cdot d\mathbf{l} = I_{\text{free}} + \frac{d\Phi_e}{dt}, \tag{8.6.46}$$

where $\Phi_m \equiv \int \mathbf{B} \cdot d\mathbf{S}$ and $\Phi_e \equiv \int \mathbf{D} \cdot d\mathbf{S}$ are the magnetic and electric fluxes linking the closed line L. These equations are generalizations of (8.2.11) to (8.2.14).

8.6.2 (*a*) Generalize the boundary conditions (8.6.5) to (8.6.8) to include the case where a surface charge density $\sigma = \Delta q_{\text{free}}/\Delta S$ and a surface current of magnitude $K = \Delta I_{\text{free}}/\Delta l$ exist on the boundary surface, establishing the conditions

$$\hat{n} \cdot (\kappa_{e2} \mathbf{E}_2 - \kappa_{e1} \mathbf{E}_1) = \frac{\sigma}{\epsilon_0}$$

$$\hat{n} \times \left(\frac{\mathbf{B}_2}{\kappa_{m2}} - \frac{\mathbf{B}_1}{\kappa_{m1}} \right) = \mu_0 \mathbf{K}.$$

(b) Show that the boundary conditions remain valid when the boundary is not plane and when the respective media are not homogeneous. (c) What are the boundary conditions on the **D** and **H** fields?

8.6.3 Consider an **E**-field *line of force*, i.e., a continuous line everywhere parallel to the local direction of **E**, deflected at the boundary between two uniform media. Show that the exit line of force lies in the plane determined by the entrance line and the normal to the boundary surface and that the angles of incidence θ_1 and exit θ_2, measured with respect to the normal, are related by the Snell's law equation

$$\frac{1}{\kappa_{e1}} \tan\theta_1 = \frac{1}{\kappa_{e2}} \tan\theta_2.$$

What are the corresponding equations for **B**, **D**, and **H**?

8.6.4 Justify (a) the minus sign in (8.6.13); (b) the final approximations given in (8.6.17) to (8.6.20).

8.6.5 Estimate the reflection coefficient of copper at normal incidence for the cases of Prob. 8.5.3b.

8.6.6 Calculate the power per unit area transmitted into the conductor of Fig. 8.6.2 by two methods. (a) From (8.6.11) and (8.6.14), compute the time-average magnitude of the Poynting vector (8.4.23) just inside the surface of the conductor. (b) Compute the power per unit area dissipated in the conductor by integrating (8.4.4) in the form

$$\int_0^\infty g\overline{E^2} \, dz.$$

Finally, compare the common result with the Poynting vector of the incident wave, using (8.6.18), to obtain the power transmission coefficient (8.6.20) by direct calculation.

★8.6.7 (a) Use Ampère's law (8.6.4) to prove that the current density $J(z)$ in the conductor of Fig. 8.6.2 is related to the net magnetic field just *outside* the conductor by

$$B_{\text{outside}} = \mu_0 \int_0^\infty J(z) \, dz \equiv \mu_0 K,$$

where the integral symbolized by K has the dimensions of a surface current density, namely, amperes per meter. (b) Consider an artificial model whereby the surface current K is distributed uniformly in the skin layer $0 < z < \delta$, so that the current density is $J_{\text{skin}} = K/\delta$ in that layer but zero everywhere else. Show that the time-average power per unit area

$$\left(\overline{\frac{J_{\text{skin}}^2}{g}} \right) \delta = \frac{\overline{K^2}}{g\delta} = \frac{\overline{B_{\text{outside}}^2}}{\mu_0^2 g\delta}$$

computed on this model is identical with that found in Prob. 8.6.6b. This model is useful for finding the power loss of waves propagating in hollow-pipe waveguides (see Prob. 8.7.14). It also suggests that the resistance of a round wire is greater for high-frequency currents than for direct current in the ratio of effective areas,

$$\frac{R_{\text{ac}}}{R_{\text{dc}}} \xrightarrow[\delta \ll a]{} \frac{2\pi a\delta}{\pi a^2} = \frac{2\delta}{a},$$

where a is the radius of the wire.

8.6.8   Show that the reflection coefficients for the *magnetic* field amplitudes (either **B** or **H**) are identical with (8.6.28) and (8.6.36), while the transmission coefficients differ from (8.6.29) and (8.6.37) by the ratio of the wave impedances of the two media, (8.5.18) or (8.5.19). Specifically, show that for the **B** field,

$$\frac{T_B}{T_E} = \frac{c_1}{c_2} = \left(\frac{\kappa_{e2}\kappa_{m2}}{\kappa_{e1}\kappa_{m1}} \right)^{1/2},$$

which is the relative refractive index for the two media; for the **H** field,

$$\frac{T_H}{T_E} = \frac{Z_{01}}{Z_{02}} = \left(\frac{\kappa_{e2}\kappa_{m1}}{\kappa_{e1}\kappa_{m2}} \right)^{1/2}.$$

Justify the cosine ratio in (8.6.39).

8.6.9   For normal incidence and nonmagnetic materials show that the power coefficients (8.6.38) and (8.6.39) reduce to

$$R_p = \left(\frac{n-1}{n+1} \right)^2$$

$$T_p = \frac{4n}{(n+1)^2},$$

where $n \equiv c_1/c_2 = Z_{01}/Z_{02}$ is the relative refractive index. Account for the difference in sign between the amplitude reflection coefficients (8.6.28) and (8.6.36) at normal incidence (see footnote, page 102). Compare with equations (1.9.6), (4.2.15), and (5.6.13).

8.6.10   Show that the general Brewster angle condition is

$$\tan(\theta_1)_{\text{Brewster}} = \left(\frac{\kappa_{e2}\kappa_{m1}/\kappa_{e1}\kappa_{m2} - 1}{1 - \kappa_{e1}\kappa_{m1}/\kappa_{e2}\kappa_{m2}} \right)^{1/2},$$

which reduces to (8.6.41) when $\kappa_{m1} = \kappa_{m2}$.

8.6.11   Consider total reflection at an interface between two nonmagnetic media, with relative refractive index $n = c_1/c_2 < 1$. For angles of incidence θ_1 exceeding the critical angle of (8.6.42), Snell's law gives

$$\sin\theta_2 = \frac{\sin\theta_1}{n} > 1,$$

which implies that θ_2 is a complex angle with an imaginary cosine,

$$\cos\theta_2 \equiv (1 - \sin^2\theta_2)^{1/2} = j\left(\frac{\sin^2\theta_1}{n^2} - 1\right)^{1/2}.$$

Substitute these relations in the case I reflection coefficient (8.6.28) to establish

$$R_{E\perp} = e^{-j2\phi_\perp},$$

where

$$\tan\phi_\perp = \frac{(\sin^2\theta_1 - n^2)^{1/2}}{\cos\theta_1}.$$

That is, the magnitude of the reflection coefficient is unity, but the phase of the reflected wave depends upon angle. Similarly show for case II from (8.6.36), that $R_{E\parallel} = e^{-j2\phi_\parallel}$ with

$$\tan\phi_\parallel = \frac{(\sin^2\theta_1 - n^2)^{1/2}}{n^2\cos\theta_1} = \frac{1}{n^2}\tan\phi_\perp.$$

Note that the two phase shifts are different, so that in general the state of polarization of an incident wave is altered.

8.6.12   Consider an electromagnetic wave E_1 incident normally on two interfaces separated by the distance t. The three media are nonconducting and nonmagnetic but have distinct permittivities. Set up the boundary conditions on the five wave amplitudes indicated in the figure and show that the reflected wave E_1' vanishes when t is a quarter-wavelength and κ_{e2} is the geometric mean $(\kappa_{e1}\kappa_{e3})^{1/2}$.

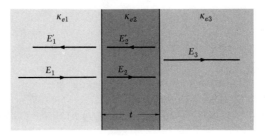

Prob. 8.6.12

8.6.13   Show that the case I reflection coefficient (8.6.28) is identical with (5.6.12) when written in terms of the characteristic impedance (8.3.12). How does the case II coefficient

(8.6.36) differ from (5.6.12)? Why is a Brewster angle associated with (5.6.12) (see Prob. 5.6.4) but not normally with (8.6.28)?

8.7 Waveguides

The propagation of electromagnetic waves in hollow conducting pipes, called *waveguides*, is fundamental to the technology of the *microwave* frequency domain, roughly from 10^9 to 10^{11} Hz (30 cm to 3 mm wavelength). At lower frequencies, two-wire transmission lines are the common method of directing the flow of electromagnetic energy. Electromagnetic waves on lines can be discussed in terms of voltages and currents without explicit reference to electromagnetic fields, as in Sec. 8.1. At higher frequencies, and in particular for visible light ($\sim 6 \times 10^{14}$ Hz, 500 nm), the model of uniform plane waves discussed in Secs. 8.3, 8.5, and 8.6 is sufficient to describe many electromagnetic processes of practical interest.† To deal with the intermediate case of hollow-pipe waveguides, we start with the vector wave equation (8.2.8) for the electric field

$$\nabla^2 \mathbf{E} = \frac{1}{c^2} \frac{\partial^2 \mathbf{E}}{\partial t^2}, \tag{8.7.1}$$

and seek non-plane-wave solutions that fit prescribed boundary conditions at the walls of the waveguide. We have seen in Sec. 7.5 that solutions to the vector wave equation can be separated into *solenoidal* and *irrotational* vector fields, i.e., into fields whose divergence or curl vanishes, respectively, and which may be thought of qualitatively as transverse or longitudinal, respectively. Only solenoidal ("transverse") solutions of (8.7.1) are consistent with Gauss' law (8.2.1),

$$\nabla \cdot \mathbf{E} = \frac{\rho}{\epsilon_0} \to 0. \tag{8.7.2}$$

(a)   The Vector Wave Equation

Equation (8.7.1) represents three scalar wave equations for the components of the vector field $\mathbf{E} = \mathbf{E}(\mathbf{r})$, where the argument $\mathbf{r}$ signifies that each component may be a function of three spatial coordinates.‡ The general solution of the *scalar* wave equation consistent with prescribed boundary conditions can be found by the direct and powerful method of separation of variables. A similar general solution of the *vector* wave equation is much more difficult since a cartesian decomposition of the vector field, necessary for the independence of the equations for the three component fields, is not compatible, in general, with the

---

† At still higher frequencies (x-rays, gamma rays), the photon or corpuscular nature of electromagnetic radiation becomes more important than its wave nature.
‡ Recall the last paragraph of Sec. 8.2.

vector boundary conditions, which are usually stated in terms of the values of the normal and tangential components of the field at the boundary surface. It turns out† that an effective strategy is to synthesize solutions of the vector wave equation by starting with a solution $\psi(\mathbf{r},t)$ of the scalar wave equation

$$\nabla^2\psi = \frac{1}{c^2}\frac{\partial^2\psi}{\partial t^2}. \tag{8.7.3}$$

If $\hat{\mathbf{a}}$ signifies a unit vector of *fixed* direction, the vector field $\hat{\mathbf{a}}\psi(\mathbf{r},t)$ is necessarily a solution of the vector wave equation, since the constant vector $\hat{\mathbf{a}}$ commutes with the derivative operators ∇^2 and $\partial^2/\partial t^2$. But the solution $\hat{\mathbf{a}}\psi$ does not, in general, satisfy the divergence condition (8.7.2). A vector field of the form

$$\mathbf{E} = \nabla \times (\hat{\mathbf{a}}\psi), \tag{8.7.4}$$

however, satisfies both the wave equation (8.7.1) and Gauss' law (8.7.2), since the curl operator also commutes with the differential operators of the wave equation and since the divergence of a curl is identically zero. A second class of solenoidal solutions is of the form

$$\mathbf{E}' = \nabla \times [\nabla \times (\hat{\mathbf{a}}\psi)], \tag{8.7.5}$$

where the prime indicates that this solution is independent of (8.7.4). Higher-order multiple curls are redundant for time-harmonic fields (Prob. 8.7.1).

The argument by which we have inferred the solutions (8.7.4) and (8.7.5) may appear arbitrary and contrived. However, it can be shown that linear combinations of solutions of the form of (8.7.4) and (8.7.5) are sufficient to provide a *general* solution for the vector wave equation (8.7.1), subject to the solenoidal condition (8.7.2).‡ The particular merit of this form of solution is that the (scalar) boundary conditions on ψ usually bear a simple relation to the prescribed (vector) boundary conditions on $\mathbf{E}$. A further advantage concerns the associated magnetic field. For time-harmonic wave fields, varying as $e^{j\omega t}$, Maxwell's curl equations (8.2.2) and (8.2.4) reduce (in vacuum) to

$$\mathbf{B} = \frac{j}{\omega}\nabla \times \mathbf{E} \tag{8.7.6}$$

$$\mathbf{E} = -\frac{jc^2}{\omega}\nabla \times \mathbf{B}. \tag{8.7.7}$$

---

† For a complete and authoritative discussion, see P. M. Morse and H. Feshbach, "Methods of Theoretical Physics," pp. 1759–1767, McGraw-Hill Book Company, New York, 1953.
‡ Similarly, a function of the form $\mathbf{E}'' = \nabla\psi$ constitutes a general solution in the irrotational case for which $\nabla \times \mathbf{E} = 0$, $\nabla \cdot \mathbf{E} \neq 0$. This alternative is not applicable to *electromagnetic* waves because of Faraday's law (8.2.2); see Morse and Feshbach, *loc. cit.*

Thus, when the electric field is of the form (8.7.4), the magnetic field is of the form (8.7.5), and vice versa.

(b)   General Solution for Waveguides

We now specialize the discussion to electromagnetic waves in hollow-pipe waveguides of constant cross section and indefinite extent in the z direction, as shown in Fig. 8.7.1. We are interested in sinusoidal waves of frequency ω traveling in the $+z$ direction. Accordingly, the scalar function ψ satisfying (8.7.3) may be written explicitly as[†]

$$\psi(\mathbf{r},t) \equiv \phi(x,y)\, e^{j(\omega t - \kappa_z z)}, \tag{8.7.8}$$

and the cylindrical symmetry leads us to adopt the unit vector $\mathbf{k}$, in the z direction, for the constant vector $\hat{\mathbf{a}}$ in (8.7.4) and (8.7.5). Working out (8.7.4), we obtain

$$\mathbf{E} = \nabla \times (\mathbf{k}\psi) = e^{j(\omega t - \kappa_z z)} \begin{vmatrix} \mathbf{i} & \mathbf{j} & \mathbf{k} \\ \dfrac{\partial}{\partial x} & \dfrac{\partial}{\partial y} & -j\kappa_z \\ 0 & 0 & \phi(x,y) \end{vmatrix} \rightarrow \begin{cases} E_x = \dfrac{\partial \phi}{\partial y}\, e^{j(\omega t - \kappa_z z)} \\[2mm] E_y = -\dfrac{\partial \phi}{\partial x}\, e^{j(\omega t - \kappa_z z)} \\[2mm] E_z = 0. \end{cases}$$

$$\tag{8.7.9}$$

[†] The arbitrary choice of cartesian coordinates for $\phi = \phi(x,y)$ and for the fields (8.7.9) to (8.7.10) and (8.7.12) to (8.7.13) is perfectly general. However, when we wish to apply boundary conditions at the waveguide walls, another coordinate system may be more convenient. For example, the circular waveguide of Fig. 8.7.1b calls for $\phi(r,\theta)$ and field components in cylindrical coordinates, as worked out in part (d) of this section.

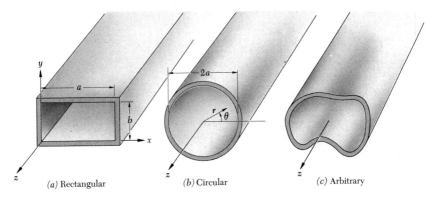

(a) Rectangular   (b) Circular   (c) Arbitrary

Fig. 8.7.1   Hollow-pipe waveguides of constant cross section.

The associated magnetic field is, by (8.7.6),

$$\mathbf{B} = \frac{j}{\omega} \nabla \times \mathbf{E} = \frac{j}{\omega} e^{j(\omega t - \kappa_z z)} \begin{vmatrix} \mathbf{i} & \mathbf{j} & \mathbf{k} \\ \dfrac{\partial}{\partial x} & \dfrac{\partial}{\partial y} & -j\kappa_z \\ \dfrac{\partial \phi}{\partial y} & -\dfrac{\partial \phi}{\partial x} & 0 \end{vmatrix}$$

$$\rightarrow \begin{cases} B_x = \dfrac{\kappa_z}{\omega} \dfrac{\partial \phi}{\partial x} e^{j(\omega t - \kappa_z z)} = -\dfrac{\kappa_z}{\omega} E_y \\ B_y = \dfrac{\kappa_z}{\omega} \dfrac{\partial \phi}{\partial y} e^{j(\omega t - \kappa_z z)} = \dfrac{\kappa_z}{\omega} E_x \\ B_z = -\dfrac{j}{\omega} \left(\dfrac{\partial^2 \phi}{\partial x^2} + \dfrac{\partial^2 \phi}{\partial y^2} \right) e^{j(\omega t - \kappa_z z)}. \end{cases} \qquad (8.7.10)$$

If the two-dimensional laplacian of ϕ vanishes, both electric and magnetic fields are transverse to the direction of propagation; such waves are called transverse electromagnetic (TEM) (see footnote, page 285). More generally, the laplacian and hence B_z are nonzero, and the resulting solutions of the **E**-field wave equation of the form (8.7.4) are known as transverse electric (TE) waves. The **E** and **B** fields, as well as their projections in the cross-sectional plane, are orthogonal since

$$\mathbf{E} \cdot \mathbf{B} = E_z B_z = 0. \qquad (8.7.11)$$

To obtain the second class of solutions, based on (8.7.5), we find it more convenient to start with the associated **B** field, which by (8.7.7) must have the *form* of (8.7.4). The results are

$$B_x = \frac{\partial \phi}{\partial y} e^{j(\omega t - \kappa_z z)}$$

$$B_y = -\frac{\partial \phi}{\partial x} e^{j(\omega t - \kappa_z z)} \qquad (8.7.12)$$

$$B_z = 0$$

and, using (8.7.7),

$$E_x = -\frac{\kappa_z c^2}{\omega} \frac{\partial \phi}{\partial x} e^{j(\omega t - \kappa_z z)} = \frac{\kappa_z c^2}{\omega} B_y$$

$$E_y = -\frac{\kappa_z c^2}{\omega} \frac{\partial \phi}{\partial y} e^{j(\omega t - \kappa_z z)} = -\frac{\kappa_z c^2}{\omega} B_x \qquad (8.7.13)$$

$$E_z = \frac{jc^2}{\omega} \left(\frac{\partial^2 \phi}{\partial x^2} + \frac{\partial^2 \phi}{\partial y^2} \right) e^{j(\omega t - \kappa_z z)}.$$

Since the two-dimensional laplacian of ϕ does not vanish, in general, this class of waves is transverse magnetic (TM).† Again, **E** and **B** are orthogonal. Formally, the roles of **E** and **B** are interchanged between the TE and the TM classes of solutions. A general solution can be written as a superposition of the TE and TM waves.

Our remaining problem is to find the function ϕ such that the solutions satisfy the boundary conditions at the waveguide walls. The full scalar wave function ψ of (8.7.8) must be a solution of the scalar wave equation (8.7.3). The laplacian operator ∇^2 in (8.7.3) can be broken into two parts

$$\nabla^2 \equiv \nabla_t^2 + \frac{\partial^2}{\partial z^2}, \tag{8.7.14}$$

where the two-dimensional transverse laplacian operator ∇_t^2 may be written out in rectangular, polar, or other suitable coordinates appropriate to the geometry of the cross section (see Fig. 8.7.1),

$$\nabla_t^2 = \frac{\partial^2}{\partial x^2} + \frac{\partial^2}{\partial y^2} = \frac{1}{r}\frac{\partial}{\partial r}\left(r\frac{\partial}{\partial r}\right) + \frac{1}{r^2}\frac{\partial^2}{\partial \theta^2} = \text{etc.} \tag{8.7.15}$$

Thus when the assumed solution (8.7.8) is substituted in the wave equation (8.7.3), the z and t derivatives can be carried out explicitly, yielding the two-dimensional Helmholtz equation for ϕ‡

$$\nabla_t^2 \phi + \kappa_c^2 \phi = 0, \tag{8.7.16}$$

where

$$\kappa_c^2 = \frac{\omega^2}{c^2} - \kappa_z^2. \tag{8.7.17}$$

Before finding solutions of (8.7.16) for specific geometries, we can anticipate the fact that only certain discrete values of the parameter κ_c permit our vector wave solution, (8.7.4) or (8.7.5), to satisfy the necessary boundary conditions at the

---

† Many writers use the terminology E wave in place of TM and H wave in place of TE, referring to the *nonzero* longitudinal component of the electric or magnetic (H) field.

‡ According to (8.7.10) and (8.7.13), TEM waves are possible only if ϕ is a solution of Laplace's equation $\nabla_t^2 \phi = 0$ rather than the Helmholtz equation (8.7.16). But then **B** in (8.7.10) and **E** in (8.7.13) are of the form of the gradient of an electrostatic potential. A well-known theorem of electrostatics, based on Gauss' law, states that the field is zero inside a boundary on which the potential is constant (or its normal derivative is zero) unless net charge is enclosed. Thus a hollow conducting pipe (Fig. 8.7.1c), cannot support a TEM mode, whereas the geometries of Fig. 8.1.2 can support a TEM mode, as well as TE and TM modes (Prob. 8.7.16). When a TEM mode is possible, the vanishing of κ_c in (8.7.16) implies that the phase velocity ω/κ_z is equal to c. TEM waves have no cutoff frequency but propagate at an arbitrarily low frequency.

walls of the waveguide.† Thus the parameter κ_c may be regarded as a property *of the shape and size of the waveguide cross section.* With this interpretation, (8.7.17) is no longer a definition of κ_c but the physical relation known as a *dispersion relation;* i.e., it relates the wave number κ_z (or wavelength $2\pi/\kappa_z$, or phase velocity c/κ_z: see Prob. 8.7.8) of a wave to the frequency ω of the wave. An important result may be seen by rewriting (8.7.17) in the form

$$\kappa_z{}^2 = \frac{\omega^2}{c^2} - \kappa_c{}^2. \tag{8.7.18}$$

If the waves are to propagate in the z direction without attenuation, as assumed in (8.7.8), κ_z must be real and the right-hand side of (8.7.18) must be positive. Thus there exists a *cutoff frequency*

$$\omega_c \equiv c\kappa_c \tag{8.7.19}$$

above which the wave travels without attenuation but below which the wave fields die out exponentially (such a wave is called *evanescent*). The dispersion relation may be put in somewhat neater form by expressing the variables as wavelengths. Let us define:

Free-space wavelength:

$$\lambda_0 \equiv \frac{2\pi c}{\omega}, \tag{8.7.20}$$

or the wavelength of a (plane) wave of frequency ω in an *unbounded* medium for which the wave speed is c (see Prob. 8.7.9);

Guide wavelength:

$$\lambda_g \equiv \frac{2\pi}{\kappa_z}, \tag{8.7.21}$$

or the wavelength of the bounded wave traveling down the waveguide;

Cutoff wavelength:

$$\lambda_c \equiv \frac{2\pi}{\kappa_c}, \tag{8.7.22}$$

which is a parameter, with dimensions of length, determined by the shape and size of the waveguide cross section [see (8.7.31) and (8.7.40)]. Then (8.7.18) may

---

† Formally, one says that the allowed values of $\kappa_c{}^2$ are the *eigenvalues* of the differential equation (8.7.16).

be written

$$\frac{1}{\lambda_g{}^2} = \frac{1}{\lambda_0{}^2} - \frac{1}{\lambda_c{}^2} \tag{8.7.23}$$

$$\lambda_g = \frac{\lambda_0}{[1 - (\lambda_0/\lambda_c)^2]^{1/2}}. \tag{8.7.24}$$

Thus the guide wavelength λ_g is always greater than the free-space wavelength λ_0, increasingly so as λ_0 approaches the cutoff wavelength λ_c. Of the three "wavelengths" just defined, only λ_g has direct physical significance as a true wavelength; λ_0 and λ_c are parameters measuring the frequency and the waveguide size and shape, respectively.†

We now examine the specific form of the function ϕ and the cutoff parameter λ_c for waveguides of rectangular and circular cross section.

(c) *Rectangular Cross Section*

Figure 8.7.1a establishes the origin of an xy coordinate system at one corner of the cross section of a rectangular waveguide of internal dimensions a by b. The general solution of (8.7.16) in cartesian coordinates, discussed in Sec. 2.1, is an infinite sum of terms of the form

$$\phi(x,y) = A \, \cos(\kappa_x x - \chi_x) \, \cos(\kappa_y y - \chi_y) \tag{8.7.25}$$

where A, χ_x, and χ_y are constants of integration, and where the separation constants $\kappa_x{}^2$ and $\kappa_y{}^2$ must satisfy the constraint

$$\kappa_x{}^2 + \kappa_y{}^2 = \kappa_c{}^2. \tag{8.7.26}$$

For the TE modes, we substitute (8.7.25) in (8.7.9) and impose the boundary condition that the tangential component of **E** must vanish at the surface of a perfect conductor, a direct extension of (8.6.7). Setting $E_x = 0$ at $y = 0$ and $y = b$ and $E_y = 0$ at $x = 0$ and $x = a$, one finds (Prob. 8.7.2) that the phase angles χ_x and χ_y must vanish and that the separation constants are restricted to the discrete values

$$\kappa_x = \frac{l\pi}{a} \qquad l = 0, 1, 2, \ldots$$

$$\kappa_y = \frac{m\pi}{b} \qquad m = 0, 1, 2, \ldots . \tag{8.7.27}$$

A particular choice of the integers l, m specifies one of the possible *modes* of propagation, identified by the notation TE_{lm}. The fields (8.7.9) and (8.7.10) become, explicitly:

† In the discussion of Sec. 2.4, waves in a narrow channel of a stretched membrane were found to have a cutoff wavelength (2.4.13) and a phase velocity (2.4.17) different from that of waves on an unbounded membrane. Equations (2.4.17) and (8.7.24) are equivalent.

TE modes, rectangular waveguide:

$$E_x = -\frac{m\pi}{b} A \cos\frac{l\pi x}{a} \sin\frac{m\pi y}{b} e^{j(\omega t - \kappa_z z)}$$

$$E_y = \frac{l\pi}{a} A \sin\frac{l\pi x}{a} \cos\frac{m\pi y}{b} e^{j(\omega t - \kappa_z z)} \tag{8.7.28}$$

$$E_z = 0$$

$$B_x = -\frac{\lambda_0}{\lambda_g}\frac{l\pi}{a}\frac{A}{c} \sin\frac{l\pi x}{a} \cos\frac{m\pi y}{b} e^{j(\omega t - \kappa_z z)}$$

$$B_y = -\frac{\lambda_0}{\lambda_g}\frac{m\pi}{b}\frac{A}{c} \cos\frac{l\pi x}{a} \sin\frac{m\pi y}{b} e^{j(\omega t - \kappa_z z)} \tag{8.7.29}$$

$$B_z = j\frac{\pi\lambda_0}{2}\left(\frac{l^2}{a^2} + \frac{m^2}{b^2}\right)\frac{A}{c} \cos\frac{l\pi x}{a} \cos\frac{m\pi y}{b} e^{j(\omega t - \kappa_z z)}.$$

From (8.7.26) and (8.7.27), we find the *cutoff frequency* (8.7.19) for the *lm* mode

$$\omega_c = \pi c \left[\left(\frac{l}{a}\right)^2 + \left(\frac{m}{b}\right)^2\right]^{1/2} \tag{8.7.30}$$

and the corresponding *cutoff wavelength* (8.7.22)

$$\lambda_c = \frac{1}{[(l/2a)^2 + (m/2b)^2]^{1/2}}. \tag{8.7.31}$$

As expected, λ_c depends only upon the dimensions a, b of the waveguide cross section and the indices l, m of the particular mode. For a TE wave l or m, but not both, may be zero.

For the TM modes, we substitute the ϕ function (8.7.25) into (8.7.12) and impose the boundary condition that the normal component of **B** must vanish at the conducting walls; that is, $B_x = 0$ at $x = 0$ and $x = a$, and $B_y = 0$ at $y = 0$ and $y = a$. The result is that $\chi_x = \chi_y = \pi/2$, while the separation constants κ_x and κ_y are again given by (8.7.27). Accordingly, the cutoff-wavelength formula (8.7.31) applies to both TE and TM solutions in the rectangular-waveguide case. The fields (8.7.12) and (8.7.13) become, explicitly:

TM modes, rectangular waveguide:

$$B_x = \frac{m\pi}{b} A \sin\frac{l\pi x}{a} \cos\frac{m\pi y}{b} e^{j(\omega t - \kappa_z z)}$$

$$B_y = -\frac{l\pi}{a} A \cos\frac{l\pi x}{a} \sin\frac{m\pi y}{b} e^{j(\omega t - \kappa_z z)} \tag{8.7.32}$$

$$B_z = 0$$

$$E_x = -\frac{\lambda_0}{\lambda_g}\frac{l\pi}{a}cA\cos\frac{l\pi x}{a}\sin\frac{m\pi y}{b}e^{j(\omega t-\kappa_z z)}$$

$$E_y = -\frac{\lambda_0}{\lambda_g}\frac{m\pi}{b}cA\sin\frac{l\pi x}{a}\cos\frac{m\pi y}{b}e^{j(\omega t-\kappa_z z)} \qquad (8.7.33)$$

$$E_z = -j\frac{\pi\lambda_0}{2}\left(\frac{l^2}{a^2}+\frac{m^2}{b^2}\right)cA\sin\frac{l\pi x}{a}\sin\frac{m\pi y}{b}e^{j(\omega t-\kappa_z z)}.$$

For a TM wave both indices l and m must be greater than zero for a nonvanishing wave.

We have thus found explicitly two families (TE$_{lm}$ and TM$_{lm}$) of wave solutions for the rectangular waveguide. Each family consists of a doubly infinite set of discrete modes, identified by the indices l, m. These solutions are *orthogonal* and *complete*.† Orthogonality means that an arbitrary wave can be represented by a unique superposition of these solutions, just as a complex periodic waveform can be expressed by a unique Fourier series. Completeness means that superposition of these solutions is sufficient to represent any possible wave field.

For a given waveguide size, there is necessarily a particular mode of lowest cutoff frequency ω_{c1}. Below ω_{c1}, waveguide propagation is not possible, except as an evanescent wave. Above ω_{c1} but below the cutoff frequency ω_{c2} of the second-lowest mode, only one mode can propagate. The lowest mode is known as the *dominant mode* and is the one normally used in practice.‡ A waveguide whose size is chosen such that only the one mode can propagate at a given frequency is called a *dominant-mode waveguide* for that frequency; if it is larger and several modes can propagate, it is often called *oversized*. A dominant-mode waveguide has the advantage that small discontinuities do not convert energy to other modes in an uncontrolled manner. For instance, it is possible to bend or twist the waveguide without causing reflections, provided the deformation takes place gradually relative to a wavelength.

In order to maximize the relative frequency range over which a rectangular waveguide is single-moded, the aspect ratio a/b is usually taken to be at least 2 (see Prob. 8.7.14). Then the dominant TE$_{10}$ mode is the only propagating mode over an octave frequency band, limited by the TE$_{20}$ mode, i.e., for free-space wavelengths in the range $2a > \lambda_0 > a$. Figure 8.7.2 is a chart of cutoff frequencies as a function of aspect ratio. Two or more modes having the same cutoff

---

† See, for example, Collin, *op. cit.*, chap. 5.

‡ Extensive discussions of the use of waveguides in microwave technology are given by T. Moreno, "Microwave Transmission Design Data," Dover Publications, Inc., New York, 1958; G. C. Southworth, "Principles and Applications of Waveguide Transmission," D. Van Nostrand Company, Inc., Princeton, N.J., 1950; C. G. Montgomery (ed.), "Technique of Microwave Measurements," McGraw-Hill Book Company, New York, 1947; and E. L. Ginzton, "Microwave Measurements," McGraw-Hill Book Company, New York, 1957.

frequencies are said to be *degenerate*. In a rectangular waveguide, but not generally, TE and TM modes of the same lm indices are degenerate. Other modes are degenerate if the aspect ratio happens to be a rational number (for instance, TE_{20} and TE_{01} when $a/b = 2$).

The field configuration of the dominant mode is shown in Fig. 8.7.3. From the distribution of surface current in the walls, also shown in the figure, it may be guessed that a narrow longitudinal slot can be cut along the middle of the broad wall without disturbing the propagating wave. This feature is often used in practical waveguide components to gain access to the interior of the guide. Similarly, a narrow transverse slot may be cut in the side wall. However, very good electric contact must be maintained across the line where top and side join and at the end when two sections of waveguide are coupled together. Higher-order modes are plotted, for instance, in the A.I.P. Handbook.†

★(d)   Circular Cross Section

To treat the circular waveguide of Fig. 8.7.1b, of diameter $2a$, we need the general solution of (8.7.16) in polar coordinates,

$$\frac{1}{r}\frac{\partial}{\partial r}\left(r\frac{\partial\phi}{\partial r}\right) + \frac{1}{r^2}\frac{\partial^2\phi}{\partial\theta^2} + \kappa_c^2\phi = 0. \tag{8.7.34}$$

This equation arose in our discussion of standing waves on a circular membrane (Sec. 2.3). Separation of variables leads to Bessel's equation for the radial function. The general solution of (8.7.34), for regions including the origin,‡ is a

† Dwight E. Gray (ed.), "American Institute of Physics Handbook," 2d ed., p. 5-61, McGraw-Hill Book Company, New York, 1963.

‡ The linearly independent second solution of Bessel's equation, the Neumann function $N_l(\kappa_c r)$, becomes infinite at the origin and must be discarded. Also we make use of our freedom to pick the θ origin in dropping the phase-angle constant of integration in the $\cos l\theta$ factor.

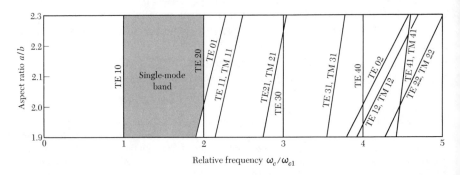

Fig. 8.7.2   Cutoff frequencies for rectangular waveguide of typical aspect ratios.

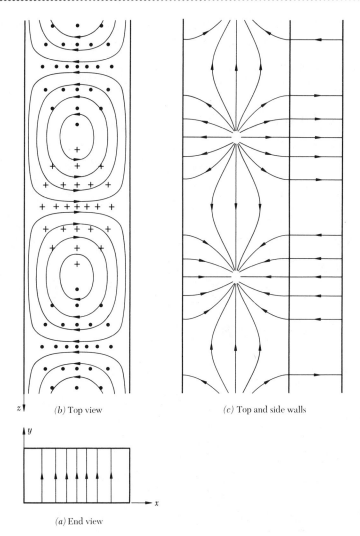

(b) Top view   *(c)* Top and side walls

(a) End view

Fig. 8.7.3 TE$_{10}$ mode in rectangular waveguide. (*a*) Electric lines of force. (*b*) Magnetic lines of force (loops); electric lines out (·) and into (**+**) figure. (*c*) Streamlines of wall current density **K**.

superposition of functions of the form

$$\phi(r,\theta) = AJ_l(\kappa_c r)\,\cos l\theta \qquad l = 0, 1, 2, \ldots, \tag{8.7.35}$$

where J_l is the usual Bessel function (2.3.9). The separation constant l is restricted to integers since the solution must be periodic in θ with period 2π; hence only integral orders of Bessel functions are involved.

For TE modes, derived from (8.7.4), we rewrite (8.7.9) and (8.7.10) in polar cylindrical coordinates and substitute (8.7.35), obtaining (Prob. 8.7.15):

TE modes, circular waveguide:

$$\mathbf{E} = \nabla \times (\mathbf{k}\psi) = e^{j(\omega t - \kappa_z z)} \begin{vmatrix} \dfrac{\hat{\mathbf{r}}}{r} & \hat{\boldsymbol{\theta}} & \dfrac{\mathbf{k}}{r} \\ \dfrac{\partial}{\partial r} & \dfrac{\partial}{\partial \theta} & -j\kappa_z \\ 0 & 0 & \phi(r,\theta) \end{vmatrix}$$

$$\rightarrow \begin{cases} E_r = \dfrac{1}{r}\dfrac{\partial \phi}{\partial \theta} e^{j(\omega t - \kappa_z r)} = -\dfrac{l}{r} A J_l(\kappa_c r) \, \sin l\theta \, e^{j(\omega t - \kappa_z z)} \\[2mm] E_\theta = -\dfrac{\partial \phi}{\partial r} e^{j(\omega t - \kappa_z z)} = -\kappa_c A J_l'(\kappa_c r) \, \cos l\theta \, e^{j(\omega t - \kappa_z z)} \\[2mm] E_z = 0 \end{cases} \quad (8.7.36)$$

$$\mathbf{B} = \dfrac{j}{\omega} \nabla \times \mathbf{E} = \dfrac{j}{\omega} e^{j(\omega t - \kappa_z z)} \begin{vmatrix} \dfrac{\hat{\mathbf{r}}}{r} & \hat{\boldsymbol{\theta}} & \dfrac{\mathbf{k}}{r} \\ \dfrac{\partial}{\partial r} & \dfrac{\partial}{\partial \theta} & -j\kappa_z \\ \dfrac{1}{r}\dfrac{\partial \phi}{\partial \theta} & -r\dfrac{\partial \phi}{\partial r} & 0 \end{vmatrix}$$

$$\rightarrow \begin{cases} B_r = \dfrac{\kappa_z}{\omega}\dfrac{\partial \phi}{\partial r} e^{j(\omega t - \kappa_z z)} = -\dfrac{\kappa_z}{\omega} E_\theta \\[2mm] \qquad = \dfrac{\lambda_0 \kappa_c}{\lambda_g}\dfrac{A}{c} J_l'(\kappa_c r) \, \cos l\theta \, e^{j(\omega t - \kappa_z z)} \\[3mm] B_\theta = \dfrac{\kappa_z}{\omega r}\dfrac{\partial \phi}{\partial \theta} e^{j(\omega t - \kappa_z z)} = \dfrac{\kappa_z}{\omega} E_r \\[2mm] \qquad = -\dfrac{\lambda_0 l}{\lambda_g r}\dfrac{A}{c} J_l(\kappa_c r) \, \sin l\theta \, e^{j(\omega t - \kappa_z z)} \\[3mm] B_z = \dfrac{j\kappa_c{}^2}{\omega} \phi e^{j(\omega t - \kappa_z z)} \\[2mm] \qquad = j\dfrac{\lambda_0 \kappa_c{}^2}{2\pi}\dfrac{A}{c} J_l(\kappa_c r) \, \cos l\theta \, e^{j(\omega t - \kappa_z z)}. \end{cases} \quad (8.7.37)$$

The primed Bessel functions, in the formulas for E_θ and B_r, are an abbreviation for the derivative,

$$J_l'(\kappa_c r) \equiv \left[\dfrac{dJ_l(u)}{du}\right]_{u = \kappa_z r}. \quad (8.7.38)$$

The TE-mode boundary condition is that $E_\theta = 0$ at $r = a$. Thus, for given radius a, only discrete values of the parameter κ_c are allowed such that

$$\frac{dJ_l(u)}{du} = 0 \qquad \text{TE modes,} \tag{8.7.39}$$

where $u = \kappa_c a$. For each order l, there is an infinite set of roots satisfying this transcendental equation. The mth positive root, in order from smallest upward, is denoted by u_{lm}. Table 8.1 gives the first few values. The cutoff wavelengths (8.7.22) for the l, m mode in circular waveguide are then given by

$$\lambda_c = \frac{2\pi}{\kappa_c} = \frac{2\pi a}{u_{lm}}, \tag{8.7.40}$$

and the cutoff frequencies (8.7.19) by

$$\omega_c = c\kappa_c = \frac{c}{a} u_{lm}. \tag{8.7.41}$$

Similarly, for the TM modes, we substitute (8.7.35) in the cylindrical coordinate version of (8.7.12) and (8.7.13), to obtain (Prob. 8.7.15):

TM modes, circular waveguide:

$$B_r = \frac{1}{r}\frac{\partial \phi}{\partial \theta} e^{j(\omega t - \kappa_z z)} = -\frac{l}{r} A J_l(\kappa_c r) \sin l\theta\, e^{j(\omega t - \kappa_z z)}$$

$$B_\theta = -\frac{\partial \phi}{\partial r} e^{j(\omega t - \kappa_z z)} = -\kappa_c A J_l'(\kappa_c r) \cos l\theta\, e^{j(\omega t - \kappa_z z)} \tag{8.7.42}$$

$$B_z = 0$$

TABLE 8.1   Bessel Function Roots u_{lm}†

Roots of $\dfrac{dJ_l(u)}{du} = 0$; TE modes					Roots of $J_l(u) = 0$; TM modes				
m \ l	0	1	2	3	m \ l	0	1	2	3
1	3.832	1.841	3.054	4.201	1	2.405	3.832	5.136	6.380
2	7.016	5.331	6.706	8.015	2	5.520	7.016	8.417	9.761
3	10.173	8.536	9.969	11.346	3	8.654	10.173	11.620	13.015

† An extensive table is given in M. Abromowitz and I. A. Stegun (eds.), "Handbook of Mathematical Functions," pp. 409, 411, Dover Publications, Inc., New York, 1965.

$$E_r = -\frac{\kappa_z c^2}{\omega}\frac{\partial\phi}{\partial r}\,e^{j(\omega t-\kappa_z z)} = \frac{\kappa_z c^2}{\omega}B_\theta$$

$$= -\frac{\lambda_0\kappa_c}{\lambda_g}\,cAJ_l'(\kappa_c r)\,\cos l\theta\,e^{j(\omega t-\kappa_z z)}$$

$$E_\theta = -\frac{\kappa_z c^2}{\omega r}\frac{\partial\phi}{\partial\theta}\,e^{j(\omega t-\kappa_z z)} = -\frac{\kappa_z c^2}{\omega}B_r$$

$$= \frac{\lambda_0 l}{\lambda_g r}\,cAJ_l(\kappa_c r)\,\sin l\theta\,e^{j(\omega t-\kappa_z z)}$$

$$E_z = -j\frac{c^2\kappa_c^2}{\omega}\,\phi e^{j(\omega t-\kappa_z z)} = -j\frac{\lambda_0\kappa_c^2}{2\pi}\,cAJ_l(\kappa_c r)\,\cos l\theta\,e^{j(\omega t-\kappa_z z)}.$$

(8.7.43)

The TM boundary condition is that $B_r = 0$ at $r = a$. Accordingly, in the TM case, the allowed values of $\kappa_c = u/a$ are given by

$$J_l(u) = 0 \qquad \text{TM modes.} \tag{8.7.44}$$

A few roots u_{lm} are listed in Table 8.1. Equations (8.7.40) and (8.7.41) again give the cutoff wavelengths and frequencies, but, of course, the roots of (8.7.44) do not coincide with those of (8.7.39) in general. An exception arises from the identity $dJ_0/du = -J_1$, so that the TE_{0m} and TM_{1m} modes are degenerate.

For circular waveguides, the cross section is determined by only one parameter, the diameter $2a$. Hence there is a unique ordering of the various modes with respect to their cutoff frequencies, as shown in Fig. 8.7.4 (compare Fig. 8.7.2). The dominant mode is TE_{11}; its electric field lines are shown in Fig. 8.7.5a.† From a practical point of view, imperfections in nominally circular waveguide cause the polarization of the dominant TE_{11} mode to wander in an uncontrolled manner; this disadvantage is removed by distorting the tubing into an elliptical cross section.‡ The most interesting modes in circular wave-

---

† The field configurations for other modes are given in the A.I.P. Handbook, *op. cit.*, p. **5**-63. A large collection of quantitative plots is given in Bell Laboratories Staff, "Radar Systems and Components," pp. 952–974, D. Van Nostrand Publishing Company, Inc., Princeton, N.J., 1949.
‡ Bell Laboratories Staff, *op. cit.*, pp. 998–1006; Moreno, *op. cit.*, pp. 119, 122, 137.

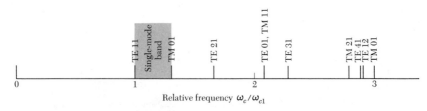

Fig. 8.7.4   Cutoff frequencies for circular waveguide.

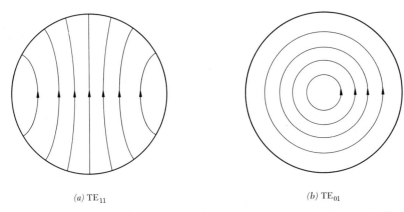

(a) TE_{11} (b) TE_{01}

Fig. 8.7.5   Electric field lines for the TE_{11} and TE_{01} modes in circular waveguide.

guide are the TE_{01} (Fig. 8.7.5b) and higher-order TE_{0m}, which have no direct analog in rectangular waveguide. For these modes $E_n = E_r$ is zero at the walls, so that no charge densities are called for. Consequently the only wall currents are those (K_θ) associated with B_z, and the finite conductivity of the walls causes relatively less attenuation for these modes.

Problems

8.7.1   Consider a solution of the vector wave equation of the form

$$\mathbf{F}(\mathbf{r},t) \equiv \nabla \times \mathbf{f}(\mathbf{r})e^{i\omega t}.$$

Show that

$$\nabla \times (\nabla \times \mathbf{F}) = \kappa^2 \mathbf{F},$$

that is, that the double curl of $\mathbf{F}$ is a redundant solution that differs from $\mathbf{F}$ only by the constant scale factor $\kappa^2 = \omega^2/c^2$.

8.7.2   Show that the general solution of the Helmholtz equation (8.7.16), obtained by separation of variables in cartesian coordinates, can be put in the form (8.7.25). Impose the boundary conditions on the electric field (8.7.9) for TE modes in rectangular waveguide to establish (8.7.27) to (8.7.29). Similarly, impose the boundary conditions on the magnetic field (8.7.12) for TM modes to establish (8.7.32) and (8.7.33).

8.7.3   (a) Extend Prob. 8.6.2 to establish that the four boundary conditions on electromagnetic fields in vacuum at the surface of a perfect conductor are

$$E_n = \sigma/\epsilon_0 \qquad B_n = 0$$
$$E_t = 0 \qquad B_t = \mu_0 \mathbf{K} \times \hat{\mathbf{n}}$$

where σ and $\mathbf{K}$ are the charge and current densities on the surface of the conductor and $\hat{\mathbf{n}}$ is a unit normal directed out of the conductor. (*b*) The function ϕ of (8.7.8) and (8.7.16) is chosen such that the E_t condition is satisfied for TE modes or the B_n condition for TM modes. Show that the orthogonality condition (8.7.11) assures that the B_n condition is also satisfied for TE modes and the E_t condition for TM modes. (*c*) By the fundamental definition of a conductor, namely, that free charge is available to move so as to cancel any interior electric field, the E_n condition is satisfied automatically so long as the inequality (8.5.12) is satisfied. Show, then, from Maxwell's equations and the equation of continuity of Prob. 8.2.6 that the B_t condition is also satisfied.

8.7.4 Consider **E** and **B** wave fields whose only dependence on z and t is included in the factor $e^{i(\omega t - \kappa_z z)}$. Further assume TE waves such that $E_z = 0$. Write out Maxwell's curl equations (8.2.2) and (8.2.4) in cartesian components and show (*a*) that all four transverse field components can be obtained from B_z by first-order partial differentiation and (*b*) that B_z must be a solution of the Helmholtz equation (8.7.16). Thus the scalar function ϕ of the text may be interpreted as proportional to B_z for TE waves or proportional to E_z for TM waves.

8.7.5 Show that parallel conducting planes of separation a can support a TE mode identical to the TE_{10} mode in rectangular waveguide with $b \to \infty$. Show further that the parallel planes can also support a TM mode that has no direct analog in rectangular waveguide (b finite) but is of the form of (8.7.32) and (8.7.33) with $m \to 0$, $\sin m\pi y/b \to 1$.

8.7.6 Consider two unbounded plane waves whose vector wave numbers κ_1 and κ_2 ($|\kappa_{1,2}| = \omega/c$) define a plane and whose electric fields are polarized normal to the plane. (*a*) Show that the superposition of these two plane waves is a wave traveling in the direction bisecting the angle α between κ_1 and κ_2 and that the **E** field vanishes on a set of nodal planes spaced $a \equiv \lambda_0/2 \sin\frac{1}{2}\alpha$ apart. (*b*) Show that plane conducting walls can be placed at two adjacent nodal planes without violating the electromagnetic boundary conditions and likewise that a second pair of conducting walls of arbitrary separation b can be introduced to construct a rectangular waveguide of cross section a by b, propagating the TE_{10} mode. Thus establish that the TE_{10} mode (more generally, the TE_{l0} modes) may be interpreted as the superposition of two plane waves making the angle $\frac{1}{2}\alpha$ with the waveguide axis and undergoing multiple reflections from the sidewalls. *Note:* The situation is directly analogous to that discussed in Sec. 2.4. Higher-order TE modes ($m > 0$) and TM modes may be described similarly as a superposition of four plane waves.

★8.7.7 Show, in general, that for TE modes the tangential-**E** boundary condition implies that the normal derivative of the scalar function ϕ must vanish at the boundary, whereas for TM modes the normal-**B** boundary condition implies that the function ϕ itself must vanish at the boundary.

8.7.8 From (8.7.18) show that the phase velocity of the wave in a waveguide is

$$c_p = \frac{\omega}{\kappa_z} = \frac{c}{[1 - (\lambda_0/\lambda_c)^2]^{1/2}}.$$

Note that this exceeds the velocity of light c! Find the group velocity $c_g = d\omega/d\kappa_z$ and show that

$$c_p c_g = c^2.$$

Explain the distinction between c_g, c, and c_p in terms of the plane-wave analysis of Prob. 8.7.6 for the TE_{10} mode in rectangular waveguide.

8.7.9 The treatment in the text tacitly assumes that the interior of the waveguide has the electromagnetic properties of vacuum. Show that if the waveguide is filled with a material of relative permittivity κ_e and permeability κ_m, all equations remain valid if c is replaced by c' of (8.2.21) and λ_0 in (8.7.20) is replaced by $\lambda' = 2\pi c'/\omega$.

8.7.10 (a) Specialize (8.7.28) and (8.7.29) to the TE_{10} dominant mode, with $E_0 \equiv \pi A/a$. (b) Integrate the time-average Poynting vector over the waveguide cross section to find the power transmitted down the waveguide,

$$\bar{P} = \frac{ab}{4}\left[1 - \left(\frac{\lambda_0}{\lambda_c}\right)^2\right]^{1/2}\frac{E_0^2}{Z_0},$$

which may be compared with (8.4.25). (c) From (8.4.6), show that the time-average electromagnetic energy per unit length of the guide is

$$\bar{W} = \frac{ab}{4}\,\epsilon_0 E_0^2,$$

(d) Use the results of Prob. 8.7.8 to show that

$$\bar{P} = c_g \bar{W}.$$

8.7.11 (a) From (8.7.10) and (8.7.13), show that the *guided wave impedance* $(E_x^2 + E_y^2)^{1/2}/(H_x^2 + H_y^2)^{1/2}$ is

$$Z_{TE} = \frac{Z_0}{[1 - (\lambda_0/\lambda_c)^2]^{1/2}} \qquad \text{TE modes}$$

$$Z_{TM} = Z_0\,[1 - (\lambda_0/\lambda_c)^2]^{1/2} \qquad \text{TM modes,}$$

where Z_0 is the unbounded wave impedance (8.3.10) or, more generally, (8.3.12). (b) For the TE_{10} dominant mode in rectangular waveguide, show that the peak potential difference between opposite points in the cross section is

$$V_0 \equiv \left[\int_0^b E_y(x = \tfrac{1}{2}a)\,dy\right]_{\text{peak}} = bE_0$$

and that the peak axial current flowing in the top wall is

$$I_0 \equiv \left[\frac{1}{\mu_0}\int_0^a B_z(y = b)\,dx\right]_{\text{peak}} = \frac{2aE_0}{\pi Z_{TE}}.$$

Since the result of Prob. 8.7.10b can be written

$$\overline{P} = \frac{ab}{4} \frac{E_0{}^2}{Z_{\mathrm{TE}}},$$

we can define three other (mode-dependent) waveguide impedances as follows:

$$Z_{V,I} = \frac{V_0}{I_0} = \frac{\pi}{2}\left(\frac{b}{a} Z_{\mathrm{TE}}\right)$$

$$Z_{P,V} = \frac{V_0{}^2}{2\overline{P}} = 2\left(\frac{b}{a} Z_{\mathrm{TE}}\right)$$

$$Z_{P,I} = \frac{2\overline{P}}{I_0{}^2} = \frac{\pi^2}{8}\left(\frac{b}{a} Z_{\mathrm{TE}}\right),$$

which differ by small numerical factors. Only systems supporting a TEM mode (e.g., Sec. 8.1), have a unique impedance.

8.7.12 For the TE$_{10}$ mode in rectangular waveguide, find the values of x at which the magnetic field is *circularly polarized;* i.e., the B_x and B_z components are equal in magnitude and 90° out of phase in time. (This feature is exploited in some waveguide devices known as *directional couplers* and *isolators*.) *Answer:* $\sin(\pi x/a) = \lambda_0/2a$.

8.7.13 Adapt the discussion at the end of Sec. 5.7 to show that the number of rectangular-waveguide modes whose cutoff frequencies are less than a given frequency $\omega_{\max}$ are approximately,

$$N = \frac{\omega_{\max}^2}{2\pi c^2} ab,$$

where N is assumed to be very large, and hence that the density of modes per unit frequency interval $dN/d\omega$ increases linearly with frequency. *Hint:* Count both TE and TM modes.

★8.7.14 Problem 8.6.7 states a model by which losses from the finite conductivity of the walls can be estimated. From the tangential components of the magnetic field, calculated on the assumption of infinite conductivity, the surface current density can be computed. This current is then assumed to flow uniformly in a surface layer of the thickness of one skin depth δ. The I^2R power loss in this layer is calculated and compared with the total power carried by the guided wave.

   Apply this model to the TE$_{10}$ rectangular-waveguide mode, to show that the attenuation is

$$\frac{2(8.686)}{2\delta g Z_0} \frac{[a/2b + (\lambda_0/2a)^2]}{[1 - (\lambda_0/2a)^2]^{1/2}} \quad \text{dB/unit length,}$$

where $\delta g Z_0 = (2g/\omega\epsilon_0)^{1/2}$. Note that the loss is reduced for small aspect ratios a/b; thus the condition for low loss is in conflict with that for maximum single-mode bandwidth (Fig.

8.7.2). Make a sketch of the attenuation as a function of frequency (implicit in δ and λ_0) and comment on the practical implications.

8.7.15 Establish the field components in (polar) cylindrical coordinates, (8.7.36) and (8.7.37) and (8.7.42) and (8.7.43).

8.7.16 Consider the coaxial transmission line or waveguide of Fig. 8.1.1*b*. (*a*) Show that the function

$$\phi = A \ln \frac{2r}{d_1},$$

when substituted into (8.7.34), (8.7.42), and (8.7.43), gives TEM waves traveling at the speed of light c. Compare these **E** and **B** wavefields with elementary static fields of coaxial symmetry, e.g., Prob. 8.1.1. ★(*b*) Show that the cutoff frequencies for the higher-order, non-TEM modes in coaxial waveguide are given by (8.7.41) with $a = \frac{1}{2}d_2$ and u_{lm} defined as the mth root of

$$J_l'\left(\frac{d_1 u}{d_2}\right) N_l'(u) - J_l'(u) N_l'\left(\frac{d_1 u}{d_2}\right) = 0 \qquad \text{TE modes}$$

$$J_l\left(\frac{d_1 u}{d_2}\right) N_l(u) - J_l(u) N_l\left(\frac{d_1 u}{d_2}\right) = 0 \qquad \text{TM modes.}$$

8.7.17 A waveguide becomes a *resonant cavity* upon placing conducting walls at the two ends. Show that a resonance occurs when the length L is an integral number n of guide half-wavelengths $\lambda_g/2$; specifically,

$$\left(\frac{\omega}{c}\right)^2 = \left(\frac{l\pi}{a}\right)^2 + \left(\frac{m\pi}{b}\right)^2 + \left(\frac{n\pi}{L}\right)^2 \qquad \text{rectangular parallelepiped}$$

$$\left(\frac{\omega}{c}\right)^2 = \left(\frac{u_{lm}}{a}\right)^2 + \left(\frac{n\pi}{L}\right)^2 \qquad \text{right circular cylinder.}$$

Cavity modes, requiring three integral indices, are named TE$_{lmn}$ or TM$_{lmn}$. Make a *mode chart* for cylindrical cavities by plotting loci of resonances on a graph of $(d/L)^2$ against $(fd)^2$, where $d \equiv 2a$, $f \equiv \omega/2\pi$.†

8.7.18 Sketch the magnetic field lines for the two modes whose electric field lines are shown in Fig. 8.7.5. What arrangement of slots or holes could you cut in resonant cavities using these modes without seriously perturbing them?

8.8 *Propagation in Ionized Gases*

Consider the case of an ionized gas, or *plasma*, as a medium for electromagnetic waves. We adopt a simplified model in which the free electrons are regarded as independent mobile particles whereas the heavy positive ions are assumed to

---

† Further discussion of the theory and practical application of cavities is given by Montgomery, *op. cit.*, chap. 5, and Bell Laboratories Staff, *op. cit.*, pp. 909–1020.

remain at rest. Collisions of the electrons with the positive ions and neutral molecules are represented as a continuous viscous drag force. The plasma is macroscopically neutral.

Although the electrons are to be treated as independent particles, we must still include the coulomb forces in a self-consistent manner through the collective electron space charge. For instance, if an initially uniform plasma is locally "plucked," i.e., perturbed by external forces and then released, the electrostatic space-charge forces act to restore the electrons to their equilibrium positions, but their inertia causes them to oscillate about the equilibrium. For a one-dimensional perturbation, this oscillation takes place at the (angular) frequency (Prob. 8.8.2)

$$\omega_p = \left(\frac{ne^2}{\epsilon_0 m}\right)^{1/2}, \tag{8.8.1}$$

where n is the number of electrons (mass m, charge $-e$) per unit volume. This frequency, known as the electron *plasma frequency*, is a fundamental parameter of the physics of ionized gases. We may think of it as the natural resonance of a plasma arising from the "spring constant" of coulomb space-charge forces together with the electrons' mass. Aside from universal constants, the plasma frequency depends only upon the particle density n of electrons.

A second parameter needed to specify the properties of an ionized gas as an electromagnetic medium is one that measures the viscous friction experienced by the free electrons arising from their collisions with heavy particles. The *collision frequency for momentum transfer* v denotes the average number of times per second that an electron originally moving in a certain direction is deflected through 90° as a result of collisions. This phenomenological constant depends in a complicated way on the particle density, temperature, and molecular species of the plasma.

Assume that a particular electron at position (x,y,z) is acted upon by a time-varying electric field E polarized in the x direction. Newton's second law gives the equation of motion

$$m\frac{d^2\xi}{dt^2} = -eE - vm\frac{d\xi}{dt}, \tag{8.8.2}$$

where ξ is the x-direction displacement of the electron from its reference position and the term $vm\,d\xi/dt$ represents the average rate of loss of momentum through collisions with heavy particles. If the electric field is oscillatory in time, $E = E_0 e^{j\omega t}$, the steady-state solution of (8.8.2) is easily obtained upon replacing the operator d/dt by $j\omega$,

$$\xi = \frac{eE}{m\omega(\omega - jv)}. \tag{8.8.3}$$

The fact that ξ is complex means that the electron's oscillatory motion is not in phase with the driving electric field (Prob. 8.8.1). For n electrons per unit volume, each undergoing the same steady-state motion, the current density is

$$J = -ne \frac{d\xi}{dt} = \frac{ne^2 E}{m(\nu + j\omega)} \tag{8.8.4}$$

and the effective conductivity of the plasma medium is

$$\breve{g} = \frac{J}{E} = \frac{ne^2}{m(\nu + j\omega)} = \frac{\omega_p^2 \epsilon_0}{\nu + j\omega}. \tag{8.8.5}$$

Since (8.8.4) is a linear, isotropic relation, it applies to fields of any polarization. Except for the fact that the conductivity (8.8.5) is complex and frequency-dependent, the plasma is electrically similar to the ohmic conducting medium discussed in Sec. 8.5. The wave equation (8.5.8) may now be rewritten in the Helmholtz form†

$$\nabla^2 \mathbf{E} + \frac{\omega^2}{c^2} \left[1 - \frac{ne^2}{\epsilon_0 m \omega(\omega - j\nu)} \right] \mathbf{E} = 0. \tag{8.8.6}$$

Plane wave solutions exist of the form (8.5.9)

$$\mathbf{E} = \mathbf{i} E_0 \exp j(\omega t - \breve{\kappa} z) = \mathbf{i} E_0 e^{\kappa_i z} e^{j(\omega t - \kappa_r z)}, \tag{8.8.7}$$

where the complex wave number, replacing (8.5.10), is

$$\breve{\kappa} \equiv \kappa_r + j\kappa_i = \frac{\omega}{c} \left[1 - \frac{ne^2}{\epsilon_0 m \omega(\omega - j\nu)} \right]^{1/2}$$

$$= \frac{\omega}{c} \left[1 - \frac{\omega_p^2}{\omega(\omega - j\nu)} \right]^{1/2}. \tag{8.8.8}$$

As in Sec. 8.5, a complex wave number signifies that the wave is damped.

Let us consider as limiting cases three frequency regions set apart by the two parameters that specify the properties of the plasma, namely, the collision frequency ν and the plasma frequency ω_p. For simplicity, we assume $\nu \ll \omega_p$. Table 8.2 gives the limiting forms of the wave number (8.8.8) for the three cases (Prob. 8.8.3). At low frequencies, below the collision frequency, the wave number has the form of (8.5.13), and the ionized gas behaves exactly like a metallic conductor with skin depth $\delta = (2\nu/\omega)^{1/2} c/\omega_p$. At intermediate frequencies, between ν and ω_p, the ionized gas is analogous to a waveguide below its cutoff frequency (8.7.19). The wave is evanescent; its amplitude decays with the con-

† In putting $\nabla \cdot \mathbf{E} = 0$ in expanding the double curl of $\mathbf{E}$, we have tacitly restricted the generality of (8.8.6). It is valid for plane TEM waves in a homogeneous plasma, i.e., where n is independent of position. Also we have assumed $\kappa_e = \kappa_m = 1$.

TABLE 8.2   Approximate Wave Numbers for the Characteristic Frequency Regions of a Plasma

Low $\omega \ll \nu$	Intermediate $\nu \ll \omega \ll \omega_p$	High $\omega \gg \omega_p$
$\breve{\kappa} \to (1-j)\left(\dfrac{\omega}{2\nu}\right)^{1/2}\dfrac{\omega_p}{c}$	$\breve{\kappa} \to -j\dfrac{\omega_p}{c}$	$\breve{\kappa} \to \dfrac{\omega}{c}\left(1-\dfrac{\omega_p^2}{\omega^2}\right)^{1/2}$
(Conducting)	(Evanescent)	(Dielectric)

stant attenuation length c/ω_p. At high frequencies, above the plasma frequency, the ionized gas has the properties of a low-loss dielectric of relative permittivity $\kappa_e = 1 - \omega_p^2/\omega^2$ (see Prob. 8.8.4). If the assumption $\nu \ll \omega_p$ is broken, the intermediate-frequency evanescent case disappears.

An illustration of these results is the effect on radio waves of the blanket of ionization (the Kennelly-Heaviside layer) existing in the upper atmosphere. A wave incident on this layer from below at a large angle of incidence (small glancing angle) goes faster and faster as it encounters increasing electron density with height. The wave is thus refracted back toward the earth, i.e., reflected (see Fig. 8.8.1), accounting qualitatively for the long-distance transmission of radio waves, as well as for the "skip distance" when the angle of incidence is too small for the wave to be returned to earth by the refraction. Sufficiently high-frequency waves (typically >40 MHz) are above the ionosphere's maximum plasma frequency and are not reflected. Intense bursts of solar wind (magnetic storms) cause the ionization to extend to lower altitudes, where the collision frequencies are higher, so that waves are heavily absorbed during reflection. The intense ionization surrounding a supersonic space vehicle, pro-

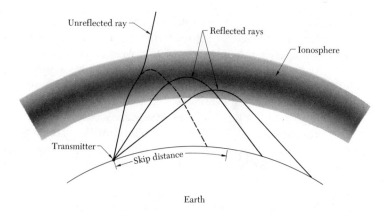

Fig. 8.8.1   Refraction of radio waves by the ionosphere.

duced by the shock phenomenon discussed in Sec. 5.10, causes the well-known reentry blackout of radio communications.

Our simplified treatment can be extended in various ways.† First, most plasmas of practical interest (including the ionosphere) are immersed in a static magnetic field, which causes the conductivity (8.8.5) to be anisotropic, i.e., different for oscillating electric fields polarized parallel and transverse to the magnetic field. The resulting problem of wave propagation is quite complex (see Prob. 8.8.6). Second, careful treatment of statistical averaging over the thermal velocity distribution of the electrons permits evaluation of the collision frequency ν from more fundamental parameters and brings in interesting effects that occur when thermal speeds are comparable with the wave phase velocity. Finally, other types of waves are possible, including an electroacoustic longitudinal (irrotational) mode and waves whose propagation characteristics are controlled largely by the positive ions rather than by the electrons.

Problems

8.8.1 Show that (8.8.3) can be written as

$$\xi = \frac{eE_0}{m\omega(\omega^2 + \nu^2)^{1/2}} \cos(\omega t + \phi)$$

for $E = E_0 \cos\omega t$, where $\tan\phi = \nu/\omega$.

8.8.2 Consider an ionized gas of uniform electron density n. Regard the positive ions as a smeared-out continuous fluid which renders the gas macroscopically neutral and through which the electrons can move without friction. Now assume that, by some external means, each electron is shifted in the x direction by the displacement $\xi = \xi(x)$, a function of its initial, unperturbed location x. (a) Show from Gauss' law (8.2.1) that the resulting electric field is $E_x = ne\xi/\epsilon_0$. (b) Show that each electron experiences a linear (Hooke's law) restoring force such that when the external forces are removed, it oscillates about the equilibrium position $\xi = 0$ with simple harmonic motion at the angular frequency

$$\omega_p = \left(\frac{ne^2}{\epsilon_0 m}\right)^{1/2},$$

which is known as the electron *plasma frequency*.

† For fuller discussion, see J. A. Ratcliffe, "The Magneto-ionic Theory and Its Applications to the Ionosphere," Cambridge University Press, New York, 1959; M. A. Heald and C. B. Wharton, "Plasma Diagnostics with Microwaves," John Wiley & Sons Inc., New York, 1965; I. P. Shkarofsky, T. W. Johnson, and M. P. Bachynski, "Particle Kinetics of Plasma," Addison-Wesley Publishing Company, Reading, Mass., 1965.

8.8.3 Carry out the approximations of the complex wave number (8.8.8) to obtain the results stated in Table 8.2. In both the low- and intermediate-frequency cases, the wave is rather strongly attenuated. How does the evanescent wave for $\nu \ll \omega < \omega_p$ differ from the lossy wave for $\omega \lesssim \nu$? *Answer:* Lossy wavelength $\ll \lambda_0 \equiv 2\pi c/\omega$; evanescent wavelength $\gtrsim \lambda_0$.

8.8.4 Show that the skin depth (attenuation distance) for a high-frequency wave ($\omega > \omega_p$) is approximately

$$\delta \equiv -\frac{1}{\kappa_i} \approx \frac{c}{\omega_p}\left(\frac{2\omega^2}{\nu\omega_p}\right)\left(1 - \frac{\omega_p{}^2}{\omega^2}\right)^{1/2}.$$

8.8.5 Use the formalism of Prob. 8.5.6a to establish that an ionized gas can be described by the complex permittivity

$$\check{\kappa}_e = 1 - \frac{ne^2}{\epsilon_0 m\omega(\omega - j\nu)}.$$

★8.8.6 Consider a plasma of electron density n immersed in a uniform static magnetic field $\mathbf{B}_0$. Let $\mathbf{B}_0$ be in the z direction. Revise the equation of motion (8.8.2) to include the Lorentz force $q(\mathbf{v} \times \mathbf{B}_0)$ on the electrons (but drop the collision term for simplicity); write out the resulting equation in cartesian components. (*a*) Show that plane waves propagating in the x direction, say, but with the electric field polarized parallel to $\mathbf{B}_0$, are unaffected by the presence of $\mathbf{B}_0$. (*b*) Show that *circularly polarized* plane waves (see Prob. 8.3.5) can propagate in the z direction with wave numbers

$$\kappa = \frac{\omega}{c}\left[1 - \frac{\omega_p{}^2}{\omega(\omega \pm \omega_b)}\right]^{1/2},$$

where $\omega_b \equiv eB_0/m$ is the cyclotron frequency. (*c*) Show that a TM wave can propagate in the x direction with the wave *magnetic* field polarized parallel to $\mathbf{B}_0$, with the wave number

$$\kappa = \frac{\omega}{c}\left[1 - \frac{\omega_p{}^2(\omega^2 - \omega_p{}^2)}{\omega^2(\omega^2 - \omega_p{}^2 - \omega_b{}^2)}\right]^{1/2}.$$

★8.8.7 (*a*) Show that the anisotropic plasma of Prob. 8.8.6 can be described by a tensor (or dyadic) conductivity such that $\mathbf{J} = \mathbf{g} \cdot \mathbf{E}$, and hence by a tensor permittivity $\varkappa_e = \mathbf{1} - j\mathbf{g}/\epsilon_0\omega$, where $\mathbf{1}$ is the unit dyadic (7.1.17). (*b*) Show from Maxwell's equations that for monochromatic plane waves in such an anisotropic medium, the fields $\mathbf{D} = \varkappa_e \cdot \mathbf{E}$, $\mathbf{B}$, and $\mathbf{H}$ are transverse to the direction of propagation but the $\mathbf{E}$ field need not be. Thus the waves are TM but not TEM, in general.

8.9 Spherical Waves

As a final example of a nontrivial solution of the electromagnetic wave equations (8.2.8) and (8.2.9), we look for functions representing waves propagating outward from a source point and having a high degree of spherical symmetry.

Recall that *scalar* spherical waves of sound have been discussed in Sec. 5.5. *Vector* spherical waves are considerably more complicated. The procedure set forth in Sec. 8.7 provides a *general* solution of the vector wave equation and can be used *in principle* to express any possible solution including spherical waves. In practice, however, it is prohibitively awkward for most spherical-wave problems.

The method of Sec. 8.7 consists of imposing appropriate boundary conditions on a linear combination of two independent families of solutions of the form (8.7.4) and (8.7.5),

$$\mathbf{E} = \nabla \times (\hat{\mathbf{a}}\psi) \qquad\qquad \text{TE modes} \qquad\qquad (8.9.1)$$

$$\mathbf{E}' = \nabla \times (\nabla \times \hat{\mathbf{a}}\psi')$$

or $\qquad\qquad\qquad\qquad\qquad$ TM modes,† $\qquad\qquad\qquad$ (8.9.2)

$$\mathbf{B}' = \nabla \times (\hat{\mathbf{a}}\psi'')$$

where $\hat{\mathbf{a}}$ is a fixed unit vector and ψ, and ψ' or ψ'', are solutions of the corresponding scalar wave equation. This formulation is particularly effective in problems having *cylindrical* symmetry, i.e., where the boundary conditions are independent of one fixed direction, as in the waveguides considered in Sec. 8.7. For waves propagating in the preferred direction, there is then a natural choice for the unit vector $\hat{\mathbf{a}}$. The required boundary conditions on $\mathbf{E}$ (and $\mathbf{B}$) reduce to simple boundary conditions on the scalar wavefunction ψ (Prob. 8.7.7). For problems with general *spherical* symmetry, however, there is no preferred fixed direction to orient $\hat{\mathbf{a}}$, and most boundary conditions do not conveniently fit solutions of the form (8.9.1) and (8.9.2). Thus another formulation is necessary for a tractable general solution of the spherical problem (Prob. 8.9.4). Nevertheless, (8.9.1) and (8.9.2) are useful for the special case of radiation from point dipole sources, to which we limit our discussion.

In Sec. 5.5, we found the particular solution (5.5.6) of the scalar wave equation representing a spherically symmetric, outward-traveling wave, namely,

$$\psi = \frac{A}{r} e^{j(\omega t - \kappa r)}, \qquad\qquad (8.9.3)$$

where $\kappa = \omega/c$. Taking $\hat{\mathbf{a}} = \mathbf{k}$ aligned with the polar axis ($\theta = 0$) of a spherical coordinate system, we may substitute (8.9.3) in the first form of (8.9.2) to obtain

$$\mathbf{E} = \nabla \times [\nabla \times \mathbf{k}\psi(r)]$$
$$= \nabla(\nabla \cdot \mathbf{k}\psi) - \nabla^2(\mathbf{k}\psi)$$
$$= \nabla \frac{\partial \psi}{\partial z} + \mathbf{k}\kappa^2\psi. \qquad\qquad (8.9.4)$$

---

† By (8.7.7), the alternate forms of (8.9.2) are equivalent for time-harmonic waves, with a simple proportionality between ψ' and ψ''.

But now using

$$\frac{\partial \psi}{\partial z} = \frac{d\psi}{dr}\frac{\partial r}{\partial z} = \frac{d\psi}{dr}\cos\theta \tag{8.9.5}$$

$$\mathbf{k} = \hat{\mathbf{r}}\cos\theta - \hat{\boldsymbol{\theta}}\sin\theta \tag{8.9.6}$$

$$\boldsymbol{\nabla} = \hat{\mathbf{r}}\frac{\partial}{\partial r} + \hat{\boldsymbol{\theta}}\frac{1}{r}\frac{\partial}{\partial \theta} + \hat{\boldsymbol{\phi}}\frac{1}{r\sin\theta}\frac{\partial}{\partial \phi}, \tag{8.9.7}$$

we obtain:

Electric dipole:

$$E_r = \left(\frac{d^2\psi}{dr^2} + \kappa^2\psi\right)\cos\theta = (1 + j\kappa r)\frac{2A}{r^3}\cos\theta\, e^{j(\omega t - \kappa r)}$$

$$E_\theta = -\left(\frac{1}{r}\frac{d\psi}{dr} + \kappa^2\psi\right)\sin\theta = (1 + j\kappa r - \kappa^2 r^2)\frac{A}{r^3}\sin\theta\, e^{j(\omega t - \kappa r)} \tag{8.9.8}$$

$$E_\phi = 0.$$

The associated magnetic field can readily be found by the following manipulations,

$$\mathbf{B} = \frac{j}{\omega}\boldsymbol{\nabla}\times\mathbf{E} = \frac{j\kappa^2}{\omega}\boldsymbol{\nabla}\times(\mathbf{k}\psi) = -\frac{j\kappa^2}{\omega}\mathbf{k}\times\boldsymbol{\nabla}\psi = -\frac{j\kappa^2}{\omega}\sin\theta\frac{d\psi}{dr}\hat{\boldsymbol{\phi}};$$

$$\tag{8.9.9}$$

that is,

$$B_r = 0$$

$$B_\theta = 0 \tag{8.9.10}$$

$$B_\phi = -\frac{j\kappa^2}{\omega}\frac{d\psi}{dr}\sin\theta = (j\kappa r - \kappa^2 r^2)\frac{A}{cr^3}\sin\theta\, e^{j(\omega t - \kappa r)}.$$

The interpretation of this particular solution becomes clear upon looking at the limiting forms for small and large distances r. In the limit $\kappa r \ll 1$, the fields reduce to the quasi-static field

$$\mathbf{E} = \frac{3(\hat{\mathbf{r}}\cdot\mathbf{p})\hat{\mathbf{r}} - \mathbf{p}}{4\pi\epsilon_0 r^3} = \hat{\mathbf{r}}\frac{2p\cos\theta}{4\pi\epsilon_0 r^3} + \hat{\boldsymbol{\theta}}\frac{p\sin\theta}{4\pi\epsilon_0 r^2} \tag{8.9.11}$$

of an oscillating electric dipole of moment

$$\mathbf{p} \equiv \mathbf{k}p = \mathbf{k}p_0 e^{j\omega t} = \mathbf{k}4\pi\epsilon_0 A\, e^{j\omega t}. \tag{8.9.12}$$

At large distances, where $\kappa r \gg 1$, the fields become

$$\mathbf{E} \rightarrow - \hat{\boldsymbol{\theta}} \frac{\kappa^2 A}{r} \sin\theta \, e^{j(\omega t - \kappa r)}$$

$$\mathbf{B} \rightarrow - \hat{\boldsymbol{\phi}} \frac{\kappa^2 A}{cr} \sin\theta \, e^{j(\omega t - \kappa r)}.$$

(8.9.13)

The *far*, or *radiation*, fields (8.9.13) fall off only as the inverse first power of the radius, so that the Poynting vector (8.4.8) is inverse square. Hence the total energy passing through a spherical surface is independent of its radius (Prob. 8.9.1). The particular solution (8.9.8) and (8.9.10) thus satisfies the boundary condition of an oscillating *electric-dipole source* at the origin and the so-called *radiation condition* of outward-traveling energy-carrying spherical waves at large distances. The special case we have examined is an example of the connection between an elementary point source of electromagnetic waves, such as an atom or a small radio antenna, and a spherical wave, which in the limit of great distances becomes the plane wave of Sec. 8.3. The companion solution of the form (8.9.1) represents radiation by an oscillating magnetic dipole (Prob. 8.9.3). The two dipole cases are identical in their $\sin^2\theta$ radiation patterns but differ in the polarization of the fields. At intermediate distances, the electric-dipole fields may be seen to be TM, while the magnetic-dipole fields are TE. A discussion of radiation by higher-order sources (oscillating quadrupoles, etc.) and of other boundary conditions (standing-wave modes in spherical cavities, for instance) requires that a more general solution be found for the vector wave equation in spherical geometry (Prob. 8.9.4 and 8.9.5).

Problems

8.9.1 Show that the time-average Poynting vector for the far fields (8.9.13) of an oscillating electric dipole (8.9.12) is

$$\bar{\mathbf{s}} = \hat{\mathbf{r}} \frac{c\kappa^4 p_0^2}{32\pi^2\epsilon_0} \frac{\sin^2\theta}{r^2}$$

and that the average total power radiated by the oscillating dipole is

$$\left(\frac{\overline{dW}}{dt} \right) = \frac{c\kappa^4 p_0^2}{12\pi\epsilon_0}.$$

Why is the sky blue and the sunset red?

8.9.2 Consider an electric dipole consisting of a charge $-e$ oscillating sinusoidally in position about a stationary charge $+e$. Show that the instantaneous total power radiated can be written in the form

$$\frac{dW}{dt} = \frac{e^2[a]^2}{6\pi\epsilon_0 c^3}$$

where $[a] = a_0 e^{j(\omega t - \kappa r)}$ is the instantaneous (retarded) acceleration of the moving charge. Since this result does not depend upon the oscillator frequency, and since by Fourier analysis, an arbitrary motion can be described by superposing many sinusoidal motions of proper frequency, amplitude, and phase, this rate-of-radiation formula has general validity for any accelerated charge (in the nonrelativistic limit $v \ll c$).

8.9.3 Substitute (8.9.3) in (8.9.1) to find the spherical wave corresponding to an oscillating magnetic dipole (current loop) of moment $m_0 e^{j\omega t}$, namely,

$$E_\phi = (-j\kappa r + \kappa^2 r^2)\frac{Z_0 m_0}{4\pi\epsilon_0 r^3}\sin\theta \; e^{j(\omega t - \kappa r)}$$

$$B_r = (1 + j\kappa r)\frac{\mu_0 m_0}{2\pi r^3}\cos\theta \; e^{j(\omega t - \kappa r)}$$

$$B_\theta = (1 + j\kappa r - \kappa^2 r^2)\frac{\mu_0 m_0}{4\pi r^3}\sin\theta \; e^{j(\omega t - \kappa r)}.$$

★8.9.4 Show that

$$\mathbf{E} = \nabla \times (\mathbf{r}\psi) = -\mathbf{r} \times \nabla\psi$$

is a solenoidal solution of the vector wave equation (8.7.1) such that $\mathbf{E}$ is everywhere tangential to a spherical boundary. Show that

$$\mathbf{E}' = \nabla \times (\nabla \times \mathbf{r}\psi') \qquad \text{or} \qquad \mathbf{B}' = \nabla \times \mathbf{r}\psi''$$

is also a solution, with tangential $\mathbf{B}$. Show that in either case the $\mathbf{E}$ and $\mathbf{B}$ fields are orthogonal. (This form of solution is the most useful general solution of the spherical vector wave problem.†)

★8.9.5 Find the general solution of the scalar wave equation in spherical coordinates by separation of variables. [The radial functions are called *spherical Bessel functions* z, related to ordinary Bessel functions Z of half-integral order by

$$z_l(\kappa r) = \left(\frac{\pi}{2\kappa r}\right)^{1/2} Z_{l+\frac{1}{2}}(\kappa r)$$

The polar-angle functions are the *associated Legendre polynomials* $P_l^m(\cos\theta)$.]

† See Panofsky and Phillips, *op. cit.*, pp. 229–233.

nine

Wave Propagation in Inhomogeneous and Obstructed Media

In most of our study of wave phenomena so far, we have considered problems that can be solved more or less exactly. Our study has perhaps given the impression that we should be able to solve any wave problem with precision, given sufficient mathematical acumen and perhaps the services of a large computer. Though this happy state of affairs may exist in principle, the task of describing wave motion in detail is a formidable one in all but a relatively few simple cases. In the real world of physics and engineering, most of the interesting problems raise such difficulties that only approximate solutions can be found.

The analysis of wave phenomena consists of first finding the appropriate partial differential equation and then solving the equation subject to the required boundary conditions. To obtain a tractable differential equation, one frequently omits terms representing minor physical processes. For instance, in our lengthy discussion of waves on a stretched string, we generally ignored the stiffness of the string and the coupling between transverse and longitudinal motion (Sec. 1.10). This kind of small effect can be put back in by *perturbation analysis*, a powerful method of successive approximations discussed in Secs. 4.4 and 4.9. Also, one usually restricts consideration to small-amplitude displacements in order to avoid a nonlinear differential equation, e.g., as in Secs. 1.1 and 6.2. Even a relatively simple differential equation may be solvable only for certain simple boundary conditions. For example, in Chap. 2 we found the normal modes for membranes with rectangular and circular boundaries, but we should be hard pressed to handle a case with an irregular boundary.

In this chapter we consider two broad classes of problems in which geometric complexities prevent an exact solution in all but very special cases. In the first, the properties of the wave-supporting structure or medium vary with position, e.g., a stretched string of varying density or an optical medium of varying refractive index. Such a medium is said to be *inhomogeneous*. In the second class, the wave propagates in a medium containing obstructions of one sort or another. In particular we are interested in the *diffraction* of waves around obstacles or through apertures. Both classes of problem help to link the concept of wave propagation to the model of *geometrical optics*, with its well-known features of rectilinear ray propagation, reflection, and refraction.

Our direct concern will be with traveling waves, propagating or *radiating* from a source to an observation point. If desired, standing-wave solutions can be constructed by superposition.

9.1 The WKB Approximation

In Sec. 1.9, we considered a wave incident upon the junction between two string segments of differing mass density. The reflected and transmitted components of this incident wave could then be found by invoking boundary conditions, which expressed the physical requirements of continuity and Newton's third law. Analogous problems were discussed in Secs. 5.6 (acoustic waves) and 8.6 (electromagnetic waves) and in Probs. 2.4.3, 4.1.2, and 4.2.4. A general conclusion is that an *abrupt discontinuity* in the properties of a wave-supporting structure or medium causes a (partial) reflection of an incident wave. By a straightforward (but rather tedious) extension of the analysis of Sec. 1.9, one may find the overall reflection and transmission coefficients for two or more abrupt discontinuities of prescribed separation. Optical thin-film interference is a familiar example, e.g., oil film on water, Newton's rings, coated lenses.

Now, however, we postulate that the properties of the structure change *continuously*. For definiteness, consider a flexible stretched string of linear density $\lambda_0(x)$. Repetition of the arguments of Sec. 1.1 shows that the displacement

η of the string from its equilibrium obeys the wave equation

$$\frac{\partial^2 \eta}{\partial x^2} = \frac{1}{[c(x)]^2} \frac{\partial^2 \eta}{\partial t^2},$$ (9.1.1)

where the velocity c is now a function of position along the string,

$$c(x) = \left[\frac{\tau_0}{\lambda_0(x)} \right]^{1/2}.$$ (9.1.2)

Equation (9.1.1) is far more difficult to solve than (1.1.3). Our present aim is to find an approximate solution to (9.1.1) for an arbitrary given function $c(x)$.

Parenthetically, we recall that in Sec. 4.3 we considered longitudinal waves in a rod of variable cross section, obtaining the somewhat different wave equation (4.3.2). Thus, we are reminded that (9.1.1) is not the only wave equation that can arise in one-dimensional inhomogeneous-medium problems (see Prob. 9.1.1). However, most such problems can be stated in such a way that (9.1.1) is indeed the equation to be solved.

Our previous discussion (Chap. 1) assumed a constant velocity and led to the traveling-wave solution

$$\eta = A \cos(\kappa x - \omega t),$$ (9.1.3)

where $\kappa = \omega/c$. If the properties of the medium are slowly varying (a qualification we soon make more precise), we may guess that the solution of (9.1.1) will look very much like (9.1.3) but perhaps with the quantities A and κ varying "slowly" with position; i.e., we assume a solution of the form

$$\eta(x,t) = A(x) \cos[S(x) - \omega t].$$ (9.1.4)

Substituting (9.1.4) in (9.1.1), we obtain

$$\left[\frac{d^2A}{dx^2} - \left(\frac{dS}{dx} \right)^2 A \right] \cos[S(x) - \omega t] - \left(2 \frac{dS}{dx} \frac{dA}{dx} + \frac{d^2S}{dx^2} A \right) \sin[S(x) - \omega t]$$
$$= -\frac{\omega^2}{c^2(x)} A \cos[S(x) - \omega t]. \quad (9.1.5)$$

But now if (9.1.5) is to hold at all positions and times, the coefficients of the sine and cosine terms must separately vanish. Thus we find that the unknown functions $A(x)$ and $S(x)$ satisfy the coupled differential equations

$$\frac{d^2A}{dx^2} + \left[\frac{\omega^2}{c^2(x)} - \left(\frac{dS}{dx} \right)^2 \right] A(x) = 0$$ (9.1.6)

$$2 \frac{dS}{dx} \frac{dA}{dx} + \frac{d^2S}{dx^2} A(x) = 0.$$ (9.1.7)

At first sight, these equations appear to be even less tractable than (9.1.1). However, d^2A/dx^2 in (9.1.6) would be zero in a uniform medium and would remain small in a slowly varying medium. Therefore, to a first approximation,

$$\frac{dS}{dx} \approx \frac{\omega}{c(x)} \equiv \kappa(x), \tag{9.1.8}$$

$$S \approx \int_{x_0}^{x} \frac{\omega}{c(x)}\, dx. \tag{9.1.9}$$

Then, substitution of (9.1.8) in (9.1.7) gives (Prob. 9.1.3)

$$A(x) = A(x_0) \left[\frac{c(x)}{c(x_0)} \right]^{1/2}, \tag{9.1.10}$$

where x_0 is simply some reference position at which the wave has a prescribed amplitude. Thus the final result of our first-approximation solution to (9.1.1) is

$$\eta(x,t) \approx A(x_0) \left[\frac{c(x)}{c(x_0)} \right]^{1/2} \cos\left[\int_{x_0}^{x} \frac{\omega}{c(x)}\, dx - \omega t \right]. \tag{9.1.11}$$

That is, the amplitude changes in proportion to the square root of the local wave velocity and, in the spatial phase factor, κx is replaced by $\int \kappa(x)\, dx$.

We may now inquire how valid this approximate solution is. Let the (unknown) exact derivative (9.1.8) be written as

$$\frac{dS}{dx} \equiv \frac{\omega}{c(x)} [1 + \epsilon(x)]. \tag{9.1.12}$$

Substitution of (9.1.10) and (9.1.12) into (9.1.6) gives

$$\epsilon \left(1 + \frac{\epsilon}{2} \right) = -\frac{1}{8\omega^2} \left(\frac{dc}{dx} \right)^2 + \frac{c}{4\omega^2} \frac{d^2c}{dx^2}. \tag{9.1.13}$$

In a slowly varying medium, the term containing the second derivative d^2c/dx^2 will generally be negligible compared with that containing $(dc/dx)^2$. Then our solution (9.1.11) is a good approximation provided

$$|\epsilon| \approx \frac{1}{8\omega^2} \left(\frac{dc}{dx} \right)^2 \ll 1. \tag{9.1.14}$$

This condition becomes more meaningful when written in terms of wavelength $\lambda(x) = 2\pi c(x)/\omega$,

$$\left(\frac{\lambda}{c} \frac{dc}{dx} \right)^2 = \left(\frac{d\lambda}{dx} \right)^2 \ll 32\pi^2 \approx 18^2. \tag{9.1.15}$$

The first two terms in (9.1.15) are squares of the *relative changes* in wave velocity and wavelength, respectively, in the distance of *one "local" wavelength*. We conclude that our approximation is good up to the point that the properties of the medium change by a large fraction of themselves within a wavelength, i.e., when the concept of wavelength becomes rather meaningless.

This method for finding approximate solutions to wave equations for inhomogeneous media is usually called the *Wentzel-Kramers-Brillouin* (WKB) *approximation*. It applies when the inhomogeneities are gradual, i.e., in the limit opposite to the abrupt-discontinuity case of Secs. 1.9, 5.6, and 8.6. The intermediate case, when the *scale length* $c/(dc/dx)$ of the inhomogeneity is neither vanishingly small nor comparable to the wavelength, is much more difficult to treat.

The WKB method predicts that the wave conserves energy as it travels through the inhomogeneous medium and thus that no reflections are set up. To justify this statement, we recall from Sec. 1.8 that the force and particle velocity of the string are related by the wave impedance $Z_0 = (\lambda_0 \tau_0)^{1/2} = \tau_0/c$ and that the power transported by a traveling wave is given by $Z_0(\partial\eta/\partial t)^2$. Then we note that the space dependence of the impedance just cancels that of the square of the amplitude factor (9.1.10). Hence the power carried by the wave is constant. The method fails to indicate the possibility of partial reflection, which is a characteristic of inhomogeneity except in this slowly varying limit.

Problems

9.1.1 In the case of a stretched string of varying density, show that the transverse force $-\tau_0(\partial\eta/\partial x)$ from (1.8.12) obeys the wave equation

$$\frac{\partial^2 F}{\partial x^2} + \frac{2}{c(x)}\frac{dc}{dx}\frac{\partial F}{\partial x} = \frac{1}{c^2(x)}\frac{\partial^2 F}{\partial t^2}.$$

9.1.2 Express the string displacement (9.1.3) in complex exponential form, substitute in (9.1.1), and show that (9.1.6) and (9.1.7) are obtained by separating the real and imaginary parts of the result.

9.1.3 Carry out the substitution and integration leading to (9.1.10). Show that (9.1.7) can be integrated directly to $A^2(ds/dx) = \text{const}$.

9.1.4 A stretched string, with ends fixed at $x = 0$ and l, has a varying lineal density

$$\lambda_0(x) = \alpha(1 + \beta x),$$

where α and β are constants. Find the frequency of the nth normal mode in the WKB approximation. How large can the inhomogeneity coefficient β be without violating the WKB limit? *Answer:* $\omega_n = \frac{3}{2}n\pi\beta c_\alpha/[(1 + \beta l)^{3/2} - 1]$ where $c_\alpha = (\tau/\alpha)^{1/2}$ and $\beta l \lesssim 3$.

9.1.5 Consider waves propagating in the xz plane of a *stratified* medium, for which $c = c(z)$ only), assuming solutions of the form

$$\psi = \phi(z)e^{i(\kappa_x x - \omega t)}.$$

Show that the WKB method can be used to find the function $\phi(z)$ when the prescribed $c(z)$ is slowly varying. What is the criterion for the validity of the WKB approximation in this case? (This geometry occurs in the propagation of radio waves in the ionosphere and acoustic waves in the ocean.)

9.1.6 From Sec. 8.8, the velocity of an electromagnetic wave in an ionized medium is (neglecting collisional damping)

$$c = \frac{c_0}{(1 - n_e/n_\omega)^{1/2}},$$

where c_0 is the velocity of light in vacuum, n_e is the number density of free electrons, and $n_\omega = \omega^2 \epsilon_0 m/e^2$ is a parameter set by the frequency ω of the wave. The figure illustrates a microwave *interferometer* for measuring the properties of a gas discharge. The signal from an oscillator is divided into two paths, one of which passes through the discharge and the other contains adjustable amplitude and phase controls. The two signals are then recombined and detected. If the controls are set for a null (destructive interference) when the gas sample is un-ionized, the shift in phase caused by ionization can be deduced from the meter reading. Show that the phase shift $\Delta\phi$ is related to the electron density by

$$\Delta\phi = \frac{\omega}{c_0} \int \left\{ 1 - \left[1 - \frac{n_e(x)}{n_\omega} \right]^{1/2} \right\} dx \xrightarrow[n_e \ll n_\omega]{} \frac{\omega}{2cn_\omega} \int n_e(x)\, dx.$$

Thus in the low-density limit the apparatus measures the total number of free electrons *per unit area* between the antennas.

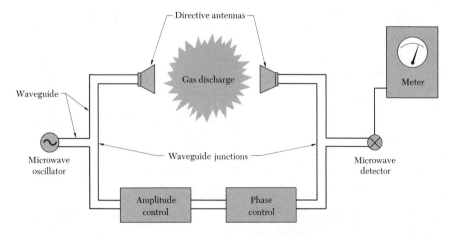

Prob. 9.1.6

★9.1.7 A uniform elastic medium is characterized by a wave impedance Z and a wave velocity c. Waves can travel in the x direction according to the ordinary one-dimensional wave equation. Now suppose that the properties of the medium vary so that both Z and c are slowly varying functions of x. (a) Show that the wave equation for the displacement becomes

$$\frac{\partial^2 \psi}{\partial x^2} + \frac{d \ln[Z(x)c(x)]}{dx} \frac{\partial \psi}{\partial x} = \frac{1}{c^2(x)} \frac{\partial^2 \psi}{\partial x^2}.$$

(b) Assume according to the WKB method a traveling-wave solution of the form

$$\psi = A(x)e^{i[S(x)-\omega t]}$$

and show that

$$A^2(x)Z(x)c(x)\frac{dS(x)}{dx} = \text{const}$$

and that approximately

$$\left(\frac{dS}{dx}\right)^2 = \frac{\omega^2}{c^2(x)} - \frac{1}{4}\left(\frac{d \ln Z}{dx}\right)^2 - \frac{1}{2}\left(\frac{d \ln Z}{dx}\right)\left(\frac{d \ln c}{dx}\right).$$

(c) Apply this result to the exponential horn discussed in Sec. 4.3 and verify that it gives the horn wave velocity (4.3.9).

9.2 Geometrical Optics

Waves that propagate in two or three dimensions may encounter inhomogeneity transverse, as well as parallel, to the direction of propagation. Parts of the wave, and the energy carried along, may then change direction. This is a more complicated problem than the purely one-dimensional case of the last section. An important and useful point of view is to describe what happens in terms of *rays*, rather than in terms of the wave motion itself. This is the model of *geometrical optics*. Although one usually associates this term with visible electromagnetic waves, we may apply it broadly to any kind of traveling wave (in two or three dimensions) in the limit that the wavelength is small compared with any dimensions of interest, such as the size of apertures or the spatial resolution of observations.

Consider a wave $\psi(x,y,z,t)$ that satisfies the scalar wave equation

$$\nabla^2 \psi = \frac{\partial^2 \psi}{\partial x^2} + \frac{\partial^2 \psi}{\partial y^2} + \frac{\partial^2 \psi}{\partial z^2} = \frac{1}{c^2(x,y,z)} \frac{\partial^2 \psi}{\partial t^2}, \tag{9.2.1}$$

where the wave speed c depends on position as the result of changes in properties of the medium. The notation can be somewhat simplified by introducing a generalized *refractive index*

$$n(x,y,z) \equiv \frac{c_0}{c(x,y,z)}, \tag{9.2.2}$$

where c_0 is some wave velocity adopted as a standard. In analogy with (9.1.4) and using the complex exponential notation, we assume a trial solution of the form

$$\psi(x,y,z,t) = A(x,y,z)\, e^{i[\kappa_0 S(x,y,z) - \omega t]},$$ (9.2.3)

where $\kappa_0 = \omega/c_0 = 2\pi/\lambda_0$ is the wave number in the standard medium and the functions $S(x,y,z)$ and $A(x,y,z)$ are to be determined.† The significance of the two spatial functions S and A may be understood by visualizing a snapshot of the wave motion at some instant of time. The points where the wave-supporting medium is at its equilibrium position, that is, $\psi = 0$, form a set of continuous surfaces, and interleaved between these surfaces is another set of surfaces where the wave disturbance is locally a maximum. These two sets of surfaces, and the continuum of surfaces that may be interpolated between them, are represented by constant values of $S(x,y,z)$. A set of *wavefronts* is simply a family of surfaces for which the values of $\kappa_0 S$ differ in increments of 2π. The A function, in contrast, is the envelope of the local maxima and in general varies over a wavefront. The distinction between envelope A and sinusoidal wiggle $e^{i\kappa_0 S}$ is clearest when A does not change greatly over the local wavelength, and it is precisely in this limit that the present line of attack is fruitful.

The function $S(x,y,z)$ is known as the *eikonal* (from Greek, for image). It has the dimensions of length and in fact may be recognized as a direct extension of the concept of optical path, familiar from elementary discussions of optics.

We now substitute our trial solution (9.2.3) into the wave equation (9.2.1), obtaining

$$\nabla^2 A + [n^2 - (\nabla S)\cdot(\nabla S)]\kappa_0^2 A = 0$$ (9.2.4)

$$2(\nabla S)\cdot(\nabla A) + (\nabla^2 S)A = 0.$$ (9.2.5)

These simultaneous equations bear a close similarity to (9.1.6) and (9.1.7). Accordingly, as a first approximation for a slowly varying medium, we may ignore the second-derivative term $\nabla^2 A$ in (9.2.4), obtaining the *eikonal equation*‡

$$|\nabla S|^2 = \left(\frac{\partial S}{\partial x}\right)^2 + \left(\frac{\partial S}{\partial y}\right)^2 + \left(\frac{\partial S}{\partial z}\right)^2 = n^2(x,y,z).$$ (9.2.6)

---

† The S function in (9.2.3) differs from that in (9.1.4) by the constant factor κ_0 simply to follow conventional notation.

‡ Note carefully that (9.2.6) is a differential equation of first order and second degree. The left-hand side is *not* the laplacian

$$\nabla^2 S = \frac{\partial^2 S}{\partial x^2} + \frac{\partial^2 S}{\partial y^2} + \frac{\partial^2 S}{\partial z^2},$$

which is of second order and first degree.

The solution $S(x,y,z)$ of (9.2.6) can in principle be found, given $n(x,y,z)$ and an initial surface $S = $ const. For example, for a uniform medium ($n = $ const) and a plane initial wave surface whose normal has the direction cosines α, β, γ, a solution is

$$S = n(\alpha x + \beta y + \gamma z). \tag{9.2.7}$$

It follows that

$$\nabla S = n\hat{\mathbf{s}}, \tag{9.2.8}$$

where $\hat{\mathbf{s}} = \mathbf{i}\alpha + \mathbf{j}\beta + \mathbf{k}\gamma$ is the unit vector with direction cosines α, β, γ. Again, for a uniform medium but an initial wave surface in the form of an infinite cylinder (or line), a solution is

$$S = n\rho \tag{9.2.9}$$
$$\nabla S = n\hat{\boldsymbol{\varrho}}, \tag{9.2.10}$$

where ρ is the radial cylindrical coordinate and $\hat{\boldsymbol{\varrho}}$ is the corresponding unit vector. Finally, for a uniform medium but a spherical (or point) initial wave surface, a solution is

$$S = nr \tag{9.2.11}$$
$$\nabla S = n\hat{\mathbf{r}}, \tag{9.2.12}$$

where r and $\hat{\mathbf{r}}$ are the radial coordinate and unit vector in spherical coordinates. More generally, we infer from (9.2.6) that

$$\nabla S = n(\mathbf{r})\hat{\mathbf{s}}(\mathbf{r}), \tag{9.2.13}$$

where $\hat{\mathbf{s}}$ is a unit vector and the functional argument $\mathbf{r}$ in place of (x,y,z) signifies dependence on three spatial coordinates but does not imply the particular choice of cartesian coordinates. It is a well-known property of the gradient operation (see Appendix A) that its vector direction is perpendicular to the surface $S = $ const. Thus we recognize the direction of the unit vector $\hat{\mathbf{s}}$ as being perpendicular to the wavefront, and continuous curves, called *rays*, may be constructed that are everywhere parallel to the local direction of $\hat{\mathbf{s}}$. In the three examples given, the rays are of course straight lines, a general result for uniform media.

Once $S(\mathbf{r})$ is known, the component of ∇A in the direction of $\hat{\mathbf{s}}$ can be found from (9.2.5). That is, we have the differential equation for $A(x,y,z)$

$$\frac{1}{A}\frac{dA}{ds} = -\frac{1}{2}\frac{\nabla \cdot (n\hat{\mathbf{s}})}{n}, \tag{9.2.14}$$

where s is the arc-length coordinate along a ray. Thus we are able to determine how the amplitude of our solution (9.2.3) changes along a ray but not tranverse to a ray. Indeed, there may even be discontinuities in A between adjacent rays.

The point of view of geometrical optics is that one deals directly with the ray trajectories, rather than finding them as a by-product of a solution of the wave equation for the eikonal function S and the resulting wavefronts. To eliminate S we look at the rate of change of the quantity $n\hat{s}$ along a ray, making repeated use of (9.2.13),

$$
\begin{aligned}
\frac{d}{ds}(n\hat{s}) &= \frac{d}{ds}(\nabla S) \\
&= \hat{s}\cdot\nabla(\nabla S) \\
&= \frac{\nabla S}{n}\cdot\nabla(\nabla S) \\
&= \frac{1}{2n}\nabla(\nabla S)^2 \\
&= \frac{1}{2n}\nabla n^2 \\
&= \nabla n.
\end{aligned}
\tag{9.2.15}
$$

Equation (9.2.15) is a differential equation that enables us to find the trajectory of a ray, given only the refractive index $n(\mathbf{r})$ and the initial direction $\hat{s}_0$ of the desired ray.

As a trivial example of the meaning of (9.2.15) consider a uniform medium, for which

$$
\frac{d\hat{s}}{ds} = 0.
\tag{9.2.16}
$$

It follows that $\hat{s} = \text{const}$ since $\hat{s}$ is a unit vector, so that the ray is simply a straight line. Next consider a more interesting example in which the refractive index varies only in the y direction and the ray lies initially in the xy plane making an angle θ_0 with the y axis (Fig. 9.2.1). The vector equation (9.2.15) resolves into the three scalar equations

$$
\frac{d}{ds}(n\sin\theta) = 0
\tag{9.2.17a}
$$

$$
\frac{d}{ds}(n\cos\theta) = \frac{dn}{dy}
\tag{9.2.17b}
$$

$$
\frac{d}{ds}(n\gamma) = 0,
\tag{9.2.17c}
$$

where $\hat{s} = \mathbf{i}\sin\theta + \mathbf{j}\cos\theta + \mathbf{k}\gamma$. Since $\hat{s}_0 = \mathbf{i}\sin\theta_0 + \mathbf{j}\cos\theta_0$, we thus obtain immediately

$$
n\sin\theta = \text{const} = n_0\sin\theta_0
\tag{9.2.18}
$$

$$
n\gamma = \text{const} = 0.
\tag{9.2.19}
$$

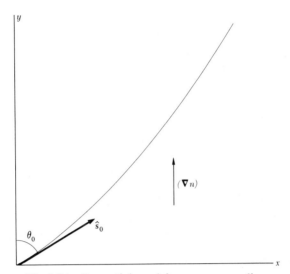

Fig. 9.2.1   Ray path in an inhomogeneous medium.

Equation (9.2.19) tells us that the ray remains in the xy plane; (9.2.18) may be recognized as Snell's law since at any point θ conforms to the usual definition of angle of incidence. Since $\hat{s}$ is a unit vector, (9.2.17b) is equivalent to (9.2.17a).

We have tried here to give a brief summary of how the familiar facts of geometrical optics can be derived as an approximate solution to a wave problem. The approximation came in neglecting the $\nabla^2 A$ term in (9.2.4). In a similar one-dimensional argument in the previous section we obtained the criterion (9.1.15), i.e., that the refractive index must change by only a small fraction of itself in the distance of a local wavelength for the approximation to be accurate. A similar criterion applies here, although the details are much more cumbersome because of the three-dimensionality. It is best summed up by saying that geometrical optics is the limiting case of wave optics as the wavelength tends toward zero.†

A final qualitative illustration is helpful. Consider the shadow cast by the edge of an opaque screen. The geometrical-optics picture is that the rays that hit the screen are absorbed while those that miss it continue on in straight lines. An observation screen beyond would thus show a uniformly illuminated region and a dark region, with a sharp boundary between the two. This description agrees with everyday experience if one does not examine the shadow boundary too closely. However, the refractive index does not vary "slowly" at the edge of the obstacle screen, and the $\nabla^2 A$ term becomes appreciable, with the result that in

† For further discussion of the connection between wave and ray optics see M. Kline and I. W. Kay, "Electromagnetic Theory and Geometrical Optics," Interscience Publishers, Inc., New York, 1965.

fact there is a gradual transition from dark to light at the observation screen, and even some interesting fine structure. We return to this problem when we discuss the Fresnel diffraction by a knife-edge in Chap. 11; we show there quantitatively how the geometrical-optics sharp shadow arises in the limiting case of vanishing wavelength. It is of interest to note here that classical (newtonian) mechanics is precisely the same limiting approximation to quantum (wave) mechanics as geometrical optics is to wave propagation.

The formulation of geometrical optics expressed by the ray equation (9.2.15) is identical with that predicted by *Fermat's principle:* the path followed by a ray in going from one point in space to another is the path that makes the time of transit of the associated wave stationary (usually a minimum). We give a brief discussion of the connection between (9.2.15) and Fermat's principle for readers familiar with the calculus of variations.†

The calculus of variations establishes that the differential equation

$$\frac{d}{d\sigma}\left(\frac{\partial F}{\partial g'}\right) - \frac{\partial F}{\partial g} = 0 \tag{9.2.20}$$

is the necessary and sufficient condition that the definite integral

$$I(g) \equiv \int_{\sigma_1}^{\sigma_2} F \, d\sigma \tag{9.2.21}$$

be an extremum, where $F \equiv F[\sigma,g(\sigma),g'(\sigma)]$ is a *given* function of the independent variable σ and of the *unknown* functions $g(\sigma)$ and $g'(\sigma) \equiv dg/d\sigma$, subject to the constraint that g has prescribed values at the limit points σ_1 and σ_2. Equation (9.2.20) is known as the *Euler differential equation* associated with stationary values of the integral (9.2.21). By an extremum or stationary value of the integral I, we mean that $I(g)$ is a minimum or maximum (or inflection point of zero slope) with respect to the choice of various functions $g(\sigma)$, having common end points. The function $g(\sigma)$ may readily be generalized to a *vector* function $\mathbf{g}(\sigma)$, with $\mathbf{g}'(\sigma) \equiv d\mathbf{g}/d\sigma$, whereupon (9.2.20) becomes a set of three differential equations, with $\partial/\partial g$ and $\partial/\partial g'$ replaced by $\partial/\partial g_x$ and $\partial/\partial g'_x$ in the first, and so on.

To summarize, the fundamental problem of the calculus of variations is to find the function $g(\sigma)$ [or $\mathbf{g}(\sigma)$] that makes the integral (9.2.21) an extremum; this goal is usually accomplished in practice by solving the Euler equation(s) (9.2.20).

Fermat's principle is of the form (9.2.21) with the integral I representing the elapsed time for a wave to go from one prescribed point to another. To

† For an introduction to the calculus of variations, see M. L. Boas, "Mathematical Methods in the Physical Sciences," chap. 8, John Wiley & Sons, Inc., New York, 1966. A thorough discussion is given by R. Courant and D. Hilbert, "Methods of Mathematical Physics," vol. 1, chap. 4, Interscience Publishers, Inc., New York, 1953.

establish the equivalence of Fermat's principle with the geometrical-optics limit of the wave equation, we must show that the ray equation (9.2.15) is the Euler equation for the Fermat integral. The transit time for a wave disturbance to pass along a path connecting point A to point B is evidently

$$\Delta t = \int_A^B dt = \int_A^B \frac{ds}{c(x,y,z)} = \frac{1}{c_0} \int_A^B n(x,y,z)\, ds, \qquad (9.2.22)$$

where $n(x,y,z)$ is the index of refraction defined by (9.2.2) and

$$ds = [(dx)^2 + (dy)^2 + (dz)^2]^{1/2}$$

is an element of length along the path. The value of Δt in general depends on the path chosen in evaluating the integral (9.2.22). Let us express the equation of the path in the parametric form

$$\begin{aligned} x &= x(\sigma) \\ y &= y(\sigma) \\ z &= z(\sigma), \end{aligned} \qquad (9.2.23)$$

where the parameter σ measures distance along any given path and is chosen, for convenience, such that $\sigma = 0$ at point A and $\sigma = 1$ at point B for all possible paths. Then (9.2.22) becomes

$$\Delta t = \int_0^1 F(x,y,z,x',y',z')\, d\sigma, \qquad (9.2.24)$$

where

$$F(x,y,z,x',y',z') = \frac{1}{c_0} n(x,y,z)(x'^2 + y'^2 + z'^2)^{1/2} \qquad (9.2.25)$$

with $x' \equiv dx/d\sigma$, $y' \equiv dy/d\sigma$, $z' \equiv dz/d\sigma$. The three Euler equations for the path that makes Δt have a stationary value are

$$\begin{aligned} \frac{d}{d\sigma}\left(\frac{\partial F}{\partial x'}\right) - \frac{\partial F}{\partial x} &= 0 \\[4pt] \frac{d}{d\sigma}\left(\frac{\partial F}{\partial y'}\right) - \frac{\partial F}{\partial y} &= 0 \\[4pt] \frac{d}{d\sigma}\left(\frac{\partial F}{\partial z'}\right) - \frac{\partial F}{\partial z} &= 0. \end{aligned} \qquad (9.2.26)$$

We find for the first Euler equation

$$\frac{d}{d\sigma}\left[\frac{nx'}{(x'^2 + y'^2 + z'^2)^{1/2}}\right] - (x'^2 + y'^2 + z'^2)^{1/2}\frac{\partial n}{\partial x} = 0, \qquad (9.2.27)$$

which may be rewritten in the form

$$\frac{d}{ds}(n\alpha) = \frac{\partial n}{\partial x}, \tag{9.2.28}$$

where $\alpha \equiv x'/(x'^2 + y'^2 + z'^2)^{1/2}$ is the direction cosine that an element of the path $ds = (x'^2 + y'^2 + z'^2)^{1/2}\,d\sigma$ makes with the x axis. We recognize (9.2.28) as being the x component of the differential equation (9.2.15) for the ray, as found from the eikonal function S. The other two Euler equations are the y and z components of (9.2.15). Thus the ray equation (9.2.15) and Fermat's principle are equivalent.

Problems

9.2.1   Obtain (9.2.15) by the following line of attack. Let $\hat{\mathbf{s}} \equiv \mathbf{i}\alpha + \mathbf{j}\beta + \mathbf{k}\gamma$. Use the operator identities $d/ds = \hat{\mathbf{s}} \cdot \nabla$ and **curl grad** $\equiv 0$ and the fact that $\hat{\mathbf{s}}$ is a unit vector to show that

$$\frac{d}{ds}(n\alpha) = \hat{\mathbf{s}} \cdot \frac{\partial}{\partial x}(n\hat{\mathbf{s}}) = \frac{\partial n}{\partial x},$$

and thus that $d(n\hat{\mathbf{s}})/ds = \nabla n$.

9.2.2   In the example sketched in Fig. 9.2.1, assume an index variation $n = (1 + ay)^{1/2}$. Find the trajectory of a ray that leaves the origin at the angle θ_0. *Answer:*

$$y + \frac{\cos^2\theta_0}{a} = \frac{a}{4\sin^2\theta_0}\left(x + \frac{2\sin\theta_0\cos\theta_0}{a}\right)^2 \qquad \text{a parabola.}$$

★**9.2.3**   Since the curl of a gradient vanishes identically, it follows from (9.2.13) that

$$\nabla \times (n\hat{\mathbf{s}}) = 0.$$

Apply Stokes' theorem (A.18) to a rectangular contour embracing the sharp interface between two media of refractive indices n_1 and n_2 to show that the components of $n\hat{\mathbf{s}}$ tangential to the interface are equal on the two sides and hence that Snell's law follows directly.

9.3   The Huygens-Fresnel Principle

We now return to the case of localized inhomogeneity in the form of opaque obstacles or apertures in an opaque screen. We assume monochromatic waves, or *radiation*, to be incident upon these discontinuities and ask what happens to the wave train after it passes them. The geometrical-optics approximation predicts that no radiation penetrates the geometrical "shadows," while otherwise the radiation is exactly that which would exist if the obstacles were not there.

However, from a wave point of view, we find it quite reasonable that the radiation can "bend around corners," although we have yet to see how to handle quantitatively this phenomenon of *diffraction*. Historically, most of the development of the theory of diffraction has been in the context of visible light. However, it must be recognized that the theory is applicable to any physical process that can be described by the ordinary wave equation in two or three spatial dimensions.

In this section, in contrast to our usual procedure, we develop the most effective way of tackling diffraction problems—the Huygens-Fresnel principle—as if it were an independent, empirical discovery. Then in the next section, we show how this principle follows deductively from the wave equation, just as in the previous section we showed how geometrical optics likewise follows from the wave equation. This reversed logic permits us to develop physical insight regarding diffraction before facing the more abstract arguments that tie Huygens' principle to the wave equation. It also happens to follow the historical development of the theory of diffraction, which played a major part in establishing the nature of light, one of the more interesting chapters in the history of physics.

Isaac Newton (1642–1727) considered, with reservations, that light consists of the flight of minute particles. The insight that rectilinear propagation (geometrical optics) is consistent with the wave equation in the limit of small wavelength is one of the smoothest pebbles that Newton failed to find on his beach.† At the time of Newton, Robert Hooke proposed that light is a wave motion, and shortly afterwards Christian Huygens developed the familiar Huygens construction, by which the position of a subsequent wavefront may be found by regarding each point of an earlier wavefront as a source of spherical secondary waves whose envelope constitutes the new wavefront. Newton's great prestige tended to block the acceptance of a wave theory of light, and it was not until 1801, when Thomas Young invoked the principle of interference to account for the colors of thin films, that the wave theory began to receive support again. Finally in 1818 Augustus Jean Fresnel combined Huygens' construction with Young's principle of interference to account both for the essentially rectilinear propagation of light and the diffraction phenomena that accompany it on a fine scale. At about this time Young proposed that light is a transverse wave, to account for polarization effects. As a result of both theoretical and experimental progress, especially by Fresnel, the wave theory of light was placed on a firm basis.

In 1861 James Clerk Maxwell discovered the possibility of electromagnetic waves and proposed that light is such a wave, rather than some sort of elastic wave in an "ether" per-

---

† "I do not know what I may appear to the world; but to myself I seem to have been only like a boy, playing on the sea-shore, and diverting myself, in now and then finding a smoother pebble or a prettier shell than ordinary, whilst the great ocean of truth lay all undiscovered before me." From manuscript of John Conduitt, Newton's nephew, in The Portsmouth Collection, University of Cambridge.

vading all space. He was able to compute the velocity of electromagnetic waves from laboratory electrical measurements and found excellent agreement with the measured velocity of light. After the origin of the quantum theory in 1900 by Max Planck, light again regained some of the particle attributes ascribed to it by Newton, especially with regard to its origin and absorption in matter. Its wave character continues to be necessary, however, for discussing diffraction and interference phenomena.

Huygens' construction is a geometric method for mapping the progress of the wavefronts, or surfaces of equal phase, of a traveling wave. Each point on a wavefront, at a particular instant, is considered to be the source of secondary wavelets that progress radially outward with a speed characteristic of the medium. At a later instant the wavefront is found by constructing the envelope of the multitude of secondary wavelets. Elementary applications of this construction include showing that the angle of incidence equals the angle of reflection for a plane mirror and deriving Snell's law of refraction at a plane boundary separating two media having different wave velocities. In this elementary form of Huygens' construction, the only physically significant portion of the secondary wavelet is that contributing to the envelope. Moreover, one simply ignores the second envelope that could be constructed on the source side of the given wavefront, an envelope which would imply a wave traveling back toward the source.

In order to modify Huygens' principle so that it is applicable to diffraction problems, which arise when the passage of waves is obstructed, it is necessary to discard the envelope construction and instead make use of all portions of each secondary wavelet. Specifically, at each observation point, one must superpose the contributions of all the Huygens wavelets emanating from points on the given wavefront, with due regard for their respective phases since the transit time to the observation point varies with the location of the Huygens source. This basic addition to Huygens' construction was made by Fresnel, and it enabled him to account quantitatively for most of the diffraction effects of importance in optics.

The Huygens-Fresnel principle can be given a mathematical formulation by considering the way in which a sinusoidal wave leaving a point source at P_s reaches a point of observation P_o, passing by obstacles or through apertures in the space between the two points. We suppose that the spherical wave leaving the point source at P_s has the form established in Sec. 5.5 for an outward-going spherical wave of frequency ω,

$$\psi = \frac{A}{r_s} e^{i(\kappa r_s - \omega t)}, \tag{9.3.1}$$

where r_s is the radial distance from P_s to any point in space and A is the amplitude of the wave at unit distance from the source. We are using ψ to designate the varying amplitude of the wave, with no stipulation as to the physical nature of the wave, whether it be an electromagnetic, acoustic, seismic wave, etc. The wave number and frequency are related as usual by $\omega/\kappa = c_m$, where c_m is the wave velocity in the medium.

Let us now suppose that all points on a closed surface S, which coincides with a wavefront of radius r_s at the point Q as shown in Fig. 9.3.1, constitute secondary sources of waves, each wavelet spreading out in all directions. We wish to add together at the observation point P_o the complex contributions arriving from each element of area dS of the surface S. It is then plausible to postulate that the contribution from any element dS, such as that at point Q in Fig. 9.3.1, will be proportional to:

(1)   The value of ψ at the surface S, $(A/r_s)e^{i(\kappa r_s - \omega t')}$, where the prime on t distinguishes the time at which the Huygens wavelets leave points on the surface S in order to reach P_o at the desired time.

(2)   The area dS at the point Q.

(3)   The spherical-wave factor $(1/r_o)e^{i[\kappa r_o - \omega(t-t')]}$, where r_o is the distance from Q to P_o and $t - t'$ is the time required for the wave to travel from Q to P_o.

(4)   Finally, and least obviously, some unknown function $f(\theta_s, \theta_o)$, called the *obliquity factor*, that takes into account the dependence of the amplitude of the secondary wave on the angles θ_s and θ_o, as defined in Fig. 9.3.1.

Accordingly, the wavelet that arrives at P_o from the secondary source element

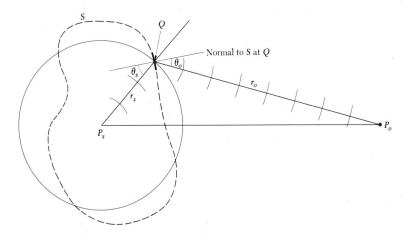

Fig. 9.3.1   The Huygens-Fresnel principle.

at Q may be written

$$d\psi = Kf(\theta_s,\theta_o) \left(\frac{A}{r_s} e^{i(\kappa r_s - \omega t')}\right)\left(\frac{1}{r_o} e^{i[\kappa r_o - \omega(t-t')]}\right) dS$$

$$= Kf(\theta_s,\theta_o) \frac{A}{r_s r_o} e^{i[\kappa(r_s+r_o)-\omega t]} \, dS. \tag{9.3.2}$$

To make definite the proportionality coefficient K, we shall take $f(\theta_s,\theta_o)$ to be normalized such that $f(0,0) = 1$. Moreover, K must have the dimensions of inverse length in order that $d\psi$ have the same dimensions as ψ in (9.3.1). Since the wavelength $\lambda = 2\pi/\kappa$ is the only characteristic length involved in the physical problem being discussed, there is a strong inference that K is proportional to $1/\lambda$. The entire disturbance at P_o is found by integrating (9.3.2) over the part ΔS of the wavefront S that is not obstructed by obstacles between P_s and P_o, that is,

$$\psi(P_o) = KAe^{-i\omega t} \int_{\Delta S} f(\theta_s,\theta_o) \frac{e^{i\kappa(r_s+r_o)}}{r_s r_o} \, dS. \tag{9.3.3}$$

If ΔS includes the entire surface S, the value of ψ computed from (9.3.3) should agree with the value given by (9.3.1), with r_s set equal to the distance $P_s P_o$. Thus if we can evaluate the integral in (9.3.3), we can then determine the proportionality constant K. The integral can be evaluated with considerable accuracy using the concepts of Fresnel zones and the vibration spiral, as we shall do in Sec. 11.1. For this calculation it is only necessary to assume that $f(\theta_s,\theta_o)$ decreases smoothly to zero as $\theta_s+\theta_o$ approaches π.

We have thus inferred from the Huygens-Fresnel postulate that the wave disturbance at the observation point P_o is given by (9.3.3). We do not as yet know the value of the proportionality constant K or the functional form of the obliquity factor $f(\theta_s,\theta_o)$. In the next section we shall discuss the method and approximations by which Kirchhoff deduced (9.3.3) from the wave equation and evaluated K and $f(\theta_s,\theta_o)$. It turns out that for many problems of practical interest, especially those involving visible light, the diffraction angles θ_s and θ_o are very small over the unobstructed area ΔS, so that one may set $f(\theta_s,\theta_o)$ equal to unity. Furthermore, one is usually interested only in relative amplitudes in the diffraction pattern, so that a knowledge of K is not needed. Most of the problems of Chap. 10 permit these simplifications.

9.4   Kirchhoff Diffraction Theory

We seek a solution to the wave equation in the presence of obstacles, i.e., a solution that satisfies the boundary conditions imposed by the obstacles. Such a solution for the general case is impossible, in the sense that overwhelmingly difficult numerical calculations would be needed. Thus, again, we shall presently need to make approximations.

To establish Huygens' principle formally, we need a relation between the known value of the wave disturbance on a surface and the unknown value at the desired observation point. This relation is found by invoking *Green's theorem*, a generalization of Gauss' theorem (Appendix A), which equates an integral of a vector function over a *closed surface* to an integral of a related function over the *volume enclosed by the surface*. The theorem is applied to a region of space surrounding, but excluding, the observation point P_o, as illustrated in Fig. 9.4.1. The volume in question is bounded by the two closed surfaces S and S'; accordingly, it is natural to break the surface integral up into two portions, one over S (the surface on which the incident wave disturbance is presumed known) and one over S' (a spherical surface of vanishing radius surrounding P_o). We shall show, first, that the volume integral vanishes, so that the two surface integrals are equal, and, second, that the surface integral over S' reduces to the desired (unknown) value of the wave disturbance at P_o. Thus the disturbance at P_o is expressed in terms of the surface integral over S, which in turn can be evaluated by making certain approximations. The closed surface S is taken in this section to enclose the observation point P_o, whereas in the previous section (Fig. 9.3.1) it was taken to enclose the source point P_s.

(a)   Green's Theorem

In boundary-value problems based on partial differential equations, the boundary conditions are normally specified in terms of the value of the function ψ (for example, the wave amplitude) and/or its normal derivative $\partial\psi/\partial n \equiv \hat{\mathbf{n}} \cdot \nabla\psi$ on the boundary, where $\hat{\mathbf{n}} \equiv d\mathbf{S}/dS$ is the unit vector locally normal to the boundary surfaces. With this motivation, we consider the integral

$$\oint_S \psi_1(\hat{\mathbf{n}} \cdot \nabla\psi_2)\, dS, \tag{9.4.1}$$

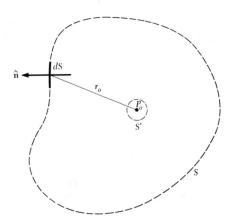

Fig. 9.4.1   Surfaces S and S' enclosing observation point P_o.

where ψ_1 and ψ_2 are scalar functions of position (perhaps two solutions of the wave equation) and the integration is carried out over the closed surface S bounding the region of interest. But the quantity $\psi_1 \nabla\psi_2$ in (9.4.1) is a vector to which Gauss' theorem (A.13) may be applied, giving

$$\oint_S (\psi_1 \nabla\psi_2) \cdot \hat{n}\, dS = \int_V \nabla \cdot (\psi_1 \nabla\psi_2)\, dV, \tag{9.4.2}$$

where V is the volume enclosed by the surface S and we must specify that the sense of the unit normal $\hat{n}$ is *outward*. The integrand on the right may be expanded

$$\nabla \cdot (\psi_1 \nabla\psi_2) = (\nabla\psi_1) \cdot (\nabla\psi_2) + \psi_1 \nabla^2\psi_2. \tag{9.4.3}$$

Finally if we interchange ψ_1 and ψ_2 in (9.4.2) and subtract the result from the original (9.4.2), using (9.4.3), we obtain *Green's theorem*

$$\oint_S (\psi_1 \nabla\psi_2 - \psi_2 \nabla\psi_1) \cdot \hat{n}\, dS = \int_V (\psi_1 \nabla^2\psi_2 - \psi_2 \nabla^2\psi_1)\, dV. \tag{9.4.4}$$

This theorem is a general relation connecting any two nonsingular scalar functions. It is not necessary that ψ_1 and ψ_2 be solutions of the wave equation, although this is the interpretation of present concern to us.[†]

(b) The Helmholtz-Kirchhoff Theorem

We choose for the two functions appearing in Green's theorem (9.4.4)

$$\psi_1 = \psi(x,y,z,t), \tag{9.4.5}$$

that is, the actual monochromatic wave present in our physical problem, and

$$\psi_2 = \frac{1}{r_o} e^{i(\kappa r_o - \omega t)}, \tag{9.4.6}$$

which has the *form* (9.3.1) of a spherically symmetric wave traveling out from the observation point P_o as origin (see Fig. 9.4.1). We assume that the sources that generate the actual wave ψ of (9.4.5) are all located outside the surface S, so that inside S the function ψ satisfies the source-free wave equation,

$$\nabla^2\psi - \frac{1}{c^2}\frac{\partial^2\psi}{\partial t^2} = \nabla^2\psi + \kappa^2\psi = 0. \tag{9.4.7}$$

Likewise, we know from Sec. 5.5 that the wave function (9.4.6) satisfies (9.4.7), except at the point P_o, which has been excluded from the volume V by the small spherical surface S' (Fig. 9.4.1). Consequently, the integrand of the right-hand

---

[†] For a mathematically sophisticated treatment of Green's theorem see, for instance, Courant and Hilbert, *op. cit.*, vol. 1, chap. 5.

side of Green's theorem (9.4.4) vanishes everywhere throughout the volume, that is,

$$\int_V (\psi_1 \nabla^2 \psi_2 - \psi_2 \nabla^2 \psi_1) \, dV = 0. \tag{9.4.8}$$

We evaluate the surface integral of Green's theorem, i.e., the left-hand side of (9.4.4), by breaking it up into two integrals, one over S and one over S'. Taking S' to be a spherical surface of radius ϵ centered on P_o and substituting the element of solid angle $d\Omega = dS/\epsilon^2$, we find

$$\oint_{S'} \left[\psi \nabla \left(\frac{1}{r_o} e^{i(\kappa r_o - \omega t)} \right) - \left(\frac{1}{r_o} e^{i(\kappa r_o - \omega t)} \right) \nabla \psi \right] \cdot \hat{\mathbf{n}} \, dS$$

$$= \oint_\Omega \left[\psi \left(\frac{1}{\epsilon^2} - \frac{i\kappa}{\epsilon} \right) e^{i(\kappa\epsilon - \omega t)} - \frac{1}{\epsilon} e^{i(\kappa\epsilon - \omega t)} \frac{\partial \psi}{\partial n} \right] \epsilon^2 \, d\Omega$$

$$\xrightarrow[\epsilon \to 0]{} 4\pi\psi(P_o,t)e^{-i\omega t}. \tag{9.4.9}$$

That is, the S' integral reduces to $4\pi e^{-i\omega t}$ times the value of ψ at the observation point, which is precisely the quantity we wish to know and which we can now evaluate in terms of the remaining integral over the outer surface S,

$$\psi(P_o,t) = -\frac{1}{4\pi} \oint_S \left[\psi \nabla \left(\frac{1}{r_o} e^{i\kappa r_o} \right) - \left(\frac{1}{r_o} e^{i\kappa r_o} \right) \nabla \psi \right] \cdot \hat{\mathbf{n}} \, dS. \tag{9.4.10}$$

This is the *Helmholtz-Kirchhoff integral theorem*. It is exact for time-harmonic scalar waves generated by sources outside the region enclosed by S. But it is not really a solution to our problem. That is, it simply shifts the unknown from the wave amplitude ψ at the desired observation point P_o to the values of ψ and $\hat{\mathbf{n}} \cdot \nabla \psi \equiv \partial\psi/\partial n$ over some enclosing surface S. Further progress rests upon finding reasonable approximations to these values on S.

(c) Kirchhoff Boundary Conditions

Let us consider an observation point P_o entirely surrounded by an opaque screen except for one or more apertures of finite extent, as suggested by Fig. 9.4.2. For definiteness, regard the wave motion as being generated by a point source of spherical waves located at P_s, so that in the absence of the screen the wavefield would be given by

$$\psi(r_s,t) = \frac{A}{r_s} e^{i(\kappa r_s - \omega t)}, \tag{9.4.11}$$

where the origin of r_s is taken at P_s and A is the amplitude at unit distance. We take the surface S to coincide with the inner surface of the screen, suitably continued across the aperture. Now in the spirit of geometrical optics we postulate that (1) over the portion of S lying in the aperture, ψ and $\partial\psi/\partial n$ have the

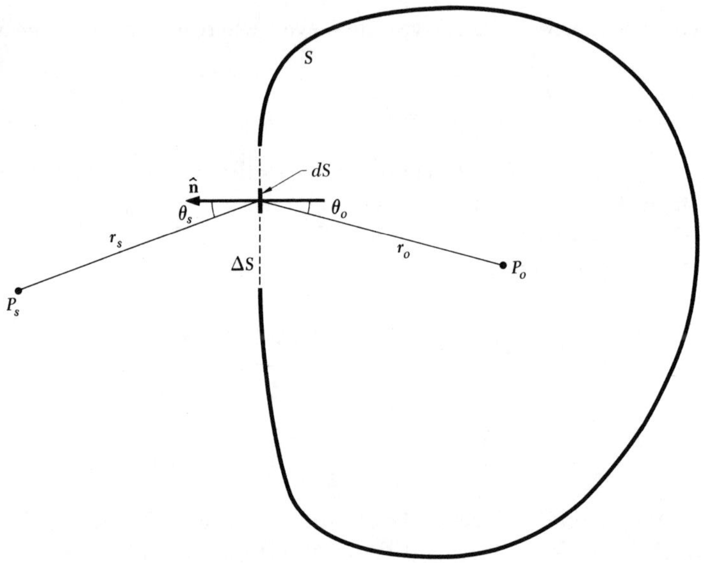

Fig. 9.4.2   Geometry involved in the Fresnel-Kirchhoff formula.

unperturbed values implied by (9.4.11), while (2) over the remainder of S, adjacent to the screen, ψ and $\partial\psi/\partial n$ are zero.† Explicitly, we make the following assumptions:

At the aperture:

$$\psi = \frac{A}{r_s}\, e^{i(\kappa r_s - \omega t)} \tag{9.4.12}$$

$$\frac{\partial \psi}{\partial n} = \left(\frac{1}{r_s} - i\kappa\right)\frac{A}{r_s}\, e^{i(\kappa r_s - \omega t)}\, \cos\theta_s; \tag{9.4.13}$$

At the screen:

$$\psi = \frac{\partial \psi}{\partial n} = 0. \tag{9.4.14}$$

These assumptions are known as *Kirchhoff boundary conditions*. They are merely physically plausible approximations. Near the edge of the aperture (within a

---

† Figure 9.4.2 shows the surface S to be closed behind P_o in a manner which suggests that a region opposite the aperture is illuminated by rays entering through the aperture and hence that the integrand of (9.4.10) should not be neglected over this portion of S. We can avoid this trouble by removing the rear surface to infinity. Even then, it is not obvious that there is no further contribution to the integral; see M. Born and E. Wolf, "Principles of Optics," 3d ed., pp. 379–380, Pergamon Press, New York, 1965.

wavelength or so), the proximity of the screen distorts the unperturbed wave in order to accommodate in detail the physical boundary conditions. Likewise, the wave amplitude does not go abruptly to zero just outside the aperture but rather diminishes gradually over a wavelength or so. Moreover, it can be shown that in specifying arbitrarily both ψ and $\partial\psi/\partial n$ we have overdetermined the boundary conditions. As a result, for instance, our treatment is not internally consistent, in that wave amplitudes near the boundary that we compute from (9.4.10) do not in general converge to our assumed amplitudes (9.4.12) to (9.4.14). However, so long as the aperture dimensions and the distances of P_s and P_o from the screen are large compared with the wavelength—the limit that we know leads to geometrical optics—the Kirchhoff assumptions give satisfactory accuracy.

Using the Kirchhoff assumptions (9.4.12) to (9.4.14) in the geometry indicated in Fig. 9.4.2, we may now write (9.4.10) in the form

$$\psi(P_o,t) = -\frac{A}{4\pi}\int_{\Delta S}\left[\left(i\kappa - \frac{1}{r_s}\right)\cos\theta_c + \left(i\kappa - \frac{1}{r_o}\right)\cos\theta_o\right]\frac{e^{i[\kappa(r_s+r_o)-\omega t]}}{r_s r_o}\,dS.$$

(9.4.15)

Kirchhoff's approximate boundary conditions lead to accurate results only when source and observation point are far from the reference surface S, relative to a wavelength. Thus we may approximate (9.4.15) in the limit $r_s, r_o \gg \lambda = 2\pi/\kappa$, obtaining the *Fresnel-Kirchhoff diffraction formula*

$$\psi(P_o,t) = -\frac{iA}{\lambda}e^{-i\omega t}\int_{\Delta S}\left(\frac{\cos\theta_s + \cos\theta_o}{2}\right)\frac{e^{i\kappa(r_s+r_o)}}{r_s r_o}\,dS.$$
(9.4.16)

The present result (9.4.16) is in exactly the same form as the integral (9.3.3) obtained from the Huygens-Fresnel postulate. Hence, we may now evaluate the proportionality coefficient

$$K = -\frac{i}{\lambda}$$
(9.4.17)

and the obliquity factor

$$f(\theta_s,\theta_o) = \frac{\cos\theta_s + \cos\theta_o}{2}$$
(9.4.18)

that remained undetermined in the earlier result. The coefficient K is indeed inversely proportional to wavelength, as surmised earlier; the factor $-i$ signifies that the Huygens wavelet is reradiated with a phase advance of 90°, a feature not anticipated in the phenomenological treatment of the previous section. The obliquity factor has a maximum value of unity in the forward direction and goes to zero for the portion of the Huygens wavelet returning toward the source.

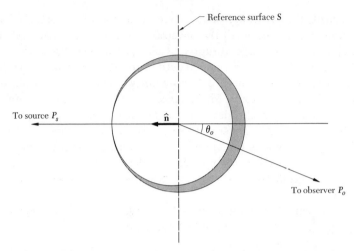

Fig. 9.4.3   Angle dependence of the amplitude of a Huygens wavelet (for $\theta_s = 0$).

Thus we visualize the wavelet as having an amplitude distribution as sketched in Fig. 9.4.3.

In summary, we have formally derived the Huygens-Fresnel-Kirchhoff principle (9.4.16) for scalar monochromatic waves from the wave equation (9.4.7) by making the approximations implicit in the Kirchhoff boundary conditions (9.4.12) to (9.4.14), which in turn assume that r_s, r_o, and the aperture dimensions are all large compared with a wavelength. The result is in full agreement with the Huygens-Fresnel hypothesis (9.3.3), the obliquity factor being given explicitly by (9.4.18).

We may easily deduce two useful theorems from (9.4.16). Since the variables r_s, θ_s and r_o, θ_o enter symmetrically in (9.4.16), a *reciprocity theorem* exists to the effect that a point source at P_s produces the same wave disturbance at P_o as a point source at P_o would produce at P_s. Another theorem, known as *Babinet's principle*, states that the wave disturbance ψ_1 at an observation point P_o for a particular opaque screen containing apertures and the disturbance ψ_2 at the same point but for the *complementary* screen, i.e., with apertures and opaque areas interchanged, sum to give the wave amplitude ψ that would exist at the point in the absence of any screen.†

Finally, we point out some important modifications of (9.4.16). If the source is one that does not radiate isotropically, the amplitude A in (9.4.11) may be a function of direction from P_s and consequently of position on the surface S.

---

† Further discussion of the limitations and interpretation of the Fresnel-Kirchhoff theory may be found in Born and Wolf, *op. cit.*, and in B. B. Baker and E. T. Copson, "The Mathematical Theory of Huygens' Principle," 2d ed., Oxford University Press, Fair Lawn, N.J., 1950.

When this variation exists, the coefficient A must be kept inside the integral in (9.4.16), where it may be termed the *aperture illumination function*. In addition, the screen forming the surface S may be of continuously varying opacity, rather than simply transparent and opaque. Similarly, the screen may introduce additional phase shift if, for instance, it consists of a dielectric of varying thickness. The Kirchhoff boundary conditions may then be readily modified (invoking the WKB approximation of Sec. 9.1, if necessary) with the result that we insert in the Fresnel-Kirchhoff integral (9.4.16) the complex *transmission function*

$$\check{T} = T e^{i\Delta\phi}, \tag{9.4.19}$$

where T is the amplitude transmission coefficient and $\Delta\phi$ the differential phase shift of the screen, both quantities being assumed to vary over the screen surface S. We thus see that by simple modifications the present theory can be applied to a number of cases of interest not directly covered by the basic diffraction formula.

Problems

9.4.1 Carry out in detail the calculation leading from the first to the last form of (9.4.9). Note that $\hat{n}$ is directed *toward* P_o on S'.

9.4.2 Verify (9.4.13), (9.4.15), and (9.4.16).

★9.4.3 Show that the analog of the Fresnel-Kirchhoff integral (9.4.16) for wave propagation in two dimensions, e.g., waves on a membrane or the surface of a lake, is, for r_s, $r_o \gg \lambda$,

$$\psi(P_o) = \frac{iA}{\pi} e^{-i\omega t} \int_{\Delta L} \left(\frac{\cos\theta_s + \cos\theta_o}{2} \right) \frac{e^{i\kappa(r_s + r_o)}}{(r_s r_o)^{1/2}} \, dL,$$

where ΔL represents the aperture(s) in an "opaque" line surrounding the observation point P_o (regard Fig. 9.4.2 as two-dimensional). *Hint:* The isotropic outgoing cylindrical wave analogous to (9.4.6) and (9.4.11) is proportional to

$$\psi = A H_0^{(1)}(\kappa r) e^{-i\omega t} \xrightarrow[\kappa r \gg 1]{} A \left(\frac{2}{\pi\kappa r} \right)^{1/2} e^{-i\pi/4} e^{i(\kappa r - \omega t)},$$

where r is now the cylindrical radial coordinate, and where $H_0^{(1)}(\kappa r)$ is the Hankel function of the first kind and zeroth order, which is related to the common Bessel functions by $H_0^{(1)} \equiv J_0 + iN_0$. The radial derivative is

$$\frac{\partial\psi}{\partial r} = A\kappa H_1^{(1)}(\kappa r) e^{-i\omega t} \xrightarrow[\kappa r \gg 1]{} A \left(\frac{2\kappa}{\pi r} \right)^{1/2} e^{-i3\pi/4} e^{i(\kappa r - \omega t)}.$$

Show that the analog of (9.4.9) is

$$\oint_{L'} (\psi_1 \nabla\psi_2 - \psi_2 \nabla\psi_1) \cdot \hat{n} \, dL \longrightarrow -i4\psi(P_0,t)e^{-i\omega t}$$

hence, that the analog of (9.4.10), called *Weber's theorem*, is

$$\psi(P_0) = -\frac{i}{4} \oint_L [\psi \nabla H_0^{(1)}(\kappa r_0) - H_0^{(1)}(\kappa r_0) \nabla\psi] \cdot \hat{n} \, dL.$$

9.5   *Diffraction of Transverse Waves*

The diffraction theory we have been developing would appear to apply only to wave motion described by a scalar wavefunction, e.g., the scalar velocity potential describing an acoustic wave in a fluid. Nevertheless, experiment clearly shows that it gives excellent results for visible light, as well as for other electromagnetic waves, provided certain conditions of a geometrical nature are met. Let us now examine qualitatively why scalar diffraction theory can be used to treat many diffraction problems involving electromagnetic (transverse, vector) waves.

First recall that even in the scalar theory certain approximations, or idealizations, have been made to achieve a useful result. In particular, the Kirchhoff boundary conditions (9.4.12) to (9.4.14) are only good approximations to the wavefield in a diffracting screen. The greatest discrepancy occurs near the edges of an aperture, which therefore should have linear dimensions large compared with a wavelength. Although the Fresnel-Kirchhoff theory fails to give an accurate description of the wavefield near an aperture, at distant points the wavefield tends to become smoothed out and not sensitive to minor inaccuracies in the form of the wavefield assumed at the aperture. The scalar theory is therefore mainly useful for predicting the relative amplitude (and phase) of the diffracted wave at observation points far (many wavelengths) from a diffracting screen. Actually, in nearly all important cases of optical diffraction, the optical aperture subtends a relatively small angle, both from the source and from an observation point—usually near the *image* of the (point) source in an optical train. We discuss these geometrical conditions more fully in Sec. 10.1.

We expect, then, that scalar diffraction theory, subject to these limitations of a geometrical nature, should hold for each (scalar) component of an electromagnetic wave, namely, the two transverse components of **E**, the electric intensity, and the two transverse components of **B**, the magnetic intensity. From our discussion of electromagnetic waves in the preceding chapter, it is clear that each component of a traveling sinusoidal electromagnetic wave has the same wavefunction factor, $\exp i(\kappa \cdot \mathbf{r} - \omega t)$, that occurs in scalar diffraction theory. Accordingly, the diffracted wave disturbance for each component can be calculated quite accurately at any observation point using the scalar Fresnel-Kirchhoff formula (9.4.16). It is then tempting to infer that the actual diffracted vector wavefield at the point, as described by **E** and **B**, is closely described by vectors formed from the diffracted scalar field components. The accuracy of this inference must somehow depend on how small an angle the aperture subtends at the observation point. For when this angle is small, the randomly polarized (transverse) contributions to the vector wavefield coming from different parts of the aperture preserve their orientation in a plane perpendicular to a line from the observation point to some point centrally located in the aperture. We then expect that a calculation based on the scalar theory should agree with one based on a more rigorous vector diffraction theory. Hence, for the geometrical conditions usually met in optical diffraction, the scalar theory gives sufficiently accurate results for randomly polarized light.

A more detailed examination of the relation of scalar diffraction theory to optics may be found in Born and Wolf.† It is shown there that when rays from various points of an optical aperture have a small inclination (<10 to $15°$) to the optical axis of an optical system, the distribution of *light intensity* in an optical image is that predicted by scalar diffraction theory. In this treatment averages are taken of the intensity contributions of randomly polarized components. It is basically intensity (average power per unit area), not amplitude, that is observed with light waves.

There exist a limited number of problems in the diffraction of electromagnetic waves for which rigorous solutions are known. Rigorous diffraction theory involves solving Maxwell's electromagnetic equations subject to (idealized) boundary conditions assumed for the aperture screen. The first such rigorous solution was obtained by Sommerfeld in 1896 for an infinitely thin, perfectly conducting half plane. Considerable mathematical ingenuity went into constructing the solution. The geometrical limitations inherent in the Fresnel-Kirchhoff theory of course do not exist in the rigorous theory, which therefore gives a complete description of the field in the vicinity of the aperture boundary, as well as at great distances. Unfortunately the number of cases that can be treated rigorously is very limited, and even the simplest case, that treated by Sommerfeld, involves very difficult mathematical analysis.‡

It is also interesting to know that one can formulate a *vector* Huygens' principle, which replaces the scalar Fresnel-Kirchhoff formula.§ As with scalar diffraction theory, the difficulty of formulating accurate boundary conditions at the aperture makes the vector theory of little use in improving the predictions

---

† *Op. cit.*, sec. 8.4.
‡ For an account of rigorous diffraction theory see Born and Wolf, *op. cit.*, chap. 11.
§ See, for example, A. Sommerfeld, "Optics," pp. 325–328, Academic Press Inc., New York, 1954; and J. D. Jackson, "Classical Electrodynamics," pp. 283–287, John Wiley & Sons, Inc., New York, 1962.

of the scalar theory. To quote Sommerfeld: "The vectorial Huygens' principle is no magic wand for the solution of boundary value problems." We shall base all our discussion of the diffraction of waves on the scalar Fresnel-Kirchhoff formula (9.4.16).

*9.6   Young's Formulation of Diffraction

When the Kirchhoff diffraction integral (9.4.16) is applied to a bounded aperture in an infinite opaque screen, the surface of integration ΔS can have an arbitrary form consistent with the given boundary. Therefore it should be possible to transform Kirchhoff's *surface integral* into an equivalent *line integral* around the edge of the aperture. Experimentally, when an aperture is viewed from an observation point within the geometrical shadow, the edges of the aperture appear luminous (the setting sun appears to "eat into" the horizon). This fact led Thomas Young to postulate (in 1802) that diffraction can be treated as an interference effect between the geometrical incident wave and a *boundary wave* generated at the edge of the aperture. In Young's model the two-dimensional array of Huygen's sources over the aperture is replaced by a one-dimensional series of Young's sources around the boundary. The quantitative form of Young's postulate was established by Rubinowicz in 1917.†

The Helmholtz-Kirchhoff integral theorem (9.4.10) is valid when the observation point P_o is inside, but the source is outside, the closed surface S (Fig. 9.6.1);

$$\psi(P_o,t) = -\frac{1}{4\pi} \oint_S \left[\psi \, \nabla \left(\frac{e^{i\kappa r_o}}{r_o} \right) - \left(\frac{e^{i\kappa r_o}}{r_o} \right) \nabla \psi \right] \cdot \hat{n} \, dS, \qquad (9.6.1)$$

† A brief and readable history is given in A. Rubinowicz, *Nature*, **180**: 160 (1957).

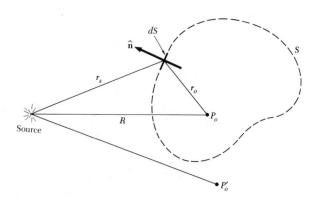

Fig. 9.6.1   Geometry of the Helmholtz-Kirchhoff integral theorem.

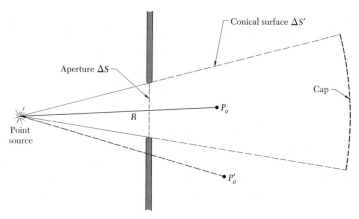

Fig. 9.6.2   Special surface of integration to establish the Young-Rubinowicz formula.

where r_o is the radial distance from P_o to dS and $\hat{n}$ is the *outward* normal at S. The theorem involves no approximation, other than the assumption of scalar waves. As a corollary, if the observation point (as well as the source) lies outside S, the integral of (9.6.1) equals zero, rather than $\psi(P_o)$, since there is then no contribution (9.4.9) arising from the surface S' of Fig. 9.4.1.

In what appears at first to be a trivial maneuver, let us apply the theorem (9.6.1) to an isotropic point source radiating waves of the form

$$\psi = \frac{A}{R} e^{i(\kappa R - \omega t)} \tag{9.6.2}$$

in an *unobstructed* medium. The integral

$$-\frac{A}{4\pi} e^{-i\omega t} \oint_S \left[\frac{e^{i\kappa r_s}}{r_s} \nabla \left(\frac{e^{i\kappa r_o}}{r_o} \right) - \frac{e^{i\kappa r_o}}{r_o} \nabla \left(\frac{e^{i\kappa r_s}}{r_s} \right) \right] \cdot \hat{n} \, dS \tag{9.6.3}$$

must then have exactly the value (9.6.2) for observation points P_o lying inside the surface S and the value zero for points P_o' lying outside (Fig. 9.6.1).

As a next step, let the given aperture ΔS in an opaque screen define a conical surface $\Delta S'$ by extending generators from the point source through the aperture boundary, as in Fig. 9.6.2. The aperture ΔS, the truncated cone $\Delta S'$, and a spherical cap at infinity constitute a closed mathematical surface, over which the integral (9.6.3) can be carried out. Symbolically,

$$\int_{\text{aperture}} \cdots dS + \int_{\text{cone}} \cdots dS + \int_{\text{cap}} \cdots dS$$

$$= \begin{cases} \dfrac{A}{R} e^{i(\kappa R - \omega t)} & P_o \text{ in direct beam} \\ 0 & P_o' \text{ in geometrical shadow.} \end{cases} \tag{9.6.4}$$

The integral over the cap vanishes.† The integral over the aperture is precisely that of the Fresnel-Kirchhoff theory, using the approximate Kirchhoff boundary conditions (9.4.12) to (9.4.14). Thus, we may rearrange (9.6.4) to obtain

$$\psi(P_o,t) \xrightarrow[\text{approx}]{\text{Kirchhoff}} \int_{\text{aperture}} \cdots dS = \left\{ \begin{array}{c} \dfrac{A}{R}\, e^{i(\kappa R - \omega t)} \\ 0 \end{array} \right\} - \int_{\text{cone}} \cdots dS. \quad (9.6.5)$$

It remains to show that the integral over the conical surface $\Delta S'$ can be transformed to a line integral around the aperture boundary.

In Fig. 9.6.3, r_s and r_o are the source and observation distances to the area element dS; they take on the limiting values ρ_s and ρ_o at the element dl of the aperture boundary. For the conical surface the normal derivatives are (Prob. 9.6.1)

$$\hat{\mathbf{n}} \cdot \nabla \left(\frac{e^{i\kappa r_s}}{r_s} \right) = \frac{\partial}{\partial n}\left(\frac{e^{i\kappa r_s}}{r_s} \right) = 0 \qquad\qquad (9.6.6)$$

$$\hat{\mathbf{n}} \cdot \nabla \left(\frac{e^{i\kappa r_o}}{r_o} \right) = \cos(r_o,n)\frac{\partial}{\partial r_o}\left(\frac{e^{i\kappa r_o}}{r_o} \right) = \cos(r_o,n)\left(\frac{i\kappa}{r_o} - \frac{1}{r_o{}^2} \right) e^{i\kappa r_o}. \qquad (9.6.7)$$

The area element is

$$dS = \sin(\rho_s,dl)\,\frac{r_s}{\rho_s}\, dl\, dr_s. \qquad\qquad (9.6.8)$$

Since the normals at dS and dl are parallel,

$$r_o \cos(r_o,n) = \rho_o \cos(\rho_o,n). \qquad\qquad (9.6.9)$$

† Although physically reasonable, this statement is difficult to prove formally; see Born and Wolf, *op. cit.*, pp. 379–380.

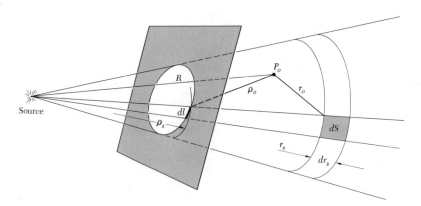

Fig. 9.6.3   Coordinates for integration over the conical surface.

Substituting (9.6.6) to (9.6.9) in (9.6.3), we obtain

$$\int_{\text{cone}} \cdots dS = -\frac{1}{4\pi} e^{-i\omega t} \oint \left[\int_{\rho_s}^{\infty} \left(i\kappa - \frac{1}{r_o} \right) \frac{e^{i\kappa(r_s+r_o)}}{r_o^2} dr_s \right]$$

$$\frac{\rho_o}{\rho_s} \cos(\rho_o,n) \sin(\rho_s,dl) \, dl. \quad (9.6.10)$$

The integrand of the integral in square brackets is an exact differential (Prob. 9.6.2)

$$\left(i\kappa - \frac{1}{r_o} \right) \frac{e^{i\kappa(r_s+r_o)}}{r_o^2} = \frac{d}{dr_s} \left\{ \frac{e^{i\kappa(r_s+r_o)}}{r_o[r_s + r_o - \rho_s + \rho_o \cos(\rho_s,\rho_o)]} \right\}, \quad (9.6.11)$$

and can thus be evaluated as

$$\int_{\rho_s}^{\infty} \left(i\kappa - \frac{1}{r_o} \right) \frac{e^{i\kappa(r_s+r_o)}}{r_o^2} dr_s = -\frac{e^{i\kappa(\rho_s+\rho_o)}}{\rho_o^2[1 + \cos(\rho_s,\rho_o]}. \quad (9.6.12)$$

We can now write (9.6.5) explicitly as the *Young-Rubinowicz diffraction formula*

$$\psi(P_o,t) = \begin{cases} \dfrac{A}{R} e^{i(\kappa R - \omega t)} & P_o \text{ in direct beam} \\ 0 & P_o \text{ in geometrical shadow} \end{cases}$$

$$- \frac{A}{4\pi} e^{-i\omega t} \oint_{\substack{\text{aperture} \\ \text{boundary}}} \left[\frac{\cos(\rho_o,n) \sin(\rho_s,dl)}{1 + \cos(\rho_s,\rho_o)} \right] \frac{e^{i\kappa(\rho_s+\rho_o)}}{\rho_s\rho_o} dl. \quad (9.6.13)$$

This formulation of diffraction is entirely equivalent to the Fresnel-Kirchhoff formula (9.4.16); the accuracy of both is limited by the approximate Kirchhoff boundary conditions (9.4.12) to (9.4.14). The first term on the right of (9.6.13) is just the wave amplitude predicted by geometrical optics, vanishing discontinuously at the geometrical shadow edge. The second term is an integral over Young's secondary sources distributed along the aperture boundary. The form of the integrand may be justified intuitively by arguments analogous to those used to deduce (9.3.2), namely, as the product of two spherical wave factors, the secondary source length dl, and an angle-dependent obliquity factor.

Physically, the total diffracted amplitude $\psi(P_o,t)$ must be continuous across the geometrical shadow edge. To compensate for the discontinuity of the geometrical-optics term in (9.6.13), the obliquity factor [square bracket in (9.6.13)] reverses sign discontinuously across the shadow edge, although the magnitude of the integral is continuous and equal to one-half the amplitude of the direct beam (Prob. 9.6.4). Consequently, the intensity at the geometrical shadow edge is one-quarter that of the direct beam, a conclusion that is difficult to derive from the Fresnel-Kirchhoff formulation.

Problems

9.6.1 Verify (9.6.6) to (9.6.10). Show that $\cos(\rho_o, n)$ is positive or negative as P_o is in the illuminated or the shadow region, respectively.

9.6.2 Verify (9.6.11) by carrying out the differentiation. *Hint:* To find dr_o/dr_s, apply the law of cosines to the triangle dl-P_o-dS in Fig. 9.6.3 to express r_o in terms of r_s and $\cos(\rho_s, \rho_o)$. The angle (ρ_s, ρ_o) is understood to be π when dl lies on the line joining source and observer.

9.6.3 Show that the Young line integral in (9.6.13) can be written in vector notation as

$$-\frac{A}{4\pi} e^{-i\omega t} \oint \frac{e^{i\kappa(\rho_s + \rho_o)}}{\rho_s \rho_o} \frac{\boldsymbol{\varrho}_s \times \boldsymbol{\varrho}_o \cdot d\mathbf{l}}{\rho_s \rho_o + \boldsymbol{\varrho}_s \cdot \boldsymbol{\varrho}_o}$$

where the aperture boundary is traversed in a counterclockwise sense when viewed from the observation point P_o and the vectors $\boldsymbol{\varrho}_s$ and $\boldsymbol{\varrho}_o$ are directed outward from the source and observation points, respectively.

★9.6.4 Show that the magnitude of the Young integral in (9.6.13) is one-half that of the direct beam at the geometrical shadow edge but that the sign (phase) of the integral reverses discontinuously across the shadow edge. *Hint:* Consider diffraction by a straight edge (see Fig. 11.4.1), evaluating the obliquity factor to obtain

$$\frac{\cos(\rho_o, n) \sin(\rho_s, dl)}{1 + \cos(\rho_s, \rho_o)} \xrightarrow[\eta, x \ll R_s, R_o]{} \frac{2x/R_o}{[(R_s + R_o)\eta/R_s R_o]^2 + (x/R_o)^2},$$

where $R_s = P_s O$, $R_o = OO'$, and η is the integration variable along the aperture edge perpendicular to the plane of Fig. 11.4.1.

★9.6.5 The Helmholtz-Kirchhoff integral (9.6.1) is of the form

$$\oint \mathbf{V} \cdot \hat{\mathbf{n}} \, dS.$$

Show that $\boldsymbol{\nabla} \cdot \mathbf{V} = 0$ in any region not containing P_0 or the sources of the wave ψ; hence there exists a vector function $\mathbf{W}$ such that

$$\mathbf{V} = \boldsymbol{\nabla} \times \mathbf{W}.$$

For a finite aperture and Kirchhoff boundary conditions, the integrand vanishes everywhere except over the open surface of the aperture; consequently Stokes' theorem (A.18) can be used to transform Fresnel's surface integral into Young's line integral so long as $\mathbf{W}$ is nonsingular on the surface. Use this approach to establish the Young-Rubinowicz formula (9.6.13).†

† See Baker and Copson, *op. cit.*, pp. 74–79; A. Rubinowicz, "Progress in Optics," vol. 4, pp. 199–240, John Wiley & Sons, Inc., New York, 1965.

ten

Fraunhofer Diffraction

In the preceding chapter we established the Fresnel-Kirchhoff diffraction integral (9.4.16),

$$\psi(P_o) = -\frac{iA}{\lambda} e^{-i\omega t} \int_{\Delta S} \frac{\frac{1}{2}(\cos\theta_s + \cos\theta_o)}{r_s r_o} e^{i\kappa(r_s + r_o)} \, dS, \qquad (10.0.1)$$

which enables us to treat, to good approximation, the diffraction of a wave through apertures in a screen in the limit that the source distance r_s, the observation distance r_o, and the aperture dimensions are all large compared with the wavelength. In spite of the simplifications already made, this integral is still cumbersome, and most diffraction problems of general interest involve further approximations.

The Kirchhoff theory applies strictly to scalar (longitudinal) waves, such as acoustic waves. However, it turns out to be generally effective for vector (transverse) waves also, except for features directly associated with polarization. Thus, within the limitations on dimensions relative to wavelength, we think of the theory as applying to all types of three-dimensional wave motion, including seismic waves and electromagnetic waves in the radio, optical, and other wavelength regions. The two-dimensional analog, referred to in Prob. 9.4.3, applies to waves on a stretched membrane or on the surface of a lake. The theory of diffraction arose historically in the context of visible-light optics. Most of our examples are drawn from this field.

10.1   The Paraxial Approximation

Before discussing the most important special case of diffraction, that of Fraunhofer, we need to introduce the *paraxial approximation*, according to which the aperture (or set of apertures) in an opaque screen separating source from observer is localized such that the greatest lateral extent d of the aperture is small compared with the distances r_s and r_o (see Fig. 10.1.1),

$$d^2 \ll r_s{}^2 \qquad d^2 \ll r_o{}^2, \tag{10.1.1}$$

that is, that the angular size of the aperture is small when viewed from either source or observer. This restriction ensures that r_s and r_o are essentially constant over the aperture, so that the average value of the factor $1/r_s r_o$ in (10.0.1) can be taken outside the integral. (Note, however, that since r_s and r_o are large compared with the wavelength, the quantity $\kappa(r_s + r_o)$ may vary by an increment comparable to π and the exponential term must be kept inside the integral.) The

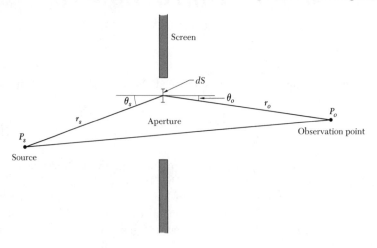

Fig. 10.1.1   Diffraction by an aperture.

paraxial approximation also ensures that the cosines are nearly constant, so that the obliquity factor can also be taken outside the integral. With these assumptions, the diffraction integral can be written

$$\psi(P_o) = -\frac{iA\,\frac{1}{2}(\cos\theta_s + \cos\theta_o)}{\lambda}\,\frac{}{r_s r_o}\,e^{-i\omega t}\int_{\Delta S} e^{i\kappa(r_s+r_o)}\,dS$$

$$\rightarrow \check{C}\int_{\Delta S} e^{i\kappa(r_s+r_o)}\,dS. \quad (10.1.2)$$

The latter, simplified form follows if we compress the factors outside the integral into the single complex coefficient $\check{C} \equiv -iA(\cos\theta_s + \cos\theta_o)e^{-i\omega t}/2\lambda r_s r_o$. Thus the problem of diffraction reduces in the paraxial approximation to the evaluation of the relatively simple integral (10.1.2).

We have used the term "paraxial approximation" so far to denote a limit on aperture size. The term can also be applied to analogous geometrical restrictions on the angular size of an extended source, e.g., the length of a line source, or on the angular extent of observations in an observation plane subtended at the aperture.

10.2 *The Fraunhofer Limit*

The special case of *Fraunhofer diffraction* arises when the source and observer are effectively at an infinite distance from a finite aperture. This simple statement needs clarification. Let us set up the paraxial diffraction integral (10.1.2) in a manner suggested by Fig. 10.2.1. A spherical reference surface of radius R_s is

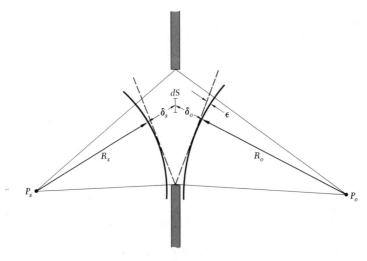

Fig. 10.2.1 Reference surfaces for computing incremental paths δ_s and δ_o.

constructed at the aperture about the source point P_s as center; a similar reference surface of radius R_o is constructed about the observation point P_o. We may now write

$$r_s \equiv R_s + \delta_s$$
$$r_o \equiv R_o + \delta_o, \tag{10.2.1}$$

and the diffraction integral becomes

$$\psi = \check{C} e^{i\kappa(R_s+R_o)} \int_{\Delta S} e^{i\kappa(\delta_s+\delta_o)} \, dS = \check{C}' \int_{\Delta S} e^{i\kappa\delta} \, dS, \tag{10.2.2}$$

where the new coefficient $\check{C}'$ absorbs the constant phase factor $e^{i\kappa(R_s+R_o)}$ and the quantity $\delta = \delta_s + \delta_o$ is termed the *path-difference function*. What we have accomplished is simply to shift our attention from the total path $r_s + r_o$ to the incremental path $\delta_s + \delta_o$ in the neighborhood of the aperture as the variable of integration.

We may now remove the source and observer to infinity ($R_s \to \infty$, $R_o \to \infty$), whereupon the reference surfaces from which we measure δ_s and δ_o become *planes*. It is easy to see that for plane reference surfaces, the path-difference function δ is a *linear function* of the (two-dimensional) coordinates of a point in the aperture plane. Thus the form of the integral (10.2.2) is no more complicated than

$$\int_{\Delta S} e^{i\kappa(\alpha x+\beta y)} \, dx \, dy, \tag{10.2.3}$$

where α and β are constants. The *linearity of the exponent* with respect to the aperture coordinates is in fact the definitive characteristic of *Fraunhofer* diffraction.

The Fraunhofer condition is fulfilled in the real world of the laboratory in two ways: (1) by placing the source and observer at sufficiently large (but still finite) distances and (2) by the use of lenses. We consider these alternatives in turn.

(1) If the spherical reference surfaces of Fig. 10.2.1 are approximated by planes, the incremental paths δ_s and δ_o computed from the planes are in error by the small distances between tangent plane and sphere, such as ϵ in Fig. 10.2.1. Let d be the maximum linear width of the aperture, as viewed from either P_s or P_o. Then the greatest error in δ, which occurs at the edges of the aperture, is (Prob. 10.2.1)

$$\left[R^2 + \left(\frac{d}{2}\right)^2 \right]^{1/2} - R \xrightarrow[d \ll R]{} \frac{d^2}{8R}, \tag{10.2.4}$$

where R represents the smaller of R_s and R_o. This error is negligible in the integral (10.2.2) if its maximum value is very small compared with a wavelength.

Thus the *Fraunhofer criterion* is that the source and observer distances are in the limit

$$R \gg \frac{d^2}{8\lambda}.$$ (10.2.5a)

In a practical engineering sense, this criterion is often written

$$R > \frac{d^2}{\lambda}.$$ (10.2.5b)

The Fraunhofer criterion differs from the assumptions of the Kirchhoff diffraction theory,

$$R > \lambda \qquad d > \lambda,$$ (10.2.6)

and of the paraxial approximation (10.1.1)

$$R^2 \gg d^2.$$ (10.2.7)

(2) According to geometrical (ray) optics, the converging lens of Fig. 10.2.2 renders parallel the rays emerging from a point source at its focus. According to wave optics, the lens converts the spherical wavefronts, radiated by the point source, into plane wavefronts. Thus a pair of lenses may be used to provide Fraunhofer's plane reference surfaces at the diffraction aperture, as shown in Fig. 10.2.3, even though P_s and P_o are not at "infinite" distances from the aperture (see Prob. 10.2.2). In practice, only one lens need be used, of suitable focal length to produce an image at P_o of an object at P_s (see Prob. 10.4.6). A spherical mirror is equivalent to a lens. Telescope objective mirrors and radar "dish" antennas, for instance, fulfill the Fraunhofer condition.

To summarize, we have defined Fraunhofer diffraction to be the limiting case where the reference wavefronts at the aperture are essentially plane, with the result that the diffraction integral reduces to the exponential of a linear function of the aperture coordinates. The case arises when source and observation distances are sufficiently great to satisfy the criterion (10.2.5) or when

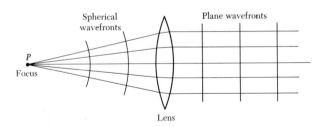

Fig. 10.2.2 Transformation between spherical and plane wavefronts by a lens.

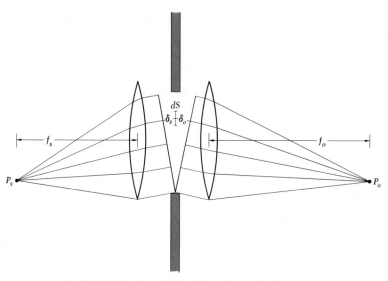

Fig. 10.2.3   Fraunhofer diffraction using lenses.

lenses are used. For visible light waves, the Fraunhofer condition is most conveniently obtained with lenses since the small wavelength makes the d^2/λ distance very large and hence the wave amplitude (proportional to $1/r_s r_o$) awkwardly low. The present diffraction theory, however, is not at all restricted to visible light waves, and the Fraunhofer condition can be well satisfied in practice, with or without lenses, for many other interesting cases, including radio-frequency electromagnetic waves, acoustic waves, and waves on the surface of a lake.

Problems

10.2.1 Prove the limiting approximation stated in (10.2.4).

10.2.2 The distances of the source and observation points from the aperture in Fig. 10.2.3 are essentially the focal lengths f_s and f_o of the respective lenses. Review the geometrical-optics theory of lenses to establish that the aberrations inherent in practical lenses limit their use to the paraxial approximation (10.2.7), in the form $f^2 \gg d^2$, where d is the maximum aperture dimension.

10.2.3 We have introduced three limiting assumptions, those of Kirchhoff (10.2.6), Fraunhofer (10.2.5), and the paraxial approximation (10.2.7). Are these all independent? Are any two sufficient to imply the third? *Answer:* Kirchhoff plus Fraunhofer implies paraxial.

10.2.4 A radio-frequency antenna (such as a TV antenna) may be thought of semiquantitatively as an aperture with dimensions comparable to the physical dimensions (usually of the order of a wavelength). How far away from the antenna, expressed in wavelengths, must one be for the Fraunhofer case to be valid? Compare with the case for a pinhole of diameter 0.1 mm for visible wavelengths.

10.3 The Rectangular Aperture

We consider a rectangular aperture, of dimensions a by b, in a plane opaque screen (Fig. 10.3.1). The screen lies in the xy plane of a cartesian coordinate system with origin at the center of the aperture. We propose to evaluate the diffraction integral in the form (10.2.2), which assumes the paraxial approximation in order to suppress the variation of obliquity factor $\frac{1}{2}(\cos\theta_s + \cos\theta_o)$ and inverse-square factor $(1/r_s r_o)$ in the full Kirchhoff integral (9.4.16). Furthermore, the Fraunhofer assumption implies that the reference surfaces, from which the incremental paths δ_s and δ_o are reckoned, are planes. That is, the Fraunhofer limit allows us to write the phase-factor exponential as a linear function of the aperture coordinates x and y, as indicated in (10.2.3). But for a rectangular aperture the limits of integration of the two variables are not coupled; i.e., the diffraction integral can be factored into two independent integrals,

$$\int_{\Delta S} e^{i\kappa(r_s+r_o)}\, dS \rightarrow e^{i\kappa(R_s+R_o)} \int_{\Delta x} e^{i\kappa\alpha x}\, dx \int_{\Delta y} e^{i\kappa\beta y}\, dy, \tag{10.3.1}$$

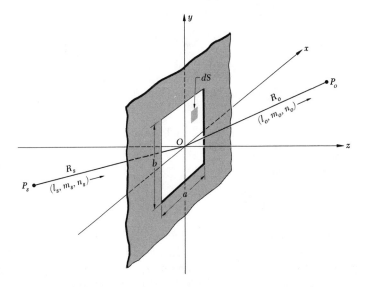

Fig. 10.3.1 Fraunhofer diffraction by a rectangular aperture.

where R_s, R_o are average source and observation distances and α, β are constants depending on the location of source and observer.

In the Fraunhofer limit the locations of source and observer, relative to the aperture, can best be specified by the *direction cosines* of the parallel rays (wavefront normals) arriving at the aperture from the source and leaving the aperture toward the observation point. We suppose, then, that plane waves of a single frequency, either from a sufficiently distant point P_s or from a point source at the focus of a lens, are incident upon the aperture in a direction specified by the direction cosines l_s, m_s, n_s. Plane wavefronts diffracted in the direction l_o, m_o, n_o by the aperture are observed at a point P_o, which is either at a great distance or at the focus of a second lens. Since $l^2 + m^2 + n^2 \equiv 1$, only two direction cosines are independent in each case.

It is convenient to compute the incremental paths δ_s and δ_o with respect to reference planes passing through the origin at the center of the aperture. For the element dS of aperture area at coordinate position (x,y), the incremental paths are (Prob. 10.3.1)

$$\delta_s = l_s x + m_s y \tag{10.3.2}$$
$$\delta_o = -(l_o x + m_o y). \tag{10.3.3}$$

Consequently, the *path-difference function* is

$$\delta(x,y) \equiv \delta_s + \delta_o = (l_s - l_o)x + (m_s - m_o)y, \tag{10.3.4}$$

and the amplitude of the diffracted wave at the observation point P_o is indeed given by an integral of the form (10.3.1), namely (Prob. 10.3.2),

$$
\begin{aligned}
\psi &= \check{C}' \int_{-b/2}^{b/2} \int_{-a/2}^{a/2} e^{i\kappa[(l_s-l_o)x+(m_s-m_o)y]} \, dx \, dy \\
&= \check{C}' \int_{-a/2}^{a/2} e^{i\kappa(l_s-l_o)x} \, dx \cdot \int_{-b/2}^{b/2} e^{i\kappa(m_s-m_o)y} \, dy \\
&= \check{C}' ab \, \frac{\sin u}{u} \frac{\sin v}{v},
\end{aligned} \tag{10.3.5}
$$

where for convenience we have introduced the new variables

$$u \equiv \frac{\kappa a}{2}(l_s - l_o) = \frac{\pi a}{\lambda}(l_s - l_o) \tag{10.3.6}$$

$$v \equiv \frac{\kappa b}{2}(m_s - m_o) = \frac{\pi b}{\lambda}(m_s - m_o). \tag{10.3.7}$$

The meaning of the variables u, v, which specify the location of source and observer relative to the aperture, can be understood most easily if we take the special case of normal incidence (whereupon $l_s = m_s = 0$, $n_s = 1$). Imagine an observation screen parallel to, and displaced from, the aperture plane by the large distance $R_o > (a^2 + b^2)/\lambda$ (or alternatively located in the focal plane of a lens of focal length f_o). Introduce coordinates x_o, y_o in this observation plane, with origin on the z axis, as shown in Fig. 10.3.2. For convenience, let us further restrict ourselves to observation points close to the origin, such that

$$l_o, m_o \ll n_o \approx 1;$$

then

$$l_o \approx \frac{x_o}{R_o} \quad \left(\text{or } \frac{x_o}{f_o}\right) \tag{10.3.8}$$

$$m_o \approx \frac{y_o}{R_o} \quad \left(\text{or } \frac{y_o}{f_o}\right). \tag{10.3.9}$$

The diffracted-wave amplitude at the observation coordinates x_o, y_o is then given by (10.3.5) with

$$u = -\frac{\pi a}{\lambda R_o} x_o \quad \left(\text{or } -\frac{\pi a}{\lambda f_o} x_o\right) \tag{10.3.10}$$

$$v = -\frac{\pi b}{\lambda R_o} y_o \quad \left(\text{or } -\frac{\pi b}{\lambda f_o} y_o\right). \tag{10.3.11}$$

Thus we may think of u, v simply as being normalized coordinates in the observation plane. The extension to off-normal incidence and large diffraction angles is straightforward (see, for instance, Probs. 10.4.5 and 10.4.6).

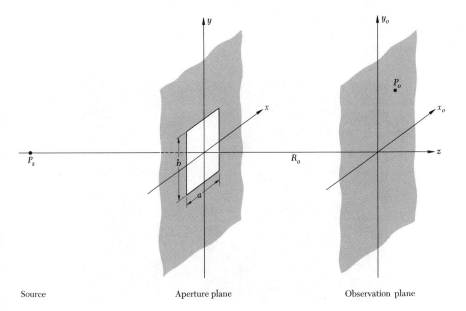

Source   Aperture plane   Observation plane

Fig. 10.3.2   Observation-plane coordinate system.

We have seen in earlier chapters that the intensity of a wave of any sort is proportional to the square of the amplitude,

$$I \propto |\psi|^2. \tag{10.3.12}$$

If we let I_0 be the intensity at the center of the diffraction pattern, where $u = v = 0$, then (see Prob. 10.3.4)

$$I = I_0 \left(\frac{\sin u}{u}\right)^2 \left(\frac{\sin v}{v}\right)^2. \tag{10.3.13}$$

A photograph of the intensity pattern in the observation plane is shown in Fig. 10.3.3 for a rectangular aperture of aspect ratio 3 by 4.

We observe that the rectangular-aperture diffraction formula (10.3.5) consists of the product of two similar terms. Each term is a function of only the aperture dimension and direction cosines associated with *one* of the coordinates, x or y. That is, decreasing (or increasing) the width a causes the diffraction pattern of Fig. 10.3.3 to expand (or contract, respectively) in the x_o direction but does not alter it in the y_o direction. Similarly, a change in the length b alters the pattern in the y_o direction but not in the x_o direction. Because of the independence of the two cartesian aperture dimensions a and b in controlling different aspects of the diffraction pattern, we can simplify the discussion by limiting our attention to the xz plane and studying the one-dimensional diffraction pattern along the x_o observation axis. Thus, in the next section we consider the *single-slit* diffraction pattern in detail. Once we have done that, we can think of the rectangular aperture case as being simply the product of two orthogonal single-slit diffraction patterns.

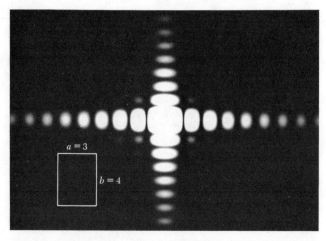

Fig. 10.3.3   Diffraction pattern of a rectangular aperture of aspect ratio 3 by 4.

Problems

10.3.1 Establish (10.3.2) and (10.3.3).

10.3.2 Verify the final form of (10.3.5).

10.3.3 A hi-fi tweeter has a rectangular aperture 5 by 12 cm. Which way would you mount this so that the sound pattern is broad in the horizontal plane and narrow in the vertical? What is the approximate radiation pattern at 10,000 Hz?

10.3.4 How does the central intensity I_0 in (10.3.13) depend upon (*a*) aperture area $S = ab$; (*b*) source wavelength λ? *Answer:* $I_0 \propto S^2/\lambda^2$. Account physically for the fact that the intensity is proportional to the square of the aperture area whereas the power in the wave incident on the aperture is, of course, proportional to the first power of the area.

10.3.5 Suppose Fraunhofer conditions are achieved using two lenses of focal lengths f_s and f_o, as shown in the sketch. How does the central intensity I_0 change if one of the lenses is replaced by a lens of different focal length (positioned to preserve the Fraunhofer condition)? *Hint:* Use conservation of energy. *Answer:* $I_0 \propto (f_s f_o)^{-2}$.

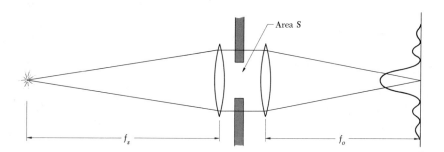

10.4 The Single Slit

Figure 10.4.1 depicts a configuration that often occurs in optical instruments. A long, narrow aperture, or *slit*, is illuminated by an incoherent *line source*, which may be thought of as a large number of randomly phased point sources arranged next to each other in a row, parallel to the long dimension of the slit. We wish to discuss the diffraction pattern in the observation plane.

Initially, we take the slit to be a rectangular aperture with $b \gg a$, and we consider only the element dy_s of the line source located at coordinates (x_s, y_s).

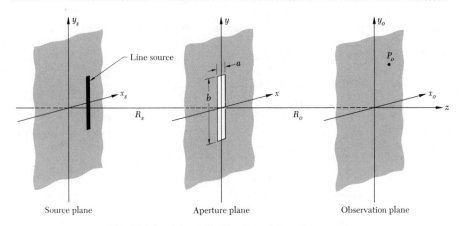

Fig. 10.4.1   A long slit illuminated by a line source.

The total diffraction pattern can then be found by invoking the principle of superposition and integrating along the line source. Furthermore, we assume that the three planes of Fig. 10.4.1 are separated by R_s, $R_o > (a^2 + b^2)/\lambda \approx b^2/\lambda$, or that lenses are used, to ensure Fraunhofer plane wavefronts at the aperture.

In the last section we solved precisely this problem, obtaining (10.3.5),

$$\psi = \check{C}'ab\,\frac{\sin u}{u}\,\frac{\sin v}{v},\tag{10.4.1}$$

where

$$u = \frac{\pi a}{\lambda}\,(l_s - l_o)\tag{10.4.2}$$

$$v = \frac{\pi b}{\lambda}\,(m_s - m_o).\tag{10.4.3}$$

To keep the geometry relatively simple, we confine our attention to source lengths and observation regions that subtend small angles at the aperture, a direct extension of the paraxial approximation. Then the respective direction cosines can be related to the respective source and observation coordinates,

$$l_s = l_s(x_s)\qquad m_s = m_s(y_s)\tag{10.4.4}$$
$$l_o = l_o(x_o)\qquad m_o = m_o(y_o).\tag{10.4.5}$$

We are now ready to carry out the integration over the line source. Qualitatively, we can see that superposing the diffraction patterns of the many "point" elements of the line source smears out the pattern of Fig. 10.3.3 in the y_o direction without affecting the structure in the x_o direction. Analytically, since the elements of the line source are assumed to radiate incoherently, we see that we must integrate the *square* of (10.4.1) with respect to y_s. Now since $b \gg a > \lambda$, a line source that is long enough to make $m_s - m_o$ reach values appreciably different from zero produces very large values of the parameter v. Thus for a sufficiently long source, the integral over the line source approaches the limit

$$\int_{-\infty}^{\infty} \left(\frac{\sin v}{v} \right)^2 dv = \pi = \text{const},$$

$(10.4.6)$

and the intensity pattern in the observation plane is

$$|\psi|^2 = C'' \left(\frac{\sin u}{u} \right)^2,$$

$(10.4.7)$

where the real coefficient C'' collects the constant factors and u is a function of x_o but not of y_o.

A long slit illuminated by a parallel-line source thus has a one-dimensional diffraction pattern that is insensitive to the length b of the slit (except as to absolute intensity). In fact, the length b need not be bounded so as to fulfill the Fraunhofer (or even the paraxial) condition (Prob. 10.4.10). Thus we can treat the single slit as a two-dimensional problem with all relevant distances and angles lying in the plane of a figure such as Fig. 10.4.2 and with merely a one-dimensional aperture integration. We could have arrived at the same conclusion by using the two-dimensional diffraction theory for cylindrical waves developed in Prob. 9.4.3 (Prob. 10.4.7). That theory assumes a *coherent* line

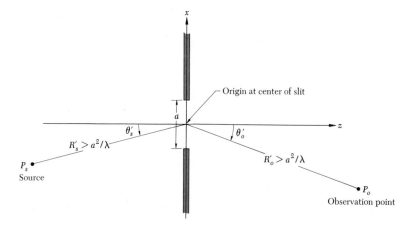

Fig. 10.4.2   Geometry of single-slit Fraunhofer diffraction. The source and slit are of indefinite extent in the y direction. The primed distances and angles and the observation point are in the plane of the figure. Note the sign convention for the angles.

source and thus differs from the incoherent case only in respect to absolute intensity.

In the notation of Fig. 10.4.2, the definition (10.4.2) of the parameter u may be written

$$u = \frac{\pi a}{\lambda} (\sin\theta_s' + \sin\theta_o').$$

(10.4.8)

For normal incidence

$$u = \frac{\pi a}{\lambda} \sin\theta_o' \xrightarrow[\theta_o' \ll 1]{} \frac{\pi a}{\lambda} \theta_o' \approx \frac{\pi a}{\lambda R_o} x_o \quad \left(\text{or } \frac{\pi a}{\lambda f_o} x_o\right),$$

(10.4.9)

where x_o is the coordinate in an observation plane at a distance R_o (or in the focal plane of a lens of focal length f_o).

Graphs of the amplitude function $\sin u/u$ and of the *single-slit intensity function*

$$w(u) \equiv \left(\frac{\sin u}{u}\right)^2$$

(10.4.10)

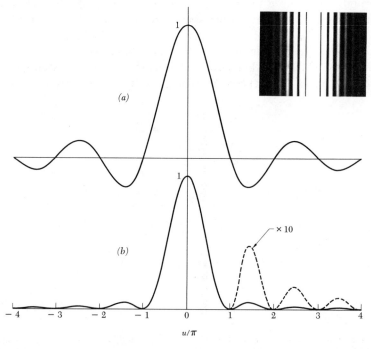

Fig. 10.4.3 Single-slit diffraction functions. (*a*) Amplitude, $\sin u/u$ and (*b*) intensity, $w(u) = (\sin u/u)^2$.

TABLE 10.1   Coordinates and Heights of Maxima for Single-slit Diffraction Pattern

u	$\dfrac{\sin u}{u}$	$w \equiv \left(\dfrac{\sin u}{u}\right)^2$	Fraction of total illumination between adjacent nulls $\dfrac{1}{\pi}\int\left(\dfrac{\sin u}{u}\right)^2 du$
0	1	1	0.9028
$4.49 = 1.43\pi$	-0.217	0.0472	$0.0236 \times 2 = 0.0471$
$7.73 = 2.46\pi$	$+0.128$	0.0169	$0.0082 \times 2 = 0.0165$
$10.90 = 3.47\pi$	-0.091	0.0083	$0.0042 \times 2 = 0.0083$
$14.07 = 4.48\pi$	$+0.071$	0.0050	$0.0025 \times 2 = 0.0050$
			$\cdots$
			1.0000

are shown in Fig. 10.4.3. Zeros, which are the minima of w, occur precisely at $u = \pm\pi, \pm2\pi, \ldots$, but at $u = 0$ the function has its *principal maximum*. *Secondary maxima* of half the width of the principal maximum occur approximately midway between the zeros, at the roots of the equation

$$\tan u = u, \tag{10.4.11}$$

which we get by setting $dw/du = 0$. Table 10.1 shows some of the roots of (10.4.11) and the corresponding values of w and its integral. The first secondary maximum is less than 5 percent of the principal maximum. The area under the principal maximum, between the zeros at $u = \pm\pi$, is 90 percent of the total area under the entire single-slit intensity pattern (10.4.6),

$$\int_{-\infty}^{\infty}\left(\frac{\sin u}{u}\right)^2 du = \pi. \tag{10.4.12}$$

A polar plot of the intensity function is shown in Fig. 10.4.4. To the extent that the slit width is very large compared with the wavelength, $a \gg \lambda$, the diffraction pattern closely surrounds the central maximum at the observation angle $\theta_o' = -\theta_s'$ (Fig. 10.4.2). The angular width $\Delta\theta_o'$ of the central maximum can be found by taking the differential of u

$$\Delta u = \frac{\kappa a}{2}\cos\theta_o' \, \Delta\theta_o' \tag{10.4.13}$$

and setting $\Delta u = 2\pi$, with the result that

$$(\Delta\theta_o')_{\substack{\text{central}\\\text{max}}} = \frac{2\lambda}{a\,\cos\theta_o'} \xrightarrow[\theta_o'\ll1]{} \frac{2\lambda}{a}. \tag{10.4.14}$$

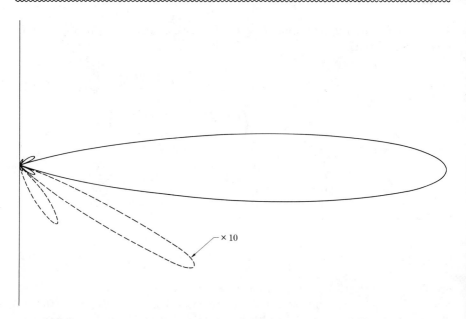

Fig. 10.4.4   Polar plot of the intensity function $(\sin u/u)^2$ for $a/\lambda = 3$. The secondary maxima are sometimes called *side lobes*.

The angular width of the central peak of intensity determines the resolution of an optical instrument having a slit aperture. Two adjacent point or line sources of equal strength are said to be resolved according to *Rayleigh's criterion* when the central peak of one diffraction pattern falls on the first minimum of the other pattern, as shown in Fig. 10.4.5. The central valley of the

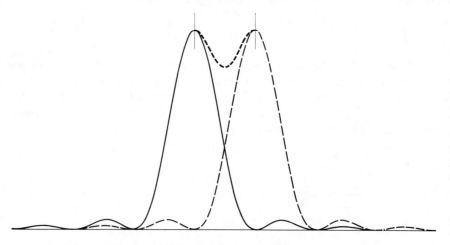

Fig. 10.4.5   Rayleigh's criterion for resolution of two sources.

total intensity curve in this case has the relative height

$$2\left(\frac{\sin\frac{1}{2}\pi}{\frac{1}{2}\pi}\right)^2 = \frac{8}{\pi^2} = 0.811. \tag{10.4.15}$$

For small θ_o', the angular separation of two sources just resolved is then

$$\Delta\theta_{\text{Rayleigh}} = \frac{\lambda}{a} \qquad \text{linear slit.} \tag{10.4.16}$$

We show in Sec. 10.5 that the analogous angular resolution of a circular aperture of diameter d is

$$\Delta\theta_{\text{Rayleigh}} = 1.22\frac{\lambda}{d} \qquad \text{circular aperture.} \tag{10.4.17}$$

Problems

10.4.1 Review the elementary arguments for discussing single-slit Fraunhofer diffraction by pairing up portions of the wavefront in the aperture that are 180° out of phase to show that (a) a null is observed when the slit may be divided into two equal portions such that

$$\frac{\delta}{2} \equiv \frac{a}{2}\sin\theta = \frac{\lambda}{2};$$

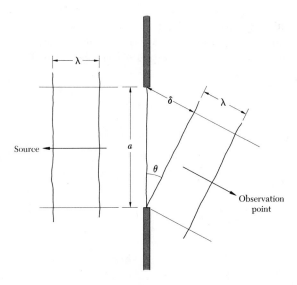

Prob. 10.4.1

(b) Similarly, nulls result when the aperture is divided up into $2n$ portions such that

$$\frac{a}{2n} \sin\theta = \frac{\lambda}{2},$$

where n is an integer; (c) the secondary maxima occur *approximately* when the aperture is divided up into $2n + 1$ portions such that

$$\frac{a}{2n + 1} \sin\theta = \frac{\lambda}{2}$$

and that the intensity of these maxima, relative to the principal maximum, is approximately

$$\frac{1}{2}\left(\frac{1}{2n + 1}\right)^2.$$

How do you justify the factor of $\frac{1}{2}$?

10.4.2   A line source at visible-light wavelengths is often made by imaging a gas-discharge lamp on a narrow slit, which then serves as the "line" source. This secondary source is placed at the focus of a lens of focal length f_s to provide collimated light. How small must the width w of the source slit be so as not to affect significantly the Fraunhofer diffraction pattern of a slit of width a? What if the lens is not used and the distance between source and diffracting slits is $R > a^2/\lambda$? *Answer:* $w \lesssim \lambda f/4a$; $a/4$.

10.4.3   An adjustable slit, of variable width a, is set up on a spectroscope table (see sketch). Collimated light from a monochromatic line source is incident normally; the diffraction pattern is observed with the telescope, which has an angular field of view of ± 0.1 rad. What does the observer see as the slit width is varied—how many secondary maxima and what relative intensity of the principal maximum? *Answer:* $a/10\lambda - 1$ on each side; $I_0 \propto a^2$.

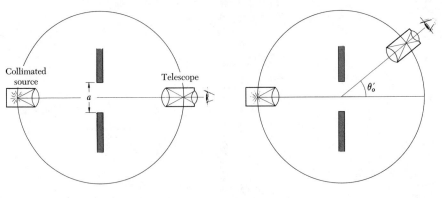

Prob. 10.4.3   Prob. 10.4.4

10.4.4 A narrow slit, a few wavelengths in width, is set up on the spectroscope table of Prob. 10.4.3. The diffraction pattern is observed by swinging the telescope over the range $-\pi/2 < \theta_o' < \pi/2$ (see sketch). What is the intensity as a function of θ_o'? *Answer:*

$$I_0 \left(\frac{1 + \cos\theta_o'}{2}\right)^2 \left(\frac{\sin u}{u}\right)^2 \qquad u = \frac{\pi a}{\lambda} \sin\theta_o'.$$

10.4.5 The telescope of Prob. 10.4.3 is again set opposite the source ($\theta_o' = 0$). A slit of width $a = 30\lambda$ is arranged to pivot about its center by the angle θ (see sketch). What is seen by the observer as θ is increased from 0 to $\pi/2$? Consider both the number of secondary maxima and the intensity of the principal maximum. *Answer:* $3 \cos\theta - 1$ secondary maxima; $I_0 \propto a^2 \cos^2\theta$; same as Prob. 10.4.3 with $a \rightarrow a \cos\theta$.

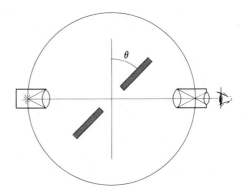

Prob. 10.4.5

★10.4.6 The statement may be made that Fraunhofer diffraction pertains to situations in which source and observer are focally conjugate, i.e., represent object and image of a focused geometrical-optics system. Consider, in the figure, an aperture placed in diverging light (plane A) or in converging light (plane B). Assume a line source and slit aperture, so that the aperture integration is one-dimensional. Show that Fraunhofer theory does indeed describe the diffraction pattern in the plane containing O, in the paraxial approximation. Discuss how the pattern changes as a slit of constant width is moved along the axis from O to S, the positions of S, L, and O remaining fixed.

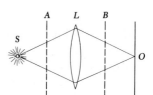

Prob. 10.4.6.

★**10.4.7**   Use the cylindrical-wave form of the Fresnel-Kirchhoff integral developed in Prob. 9.4.3 to calculate the Fraunhofer diffraction pattern (10.3.5) of an infinite slit of constant width a. This approach implies a line source that emits coherently along its length. How does the diffraction pattern predicted by this model differ from that of the incoherent line source assumed in the text?

10.4.8   Calculate the Fraunhofer diffraction pattern of a single slit which is illuminated (at normal incidence) by a wave whose amplitude varies parabolically across the slit (zero at edges, rising to a maximum at the center). Show that, relative to the case of uniform illumination, the central maximum is broadened and the secondary maxima are depressed. *Answer:* $(\sin u - u \cos u)/(u^3/3)$.

10.4.9   Repeat Prob. 10.4.8 for half-period cosinusoidal illumination (zero at edges, single maximum at the center). *Answer:* $\cos u/(1 - 4u^2/\pi^2)$.

★**10.4.10**   Show that the integration of the Kirchhoff integral (9.4.16) can be formally carried out over y for an arbitrarily long slit (Fig. 10.4.2) to obtain

$$\int_L \frac{\frac{1}{2}(\cos\theta_s + \cos\theta_0)}{r_s r_0} e^{i\kappa(r_s+r_0)} \, dy = \check{C}(L,\theta_s',\theta_0') \, e^{i\kappa(\sin\theta_s'+\sin\theta_0')x}$$

where $\check{C}$ is a complex constant whose value depends upon the limits L of the slit's extent in the y direction but not upon x. *Hint:* The reference surface for computing incremental paths may

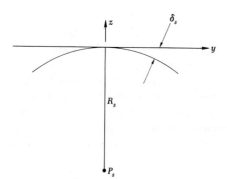

Prob. 10.4.10

be visualized as a circular cylinder. Define a function that represents the variation of δ_s with y for the elementary strip of width dx passing through the origin (see sketch). Then show that for the elementary strip passing through x the *same* function represents the incremental path with respect to a *different* reference cylinder but nevertheless one whose radius is known relative to the former reference cylinder.

★*10.4.11* Find an aperture illumination function such that the diffraction pattern has no secondary maxima (no side lobes) but rather falls to zero monotonically on either side of the principal maximum.

★*10.4.12* Adapt the Young-Rubinowicz formula (9.6.13) to the single slit in the Fraunhofer limit, showing that the diffraction integral reduces to a discrete sum of two terms.

10.5 The Circular Aperture

Fraunhofer diffraction at a circular aperture is of considerable practical importance since most optical instruments (including the eye) have circular apertures. We assume that the diameter $d = 2a$ of the aperture is large compared with the wavelength λ, and we shall simplify the analysis by putting the monochromatic point source on the axis of the aperture, as in Fig. 10.5.1. We assume that both the point source and the observation point are effectively at infinity. The integration of the diffraction integral (10.2.2) over a circular aperture is most conveniently carried out in polar coordinates. Hence let us define a polar coordinate system ρ, ϕ, equivalent to the x, y coordinate system, with its origin O at the center of the aperture and with the radius $\phi = 0$ coincident with the x axis. If we choose the direction of the x axis to lie in the plane defined by the three points P_s, O, and P_o, the path-difference function $\delta(x,y)$ in the diffraction integral depends only on the coordinate x. Hence the expression for δ, as given by (10.3.4), becomes

$$\delta(\rho,\phi) = \rho \sin\theta_o \cos\phi \tag{10.5.1}$$

since $l_s = m_s = m_o = 0$, $l_o = -\sin\theta$ and $x = \rho \cos\phi$. The Fraunhofer diffraction integral then becomes

$$\psi = \breve{C}' \int_0^{2\pi} \int_0^a e^{ik\rho\sin\theta_o\cos\phi} \, \rho \, d\rho \, d\phi \tag{10.5.2}$$

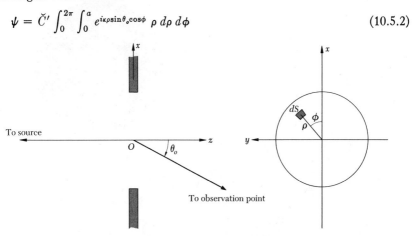

Fig. 10.5.1 Diffraction by a circular aperture.

with the element of area $dS = \rho\,d\rho\,d\phi$. The integral can be evaluated by expanding the exponential in an infinite series and integrating the series term by term. The series so obtained is recognized as being closely related to a Bessel function of the first order, as given by (2.3.9) with $m = 1$ (Fig. 2.3.1). If we introduce the variable

$$u = a\kappa\sin\theta_o = \frac{\pi d}{\lambda}\sin\theta_o,$$ (10.5.3)

the amplitude at P_o is found to be

$$\psi = \check{C}'\pi a^2\,\frac{2J_1(u)}{u},$$ (10.5.4)

where

$$J_1(u) \equiv \frac{u}{2}\left[1 - \frac{1}{1!2!}\left(\frac{u}{2}\right)^2 + \frac{1}{2!3!}\left(\frac{u}{2}\right)^4 - \frac{1}{3!4!}\left(\frac{u}{2}\right)^6 + \cdots\right]$$ (10.5.5)

is the Bessel function of first order. The intensity, which is proportional to $|\psi|^2$, may be written

$$I = I_0\left[\frac{2J_1(u)}{u}\right]^2$$ (10.5.6)

where I_0 is the intensity at the center of the diffraction pattern. The circular-aperture pattern (10.5.6) is directly analogous to the rectangular-aperture pattern (10.3.13).

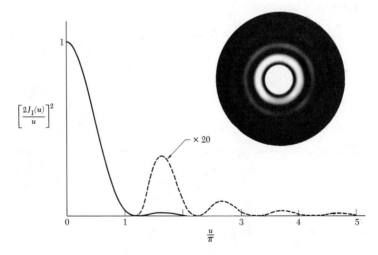

Fig. 10.5.2   Plot of circular-aperture intensity function.

TABLE 10.2   Coordinates of Maxima and Minima for
Circular-aperture Diffraction Pattern

u	$\dfrac{I}{I_0} = \left[\dfrac{2J_1(u)}{u}\right]^2$	Fraction of total illumination between adjacent nulls $\dfrac{1}{2}\displaystyle\int \left[\dfrac{2J_1(u)}{u}\right]^2 u\,du$
0	1	0.838
$3.832 = 1.220\pi$	0	
$5.136 = 1.635\pi$	0.0175	0.072
$7.016 = 2.233\pi$	0	
$8.417 = 2.679\pi$	0.0042	0.028
$10.173 = 3.238\pi$	0	
$11.620 = 3.699\pi$	0.0016	0.014
		$\cdots$
		1.000

A plot of the intensity function $[2J_1(u)/u]^2$ for the circular aperture is given in Fig. 10.5.2. It resembles the single-slit function $(\sin u/u)^2$ but differs in details. The diffraction pattern of a point source thus consists of a bright central disk, sometimes called the *Airy disk*, surrounded by a system of dark and light rings. In Prob. 10.5.2 it is established that the intensity maxima occur at the roots of the Bessel function of second order, $J_2(u)$. Table 10.2 gives the values of u making the intensity zero and maximum, the relative intensities at the maxima, and the total illumination in various parts of the pattern.

The angular width of the central disk can be found from the value $u = 3.832$ for the first zero of the intensity function. For $a \gg \lambda$, the angular width is a small angle, and we may replace $\sin\theta_o$ in (10.5.3) by θ_o. Since the separation between the first-order nulls is

$$\Delta u = \kappa a\, \Delta\theta_o = 2 \times 3.832,$$

the angular diameter of the central disk is

$$(\Delta\theta)_{\substack{\text{central} \\ \text{max}}} = \frac{3.832\,\lambda}{\pi\quad a} = 1.22\,\frac{\lambda}{a}. \tag{10.5.7}$$

The Rayleigh criterion for the resolution of two point sources requires that the central disk of one diffraction pattern fall on the first zero of the other pattern. This occurs for an angular separation of two point sources that is half the angular width of the central disk. Hence, as quoted in (10.4.17),

$$(\Delta\theta)_{\text{Rayleigh}} = 0.61\frac{\lambda}{a} = 1.22\frac{\lambda}{d}, \qquad (10.5.8)$$

where $d = 2a$ is the diameter of the aperture.

It is of interest to apply this result to estimate the theoretical resolution of an astronomical telescope, whose objective lens, or mirror, serves as the aperture. For the 200-in. Hale telescope on Mount Palomar, $\Delta\theta_{\text{Rayleigh}} = 0.03$ second of arc (the angular diameter of Mars viewed from the earth at closest approach is about 18 seconds of arc). This resolution is not realized in practice because of image distortions caused by atmospheric turbulence.

The resolution of the eye is also limited in a fundamental way by diffraction. The diameter of the pupil varies from approximately 1.5 to 6 mm, depending on the intensity of light. At a wavelength of 560 nm, the angular resolution of the eye should then lie between approximately 1.6 and 0.4 minute of arc. The eye has a diameter of about 1.5 cm, so that an angular separation of 1 minute of arc corresponds to a distance of about 0.005 mm on the retina. This is in fact the order of magnitude of the diameter of the light-sensitive cells (*cones*) in the *fovea*, the small region in the eye of most distinct vision.

The measured resolution of a normal eye, of about 1 minute of arc, agrees fairly closely with the limit set by diffraction theory. With the pupil wide open, the angular resolution suffers somewhat because of aberrations of the lens of the eye. Visual optical instruments are usually designed, therefore, so that the pencil of rays entering the eye has a diameter of 4 to 5 mm. Except for high light intensities, the eye is then used to its best advantage so far as resolution is concerned.

Problems

10.5.1 Evaluate the diffraction integral (10.5.2) by the method described and obtain the expression (10.5.4) for the wave amplitude at the observation point.

10.5.2 Use the Bessel function recursion relations (2.3.10) and (2.3.11) to establish that the maxima of $[J_1(u)/u]^2$ occur at the roots of $J_2(u) = 0$.

10.5.3 Carry out the numerical calculations for the resolution of the eye to verify the results quoted in the text.

★10.5.4 Show that the Fraunhofer diffraction pattern of an elliptical aperture, of semiaxes a and b, is identical to that of a circular aperture except for a linear expansion by the ratio a/b in the direction parallel to the minor axis b.

★10.5.5   Show that the fraction of the total energy in the circular-aperture diffraction pattern out to the radius specified by u_{max} is

$$\frac{1}{2} \int_0^{u_{max}} \left[\frac{2J_1(u)}{u} \right]^2 u \, du = 1 - J_0{}^2(u_{max}) - J_1{}^2(u_{max}).$$

Hence, verify the final column of Table 10.2.

10.5.6   The largest fully steerable radio telescope at the National Radio Astronomy Observatory (Green Bank, W.Va.) has a parabolic mirror 140 ft in diameter. Approximately what angular resolution does it have for the 1,420-MHz (21-cm) line radiated by interstellar atomic hydrogen? (In practice, the sensitivity of the detector at the parabola focus is not isotropic. A careful analysis would include a nonuniform aperture-illumination function analogous to Probs. 10.4.8 and 10.4.9.) The mirror was constructed to conform to an ideal paraboloid within a tolerance of 0.030 in. What minimum wavelength can be used with this instrument if the criterion is that the construction errors not exceed $\lambda/8$? What is the approximate resolution for the minimum wavelength? *Answer:* 21 minutes of arc; $\lambda = 6$ mm, 0.6 minute of arc.

★10.5.7   The reflection cross section of a radar target is defined by

$$4\pi \frac{\text{power/unit solid angle reflected back toward radar antenna}}{\text{power/unit area incident upon target}}.$$

Show that for a circular plane mirror of diameter d at normal incidence, the cross section is of the order of magnitude d^4/λ^2. If possible, find the numerical coefficient.

★10.5.8   Show that when a circular aperture is illuminated by a plane wave whose amplitude varies with radius as $T(\rho)$, from (9.4.19), the diffraction pattern is given by

$$\psi(u) = 2\pi \breve{C}' \int_0^a T(\rho) J_0\left(\frac{u\rho}{a}\right) \rho \, d\rho,$$

which reduces to (10.5.4) when $T(\rho) = 1$.

10.6   The Double Slit

The Fraunhofer diffraction at a double slit serves to introduce several features that characterize diffraction at aperture screens having a number of regularly spaced openings. In addition, this case has an intrinsic interest as the basis of the Michelson stellar interferometer, which has been used to establish the angular size of nearby giant stars. Let us designate the width of each slit by a and the center-to-center spacing by b.

   In analogy with the single-slit case of Sec. 10.4, we can treat the diffraction at two parallel slits as a one-dimensional problem, using the geometry shown in

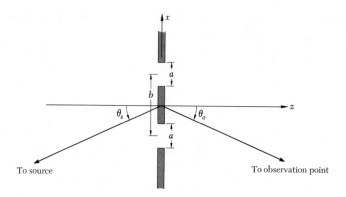

Fig. 10.6.1   Diffraction by a double slit.

Fig. 10.6.1. In terms of the angle of incidence θ_s and the angle of diffraction θ_o, the path-difference function $\delta(x,y)$ takes the form

$$\delta(x) = (\sin\theta_s + \sin\theta_o)x. \tag{10.6.1}$$

The Fraunhofer diffraction integral thus becomes

$$\psi = \check{C}\left(\int_{(-b-a)/2}^{(-b+a)/2} e^{i\kappa\delta(x)}\, dx + \int_{(b-a)/2}^{(b+a)/2} e^{i\kappa\delta(x)}\, dx\right). \tag{10.6.2}$$

On carrying out the integration and introducing the notation

$$\begin{aligned} u &\equiv \frac{a\kappa}{2}\,(\sin\theta_s + \sin\theta_o) \\ v &\equiv \frac{b}{a}\,u = \frac{b\kappa}{2}\,(\sin\theta_s + \sin\theta_o), \end{aligned} \tag{10.6.3}$$

the amplitude of the wave at the observation point is found to be

$$\psi = \check{C}2a\,\frac{\sin u}{u}\,\cos v. \tag{10.6.4}$$

The intensity distribution in the diffraction pattern is therefore

$$I = I_0\left(\frac{\sin u}{u}\right)^2 \cos^2 v, \tag{10.6.5}$$

where I_0 is the central intensity.

Equation (10.6.5) for the intensity contains as factors the single-slit diffraction factor $(\sin u/u)^2$ and what we may call the *double-slit interference factor* $\cos^2 v$. Since necessarily $b > a$, the zeros of the single-slit factor are more widely spaced in angle than the zeros of the interference factor. Hence a plot of the intensity function has the enveloping shape of a single-slit pattern, with a finer structure due to the interference of the waves coming from the two slits. A typical double-slit pattern is shown in Fig. 10.6.2. Note that if either slit were to be covered up, the two single-slit patterns would be identical, occupying the same angular position as the pattern drawn with dashed lines in Fig. 10.6.2 but having only one-fourth the intensity amplitude. The angular width of the narrow lines, between $\cos v$ zeros, is easily found to be

$$\Delta\theta = \frac{\lambda}{b \ \cos\theta_o}. \tag{10.6.6}$$

Two neighboring point sources of equal strength are resolved, according to the Rayleigh criterion, when they are separated by half the angular width of a narrow line. For normal incidence, this angle is

$$(\Delta\theta)_{\text{Rayleigh}} = \frac{\lambda}{2b}. \tag{10.6.7}$$

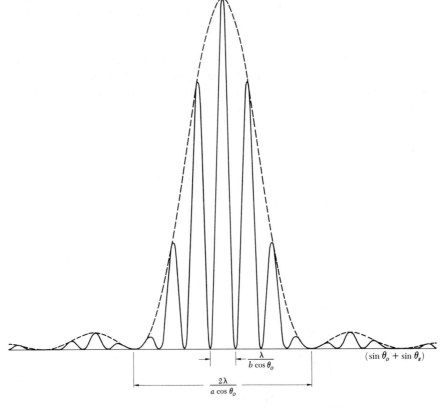

Fig. 10.6.2   Double-slit intensity pattern $(b/a = 3\frac{1}{2})$.

By making b large, a high resolution can be obtained. A comparison of (10.6.7) with the corresponding expression for the angular resolution of a circular aperture (10.5.8) indicates that the resolution of an optical instrument can be nearly doubled, at the expense of light intensity, by placing an opaque screen with two widely spaced slits in front of the objective lens.

An aperture consisting of two slits with an adjustable spacing mounted at the objective of an astronomical telescope has been used to measure the angular separation of certain double stars. Such an arrangement can also be used to measure the angular diameter of certain planetary satellites but fails for single stars because of their much smaller angular diameter. Michelson, however, managed to measure the angular diameter of the star Betelgeuse by the arrangement illustrated in Fig. 10.6.3, where mirrors are used to increase the effective separation of the two slit apertures at the objective of the telescope. The mirrors M_1 and M_2, whose separation could be made as large as 20 ft (6 m) permitted an angular resolution (at $\lambda = 560$ nm) of about 0.01 second of arc. If we think of the light from the disk of a star as coming from two point sources separated somewhat less than one radius, Michelson's original *stellar interferometer* should be able to measure angular diameters of stars as small as about 0.02 second of arc. On the basis of observation and a more exact analysis, Betelgeuse was found to have an angular diameter of 0.047 second of arc. Since the distance of this star is known by triangulation, using the diameter of the earth's orbit as a base line, its actual diameter can be computed; it is found to be 300 times that of the sun, making it greater than the diameter of the earth's orbit. The angular diameters of a few other nearby giant stars have also been measured, but most stars subtend too small an angle to be measured by Michelson's method.

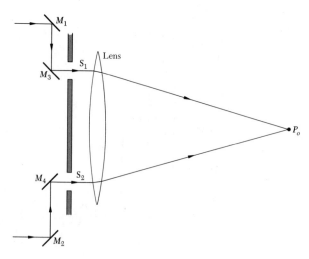

Fig. 10.6.3 Michelson's stellar interferometer.

Problems

10.6.1 Plane waves are incident normally on a double slit. Assume that the amplitude of the wave diffracted at an angle θ by each slit is proportional to some function $F(u)$ of $u = \frac{1}{2}a\kappa\sin\theta$. Show that the amplitude of the wave found by combining the two waves from the slits, assuming Fraunhofer geometry, is proportional to $F(u)\cos v$, where $v = \frac{1}{2}b\kappa\sin\theta$.

10.6.2 Investigate what happens to the double-slit pattern in the limit when b approaches a so that the two slits merge into a single slit of width $2a$.

★10.6.3 Work out the theory of double-slit Fraunhofer diffraction when the two slits have different widths $a_1 = 2a_2$ and b, the distance between slit centers, is $4a_2$.

10.7 Multiple Slits

Fraunhofer diffraction at apertures having many evenly spaced parallel slits is an extension of the double-slit case that we have just considered. Its study serves to introduce the theory of diffraction gratings, which have great importance in the measurement of electromagnetic radiations emitted by atoms and molecules. As in the double-slit case, we find that the intensity of the diffracted wave is given by the product of a *diffraction factor* that depends on the width a of the individual slits and an *interference factor* that depends on the slit spacing b and on the total number of slits N. Both factors depend, of course, on the angle of incidence θ_s and the angle of diffraction θ_o. In the present section we look into the theory of the idealized "picket-fence" diffraction grating, and in the following section we extend the analysis to the practical diffraction gratings used in spectral analysis.

The path-difference function depends only upon the one-dimensional aperture coordinate x and is identical to (10.6.1) used for the double slit,

$$\delta(x) = (\sin\theta_s + \sin\theta_o)x. \tag{10.7.1}$$

It is convenient to place the origin at the center of the first slit, as in Fig. 10.7.1. The Fraunhofer diffraction integral for a grating of N evenly spaced slits is then the sum of N integrals, one for each slit,

$$\psi = \check{C}\left(\int_{-a/2}^{a/2} + \int_{b-a/2}^{b+a/2} + \int_{2b-a/2}^{2b+a/2} + \cdots + \int_{(N-1)b-a/2}^{(N-1)b+a/2}\right) e^{i\kappa\delta(x)}\,dx. \tag{10.7.2}$$

By changing the variable of integration in the nth integral from x to $x' \equiv x - (n-1)b$ and introducing the notation found convenient in discuss-

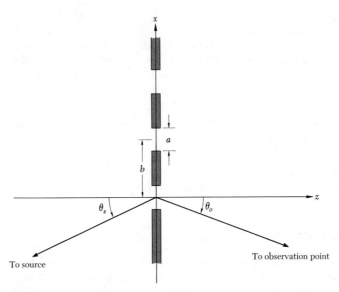

Fig. 10.7.1   Diffraction by a picket-fence grating, consisting of N evenly spaced parallel slits.

ing the double slit

$$u \equiv \frac{a\kappa}{2}\,(\sin\theta_s + \sin\theta_o)$$

$$v \equiv \frac{b}{a}\,u = \frac{b\kappa}{2}\,(\sin\theta_s + \sin\theta_o), \tag{10.7.3}$$

the diffraction integral becomes (Prob. 10.7.1)

$$\psi = \check{C}\left(\int_{-a/2}^{a/2} e^{i2ux'/a}\,dx'\right)\left(1 + e^{i(2v)} + e^{2i(2v)} + \cdots + e^{(N-1)i(2v)}\right). \tag{10.7.4}$$

The integral that is a common factor for all the terms in (10.7.4) is the single-slit diffraction integral having the value $(a\,\sin u)/u$. The series multiplying this integral constitutes a geometric progression that can be summed using the identity

$$1 + x + x^2 + \cdots + x^{n-1} = \frac{x^n - 1}{x - 1}. \tag{10.7.5}$$

Accordingly we find that

$$\psi = \check{C}a\,\frac{\sin u}{u}\,\frac{e^{iN(2v)} - 1}{e^{i(2v)} - 1}$$

$$= \check{C}ae^{i(N-1)v}\,\frac{\sin u}{u}\,\frac{e^{iNv} - e^{-iNv}}{e^{iv} - e^{-iv}}$$

$$= \check{C}ae^{i(N-1)v}\,\frac{\sin u}{u}\,\frac{\sin Nv}{\sin v}. \tag{10.7.6}$$

The intensity in the diffraction pattern may be written

$$I = I_0\left(\frac{\sin u}{u}\right)^2\left(\frac{\sin Nv}{N\,\sin v}\right)^2, \tag{10.7.7}$$

where I_0 is the intensity when $u = v = 0$. The factor N^2 has been placed in the denominator of (10.7.7) so that

$$\lim_{v\to 0}\left(\frac{\sin Nv}{N\,\sin v}\right)^2 = 1.$$

Let us now examine the significance of what we have found. As mentioned earlier, the intensity function is the product of the single-slit diffraction factor $(\sin u/u)^2$ and what we may call the *grating interference factor*,

$$\left(\frac{\sin Nv}{N\,\sin v}\right)^2. \tag{10.7.8}$$

A plot of the intensity function for $N = 5$ and $b/a = v/u = 2\frac{1}{2}$ appears in Fig. 10.7.2. The interference factor gives rise to a fine-scaled pattern having the single-slit pattern as an envelope. The pattern is symmetrical about the v origin. The central peak occurs, of course, at the angle $\theta_o = -\theta_s$.

In Sec. 10.4 we examined the mathematical properties of the single-slit diffraction factor $(\sin u/u)^2$. Now we must investigate the N-slit interference factor $(\sin Nv/N\,\sin v)^2$. Several examples are shown in Fig. 10.7.3. First note that it has zeros when $Nv = n\pi$, where n is any positive or negative integer *that is not an integral multiple of N*. When $n = mN$ ($m = 0, \pm 1, \pm 2, \ldots$), v takes on the values $v = m\pi$, which make $\sin v = 0$. For these values of v the interference factor has the value

$$\lim_{v\to m\pi}\left(\frac{\sin Nv}{N\,\sin v}\right)^2 = 1 \tag{10.7.9}$$

instead of zero, and maxima, known as *principal maxima*, occur in the diffraction pattern. The intensities at the principal maxima touch the enveloping

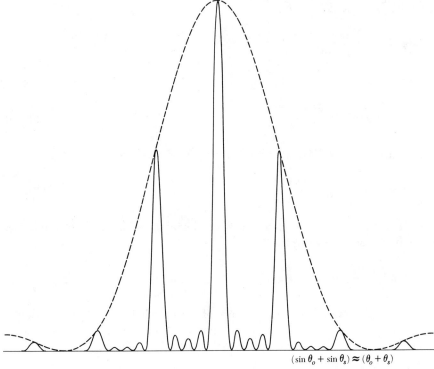

$$(\sin\theta_o + \sin\theta_s) \approx (\theta_o + \theta_s)$$

Fig. 10.7.2   The Fraunhofer diffraction pattern for five slits ($b/a = 2\frac{1}{2}$).

single-slit pattern

$$I = I_0\left(\frac{\sin u}{u}\right)^2.$$

Between pairs of adjacent principal maxima, e.g., between those at $v = 0$ and $v = \pi$, the interference factor has $N - 1$ zeros, occurring precisely at

$$v = \frac{\pi}{N}, \frac{2\pi}{N}, \cdots, \frac{N-1}{N}\pi \quad \text{(zeros)}. \tag{10.7.10}$$

In the range $\pi < v < 2\pi$, the zeros are at the values (10.7.10) increased by π, and so forth.

If we neglect the very slight effect of the diffraction factor $(\sin u/u)^2$ in displacing the location of maxima in the diffraction pattern, i.e., if we assume $a \ll b$, we can find the position of all the maxima (and minima) by setting $d(\sin Nv/N \sin v)^2/dv = 0$. The maxima are found to occur at the roots of the equation

$$N \tan v = \tan Nv. \tag{10.7.11}$$

The roots $v_m = m\pi$ ($m = 0, \pm 1, \pm 2, \ldots$), which make each side of (10.7.11) vanish, give the principal maxima already described. In addition to these principal maxima, there are $N - 2$ *subsidiary maxima* at the roots of (10.7.11) occurring nearly midway between adjacent zeros. In the range $0 < v < \pi$, the subsidiary maxima therefore occur approximately at

$$v \approx \frac{3\pi}{2N}, \frac{5\pi}{2N}, \cdots, \frac{(2N - 3)\pi}{2N} \qquad \text{(subsidiary maxima)} \tag{10.7.12}$$

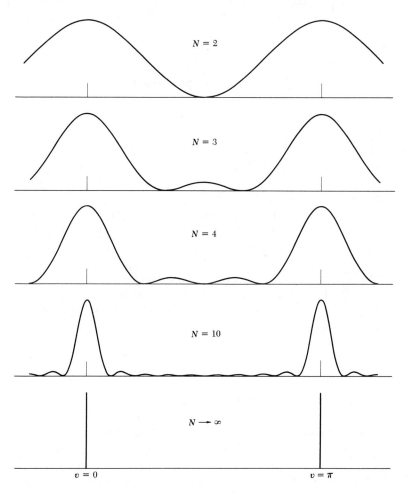

Fig. 10.7.3   The interference factor $(\sin Nv/N \sin v)^2$ for $N = 2, 3, 4, 10$, and ∞.

and at corresponding positions in the intervals between other adjacent principal maxima. The subsidiary maxima have a width $\Delta v = \pi/N$ between adjacent zeros, whereas the principal maxima have twice this width.

If v_m is the value of v at a subsidiary maximum of the interference factor, the intensity at this maximum is given by

$$I_m = I_0 \left(\frac{\sin N v_m}{N \sin v_m}\right)^2 = I_0 \left(\frac{\cos N v_m}{\cos v_m}\right)^2. \tag{10.7.13}$$

We may find the approximate height of the subsidiary maximum next to a principal maximum for large N by setting $v_m = 3\pi/2N$, the first of the values listed in (10.7.12), and making use of the fact that $3\pi/2N \ll 1$, so that the approximation $\sin\theta \approx \theta$ may be used. We find (see also Prob. 10.7.2) that

$$I_m \approx \frac{I_0}{(3\pi/2)^2} \approx \frac{I_0}{22}. \tag{10.7.14}$$

Near the middle of the interval between principal maxima,

$$I_m \approx \frac{I_0}{N^2}. \tag{10.7.15}$$

Except for the effect of the single-slit factor (neglected here) the subsidiary maxima form a symmetrical array between the principal maxima, as is evident in Fig. 10.7.3.

When N is very large, the subsidiary maxima of appreciable height are crowded in so close to the principal maxima that they cannot be resolved, and only the extremely narrow principal maxima are observed. The subsidiary maxima give rise to only a slight background illumination between the principal maxima. When the slit opening a is very small, comparable with a wavelength λ, the Kirchhoff diffraction theory is less accurate, and it is to be expected that the single-slit diffraction factor will no longer give a good description of the envelope. The interference factor (10.7.8), however, continues to give the positions and the widths of the principal maxima quite accurately.

Problems

10.7.1 Establish the diffraction integral in the form (10.7.4).

10.7.2 Show that in the limit of many slits ($N \to \infty$), the interference pattern in the vicinity of the principal maxima takes on the form of the single-slit diffraction pattern.

10.7.3 The width of a curve having a maximum (with zeros on each side) is often specified by physicists by stating its *full width at half height*. Prove that the angular width at half intensity of a principal maximum of an N-slit diffraction pattern is 0.44 times the angular width measured between adjacent zeros (which has the value $\Delta\theta = 2\lambda/Nb \cos\theta$).

*10.8   Practical Diffraction Gratings for Spectral Analysis

(a)   Gratings of Arbitrary Periodic Structure

Most diffraction gratings used for spectral analysis differ considerably from the ideal picket-fence grating we have been discussing. A more general sort of diffraction grating consists of any *periodic* structure that can alter the amplitude, or phase, or both of an incident wave. Practical gratings are usually made by a *ruling engine,* which cuts a large number of identical evenly spaced parallel grooves on a glass plate or a metallic mirror surface.

Consider a general periodic variation of the transmission properties of a plane screen as suggested in Fig. 10.8.1. We suppose that both the transmitted amplitude and the phase shift have a periodicity b in the x direction and that the total aperture comprises N such periods or cycles.

We can express the path-difference function in the form

$$\delta(x) = (\sin\theta_s + \sin\theta_o)x + \breve{P}(x), \qquad (10.8.1)$$

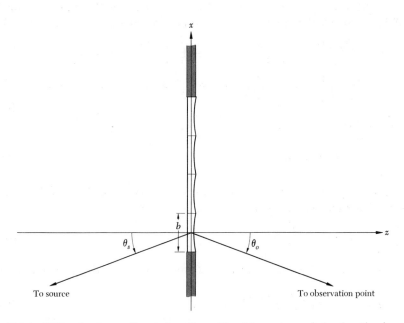

Fig. 10.8.1   Diffraction by an N-period grating with arbitrary transmission function in each period of width b.

where $\check{P}(x)$ is a periodic function of x, of period b. This form is clearly valid for a periodic *phase shift* since a variation in optical path caused by altered wave velocity is equivalent to a variation in geometrical path length between source and observation point. It is perhaps less obvious that taking $\check{P}(x)$ to be complex,

$$\check{P}(x) \equiv P_r(x) + iP_i(x), \tag{10.8.2}$$

permits taking a periodic variation of *opacity* into account. To see that the imaginary component of $\check{P}(x)$ expresses a variation in the amplitude of the wave passing through the diffraction grating, we substitute (10.8.1), with $\check{P}(x)$ complex, into the Fraunhofer diffraction integral (10.2.2), to obtain

$$\psi = \check{C} \int_{\text{grating}} e^{i\kappa[(\sin\theta_s + \sin\theta_o)x + P_r(x)]} \, e^{-\kappa P_i(x)} \, dx. \tag{10.8.3}$$

We now see that the periodic factor $e^{-\kappa P_i(x)}$ can describe an arbitrary periodic variation in transmitted wave *amplitude* with position along the grating. Hence we have a simple formalism for writing the diffraction integral for a grating of an arbitrary periodic structure, regardless of the nature of the structure.

Considerable progress can be made in simplifying the diffraction integral (10.8.3) without knowing anything about $\check{P}(x)$ except that it has a period b and that the grating contains N of these periods across its width. For convenience let us put the x origin at a distance $b/2$ from the edge of the grating, as in Fig. 10.8.1, and break the diffraction integral into N parts as we did in (10.7.4),

$$\psi = \check{C} \left(\int_{-b/2}^{b/2} + \int_{b/2}^{3b/2} + \cdots + \int_{[(2N-3)b]/2}^{[(2N-1)b]/2} \right) e^{i\kappa\delta(x)} \, dx$$

$$= \check{C} \left(\int_{-b/2}^{b/2} e^{i\kappa\delta(x)} \, dx \right) \left(1 + e^{i(2v)} + e^{2i(2v)} + \cdots + e^{(N-1)i(2v)} \right), \tag{10.8.4}$$

where, as before,

$$v = \frac{\kappa b}{2} (\sin\theta_s + \sin\theta_o) \tag{10.8.5}$$

and $\delta(x)$ is given by (10.8.1). Since we do not know the form of $\check{P}(x)$, let us denote the unknown value of the integral in (10.8.4) by

$$\check{F}(\theta_s, \theta_o) \equiv \check{C} \int_{-b/2}^{b/2} e^{i\kappa\delta(x)} \, dx. \tag{10.8.6}$$

The amplitude of the diffracted wave at the observation point then becomes

$$\psi = e^{i(N-1)v} \, \check{F}(\theta_s, \theta_o) \, \frac{\sin Nv}{\sin v}, \tag{10.8.7}$$

where the series in (10.8.4) has been summed exactly as was done in obtaining

(10.7.6). The intensity of the diffracted wave is clearly

$$I \propto N^2 |\breve{F}|^2 \left(\frac{\sin Nv}{N \sin v} \right)^2. \tag{10.8.8}$$

Hence, for the general grating, the intensity is proportional to (1) a *diffraction factor* $|\breve{F}|^2$ that depends on the nature of the individual rulings of the grating and, of course, on the angles θ_s and θ_o and (2) to the same *grating interference factor* $(\sin Nv/N \sin v)^2$ that occurred for the picket-fence grating. Accordingly the discussion of the properties of the interference factor in the last section applies to the diffraction pattern made by a grating having rulings of arbitrary form. The only important difference between such an arbitrary grating and the picket-fence grating is in the diffraction factor, which describes the slowly varying distribution of light intensity among the principal maxima.

(b)   The Grating Equation

If the source of radiation for a grating contains atoms or molecules that emit radiation at several distinct frequencies, the diffraction pattern consists of the superposition of several of the monochromatic diffraction patterns we have been discussing, one for each frequency. We recall that when N is large, only the principal maxima in these patterns have appreciable intensity and they occur at the angles given by

$$v = \frac{\kappa b}{2} (\sin\theta_s + \sin\theta_o) = m\pi \qquad m = 0, \pm 1, \pm 2, \ldots, \tag{10.8.9}$$

which may be rearranged as the *grating equation*

$$\sin\theta_o = \frac{m\lambda}{b} - \sin\theta_s. \tag{10.8.10}$$

When $m = 0$, $\theta_o = -\theta_s$, regardless of the value of λ, and all the spectral components are superposed on the central, or *zero-order* maximum. When $m = \pm 1$, the various principal maxima of the spectral components are said to comprise the *first-order spectrum* of the grating, and so on. Each principal maximum is an image of the (line) source formed by radiation of a particular wavelength λ, and when $m = \pm 1$, it is referred to as a first-order *spectral line*. The higher-order spectral lines ($|m| > 1$) occur to the extent permitted by the grating equation (10.8.10). It is evident that a measurement of θ_s, θ_o, b and a knowledge of the spectral order m enable λ to be calculated. Since the visible spectral range occupies slightly less than a factor of 2 in wavelength (approximately from 400 to 750 nm), only the first-order visible spectrum is free from an overlap with

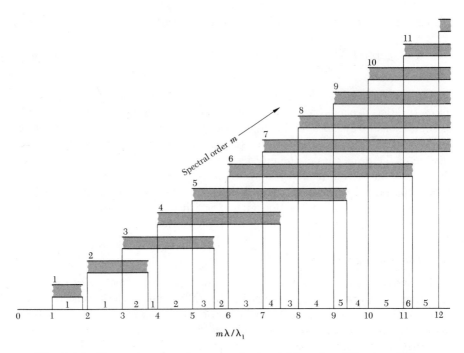

Fig. 10.8.2   The overlap of grating spectra in the visible region, $400 < \lambda < 750$ nm.

other orders. Evidently if spectra are to be measured when m is large, a certain amount of care is needed to establish the order of a given spectral line. The extent of the overlap in the first few spectral orders is shown in Fig. 10.8.2.

The form of the grating rulings does not enter into the grating equation (10.8.10). The diffraction factor $|\breve{F}|^2$, however, may cause certain spectral lines to be weak or missing if they happen to fall near (or on) minima or zeros of the diffraction factor. The effect of diffraction at the individual rulings of a grating can, in fact, be exploited by controlling the shape of the rulings when a grating is made. By this means it is possible to give $|\breve{F}|^2$ a broad maximum in a particular range of angles, e.g., where the second-order spectrum lies on one side of the central peak. Such *blazed gratings* enable weak spectral sources to be studied since most of the diffracted light falls in one spectral order rather than being spread out over many orders (see Prob. 10.8.3). Blazed gratings are particularly useful for studying the spectra of stars.

(c)   *Dispersion*

The *dispersion* of a grating, $D \equiv |d\theta_o/d\lambda|$, expresses the angular spread of spectral lines with respect to wavelength. From the grating equation (10.8.10) we find that

$$D = \left| \frac{d\theta_o}{d\lambda} \right| = \frac{|m|}{b \cos\theta_o}. \tag{10.8.11}$$

Hence the dispersion is proportional to the spectral order m and varies inversely with the grating space b and the cosine of the angle of diffraction. We note that if spectra are observed approximately at right angles to the grating, so that $\cos\theta_o \approx 1$, the dispersion has the constant value $|m|/b$. Hence in the vicinity of $\theta_o = 0$, the wavelength scale is linear in angle. The dispersion is considerably increased, but with a nonlinear wavelength scale, if spectra are viewed near *grazing incidence*, i.e., with θ approaching $\pi/2$.

(d)   Resolving Power

We have already pointed out that the principal maxima have a width, between adjacent zeros, of $\Delta v = 2\pi/N$. Since the corresponding angular width $\Delta\theta_o$ is a very small angle, we may take the differential of v, as defined in (10.8.5), to find the angular width. Thus from

$$\Delta v = \frac{2\pi}{N} = \frac{\kappa b}{2} \cos\theta_o \, \Delta\theta_o,$$

we deduce that

$$\Delta\theta_{\substack{\text{principal} \\ \text{max}}} = \frac{2\lambda}{Nb \cos\theta_o}. \tag{10.8.12}$$

According to Rayleigh's criterion for resolution, two monochromatic spectral lines emitted from the same source with equal intensity but with slightly different wavelengths are resolved if the principal maximum of one falls on the adjacent zero of the other. The two spectral lines are then separated in angle by half the angular width (10.8.12) of each line. Hence the minimum angle for which two such lines are resolved is

$$\Delta\theta_{\text{Rayleigh}} = \frac{\lambda}{Nb \cos\theta_o}. \tag{10.8.13}$$

The dispersion relation (10.8.11) shows that the two lines just resolved are separated in wavelength an amount

$$\Delta\lambda_{\text{Rayleigh}} = \frac{\Delta\theta_{\text{Rayleigh}}}{D} = \frac{b \cos\theta_o}{|m|} \Delta\theta_{\text{Rayleigh}}. \tag{10.8.14}$$

Hence, we find the *spectroscopic resolving power*

$$R \equiv \frac{\lambda}{\Delta\lambda_{\text{Rayleigh}}} = |m|N. \tag{10.8.15}$$

This important result shows that the resolving power of a grating for analyzing spectra equals the product of the number of rulings on the grating and the spectral order in which the spectra are observed.

There is a definite limit to the value of m, since $\sin\theta_o$ in the grating equation must be less than unity. If this limitation is put on m, we find (see Prob. 10.8.5) that $R_{max} \leq 2B/\lambda$, where $B = Nb$ is the total width of the grating. Thus a 10-cm grating has a maximum resolution of 400,000 for waves having a wavelength of 500 nm. Gratings have been constructed with a useful resolution as high as 1 million in the visible spectral region.

Problems

10.8.1 A particular transparent diffraction grating is ruled with 5,500 lines per centimeter, and 2.5 cm of it is used in a simple spectrometer having a source slit, collimating lens, and telescope. At what angles are the sodium D lines ($\lambda = 589.0$ and 589.6 nm) found if the grating is illuminated at normal incidence? What is the dispersion in minutes of arc per nanometer and the theoretical resolution at each of these angular positions?

10.8.2 The telescope of the spectrometer of Prob. 10.8.1 has a magnifying power of 5; that is, rays from two distant point objects subtending an angle $\Delta\theta$ emerge as if from two distant virtual images subtending an angle $5\Delta\theta$. With what angular separation do the D lines appear to the eye in each order present? The eye has an angular resolution of about 1 minute of arc. Are the D lines clearly resolved? Can the eye make use of the theoretical resolving power of the grating with the 5-power telescope? What minimum power should it have to match the grating resolution in the first order?

10.8.3 Try to devise, i.e., invent, a type of ruling that throws much of the intensity of spectral lines into the first order on one side of the diffraction pattern (when $\theta_s = 0$). Justify your design by a qualitative (or quantitative, if possible) argument. *Hint:* It is only necessary to consider the diffraction pattern of a single ruling.

10.8.4 A linear array of N identical point sources of waves a distance b apart is shown in the figure. The wave disturbance produced by each source at a distance r from that source is

$$\psi = \frac{A}{r} \exp i(\kappa \cdot \mathbf{r} - \omega t).$$

Find the relative intensity of the wave disturbance at a far-off, i.e., Fraunhofer, observation point P_o situated at a distance R_o from the center of the array on a line making an angle $\frac{1}{2}\pi - \theta$ with the array. Relate the result to the theory of diffraction gratings. Investigate the distant radiation pattern when $b < \lambda$, for example, when $b = \frac{1}{2}\lambda$. Explain how the radiation pattern can be "steered" by altering the relative phases of the N point sources in some fashion.

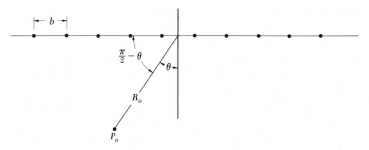

Prob. 10.8.4

10.8.5   Show that the spectroscopic resolving power of a grating has an upper limit of $2B/\lambda$, where $B = Nb$ is the width of the grating.

10.8.6   The *deviation* $\Theta = \theta_s + \theta_o$ is the angle between the direction of the incident light and the direction at which a principal maximum (or spectral line) is observed. Prove that for a particular spectral line, Θ is a minimum when $\theta_s = \theta_o$. Why would an experimentalist probably choose to operate his spectrograph with the grating oriented for minimum deviation?

10.8.7   The spectrum of mercury has a blue line at $\lambda = 435.8$ nm, a green line at 546.1, and a yellow doublet at 577.0 and 579.1. It is observed with a grating consisting of 40 slits. Discuss the appearance of the yellow doublet in the third, seventh, and seventeenth orders. Consider both resolution and overlapping of orders.

10.8.8   For the N-slit grating at normal incidence, devise elementary arguments, not involving integrals or a formal summation like (10.7.5), to show that (a) the principal interference maxima occur for

$$b \sin\theta_o = m\lambda$$

where m is an integer; (b) the nulls adjacent to a particular principal maximum are displaced in angle from the maximum by $\Delta\theta_o$ such that

$$\left(\frac{N}{2} b \cos\theta_o\right) \Delta\theta_o = \pm\frac{\lambda}{2};$$

(c) hence, the resolving power is mN.

*10.9   Two-dimensional Gratings

We have found that the intensity distribution in a Fraunhofer diffraction pattern of a one-dimensional or linear grating always can be separated into a diffraction factor and an interference factor. The diffraction factor depends on the properties of an individual element or ruling of the grating and specifies the fairly broad distribution of intensity in various orders of the diffraction pattern. The interference factor depends upon the number N of repeated elements and specifies the finer structure of the diffraction pattern, in particular, the position and the angular width of the narrow principal maxima of the pattern, as given by the grating equation (10.8.10).

The grating interference factor $(\sin Nv/N \sin v)^2$ can be derived without knowing the form of the diffraction factor. Indeed, the results of Prob. 10.8.4 lead us to conclude that whenever there exists a linear array of identical sources of waves having either identical phases or phases that progress linearly with position along the array, the relative wave intensity at a distance is given by a grating interference factor. We conclude that it is possible to discuss the essential features of the diffraction pattern of a linear grating by considering that the waves from each ruling come from an ideal line (or point) source located, for example, at the center of each ruling. As Prob. 10.8.4 teaches us, the grating interference factor arises from the superposition of the wavelets from this discrete set of sources.

To discuss Fraunhofer diffraction by two-dimensional gratings (and by three dimensional gratings in the following section) we adopt the viewpoint just expressed and consider that each diffracting aperture in the grating acts as if it were a point source of secondary wavelets whose phase is established by the direction of the incident plane waves. The same model also applies to a two-dimensional array of diffracting obstacles (rather than apertures), which are often termed *scattering centers*. In representing finite-size apertures or obstacles by point sources we discard the broad diffraction factor, characteristic of each scatterer, which modulates the more interesting interference pattern.

We can use Fig. 10.3.1 to define the direction cosines l_s, m_s, n_s specifying the direction from the source point and the direction cosines l_o, m_o, n_o specifying the direction to the observation point. We assume either that the source and observation points are sufficiently distant or that lenses are used, as described in Sec. 10.2, to achieve the Fraunhofer limit.

For simplicity we consider here only two-dimensional gratings having a unit cell in the form of a rectangle of dimensions b_1 and b_2. Such a grating can be formed, for example, by placing in contact two linear gratings having these spacings, with the direction of the rulings at right angles. We suppose that there are N_1 unit cells of width b_1 in the x direction and N_2 unit cells of width b_2 in the y direction.

The contributions to the wave amplitude reaching the observation point from each of the N_1N_2 cells, considered as diffracting centers, have identical magnitudes but have phases set by the path-difference function (10.3.4) that we introduced in discussing the rectangular aperture. Here the aperture position coordinates, x and y, take on the discrete values that specify the position of each of the unit cells of the grating. If we place the xy origin at one corner of the grating, these values of x and y are

$$x = n_1 b_1 \qquad 0 \leq n_1 \leq N_1 - 1$$
$$y = n_2 b_2 \qquad 0 \leq n_2 \leq N_2 - 1. \qquad (10.9.1)$$

According to (10.3.4), the path-difference function for the n_1, n_2 cell then takes the form

$$\delta(n_1, n_2) = n_1(l_s - l_o)b_1 + n_2(m_s - m_o)b_2. \qquad (10.9.2)$$

The amplitude of the diffracted wave at the observation point is the sum of the contributions from each of the cells of the grating. If $\breve{F}$ is the amplitude of the contribution from one of the cells, the entire amplitude may be written

$$\psi = \breve{F} \sum_{n_1=0}^{N_1-1} \sum_{n_2=0}^{N_2-1} e^{ik[n_1(l_s-l_o)b_1+n_2(m_s-m_o)b_2]} \qquad (10.9.3)$$

The amplitude factor $\breve{F} = \breve{F}(l_s - l_o, m_s - m_o)$ varies slowly with the directions of the source and observation points, expressing the diffraction of waves passing through one elemental cell of the grating.

The two sums in (10.9.3) give rise to the interference factor of the two-dimensional grating. If we introduce the variables

$$v_1 \equiv \frac{\pi b_1}{\lambda}(l_s - l_o)$$
$$v_2 \equiv \frac{\pi b_2}{\lambda}(m_s - m_o) \qquad (10.9.4)$$

and carry out the indicated sums, we find that the interference factor for the intensity at the observation point is just the product of two linear (one-dimensional) grating interference factors (Prob. 10.9.1). The intensity in the diffraction pattern is found to take the form

$$I \propto |\breve{F}|^2 (N_1 N_2)^2 \left(\frac{\sin N_1 v_1}{N_1 \sin v_1}\right)^2 \left(\frac{\sin N_2 v_2}{N_2 \sin v_2}\right)^2. \qquad (10.9.5)$$

Hence, just as with the rectangular aperture discussed in Sec. 10.3, we can think of the two-dimensional grating case as being the product of two orthogonal linear diffraction patterns, at least insofar as the interference factors are concerned. Accordingly, much of what we have learned about the diffraction of waves by a linear grating can be applied to the present case.

Principal maxima occur when $v_1 = m_1 \pi$, $v_2 = m_2 \pi$ ($m_1, m_2 = 0, \pm 1, \pm 2,$ . . .). The two equations

$$m_1 \lambda = b_1(l_s - l_o)$$
$$m_2 \lambda = b_2(m_s - m_o) \qquad (10.9.6)$$

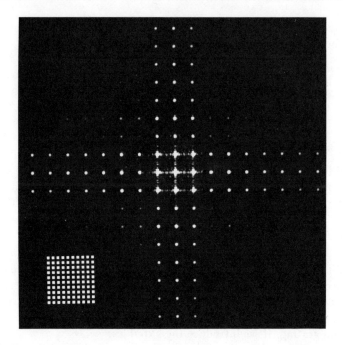

Fig. 10.9.1   Fraunhofer diffraction pattern for a two-dimensional array of 10 by 10 rectangular apertures (*inset*).

prescribe the direction cosines of principal maxima. For a fixed position of the point source, the first of these equations defines a set of conical surfaces surrounding the x axis, of half-angle $\cos^{-1}l_o$, corresponding to the discrete set of values that the direction cosine l_o can take on for various values of the integer m_1. Similarly, the second of the equations defines another set of conical surfaces surrounding the y axis, of half-angle $\cos^{-1}m_o$, corresponding to the discrete set of values that the direction cosine m_o can take for various values of the integer m_2. The principal maxima of the two-dimensional grating pattern occur in directions that are given by the intersection of the two sets of conical surfaces. The angular size of a maximum given by (10.9.6) can be found by adapting the procedure used in Sec. 10.8 for linear gratings. A diffraction pattern of a two-dimensional grating is illustrated in Fig. 10.9.1.

Problems

10.9.1   Carry out the summations indicated in (10.9.3) and obtain the intensity formula (10.9.5) for a two-dimensional grating.

10.9.2   A plane grating consists of an opaque screen perforated by rectangular openings having the dimensions a_1 and a_2 in the x and y directions, respectively. The center-to-center spacing of the N_1 openings in the x direction is b_1, and that of the N_2 openings in the y direction is b_2, with $a_1 < b_1$ and $a_2 < b_2$. Investigate the Fraunhofer diffraction through such an aperture screen. In particular, evaluate the form of the single-cell amplitude function $\breve{F}$ occurring in (10.9.3).

10.9.3   Repeat the calculation of the previous problem on the assumption that the screen is perforated by circular holes of diameter $d = 2a < b_1, b_2$.

★10.9.4   Find the interference pattern for the crossed array shown.

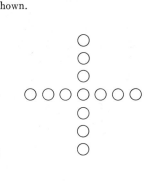

Prob. 10.9.4

10.10   Three-dimensional Gratings

As a final example of Fraunhofer diffraction, let us examine certain aspects of the diffraction of plane waves by a three-dimensional grating, or *lattice*. We limit the discussion to a simple lattice of scattering centers, forming a rectangular array. We assume that each center diffracts, or scatters, the same tiny fraction of the energy of an incident wave, so that the amplitude of a wave passing through the lattice has very nearly a constant magnitude at all points in the lattice. In addition this assumption ensures that the secondary scattering of waves already scattered by other centers can be neglected.

An important example of diffraction of waves by a three-dimensional lattice is afforded by the diffraction of x-rays (short-wavelength electromagnetic waves, $\lambda \sim 10^{-10}$ m) by crystalline solids. The regular arrangement of identical atoms (or molecules) into crystalline lattice structures is a distinguishing feature of most solids.† Single crystals occur in nature or can be grown in the laboratory by various techniques. Not only can x-rays pass through a small crystal with only a moderate amount of scattering, but the x-ray wavelength can easily be made somewhat less than twice the crystal-lattice spacing, a necessary condition for the occurrence of principal diffraction maxima.

† Exceptions include glasslike solids, which can be regarded as highly viscous liquids having a random arrangement of molecules.

The use of crystals as diffraction gratings for x-rays was first proposed by von Laue in 1912, and the first successful diffraction pattern was observed by Friedrich and Knipping.† Their experiment conclusively established the wave nature of x-rays, formed by the bombardment of a metal target by energetic electrons, and led to two very fruitful areas of research, namely, the study of crystal structures, using x-rays as a tool, and the study of x-ray spectra of atoms, using crystals as diffraction gratings.

Since the wavelength of an x-ray spectral line can be measured in absolute units using a ruled-line diffraction grating near grazing incidence, the spacing of atoms in a particular crystal can be established with considerable accuracy. By combining this knowledge with that of the atomic weight and the density of the crystal, an accurate value of Avogadro's number can be computed. The literature on determination of crystal structures using x-ray diffraction techniques is enormous. We must be content here to examine only a few of the characteristic features of the diffraction of waves by a simple three-dimensional lattice.

Let us place the origin of an xyz cartesian frame at one corner of the lattice, as indicated in Fig. 10.10.1. The coordinates of the diffracting centers of the lattice are then given by

$$
\begin{aligned}
x &= n_1 b_1 & 0 \leq n_1 \leq N_1 - 1 \\
y &= n_2 b_2 & 0 \leq n_2 \leq N_2 - 1 \\
z &= n_3 b_3 & 0 \leq n_3 \leq N_3 - 1,
\end{aligned}
\tag{10.10.1}
$$

† W. Friedrich, P. Knipping, and M. von Laue, *Ann. Phys.*, **41**: 971 (1913). See also A. H. Compton and S. K. Allison, "X-rays in Theory and Experiment," D. Van Nostrand Company, Inc., Princeton, N.J., 1935.

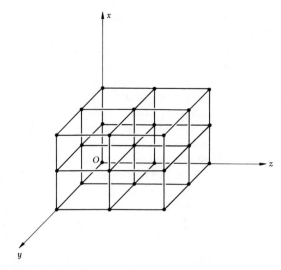

Fig. 10.10.1   A three-dimensional rectangular array of scattering centers.

where the b's are the lattice spacings and the N's are the number of scatterers in the respective directions. Since the lattice, considered as a diffraction aperture, has an appreciable extension in the z direction, as well as in the xy plane, where we have hitherto placed plane apertures, the path-difference function connecting a (distant) source point with a (distant) observation point via the n_1, n_2, n_3 center takes the form (Prob. 10.10.1)

$$\delta(n_1,n_2,n_3) = n_1(l_s - l_o)b_1 + n_2(m_s - m_o)b_2 + n_3(n_s - n_o)b_3. \qquad (10.10.2)$$

The direction cosines l_s, m_s, n_s and l_o, m_o, n_o are those defined in Fig. 10.3.1.

The amplitude of the diffracted wave reaching the observation point is the sum of the many ($N_1 \times N_2 \times N_3$) contributions having equal amplitude but having phases set by the path-difference function (10.10.2). If we let $\breve{F}$ be the amplitude of one of these elementary contributions, then

$$\psi = \breve{F} \sum_{n_1=0}^{N_1-1} \sum_{n_2=0}^{N_2-1} \sum_{n_3=0}^{N_3-1} e^{ik\delta(n_1,n_2,n_3)} \qquad (10.10.3)$$

expresses the wave amplitude at the observation point. The elementary wave amplitude $\breve{F}$, which is dependent upon the nature of an individual scattering (diffracting) center, is a slowly varying function of the direction cosines l_s, m_s, n_s and l_o, m_o, n_o.

The three summations in (10.10.3) can be carried out independently, each being similar to that occurring for the linear grating. Accordingly we find that the intensity of the diffracted wave contains a product of three one-dimensional grating interference factors, as well as the diffraction factor $|\breve{F}|^2$ characteristic of each scatterer;

$$I \propto |\breve{F}|^2 (N_1 N_2 N_3)^2 \left(\frac{\sin N_1 v_1}{N_1 \sin v_1}\right)^2 \left(\frac{\sin N_2 v_2}{N_2 \sin v_2}\right)^2 \left(\frac{\sin N_3 v_3}{N_3 \sin v_3}\right)^2, \qquad (10.10.4)$$

where now, extending (10.9.4),

$$v_1 \equiv \frac{\pi b_1}{\lambda} (l_s - l_o)$$

$$v_2 \equiv \frac{\pi b_2}{\lambda} (m_s - m_o) \qquad (10.10.5)$$

$$v_3 \equiv \frac{\pi b_3}{\lambda} (n_s - n_o).$$

The three interference factors have a common principal maximum provided that

$$\begin{aligned} v_1 &= m_1\pi & m_1 &= 0, \pm 1, \pm 2, \ldots \\ v_2 &= m_2\pi & m_2 &= 0, \pm 1, \pm 2, \ldots \\ v_3 &= m_3\pi & m_3 &= 0, \pm 1, \pm 2, \ldots, \end{aligned} \qquad (10.10.6)$$

that is, provided that

$$m_1\lambda = b_1(l_s - l_o) \tag{10.10.7}$$
$$m_2\lambda = b_2(m_s - m_o) \tag{10.10.8}$$
$$m_3\lambda = b_3(n_s - n_o). \tag{10.10.9}$$

These *three* simultaneous equations do not, in general, have a common solution subject to the constraints that

$$l_s{}^2 + m_s{}^2 + n_s{}^2 = 1 \tag{10.10.10}$$
$$l_o{}^2 + m_o{}^2 + n_o{}^2 = 1. \tag{10.10.11}$$

The geometrical significance of this fact is easily seen if we note that given values of λ and l_s in (10.10.7), for various values of the integer m_1, give rise to a set of cones of half-angle $\cos^{-1}l_o$ about the x axis. A second set of cones about the y axis having half angles $\cos^{-1}m_o$, as given by (10.10.8), intersect the first set in a limited number of directions, precisely as in the diffraction pattern of a two-dimensional grating. A third set of cones about the z axis having half angles $\cos^{-1}n_o$, as given by (10.10.9), in general do not intersect the first two sets in any direction common to all three sets of cones. *Hence the diffraction pattern of a three-dimensional grating is subject to more severe constraints than those applying to one- and two-dimensional gratings.* Only when $m_1 = m_2 = m_3 = 0$ do the equations always have a common solution, corresponding to a diffracted wave in the original direction.

Evidently for waves that are incident in a given direction to give rise to principal maxima, (10.10.11) requires that the equation

$$\left(\frac{m_1\lambda}{b_1} - l_s\right)^2 + \left(\frac{m_2\lambda}{b_2} - m_s\right)^2 + \left(\frac{m_3\lambda}{b_3} - n_s\right)^2 = 1 \tag{10.10.12}$$

be satisfied. This equation shows that for a particular value of λ, provided it is not too large, there are only certain directions for the incident wave that lead to an intense diffracted wave. Alternatively, for a given direction of incidence, there is a discrete set of λ's that give intense diffracted waves.

Equations such as (10.10.7) to (10.10.11) are not very convenient for analyzing the diffraction of x-rays by crystals. Instead it is customary to make use of the Bragg formula (Prob. 10.10.2)

$$n\lambda = 2d\,\sin\theta, \tag{10.10.13}$$

which expresses the condition for a strong reflection (a principal maximum) arising from the reflection of waves from one of the sets of parallel planes of diffracting centers into which the lattice may be subdivided; d is the separation of adjacent planes, and n is the order of the interference. The angle θ in (10.10.13) is the *glancing angle*, measured from the *plane itself* rather than from the *normal* to the plane (see Prob. 10.10.3).

We can show that the Bragg formula (10.10.13) follows from the present more general analysis. Let

$$\hat{\boldsymbol{\kappa}}_s \equiv l_s \mathbf{i} + m_s \mathbf{j} + n_s \mathbf{k} \tag{10.10.14}$$
$$\hat{\boldsymbol{\kappa}}_o \equiv l_o \mathbf{i} + m_o \mathbf{j} + n_o \mathbf{k} \tag{10.10.15}$$

be unit vectors in the direction of the incident waves and in the direction of one of the principal maxima. If we multiply Eqs. (10.10.7) to (10.10.9) by the respective positive integers n_1, n_2, n_3 of an arbitrary diffracting center, we find that the resulting three equations may be written as the single vector equation

$$N\lambda = \mathbf{r}_a \cdot (\hat{\boldsymbol{\kappa}}_s - \hat{\boldsymbol{\kappa}}_o), \tag{10.10.16}$$

where $\mathbf{r}_a$ is the vector position of the diffracting center specified by n_1, n_2, n_3, and where N is the integer

$$N \equiv m_1 n_1 + m_2 n_2 + m_3 n_3. \tag{10.10.17}$$

For a given integer N, and set of integers m_1, m_2, m_3, there are many integral choices for n_1, n_2, n_3 giving the coordinates of the diffracting centers. Equation (10.10.16) shows that the diffracting centers involved all lie in a plane perpendicular to the vector $\hat{\boldsymbol{\kappa}}_s - \hat{\boldsymbol{\kappa}}_o$. Different choices for N correspond to other parallel planes of diffracting centers perpendicular to $\hat{\boldsymbol{\kappa}}_s - \hat{\boldsymbol{\kappa}}_o$. The geometry involved is illustrated in Fig. 10.10.2. It is evident from the figure that a particular plane specified by a given value of N is located at a distance

$$D \equiv \frac{\mathbf{r}_a \cdot (\hat{\boldsymbol{\kappa}}_s - \hat{\boldsymbol{\kappa}}_o)}{|\hat{\boldsymbol{\kappa}}_s - \hat{\boldsymbol{\kappa}}_o|} \tag{10.10.18}$$

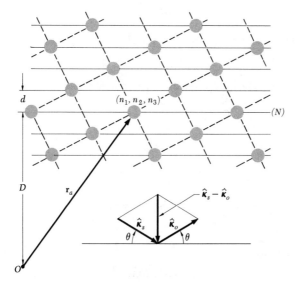

Fig. 10.10.2   Diagram illustrating (10.10.16) and (10.10.18).

from the origin and, furthermore, that

$$|\hat{\kappa}_s - \hat{\kappa}_o| = 2 \sin\theta,$$

where θ is the angle appearing in Bragg's formula. Hence (10.10.18) becomes

$$N\lambda = 2D \sin\theta. \tag{10.10.19}$$

The various integral values that N can assume depend on the integers m_1, m_2, m_3. For example, if $m_1 = m_2 = m_3 = 1$, N can be any integer, whereas if $m_1 = m_2 = m_3 = 2$, N is necessarily an even integer, etc. In general, for a given choice of m_1, m_2, m_3, the possible values of N are integral multiples of the largest common factor n of the three m's. Planes corresponding to successive multiples of n have a minimum spacing d such that

$$n\lambda = 2d \sin\theta. \tag{10.10.20}$$

Hence we have shown how the Bragg formula can be obtained by combining the waves diffracted, or scattered, by individual diffracting centers and relating principal maxima to the sets of planes of diffracting centers into which the lattice may be subdivided. The Bragg formula evidently can be used for sets of planes in crystals having nonorthogonal axes, although the analysis leading to an equation equivalent to (10.10.12), for example, then requires considerable revision.

Problems

10.10.1   Establish the path-difference function (10.10.2) that applies to the present case of Fraunhofer diffraction.

10.10.2   Supply the derivation of the Bragg formula (10.10.13) on the basis of the figure.

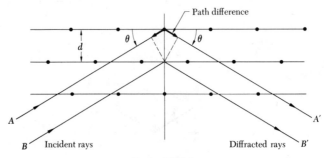

Prob. 10.10.2

10.10.3   The Bragg equation (10.10.13) is similar in form to the grating equation (10.8.10) with $\theta_o = \theta_s$. However, for the grating the angles θ_s, θ_o are measured from the normal, whereas in the Bragg case the angle θ is measured, by convention, from the diffracting planes in question (rather than from the normal to the planes). What is the relation between the two situations? Why not use the cosine in place of the sine in one equation, since the angles appear to be complements?

eleven

Fresnel Diffraction

The term *Fresnel diffraction* refers to applications of the Kirchhoff theory in which source and observer are closer to the diffracting aperture than Fraunhofer's "infinite" distances. This more general case can be described semiquantitatively by the useful concept of *Fresnel zones*. The formal treatment of rectangular apertures introduces the *Fresnel integrals* and the *Cornu spiral*. A fundamental and characteristic Fresnel problem is diffraction by a knife-edge, i.e., by the edge of a semi-infinite aperture.

11.1 Fresnel Zones

(a) Circular Zones

We wish to investigate the wave disturbance reaching the observation point P_o from the point source P_s through a circular aperture of radius a, as in Fig.

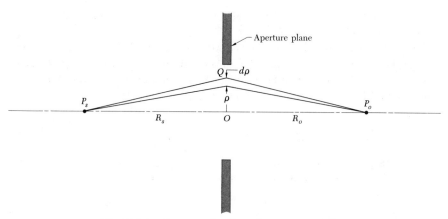

Fig. 11.1.1   Geometry for Fresnel zone construction.

11.1.1. Points P_s and P_o lie on the axis of rotational symmetry, perpendicular to the aperture plane. This case is more general than the Fraunhofer case in that the distances R_s and R_o may now be finite.

To apply the Kirchhoff diffraction formula (9.4.16),

$$\psi(P_o) = -\frac{iA}{\lambda} e^{-i\omega t} \int_{\Delta S} \left[\frac{\frac{1}{2}(\cos\theta_s + \cos\theta_o)}{r_s r_o} \right] e^{i\kappa(r_s+r_o)} \, dS, \tag{11.1.1}$$

we consider the contribution to the wave at P_o that passes through the infinitesimal annular strip at Q, of radius ρ, width $d\rho$, and area $dS = 2\pi\rho \, d\rho$ (Fig. 11.1.1).† The phase factor may be written in the form used in (10.2.2) as

$$e^{i\kappa(r_s+r_o)} \equiv e^{i\kappa(R_s+R_o)} e^{i\kappa\delta(\rho)}, \tag{11.1.2}$$

where the first factor on the right is a constant. The path-difference function $\delta(\rho)$ is conveniently expressed in terms of the number n of *half*-wavelengths by which the path P_sQP_o is longer than P_sOP_o, that is,

$$n\frac{\lambda}{2} \equiv \delta(\rho) = (R_s^2 + \rho^2)^{1/2} + (R_o^2 + \rho^2)^{1/2} - (R_s + R_o). \tag{11.1.3}$$

The aperture coordinate ρ is thereby replaced by n, a continuous variable, though we shall mainly be concerned with *integral* values of n. In particular, the

† Many texts apply the Huygens-Kirchhoff integral to a *spherical* surface with center at P_s and bounded by the aperture edge rather than to a *plane* surface extended across the aperture. The two conventions coincide in the paraxial limit; beyond that, they are simply alternative methods of bookkeeping since a circular aperture uncovers the same number of zones in each model.

aperture area between circles of radii $\rho(n-1)$ and $\rho(n)$ with n integral is known as the nth *Fresnel (half-period) zone.*[†]

The paraxial approximation of Sec. 10.1, whereby the lateral extent of the aperture is assumed small compared with the distances R_s and R_o ($a^2 \ll R_s{}^2, R_o{}^2$), permits a considerable simplification of the integrand of (11.1.1). First, the obliquity and inverse-distance factors reduce to a constant

$$\frac{\frac{1}{2}(\cos\theta_s + \cos\theta_o)}{r_s r_o} \to \frac{1}{R_s R_o}. \tag{11.1.4}$$

Second, the path-difference function (11.1.3) can be approximated by the first term of a series expansion in powers of ρ^2 (Prob. 11.1.1),

$$\delta(\rho) \equiv n\frac{\lambda}{2} \to \frac{1}{2}\left(\frac{1}{R_s} + \frac{1}{R_o}\right)\rho^2. \tag{11.1.5}$$

Finally, using (11.1.5), the area element can be written in terms of the zone parameter n as

$$dS = 2\pi\rho\,\frac{d\rho}{dn}\,dn \to \pi\lambda\,\frac{R_s R_o}{R_s + R_o}\,dn, \tag{11.1.6}$$

from which it follows that the area of a full zone is a constant,

$$\Delta S = \pi\lambda\,\frac{R_s R_o}{R_s + R_o}, \tag{11.1.7}$$

in the paraxial approximation.

Before evaluating the Kirchhoff integral (11.1.1) for the entire circular aperture, we evaluate the contribution from one full (half-period) zone, in the paraxial approximation,

$$
\begin{aligned}
\psi_n &= -\frac{iA}{\lambda}\,\frac{e^{i[\kappa(R_s+R_o)-\omega t]}}{R_s R_o}\int_{\rho(n-1)}^{\rho(n)} e^{i\kappa\delta(\rho)}2\pi\rho\,d\rho \\
&= -i\pi A\,\frac{e^{i[\kappa(R_s+R_o)-\omega t]}}{R_s + R_o}\int_{n-1}^{n} e^{i\pi n}\,dn \\
&= (-1)^{n-1}2A\,\frac{e^{i[\kappa(R_s+R_o)-\omega t]}}{R_s + R_o},
\end{aligned} \tag{11.1.8}
$$

which is of constant magnitude and of positive or negative sign as n is odd or even. Thus all zones contribute equal amplitudes, and the contributions from

[†] A Fresnel zone is called a half-period zone to emphasize that the zones are defined for a *half*-wavelength path increment, as in (11.1.3). This terminology is misleading unless one understands that a half-period zone is a full (not a half) zone! The term "full-period" zone, rarely used, would denote *two* adjacent Fresnel zones.

adjacent zones are precisely 180° out of phase. We reach the important conclusion that if a circular aperture uncovers an *even* integral number of Fresnel
zones, the intensity at P_o goes to zero.

Finally, if a circular aperture of radius a uncovers a *total* of n zones, so that
(11.1.5) becomes

$$n = \left(\frac{1}{R_s} + \frac{1}{R_o}\right)\frac{a^2}{\lambda},$$

(11.1.9)

the amplitude is

$$\psi = -i\pi A \frac{e^{i[\kappa(R_s+R_o)-\omega t]}}{R_s + R_o} \int_0^n e^{i\pi n}\, dn$$

$$= i^{n-1}2A \frac{e^{i[\kappa(R_s+R_o)-\omega t]}}{R_s + R_o} \sin\frac{\pi n}{2}.$$

(11.1.10)

It follows from (11.1.10) and (9.4.11) that the intensity at P_o is

$$I = 2I_0(1 - \cos\pi n),$$

(11.1.11)

where I_0 is the intensity that would be observed at P_o in the absence of any
obstruction (see Prob. 11.1.2). Thus the intensity has the maximum value $4I_0$
when n is an *odd* integer. In (11.1.5) the parameter n, representing the number
of zones unmasked by the aperture, may be changed by changing either the
aperture radius a or the axial distances R_s and/or R_o.

A graphical description of this analysis may be made as follows. In Sec. 1.3
we established that a wave may be represented by a rotating vector, the projection of which on some reference axis, e.g., the real axis of the complex plane,
yields the actual physical displacement of the wave at a particular instant of
time and position in space. When two or more coherent waves interfere, the
amplitude and phase of the net wave are given by the vector sum of the component wave vectors. Now imagine the first Fresnel zone to be subdivided into
a large number of elements (eight, say, for purposes of illustration), each representing a constant increment in path length. Then the contribution from the
first full zone is given by the resultant in Fig. 11.1.2*a*. The vectors representing
the eight component waves are of equal length on account of the direct proportionality between dS and dn in (11.1.6). The smooth curve representing an
infinite number of infinitesimal component waves is known as a *vibration curve*.
If, for instance, the contribution from the second Fresnel zone is added, the
component vectors complete a circle and the net wave amplitude goes to zero
(Fig. 11.1.2*b*).

In the paraxial limit assumed in (11.1.4) to (11.1.11), the vibration curve is
a multiply traced circle, completed *once* for each additional *two* Fresnel zones.
Eventually, the effects neglected in the Kirchhoff integral, namely, the obliquity

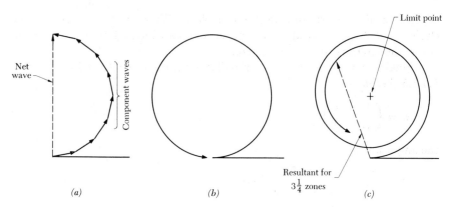

Fig. 11.1.2   (*a*) Vector diagram for first Fresnel zone, divided into eight subzones; (*b*) vibration curve for two Fresnel zones (paraxial limit); (*c*) vibration spiral for large aperture violating the paraxial approximation.

factor $\frac{1}{2}(\cos\theta_s + \cos\theta_o)$, the inverse-square law, and the higher-order terms in the expansion of (11.1.3), cause the amplitude contributed by successive zones to fall off. The vibration curve then becomes a spiral, gradually converging to the point at the center of the original circle (Prob. 11.1.3). The effect is shown exaggerated in Fig. 11.1.2*c*, in which the net wave vector for an aperture of $3\frac{1}{4}$ zones is also illustrated. It may be shown that the vibration curve remains a circle so long as $n \ll 4R_sR_o/(R_s + R_o)\lambda$ (Prob. 11.1.4). Since this latter quantity may be gigantic for visible wavelengths, the paraxial approximation and the resulting cancellation of adjacent zones remain accurate up to large values of n.

The Fresnel zone analysis is immediately applicable to a circular *obstacle*, as well as to a circular aperture. In this case the net wave vector is drawn to the limit point from the appropriate point on the circle of Fig. 11.1.2*c*. The surprising conclusion is that the intensity observed on the axis of the obstacle, in the middle of the geometrical shadow, is equal to that of the unimpeded wave! This effect, known as *Poisson's bright spot*, was of considerable significance in the history of physical optics theory.

(b)   Off-axis Diffraction

The preceding discussion has shown the usefulness of the Fresnel zone concept in computing the diffracted amplitude *on the axis* of a circular aperture or obstacle. It is also useful for a semiquantitative analysis of off-axis and irregular-aperture diffraction. Figure 11.1.3*a* represents a three-zone aperture viewed from a particular observation point off the axis. The boundaries of the Fresnel zones, in the plane of the aperture, are constructed with respect to the line of sight (direct path) from source to observation point. In this example, all of zones 1 and 2, but only portions of zones 3 to 5, are unmasked. The amplitude of the net contribution from a partially uncovered zone can be estimated semiquantitatively from the fraction of its area exposed. The phase may be taken to be that characteristic of the entire zone. (But that is not really a very good approximation here. Why?) Thus a one-dimensional vector diagram may be drawn with five collinear (parallel and antiparallel) vectors (Fig. 11.1.3*b*). The magnitude of the resultant vector is (approximately) proportional to the amplitude of the net diffracted wave. By point-by-point numerical estimates of this sort the off-axis intensity pattern can be mapped out. A similar procedure can be applied to the irregular aperture of Fig. 11.1.3*c*.

A careful analysis of the off-axis diffraction of a circular aperture requires a two-coordinate integration over the aperture, analogous to that carried out in Sec. 10.5 for the Fraunhofer case. In the more general Fresnel case but retaining the paraxial approximation, the resulting integrals are called *Lommel functions*.† Two variables are needed to specify the complete diffraction pattern: (1) the number *n* of unmasked zones, from (11.1.9), measures the aperture radius *a* relative to the source and observation distances R_s and R_o, and (2) the parameter

$$u = \frac{2\pi a}{\lambda} \sin\theta \approx \frac{2\pi a}{\lambda} \theta \qquad (11.1.12)$$

of (10.5.3) measures the (angular) displacement of the observation point from the axis. This same type of analysis enables one to study the three-dimensional diffraction pattern that forms the "point image" of a point source produced by a focused geometrical-optics lens system.‡

† See A. Gray, G. B. Mathews, and T. M. MacRobert, "A Treatise on Bessel Functions and Their Applications to Physics," 2d ed., chap. 14, Dover Publications, Inc., New York, 1966.
‡ M. Born and E. Wolf, "Principles of Optics," 3d ed., sec. 8.8, Pergamon Press, New York, 1965.

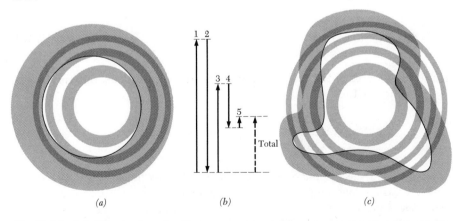

Fig. 11.1.3   (*a*) Off-axis view of a three-zone aperture; (*b*) approximate vector diagram for case (*a*); (*c*) zones exposed by an irregular aperture.

From Sec. 9.4 we recall the theorem, known as *Babinet's principle*, stating that the diffraction patterns of two *complementary* screens, i.e., one has apertures where the other is opaque, and vice versa, are such that the vector sum of the respective diffracted *amplitudes* (not intensities) at any point is equal to the amplitude of the original wave when no screen is present. Thus if the diffraction pattern of a certain aperture is known, the pattern of the corresponding obstacle can be obtained readily.

Fraunhofer diffraction is the limit of Fresnel diffraction when only a small fraction of the first Fresnel zone is unmasked by the aperture. Thus for fixed wavelength, the Fraunhofer case can be achieved by increasing R_s and R_o (perhaps with the aid of lenses) or by decreasing the aperture radius a. The limit of *geometrical* optics corresponds to $a/\lambda \to \infty$ with R_s and R_o bounded such that $n \to \infty$, which is simply the limit of small wavelength relative to other dimensions. According to quantum mechanics a particle of momentum mv has associated with it a wavelength $\lambda = h/mv$, where h is Planck's constant. Classical mechanics, i.e., rectilinear propagation of particles, is valid in the limit when λ is small compared with the dimensions of the relevant apertures or obstacles, such as an atom.

(c) Linear Zones

Since a simple but important geometry for a diffraction aperture is the linear slit, it is of interest to apply the Fresnel zone argument to the case of a slit illuminated by a line source. The diagram of Fig. 11.1.1 can be used with the understanding that the center line now represents a plane, rather than an axis, of symmetry. We include in the nth zone the *pair* of linear strips, one on each side of the centerline, bounded by $\rho(n - 1)$ and $\rho(n)$, where from (11.1.5), in the paraxial approximation,

$$n = \left(\frac{1}{R_s} + \frac{1}{R_o}\right)\frac{\rho^2}{\lambda}.$$
(11.1.13)

However, the area of the pair of infinitesimal strips of width dn and unit length is, replacing (11.1.6),

$$dS = 2\frac{d\rho}{dn}\,dn = \left[\frac{\lambda R_s R_o}{n(R_s + R_o)}\right]^{1/2} dn.$$
(11.1.14)

In contrast to the circular-zone case, the area of the linear zones goes down with increasing zone number even in the paraxial approximation; i.e., higher-order zones make a smaller contribution to the diffracted amplitude at P_o.

If we attempt to evaluate the contribution from a full linear zone, in anal-

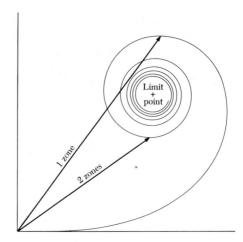

Fig. 11.1.4 Vibration spiral for linear zones.

ogy with (11.1.8), we must deal with an integral of the form

$$\int_{n-1}^{n} \frac{e^{i\pi n}}{(2n)^{1/2}} \, dn \equiv \int_{[2(n-1)]^{1/2}}^{(2n)^{1/2}} e^{i(\pi/2)v^2} \, dv, \qquad (11.1.15)$$

which is known as Fresnel's integral and is conventionally written in terms of the variable $v \equiv (2n)^{1/2}$. A full discussion of the linear-slit Fresnel diffraction problem is taken up in the following sections. Our simple argument with linear half-period zones fails to be as useful as the circular-zone argument because of the change of area of successive zones. Qualitatively we can say that the vibration curve is a spiral (Fig. 11.1.4), converging rather rapidly even though we are ignoring the second-order effects (obliquity factor, etc.) in accordance with the paraxial approximation. We return to this elementary linear-zone argument at the end of Sec. 11.3.

Problems

11.1.1 Carry out an expansion of (11.1.3) in the paraxial limit to obtain (11.1.5).

11.1.2 Verify the final forms of (11.1.8) and (11.1.10) and justify the coefficient $2I_0$ in (11.1.11).

11.1.3 Consider a circular aperture of radius a such that $a = R_s = R_o$, thus violating the paraxial approximation. (a) Show that the width of the Fresnel half-period zone closest to the aperture edge is

$$\Delta\rho \approx \frac{\lambda}{2\sqrt{2}}.$$

(b) Show that the amplitude of the wave component contributed by this zone, i.e., the product of $dS = 2\pi a\,\Delta\rho$ and the square-bracketed quantity in (11.1.1), is one-half the amplitude con tributed by zones near the axis. (c) Show that the amplitude contributed by successive zones-goes monotonically to zero as the radius a goes to infinity and thus that the vibration spiral of Fig. 11.1.2c converges to a limit point. ★(d) Show that the limit point equals one-half the amplitude contributed by the first zone.

★**11.1.4** The theory of circular Fresnel zones leading to (11.1.8) and (11.1.10) assumed the paraxial approximation, $\rho^2 \ll R_s^2,\ R_o^2$. Now, without making the paraxial simplification, show that

$$\frac{\tfrac{1}{2}(\cos\theta_s + \cos\theta_o)}{r_s r_o}\,dS = \frac{\pi\lambda}{R_s + R_o}\,dn\,\frac{\tfrac{1}{2}[R_s/(R_s^2 + \rho^2)^{1/2} + R_o/(R_o^2 + \rho^2)^{1/2}]}{[(R_s^2 + \rho^2)^{1/2} + (R_o^2 + \rho^2)^{1/2}]/(R_s + R_o)}$$

$$= \frac{\pi\lambda}{R_s + R_o}\,dn\,\frac{1 - \tfrac{1}{4}(1/R_s^2 + 1/R_o^2)\rho^2 + \cdots}{1 + \tfrac{1}{2}(1/R_s R_o)\rho^2 + \cdots}$$

$$= \frac{\pi\lambda}{R_s + R_o}\,dn\left[1 - \frac{1}{4}\left(\frac{1}{R_s} + \frac{1}{R_o}\right)^2\rho^2 + \cdots\right].$$

Thus confirm that the paraxial approximation is valid so long as

$$n \ll \frac{4}{\lambda}\frac{R_s R_o}{R_s + R_o} \sim \frac{R_{\text{smaller}}}{\lambda}.$$

11.1.5 A *zone plate* is a screen made by blackening even-numbered (alternatively, odd-numbered) zones whose outer radii are given by $\rho_n^2 = n\lambda R_o$ ($n = 1, 2, \ldots$). Plane monochromatic waves incident on it are in effect focused on a point P_o, distant R_o from the zone plate, since the wavefronts arriving at P_o from the odd-numbered zones are all in phase. Show that a sequence of such focus points exist. Show also that the zone plate focuses a point source at a distance p into a point image at a distance q, where p and q are related by the familiar lens equation

$$\frac{1}{p} + \frac{1}{q} = \frac{1}{f} = \frac{\lambda}{\rho_1^2}.$$

Does an equation of this form hold for each of the other images?

11.1.6 In the paraxial approximation, the intensity along the axis of a circular aperture varies between zero and four times that of the primary illumination. The intensity along the axis of a circular obstacle is constant and equal to that of the primary illumination. Reconcile these two statements with Babinet's principle.

11.1.7 You wish to construct a pinhole camera (no lens) with 10 cm separation between pinhole and film plane. The film is sensitive to visible light ($\lambda \sim 500$ nm). What pinhole diameter would you choose for optimum resolution, i.e., smallest photographic image of a point object, and what order of magnitude of angular resolution would be obtained?

11.2   The Rectangular Aperture

(a)   Geometry and Notation

The geometry involved in Fresnel diffraction by a rectangular aperture is cumbersome and must be set up with care. In Fig. 11.2.1, P_s is the location of a point source emitting monochromatic waves. A plane opaque screen, containing the rectangular aperture, is located a distance R_s from the source. The observation plane is parallel to the aperture plane and separated from it by the distance R_o. The line P_sOO' coincides with the z axis, perpendicular to the aperture and observation planes.

The observation point P_o is specified by the cartesian coordinates x, y, with origin at O'. An element of aperture area at Q is located by the cartesian coordinates ξ, η with respect to the origin Ω, defined by the intersection of the diagonal P_sP_o with the aperture plane. The point O then has the coordinates $\xi_0 \equiv xR_s/(R_s + R_o)$, $\eta_0 \equiv yR_s/(R_s + R_o)$. The boundaries of the rectangular aperture (which need not be centered on O) are specified by ξ_1, ξ_2, η_1, η_2. The origin Ω "moves around" with P_o, and consequently ξ_1, ξ_2, η_1, η_2, as well as ξ_0, η_0, are functions of x, y. However, the aperture dimensions $a \equiv \xi_2 - \xi_1$ and $b \equiv \eta_2 - \eta_1$ are necessarily constant.

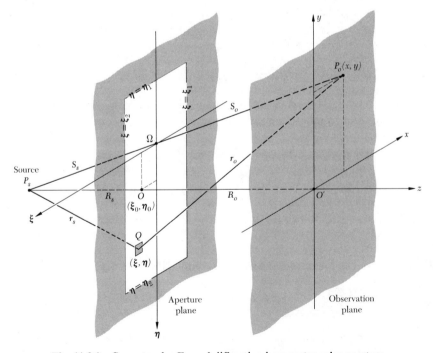

Fig. 11.2.1   Geometry for Fresnel diffraction by a rectangular aperture.

The only nontrivial idea in establishing this system of notation is the choice of the moving origin Ω for the aperture coordinates ξ, η (see Prob. 11.2.3).

In order to make the Kirchhoff integral (11.1.1) tractable, we again assume the paraxial approximation, i.e. (see Prob. 11.2.2),

$$\left.\begin{array}{c} a^2 + b^2 \\ x^2 + y^2 \end{array}\right\} \ll \left\{\begin{array}{c} R_s{}^2 \\ R_o{}^2. \end{array}\right. \tag{11.2.1}$$

The integral becomes

$$\psi = -\frac{iA}{\lambda}\, \frac{e^{i[\kappa(S_o+S_s)-\omega t]}}{R_s R_o} \int_{\eta_1}^{\eta_2}\int_{\xi_1}^{\xi_2} e^{i\kappa\delta(\xi,\eta)}\, d\xi\, d\eta, \tag{11.2.2}$$

where the path-difference function is

$$\delta(\xi,\eta) \equiv (r_s + r_o) - (S_s + S_o), \tag{11.2.3}$$

and where S_s and S_o are the lengths of the line segments $P_s\Omega$ and ΩP_o, respectively (Fig. 11.2.1). Our task is to obtain δ as a simple function of ξ, η. From the figure,

$$r_s{}^2 = R_s{}^2 + (\xi - \xi_0)^2 + (\eta - \eta_0)^2 \tag{11.2.4}$$
$$r_o{}^2 = R_o{}^2 + (x + \xi - \xi_0)^2 + (y + \eta - \eta_0)^2 \tag{11.2.5}$$
$$S_s{}^2 = R_s{}^2 + \xi_0{}^2 + \eta_0{}^2 \tag{11.2.6}$$
$$S_o{}^2 = R_o{}^2 + (x - \xi_0)^2 + (y - \eta_0)^2 \tag{11.2.7}$$
$$\xi_0 = \frac{R_s}{R_s + R_o}\, x \qquad \eta_0 = \frac{R_s}{R_s + R_o}\, y. \tag{11.2.8}$$

Now, in the paraxial limit, we may expand the right-hand sides of (11.2.4) to (11.2.7) by the binomial theorem; for instance,

$$\begin{aligned} r_s &= R_s\left[1 + \frac{(\xi - \xi_0)^2}{R_s{}^2} + \frac{(\eta - \eta_0)^2}{R_s{}^2}\right]^{1/2} \\ &= R_s\left[1 + \frac{1}{2}\frac{(\xi - \xi_0)^2}{R_s{}^2} + \frac{1}{2}\frac{(\eta - \eta_0)^2}{R_s{}^2} + \cdots\right] \\ &= R_s + \frac{1}{2}\frac{(\xi - \xi_0)^2}{R_s} + \frac{1}{2}\frac{(\eta - \eta_0)^2}{R_s} + \cdots \end{aligned} \tag{11.2.9}$$

Keeping only the lowest-order terms, one finds (Prob. 11.2.1)

$$\delta(\xi,\eta) = \frac{1}{2}\left(\frac{1}{R_s} + \frac{1}{R_o}\right)(\xi^2 + \eta^2), \tag{11.2.10}$$

which is directly analogous with (11.1.5). Indeed, we could have written (11.2.10) down directly from (11.1.5), although the expansions of (11.2.4) to (11.2.7) make explicit the nature of the paraxial assumptions (Prob. 11.2.2).

The resulting integral in (11.2.2),

$$\int_{\eta_1}^{\eta_2} \int_{\xi_1}^{\xi_2} \exp\left[i\frac{\kappa}{2}\left(\frac{1}{R_s} + \frac{1}{R_o}\right)(\xi^2 + \eta^2)\right] d\xi\, d\eta$$
$$= \int_{\xi_1}^{\xi_2} \exp\left[i\frac{\kappa}{2}\left(\frac{1}{R_s} + \frac{1}{R_o}\right)\xi^2\right] d\xi \cdot \int_{\eta_1}^{\eta_2} \exp\left[i\frac{\kappa}{2}\left(\frac{1}{R_s} + \frac{1}{R_o}\right)\eta^2\right] d\eta,$$

$$(11.2.11)$$

is conventionally written in a form obtained by the elementary substitutions

$$\frac{\pi}{2} u^2 \equiv \frac{\kappa}{2}\left(\frac{1}{R_s} + \frac{1}{R_o}\right)\xi^2$$
$$\frac{\pi}{2} v^2 \equiv \frac{\kappa}{2}\left(\frac{1}{R_s} + \frac{1}{R_o}\right)\eta^2,$$

that is,

$$u \equiv \left[\frac{2}{\lambda}\left(\frac{1}{R_s} + \frac{1}{R_o}\right)\right]^{1/2} \xi \qquad (11.2.12)$$

$$v \equiv \left[\frac{2}{\lambda}\left(\frac{1}{R_s} + \frac{1}{R_o}\right)\right]^{1/2} \eta. \qquad (11.2.13)$$

These new variables u, v are simply dimensionless, or *normalized*, aperture coordinates, replacing ξ, η. Accordingly, (11.2.2) becomes

$$\psi = -\frac{iA}{2}\frac{e^{i[\kappa(S_o+S_s)-\omega t]}}{R_s + R_o}\int_{u_1}^{u_2} e^{i(\pi/2)u^2}\, du \int_{v_1}^{v_2} e^{i(\pi/2)v^2}\, dv. \qquad (11.2.14)$$

As in the Fraunhofer treatment of the rectangular aperture (Sec. 10.3), the diffraction integral factors into two independent integrals, each controlled by only one of the aperture dimensions. Thus it is sufficient for us to study the properties of one of the integrals in (11.2.14); the complete diffraction pattern is simply the product of two evaluations of this integral for the respective pairs of limits. Furthermore, as in Sec. 10.4, we may allow one of the aperture dimensions to become very large, whereupon the corresponding integral goes to a (constant) limiting value. The *Fresnel single-slit pattern* is then given by the remaining integral, which is a function of the narrow dimension of the "infinite" slit. In summary, we reach the same conclusion as in the Fraunhofer case, namely, that the diffraction pattern of the rectangular aperture is simply the product of the two corresponding single-slit patterns.

(b) *The Cornu Spiral*

By Euler's identity (1.3.6),

$$\int_0^u e^{i(\pi/2)u^2}\, du \equiv \int_0^u \cos\frac{\pi}{2}u^2\, du + i\int_0^u \sin\frac{\pi}{2}u^2\, du. \qquad (11.2.15)$$

The definite integrals

$$C(u) \equiv \int_0^u \cos \frac{\pi}{2} u^2 \, du \tag{11.2.16}$$

$$S(u) \equiv \int_0^u \sin \frac{\pi}{2} u^2 \, du \tag{11.2.17}$$

are known as the *Fresnel integrals*, for which extensive tables exist.† The complex integral (11.2.15) can be represented graphically in the complex plane by plotting $S(u)$ against $C(u)$, a figure known as the *Cornu spiral* (Fig. 11.2.2).

† See, for instance, E. Jahnke and F. Emde, "Tables of Functions," 4th ed., pp. 34–37, Dover Publications, Inc., New York, 1945; M. Abramowitz and I. A. Stegun (eds.), "Handbook of Mathematical Functions," pp. 321–324, Dover Publications, Inc., New York, 1965.

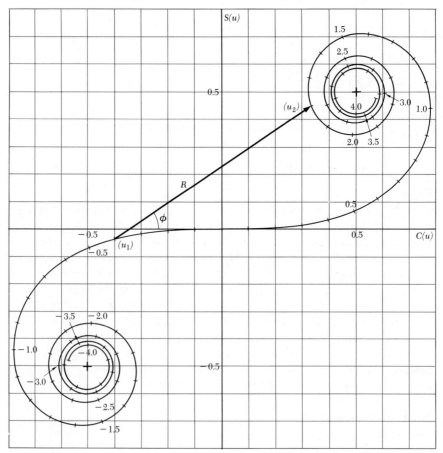

Fig. 11.2.2 The Cornu spiral.

The variable u appears as a parameter, each point on the spiral being identified with a specific value of u. For instance, the curve passes through the origin when $u = 0$. In fact, since an element of arc length along the spiral is (Prob. 11.2.4)

$$[(dC)^2 + (dS)^2]^{1/2} = du, \tag{11.2.18}$$

equal increments in u correspond to equal arc segments. Thus u may be thought of as simply the coordinate measuring arc length along the spiral, as shown by hatch marks in Fig. 11.2.2. As $u \rightarrow \pm \infty$, the spiral converges to the two limit points $\pm(\frac{1}{2} + i\frac{1}{2})$ (Prob. 11.2.5). The slope of the curve at any point is

$$\frac{dS}{dC} = \frac{\sin(\pi/2)u^2}{\cos(\pi/2)u^2} = \tan\frac{\pi}{2}u^2; \tag{11.2.19}$$

that is, at every point the tangent to the curve makes the angle $\frac{1}{2}\pi u^2$ with the real (C) axis. The radius of curvature of the spiral is given by

$$\left[\frac{d(\frac{1}{2}\pi u^2)}{du}\right]^{-1} = \frac{1}{\pi u}, \tag{11.2.20}$$

showing clearly that the curve spirals in tightly toward the limit points as $|u|$ becomes large.

The value of the integral

$$\int_{u_1}^{u_2} e^{i(\pi/2)u^2}\, du = [C(u_2) - C(u_1)] + i[S(u_2) - S(u_1)]$$
$$\equiv Re^{i\phi} \tag{11.2.21}$$

is given by the straight line, or chord, connecting the two points on the Cornu spiral specified by the limits u_1 and u_2, as indicated in Fig. 11.2.2. The length of the line is the *magnitude* R of this complex integral; the angle the line makes with the real axis is the *phase* ϕ. Thus the value of the integral can be quickly read off the graph in the form $Re^{i\phi}$, and the effect of changing the limits u_1 and u_2, as the observation point P_o is changed, can be readily studied. It is usually only the diffracted *intensity*, proportional to $|Re^{i\phi}|^2 = R^2$, that is of interest, in which case only the length of the chord between the two u values need be read from the graph.

Problems

11.2.1   Expand (11.2.4) to (11.2.7) binomially to obtain (11.2.10).

11.2.2   The prescription of the paraxial approximation given by the inequalities of (11.2.1) does not exactly justify the expansions used in obtaining (11.2.10). What are the more rigorous (and more cumbersome) inequalities? Can you imagine a situation in which (11.2.1) is satisfied but (11.2.10) is not valid? Under what conditions would the distinction between the two prescriptions be important in practical laboratory work?

★11.2.3 Develop the Fresnel diffraction of a slit using a coordinate system fixed to the center of the slit (but source and observation point may be off axis). Show that the phase exponent $\kappa(r_s + r_o)$ in the integrand of the Kirchhoff integral contains both a linear and a quadratic term in the aperture coordinate. (Under what conditions can the quadratic term be neglected, giving the Fraunhofer limit?) By completing the square, show that the "moving" coordinate system used in this section (Fig. 11.2.1) eliminates the linear term, i.e., results in a pure quadratic phase exponent and the conventional Fresnel integrals.

11.2.4 Confirm the results stated in (11.2.18) to (11.2.20).

★11.2.5 Prove formally the limiting values of the Fresnel integrals for $u \to \infty$, that is,

$$\int_0^\infty \cos \frac{\pi}{2} u^2 \, du = \int_0^\infty \sin \frac{\pi}{2} u^2 \, du = \tfrac{1}{2}.$$

★11.2.6 Show that the Fresnel integrals are related to Bessel functions of half-integral order by

$$C(u) = J_{1/2}\left(\frac{\pi}{2} u^2\right) + J_{5/2}\left(\frac{\pi}{2} u^2\right) + J_{9/2}\left(\frac{\pi}{2} u^2\right) + \cdots$$

$$S(u) = J_{3/2}\left(\frac{\pi}{2} u^2\right) + J_{7/2}\left(\frac{\pi}{2} u^2\right) + J_{11/2}\left(\frac{\pi}{2} u^2\right) + \cdots .$$

11.3   The Linear Slit

For simplicity we consider the case where the point (or line) source lies on the perpendicular bisector of a slit of width a. The geometry is specified in Fig. 11.3.1. According to geometrical optics, sharp shadow edges would occur at

$$\pm x_0 \equiv \pm \frac{R_s + R_o}{R_s} \frac{a}{2}. \tag{11.3.1}$$

With respect to the "moving" origin Ω, the aperture center O is at

$$\xi_0 = \frac{R_s}{R_s + R_o} x, \tag{11.3.2}$$

and the aperture boundaries are at

$$\xi_1 = \xi_0 - \frac{a}{2} \qquad \xi_2 = \xi_0 + \frac{a}{2}. \tag{11.3.3}$$

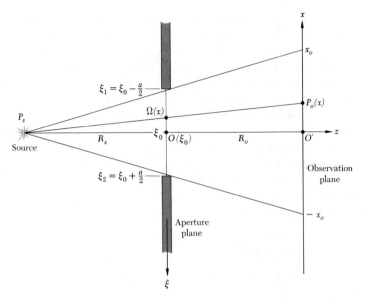

Fig. 11.3.1   Geometry of Fresnel diffraction by a long slit of width a.

In terms of the normalized variable u of (11.2.12), the aperture boundaries are

$$u_1(x) = \left[\frac{2}{\lambda}\left(\frac{1}{R_s} + \frac{1}{R_o}\right)\right]^{1/2} \left(\frac{R_s}{R_s + R_o} x - \frac{a}{2}\right) \tag{11.3.4}$$

$$u_2(x) = \left[\frac{2}{\lambda}\left(\frac{1}{R_s} + \frac{1}{R_o}\right)\right]^{1/2} \left(\frac{R_s}{R_s + R_o} x + \frac{a}{2}\right) \tag{11.3.5}$$

$$\Delta u \equiv u_2 - u_1 = \left[\frac{2}{\lambda}\left(\frac{1}{R_s} + \frac{1}{R_o}\right)\right]^{1/2} a. \tag{11.3.6}$$

From (11.2.14), allowing the length of the slit to become very large $(v_1, v_2 \to \mp \infty)$, the intensity of the Fresnel single-slit pattern is

$$I(x) = \tfrac{1}{2}I_0 \left|\int_{u_1}^{u_2} e^{i(\pi/2)u^2} du\right|^2 = \tfrac{1}{2}I_0[R(\Delta u, x)]^2, \tag{11.3.7}$$

where as usual I_0 is the intensity in the absence of any obstruction and $R(\Delta u, x)$ is the length of the chord connecting the points $u_1(x)$ and $u_2(x) = u_1(x) + \Delta u$ on the Cornu spiral.

Imagine a model of the Cornu spiral constructed of stiff wire and a piece of flexible tubing, of the constant length Δu given by (11.3.6), that slides along the spiral. The position of the center of the tubing is related to the observation coordinate x by

$$u_{\text{center}} \equiv \left[\frac{2}{\lambda}\left(\frac{1}{R_s} + \frac{1}{R_o}\right)\right]^{1/2} \frac{R_s}{R_s + R_o} x = \left[\frac{2R_s}{\lambda R_o(R_s + R_o)}\right]^{1/2} x. \tag{11.3.8}$$

As x is changed, the center of the tubing moves along the spiral proportionately and the intensity at x is proportional to the square of the chord connecting the two ends of the tube. The intensity pattern differs from that of the Fraunhofer case in that no true zeros of intensity occur within the pattern. However there exist positions where the ends of the tubing are quite near each other on neighboring turns of the spiral, giving intensity minima. At the geometrical shadow edges, one end of the tubing is at $u = 0$, the other at $u = \pm \Delta u$ (Prob. 11.3.1). The intensity approaches zero in the shadow regions as the tubing winds up tightly at one or the other of the two limit points.

Let us now relate the formal Cornu spiral treatment of the single slit to the elementary Fresnel linear-zone model introduced in Sec. 11.1c. In that discussion, the nth Fresnel (linear) zone denoted the *pair* of strips in the aperture between coordinates $\xi(n-1)$ and $\xi(n)$, where from (11.1.13)

$$\xi(n) = \pm \left(\frac{n \lambda R_s R_o}{R_s + R_o} \right)^{1/2}. \tag{11.3.9}$$

The zone number n and the Cornu spiral variable u are both normalized, or dimensionless, substitutes for the aperture coordinate ξ and are related by

$$u_1, u_2 = \mp (2n)^{1/2} \tag{11.3.10}$$

when the observation point is on the axis from source to aperture center. An aperture of width Δu, given by (11.3.6), exposes

$$n = \tfrac{1}{8}(\Delta u)^2 \tag{11.3.11}$$

zones on axis. When viewed from off axis, the slit width Δu remains fixed but the number of zones exposed changes since the area (11.1.14) is not constant. Thus the Fresnel zone argument for *linear* zones is generally useful only for discussing the intensity distribution on the axis.

In Sec. 11.1a we showed for a circular aperture using *circular* zones that the on-axis diffracted intensity is a maximum or zero as n is odd or even, according to the vibration circle of Fig. 11.1.2b. Similarly, in the case of a slit using *linear* zones, the on-axis intensity goes through maxima and minima in accordance with the *damped* vibration spiral of Fig. 11.1.4, which is essentially the Cornu spiral (Prob. 11.3.2). Since the Fresnel zones are defined by phase considerations (180° phase change for each half-period zone), the points on the spiral representing full integral zones are those at extrema of the function $S(u)$. The on-axis maxima and minima, however, are at extrema of $C^2(u) + S^2(u)$. From Fig. 11.1.4, it may be seen then that the intensity maxima and minima occur approximately for *one-quarter of a zone less* than integral zones (Prob. 11.3.3), i.e., for

$$n \approx k - \tfrac{1}{4} \qquad \begin{array}{l} k = 1, 3, 5, \ldots \qquad \text{maxima} \\ k = 2, 4, 6, \ldots \qquad \text{minima.} \end{array} \tag{11.3.12}$$

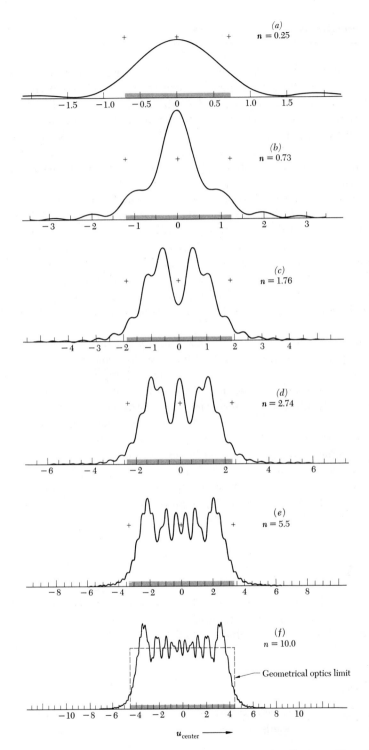

Fig. 11.3.2 Examples of Fresnel (intensity) diffraction patterns for slits unmasking n Fresnel linear zones on axis.

Similarly, for $n \approx k + \frac{1}{4}$ (k odd or even) the intensity is that of the unobstructed wave. Thus the linear-zone model together with the extremum condition (11.3.12) gives us a handy way to find the combinations of slit width a and observation distance R_o that give maximum and minimum intensity at the center of the diffraction pattern.

Representative Fresnel diffraction patterns are shown in Fig. 11.3.2. Included are special cases near $k - \frac{1}{4}$ that give maxima or minima at the center of the pattern.

Fraunhofer diffraction is the limiting case of Fresnel when the aperture unmasks *much less than one* Fresnel zone (either circular or linear). The Fraunhofer criterion (10.2.5) states precisely this limit. The "tubing" sliding on the Cornu spiral is then very short—so short that it must slide many times around the limit point before the opposite ends of the tubing come close together on adjacent turns. Since the Cornu turns converge rather slowly, this minimum intensity approaches a null. Indeed Fig. 11.3.2a, a Fresnel pattern for $n = \frac{1}{4}$, is already barely distinguishable from the Fraunhofer pattern ($n \to 0$) of Fig. 10.4.3b.

Our discussion of the rectangular aperture and the slit has assumed the paraxial approximation. Even in this limit, the vibration curve (Fig. 11.1.4) is a converging (Cornu) spiral, in contrast to the vibration *circle* of Fig. 11.1.2b holding for circular apertures. Only the first few linear zones are important in determining the diffracted intensity for a slit (Prob. 11.3.9), whereas all circular zones are important for a circular aperture until nonparaxial effects occur. When the paraxial condition is violated (Prob. 11.1.2), the obliquity and inverse-distance factors in the full Kirchhoff integral (11.1.1) cause the vibration curve for circular zones to converge to a limit point, as in Fig. 11.1.2c. For linear zones, violation of the paraxial assumptions causes the paraxial (Cornu) spiral to converge more rapidly to its limit. A straightforward analysis similar to that outlined in Prob. 11.1.3 shows that the Cornu spiral is not appreciably changed so long as

$$n = \tfrac{1}{8}(\Delta u)^2 \ll \frac{R_s R_o}{\lambda(R_s + R_o)}. \tag{11.3.13}$$

Unless the right-hand side of (11.3.13) is small, nonparaxial corrections occur only after the paraxial spiral has closely approached its limit, in which case the nonparaxial effects are negligible.

Problems

11.3.1 Show that the intensity at the shadow edge positions, $\pm x_0$ in Fig. 11.3.1, approaches the limit $\frac{1}{4}I_0$ as the slit width increases.

11.3.2 Figure 11.1.4 is the vibration spiral for the on-axis diffracted amplitude for a slit exposing n linear zones (n not necessarily integral). How is this figure related to the Cornu spiral?

11.3.3 (a) Show that the on-axis intensity maxima and minima for a slit occur approximately at one-quarter of a zone less than integral zones, as stated in (11.3.12). ★(b) Show that the error involved in this approximation is

$$n_{\text{extremum}} - k + \tfrac{1}{4} \approx \frac{(-1)^k}{2\pi^3(k - \tfrac{1}{4})^{3/2}}. \tag{11.3.14}$$

Hint: The asymptotic expansions of the Fresnel integrals (11.2.16) and (11.2.17) for $u \gg 1$ are

$$C(u) = \frac{1}{2} + \frac{\sin(\pi/2)u^2}{\pi u} - \frac{\cos(\pi/2)u^2}{\pi^2 u^3} - \cdots \tag{11.3.15}$$

$$S(u) = \frac{1}{2} - \frac{\cos(\pi/2)u^2}{\pi u} - \frac{\sin(\pi/2)u^2}{\pi^2 u^3} + \cdots . \tag{11.3.16}$$

11.3.4 For the extrema conditions given by (11.3.12), show that as the observation point is moved slightly off axis ($x \neq 0$ in Fig. 11.3.1), the intensity goes down for k odd and goes up for k even.

11.3.5 Using an accurate graph of the Cornu spiral, compute and plot the diffraction pattern for a case not shown in Fig. 11.3.2, e.g., for $\Delta u = 3$.

11.3.6 In the paraxial approximation, the Fresnel diffraction pattern of a slit may be specified uniquely by two normalized parameters: (1) the width of the slit measured as

$$\Delta u = \left[\frac{2}{\lambda} \left(\frac{1}{R_s} + \frac{1}{R_0} \right) \right]^{1/2} a,$$

or alternatively the number $n = (\Delta u)^2/8$ of Fresnel linear zones unmasked, and (2) the angle of the observation point off the axis measured as θ/θ_0 where $\theta_0 = \lambda/a$. Compare the diffraction patterns when (a) $a = 1$ mm, $R_s = R_0 = 1$ m and, (b) $a = 2$ mm, $R_s = R_0 = 4$ m.

11.3.7 Show how to obtain from the Cornu spiral the diffraction pattern of a strip obstacle of width a with the source symmetrically placed behind the obstacle.

11.3.8 What is the effect of an obstacle (as opposed to an aperture) in the *Fraunhofer* diffraction limit?

11.3.9 Using the Cornu spiral or tables of the Fresnel integrals, investigate how large the boundary coordinate u_1 or u_2 must be in order that an *intensity* error of only 1 percent be made by substituting the limit point ($|u| \to \infty$). Assuming $R_s = R_0 = 2$ m, $\lambda = 500$ nm, comment on the statement that most of the light reaching P_0 comes from the region of the aperture near Ω (Fig. 11.3.1).

11.4 The Straight Edge

Diffraction by a straight edge, or knife-edge, is a famous example of the Fresnel theory. Since this case is in fact a *semi-infinite* slit, the solution is implicit in the preceding section. It is convenient here, however, to place the origin O', from which the observation point P_o is measured, at the geometrical shadow edge, as shown in Fig. 11.4.1. The "tubing," which we imagine to be sliding on a wire model of the Cornu spiral, thus extends from the limit point $u_1 \rightarrow -\infty$ to the point

$$u_2(x) = \left[\frac{2R_s}{\lambda R_o(R_s + R_o)} \right]^{1/2} x. \tag{11.4.1}$$

The intensity observed at P_o is, from (11.3.7),

$$I(x) = \tfrac{1}{2} I_0 [R(x)]^2, \tag{11.4.2}$$

where R is the length of the chord drawn from the limit point in the third quadrant to the point on the Cornu spiral where $u = u_2$. At $x = 0$, the intensity is $\tfrac{1}{4} I_0$. As the observation point moves into the shadow region ($x < 0$), the intensity goes monotonically to zero. In the illuminated region ($x > 0$) there are

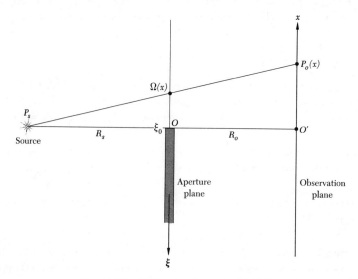

Fig. 11.4.1 Diffraction by a straight edge.

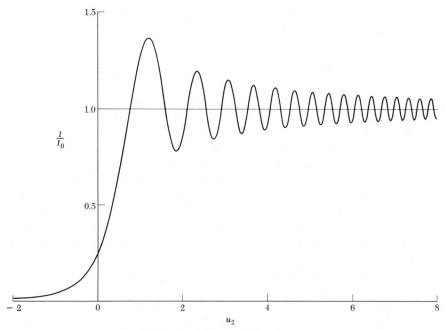

Fig. 11.4.2 Straight-edge diffraction pattern.

maxima and minima, which gradually damp out to the unobstructed intensity I_0 as $x \to +\infty$. The intensity variation is shown quantitatively in Fig. 11.4.2.

The approximate positions of the maxima and minima may readily be found from the Fresnel zone model. Instead of taking the nth linear zone to include *two* strips symmetrically located on either side of the line of sight from source to observation point, we may consider separately the set of zones on each side. This subdivision of the zones simply halves the area and hence changes the amplitude contributions; it does not alter the phase of the contributions from respective zones. Thus, for instance, at the geometrical shadow point the aperture exposes all the subdivided zones $n = 1, 2, \cdots, \infty$ on the $\xi < 0$ side of Ω (Fig. 11.4.1). Since the corresponding mirror-image zones on the $\xi > 0$ side are blocked, the amplitude is reduced to one-half and the intensity to one-quarter that of the unobstructed wave.

At an observation point $x > 0$, the aperture consists of an open half-plane from Ω to infinity, plus

$$n = \left(\frac{1}{R_s} + \frac{1}{R_o}\right)\frac{\xi_0^2}{\lambda} = \frac{R_s}{\lambda R_o(R_s + R_o)} x^2 \tag{11.4.3}$$

zones on the other side of Ω. The total amplitude at $P_o(x)$ is the vector sum of the fixed contribution from the open half-space and the variable contribution from the n zones. As we saw in the preceding section (Prob. 11.3.3), the latter contribution is an extremum when

$$n \approx k - \tfrac{1}{4} \qquad \begin{array}{ll} k = 1, 3, 5, \ldots & \text{maxima} \\ k = 2, 4, 6, \ldots & \text{minima.} \end{array} \qquad (11.4.4)$$

Moreover, inspection of the Cornu spiral shows that the phase of the contribution of $k - \tfrac{1}{4}$ zones is the same as that of the half-space. Thus, we conclude that the condition (11.4.4), identical with (11.3.12), locates the straight-edge intensity extrema in the observation plane, as well as locating the finite-slit extrema along the axis from source through the center of the slit (see Prob. 11.4.4).

As a final summary of the relation of wave to ray optics, consider the intensity distribution of light at an observation plane some distance R_o beyond a slit of width a. For a small, $\lambda < a < (\lambda R_0)^{1/2}$, Fraunhofer conditions apply and the patterns of Figs. 10.4.3 and 11.3.2a are observed. As a increases, the observed pattern *decreases* in width and gradually deforms into the sharply peaked pattern of Fig. 11.3.2b, for which $n \approx \tfrac{3}{4}$. Further increase in a produces "choppy" patterns such as Fig. 11.3.2c and d. Finally, as the aperture widens to expose many Fresnel zones (Fig. 11.3.2e and f), the pattern may be recognized as two interfering *straight-edge* patterns (Prob. 11.4.5). The geometrical-optics approximation eliminates the fine structure of these patterns, predicting instead a rectangular intensity profile. Mathematically, geometrical optics is the limit as $\lambda \to 0$, whereupon the shadow edges become very sharp (Prob. 11.4.3) and the straight-edge intensity "wiggles" crowd close to the shadow edge. Diffraction fine structure near shadow edges is not often seen in everyday life because of the finite angular size of most light sources and the fact that they usually emit a wide range of wavelengths (Prob. 11.4.2).

Problems

11.4.1   Treating the straight edge as a (semi-)*infinite* slit seems to violate the paraxial approximation. Under what conditions is our paraxial treatment accurate or inaccurate?

11.4.2   In a laboratory experiment to observe the straight-edge diffraction pattern, the effective source is a narrow slit of width a_s, parallel to the straight edge (this source slit is illuminated by a gas-discharge lamp as in Prob. 10.4.2). Let $R_s = R_o = 2$ m, $\lambda = 500$ nm. How narrow would you make the source slit? Suppose the lamp is not monochromatic but emits the range of wavelengths between 475 and 525 nm?

11.4.3   A *step function* is a function whose value changes abruptly from zero to some constant. In general, this abrupt rise takes place over a small but finite interval of the argument. Also, the function may *overshoot* before converging to its final constant value. The *rise-interval* (or *rise-time* if the abscissa happens to be time) is usually defined as the increment in the argument over which the function changes from 10 to 90 percent of its final value.

What is the rise-interval δu of the straight-edge intensity diffraction pattern? To what distance in the observation plane does this correspond for the special case of $R_s = R_o = 2$ m, $\lambda = 500$ nm? What is the percentage overshoot? *Answer:* $\delta u = 1.17$, $\delta x = 1.17$ mm; 37 percent.

★**11.4.4**   Show that the error in locating the straight-edge extrema at $n = k - \frac{1}{4}$ is one-half that of (11.3.14).

11.4.5   Show that the Fresnel diffraction pattern of a wide linear slit (exposing many Fresnel zones) can be regarded as the sum of two straight-edge patterns.

twelve

Spectrum Analysis of Waveforms

In this last chapter we examine some of the analytic tools available for discussing waves that are nonsinusoidal. We have seen in our brief study of Fourier series in Chap. 1 that a periodic but nonsinusoidal wave can always be regarded as the superposition of sinusoidal waves of suitably chosen amplitude and phase. It is not immediately obvious that all other sorts of waves can also be so regarded, provided only that they are solutions of a *linear* wave equation. If this is indeed the case, then the steady-state analysis of sinusoidal wave propagation in various media can serve as the basis for discussing the propagation of other shapes of waves such as transient disturbances, wave packets, and modulated waves. Our interest here is primarily in the mathematical description of various types of nonsinusoidal waveforms and not on the propagation of such waves in specific media. The techniques presented have far-reaching application in many areas of pure and applied physics.

12.1 *Nonsinusoidal Periodic Waves*

In Chap. 1 we found that the normal-mode analysis of the vibrations of a string segment constituted an introduction to Fourier series. In Sec. 1.7 we stated the mathematical theorem that any periodic function $f(\theta)$ of period 2π, subject to certain restrictions, can be expressed as the infinite series of sine and cosine functions (1.7.1), with coefficients given by (1.7.2). For a periodic function of time $\psi(t)$, of period $T_1 = 2\pi/\omega_1$, these formulas become, upon setting $\theta = \omega_1 t$,

$$\psi(t) = a_0 + \sum_{n=1}^{\infty} a_n \cos n\omega_1 t + \sum_{n=1}^{\infty} b_n \sin n\omega_1 t \tag{12.1.1}$$

$$a_0 = \frac{1}{T_1} \int_0^{T_1} \psi(t)\, dt$$

$$a_n = \frac{2}{T_1} \int_0^{T_1} \psi(t) \cos n\omega_1 t\, dt \tag{12.1.2}$$

$$b_n = \frac{2}{T_1} \int_0^{T_1} \psi(t) \sin n\omega_1 t\, dt.$$

By replacing t by $t \pm x/c$, such a function, of course, becomes a periodic solution of the one-dimensional wave equation

$$\frac{\partial^2 \psi}{\partial x^2} = \frac{1}{c^2} \frac{\partial^2 \psi}{\partial t^2},$$

where c is the constant wave velocity.

As a result of Prob. 1.7.7, we found that the series (12.1.1) can be expressed in the more compact, alternative form

$$\psi(t) = \sum_{n=-\infty}^{\infty} \breve{A}_n e^{-in\omega_1 t}, \tag{12.1.3}$$

where the Fourier coefficient,

$$\breve{A}_n = \frac{1}{T_1} \int_{-T_1/2}^{T_1/2} \psi(t) e^{in\omega_1 t}\, dt, \tag{12.1.4}$$

replacing the real coefficients (12.1.2), may be a complex number.† In this form of the series, the summation index n runs over *all* integers, negative, zero, and positive. Hence we can speak of *negative* as well as *positive* (and *zero*) harmonic frequencies. They uniformly populate the entire frequency axis with a spacing $\Delta\omega = \omega_1$ in a bar graph showing the discrete *spectrum* of $\psi(t)$. The magnitude of

---

† We have changed the signs of i in (12.1.3) and (12.1.4), as compared with those in Prob. 1.7.7, to make (12.1.3) agree with our convention of using $e^{-i\omega t}$ (or $e^{+i\omega t}$) as a complex exponential time factor when discussing waves.

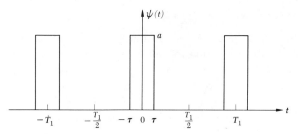

Fig. 12.1.1   A periodic waveform of rectangular pulses.

$\check{A}_n$ is evidently the amplitude of the nth harmonic; the angle that $\check{A}_n$ makes with the real axis (found from the ratio of its imaginary and real components) gives the phase angle of the nth harmonic. The complex series (12.1.3), of course, contains precisely the same information as the real series (12.1.1).

Let us now examine the spectrum of a particular periodic waveform, such as that of the sequence of rectangular pulses shown in Fig. 12.1.1. In particular we wish to discover how the spectrum depends on the fundamental period T_1 when the pulse width 2τ is held constant. When T_1 is made very large, the waveform approaches that of a single pulse. In this way we can gain an insight into the Fourier spectrum of nonrecurrent waveforms.

In real form, the Fourier expansion of the waveform of Fig. 12.1.1 is found to be (Prob. 12.1.2)

$$\psi(t) = \frac{2a\tau}{T_1} + \frac{2a}{\pi}(\sin\omega_1\tau \cos\omega_1 t + \tfrac{1}{2}\sin2\omega_1\tau \cos2\omega_1 t + \cdots). \qquad (12.1.5)$$

Before discussing the spectrum of $\psi(t)$ let us change the series (12.1.5) to complex form using the identity $\cos\theta = \tfrac{1}{2}(e^{i\theta} + e^{-i\theta})$. If we also introduce the *discrete* (or quantized) frequency variable $\omega_n \equiv n\omega_1$, which progresses in equal steps of frequency $\Delta\omega \equiv \omega_{n+1} - \omega_n = \omega_1$, the series (12.1.5) becomes

$$\psi(t) = \frac{a\tau \, \Delta\omega}{\pi} \sum_{n=-\infty}^{\infty} \frac{\sin\omega_n\tau}{\omega_n\tau} e^{-i\omega_n t}. \qquad (12.1.6)$$

The Fourier coefficients $\check{A}_n$ in this example are real.

The amplitude of the nth harmonic, at the frequency $\omega_n = n\omega_1$ (recall that n may be negative as well as positive), has as factors the constant $(a\tau \, \Delta\omega)/\pi$ and the amplitude function $\sin\omega_n\tau/\omega_n\tau$. The constant is evidently the area $2a\tau$ under one rectangular pulse times the repetition rate $1/T_1 = \Delta\omega/2\pi$. We recognize

$$f(u) = \frac{\sin u}{u} \qquad (12.1.7)$$

as the function occurring in the theory of the Fraunhofer diffraction by a single slit (Sec. 10.4). Here the variable u takes on the discrete values $u_n = \omega_n\tau$. A graph of $f(u)$ appears in Fig. 12.1.2; on separate u axes are shown the location of the values of $u_n = \omega_n \tau$ when $T_1 = 5\tau$ and when $T_1 = 50\tau$.

When the repetition period T_1 becomes very large and the frequency spacing $\Delta\omega$ of the discrete frequencies therefore becomes very small, the amplitude function $f(u_n)$ gives an almost continuous description of the spectrum. We then infer that if we proceed to the limit $T_1 \to \infty$, so that $\Delta\omega \to d\omega$, the Fourier series (12.1.6) becomes the *Fourier integral*

$$\psi(t) = \frac{a\tau}{\pi} \int_{-\infty}^{\infty} \frac{\sin\omega\tau}{\omega\tau} e^{-i\omega t} \, d\omega. \tag{12.1.8}$$

What was originally a spectrum of *discrete* frequencies, defined over the entire frequency axis $-\infty < \omega < +\infty$, now becomes a *continuous* spectrum of frequencies. In such a case we do not speak of the amplitude of a spectral component of frequency ω. Rather we speak of the (infinitesimal) amplitude of the spectral frequency components in the (infinitesimal) frequency range between ω and $\omega + d\omega$. Though plausible, this derivation of the Fourier integral lacks rigor since the various limiting processes involved are not carefully treated.

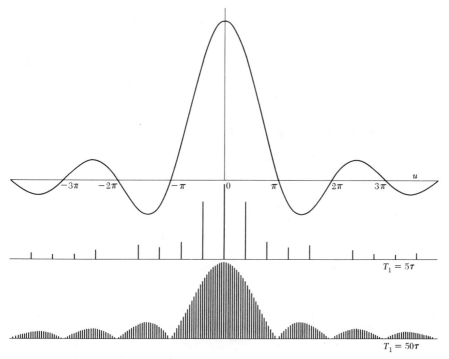

Fig. 12.1.2   The function $\sin u / u$ and the location of the harmonic frequencies when $T_1 = 5\tau$ and $T_1 = 50\tau$.

Nevertheless an analysis of the sort given here is very suggestive and, at least for waveforms of interest in physics, leads to a correct description of the frequency spectrum of a nonrecurrent waveform. Physically, we surely expect that the discrete amplitude spectrum of a periodic wave will transform smoothly into a continuous amplitude spectrum as the period of the wave becomes indefinitely great.

In many applications of wave theory, we are particularly interested in the energy (or power) spectrum rather than the amplitude spectrum. We know from our study of wave motion that the square of a wave amplitude is proportional to the energy density in the wave. If we average the square of the periodic time function (12.1.3) over one fundamental period, we find that the energy associated with the nth harmonic wave is proportional to $|\breve{A}_n|^2 = \breve{A}_n \breve{A}_n^*$ (Prob. 12.1.3). Hence it is easy to convert an amplitude spectrum into a (relative) energy spectrum. Note that information about the phase of the harmonics is present only in the complex amplitude spectrum.

Problems

12.1.1 Construct bar graphs showing the amplitude and energy spectra of several of the waveforms in Probs. 1.7.3 and 1.7.4.

12.1.2 Verify the Fourier expansion (12.1.5). Obtain (12.1.6) by the substitution $\cos\theta = \frac{1}{2}(e^{i\theta} + e^{-i\theta})$ and also by using (12.1.4).

12.1.3 Show that when $\psi(t)$ is expressed by (12.1.3),

$$\overline{|\psi(t)|^2} = \sum_{-\infty}^{\infty} \breve{A}_n \breve{A}_n^* = \sum_{-\infty}^{\infty} |\breve{A}_n|^2,$$

where the bar denotes a time average taken over the fundamental period T_1. This relation is very similar to (1.8.17), except for a factor of $\frac{1}{2}$. Account for the difference.

12.2 Nonrecurrent Waves

In the preceding section we examined the spectrum of a periodic wave expressed as a Fourier series and found it convenient to regard its spectrum as populating the entire frequency axis from $-\infty$ to $+\infty$. We also found from a study of a sequence of rectangular pulses that the spectrum becomes more and more dense as the repetition period is lengthened while keeping the pulse width constant.

This behavior strongly implies that a Fourier *series* turns into an *integral* with infinite limits when the period is allowed to increase indefinitely.

Considerations of this sort lead to the definition of the *Fourier* (integral) *transform pair*

$$f(t) = \frac{1}{2\pi} \int_{-\infty}^{\infty} F(\omega)e^{-i\omega t}\, d\omega \tag{12.2.1}$$

$$F(\omega) = \int_{-\infty}^{\infty} f(t)e^{i\omega t}\, dt. \tag{12.2.2}$$

We adopt here a common notation; in the time domain a function is expressed by a lowercase letter, for example, $f(t)$; its Fourier transform in the frequency domain is expressed by the same letter written as a capital, for example, $F(\omega)$. Either or both functions may be complex (see Prob. 12.2.1). It is possible to show that (12.2.1) implies (12.2.2), and vice versa, when

$$\int_{-\infty}^{\infty} |f(t)|\, dt < \infty$$

and when $f(t)$ possesses at most a finite number of discontinuities and maxima and minima (turning points) in any finite interval—the so-called *Dirichlet conditions* that limit the form of functions expressible as Fourier series. At a discontinuity, the value of $f(t)$ as given by the transform (12.2.1) is found to converge to $\lim \frac{1}{2}[f(t + \epsilon) + f(t - \epsilon)]$. A rigorous mathematical treatment of Fourier integrals is not simple, and as with the Fourier series discussed in Chap. 1, we must be content merely to state a few results that can serve to introduce this application of formal Fourier analysis.†

The first member of the Fourier transform pair, (12.2.1), implies that there exists a (complex) amplitude spectrum-density function $F(\omega)$ which, when multiplied by $(e^{-i\omega t}\, d\omega/2\pi)$ and integrated over the entire frequency range, gives rise to a particular function of time $f(t)$. The second member of the pair, (12.2.2), enables us to find the spectrum-density function, i.e., the *Fourier transform $F(\omega)$* for a given $f(t)$. These two equations should be compared carefully with the analogous two equations (12.1.3) and (12.1.4) for a periodic function. We see that the discrete harmonic frequencies $n\omega_1$ occurring in the Fourier series expansion (12.1.3) are replaced by a continuous frequency variable ω in the analogous

---

† For a brief introductory account, see M. L. Boas, "Mathematical Methods in the Physical Sciences" chap. 13, John Wiley & Sons, Inc., New York, 1966. A more extended introductory account is given by R. Bracewell, "The Fourier Transform and Its Applications," McGraw-Hill Book Company, New York, 1965. For an advanced treatment, with many applications of interest to physicists, see I. N. Sneddon, "Fourier Transforms," McGraw-Hill Book Company, New York, 1951. The mathematical aspects are emphasized in H. S. Carslaw, "Introduction to the Theory of Fourier's Series and Integrals," Dover Publications, Inc., New York, 1930.

Fourier transform (12.2.1). In both cases the frequencies range from $-\infty$ to $+\infty$. The finite limits of the integral (12.1.4) for the Fourier coefficients $\breve{A}_n$ are replaced by infinite limits in the integral (12.2.2) for the Fourier spectrum-density function $F(\omega)$. The factor $1/T_1$ in (12.1.4) has, in effect, been transferred to the first Fourier transform (12.2.1), where it appears as the factor $d\omega/2\pi$.†
We make no direct use of the real-part convention when employing either the Fourier series (12.1.3) and (12.1.4) or the Fourier transforms (12.2.1) and (12.2.2). The reason it is not necessary to use this convention derives from the fact that when we replaced real sines and cosines by their complex equivalents to obtain a complex form for a Fourier series, we extended the frequency range to include negative as well as positive frequencies, as an alternative to making use of the real-part convention. It is important to understand this point since real functions of time may possess real, imaginary, or complex transforms (spectra), depending on their symmetry properties of oddness and evenness (see Prob. 12.2.1). There exist extensive tables of Fourier transform pairs listing the frequency function $F(\omega)$ for a wide variety of temporal functions $f(t)$.‡

It is not difficult to establish the so-called *similarity relation*, namely, if $F(\omega)$ is the Fourier transform of $f(t)$, then $(1/a)F(\omega/a)$ is the Fourier transform of $f(at)$, where a is a positive constant scale factor which changes the duration of the waveform in time (Prob. 12.2.4). If $a > 1$, the waveform $f(at)$ is compressed in comparison with the waveform $f(t)$. The spectrum, in contrast, is then *spread out*. We can be more specific. If Δt is some consistently applied measure of the temporal width, or *duration*, of the waveform $f(t)$, then $\Delta t/a$ is a measure of the duration of the similar but compressed waveform $f(at)$. Similarly, if $\Delta\omega$ is some consistently applied measure of the spread, or *bandwidth*, of the spectrum $F(\omega)$, then $a\,\Delta\omega$ is a measure of the bandwidth of the spectrum $(1/a)F(\omega/a)$, which is also everywhere decreased in height by the scale factor $1/a$.

The dimensionless product of the waveform duration and the spectrum bandwidth is *independent of the scale factor* and hence a constant for similar waveforms. The numerical value of the constant, however, depends on how Δt and $\Delta\omega$ are defined, and it can be expected to depend on the form of the function $f(t)$. There are various ways of defining the duration Δt of the waveform (or *wave packet*) described by $f(t)$ and the bandwidth $\Delta\omega$ of the spectrum of $f(t)$. The definition having the greatest physical significance is the *root-mean-square* width of the energy (or power) associated with the waveform and with its related energy spectrum. Let $f(t)$ be a real function of time for which $\int_{-\infty}^{\infty} |f(t)|^2\,dt$ is

† The pair of transform integrals appears more symmetrical if the cyclic frequency variable $\nu \equiv \omega/2\pi$ is used in place of the radian frequency ω. Other authors achieve symmetry by assigning a factor $(2\pi)^{-1/2}$ to each of the integrals.
‡ G. A. Campbell and R. M. Foster, "Fourier Integrals for Practical Applications," D. Van Nostrand Company, Inc., Princeton, N.J., 1948; A. Erdélyi, "Tables of Integral Transforms," vol. 1, McGraw-Hill Book Company, New York, 1953.

bounded. Then

$$\bar{t} = \frac{\int_{-\infty}^{\infty} t|f(t)|^2\, dt}{\int_{-\infty}^{\infty} |f(t)|^2\, dt} \qquad (12.2.3)$$

is the *centroid* of $|f(t)|^2$, and the mean-square duration of $|f(t)|^2$ is

$$(\Delta t)^2 = \frac{\int_{-\infty}^{\infty} (t - \bar{t})^2 |f(t)|^2\, dt}{\int_{-\infty}^{\infty} |f(t)|^2\, dt}$$

$$= \frac{\int_{-\infty}^{\infty} t^2 |f(t)|^2\, dt}{\int_{-\infty}^{\infty} |f(t)|^2\, dt} - \left[\frac{\int_{-\infty}^{\infty} t|f(t)|^2\, dt}{\int_{-\infty}^{\infty} |f(t)|^2\, dt} \right]^2, \qquad (12.2.4)$$

provided the integrals converge. The related mean-square bandwidth of the energy spectrum is given by

$$(\Delta \omega)^2 = \frac{\int_{-\infty}^{\infty} \omega^2 |F(\omega)|^2\, d\omega}{\int_{-\infty}^{\infty} |F(\omega)|^2\, d\omega} \qquad (12.2.5)$$

where $F(\omega)$ is the Fourier transform of $f(t)$. If the waveform $f(t)$ is a real function of time, as we have assumed, its frequency spectrum is found to be symmetrical about zero frequency (see Prob. 12.2.6). Hence we do not need to introduce formally the centroid $\bar{\omega}$ of the relative energy spectrum $|F(\omega)|^2$ in defining the mean-square bandwidth, since $\bar{\omega} = 0$. In each case we are using what is known in statistics as the *standard deviation* as the measure of width of $|f(t)|^2$ and of $|F(\omega)|^2$. We note in passing Parseval's theorem (Appendix C)

$$\int_{-\infty}^{\infty} |f(t)|^2\, dt = \frac{1}{2\pi} \int_{-\infty}^{\infty} |F(\omega)|^2\, d\omega, \qquad (12.2.6)$$

which connects energy-related quantities in the time and frequency domains.

Now we have already established the similarity relation that the product $\Delta t\, \Delta \omega$ is independent of a scale factor. With Δt and $\Delta \omega$ defined by (12.2.4) and (12.2.5), it is possible to show (Appendix C), using a mathematical theorem known as *Schwarz's inequality*, that

$$\Delta t\, \Delta \omega \geq \tfrac{1}{2}. \qquad (12.2.7)$$

This equation is known as the *uncertainty relation* and expresses in a quantitative way, holding for functions of all shapes, that $\tfrac{1}{2}$ sets a *lower limit* for the product $\Delta t\, \Delta \omega$. The uncertainty relation, as given by (12.2.7), expresses the fact that no matter what form $f(t)$ takes, if it has an rms Δt defined by (12.2.4), the minimum possible rms width of its energy spectrum, as defined by (12.2.5), is $\Delta \omega = 1/2\Delta t$.

An example of a waveform having a minimum spectral width is the gaussian function, $f(t) = e^{-(t^2/\tau^2)}$ (Prob. 12.2.5). At the other extreme, the spectral width of a rectangular pulse is found to be infinite! The uncertainty relation expresses a fundamental connection between the temporal duration of a waveform $f(t)$ and the bandwidth of its energy spectrum in the related frequency domain. Although the uncertainty relation is a deduction of pure mathematics, it is given an important physical interpretation in quantum mechanics, where it bears the name of *Heisenberg's uncertainty principle* (see Prob. 12.2.9).

We have endeavored to introduce several important ideas in the present section: (1) Any nonrecurrent waveform $f(t)$ of reasonable shape in the temporal domain can be considered to have a continuous harmonic amplitude spectrum $F(\omega)$ in the frequency domain. (2) If we are given an amplitude spectrum $F(\omega)$, it is possible to compute the waveform $f(t)$ to which it corresponds. (3) There exist important relations between the time and frequency domains, such as those expressed by the similarity relation, Parseval's theorem (12.2.6), and the uncertainty relation (12.2.7).

The possibility of *spectrum analysis* and *spectrum synthesis* by Fourier series and Fourier integrals again reminds us that the theory of steady sinusoidal waves, which never end in space or time, is far more than an idealized special case. Using the concepts of Fourier analysis and synthesis, the theory of continuous waves constitutes the basis for the description of waves of arbitrary shape without further reference to the wave equation or to particular non-sinusoidal solutions that it may have. The only basic requirement is that the wave equation be *linear*, so that the principle of superposition applies. The technique, of course, can be used even though the wave velocity is frequency-dependent; i.e., the wave medium is dispersive. A problem of this sort is discussed in Sec. 12.5.

Problems

12.2.1 By examining the Fourier transform (12.2.1), establish the following diagram

$$f(t) = o(t) + e(t) = \text{Re}[o(t)] + i\,\text{Im}[o(t)] + \text{Re}[e(t)] + i\,\text{Im}[e(t)]$$

$$F(\omega) = O(\omega) + E(\omega) = \text{Re}[O(\omega)] + i\,\text{Im}[O(\omega)] + \text{Re}[E(\omega)] + i\,\text{Im}[E(\omega)]$$

which relates the real and imaginary parts of $f(t)$, expressed as the sum of odd and even parts, to its transform broken into the corresponding four parts or categories. For example, if the transform of $f(t)$ is a real odd function of ω, then $f(t)$ is an imaginary odd function of t.

12.2.2 Find, using (12.2.2), the Fourier transform $F(\omega)$ of a single rectangular pulse of amplitude a extending from $-\tau$ to $+\tau$. Substitute this $F(\omega)$ in (12.2.1) and compare the resulting expression with (12.1.8), obtained from the Fourier series expansion of a sequence of rectangular pulses.

12.2.3 A pulse described by the *Dirac delta function* $\delta(t)$ has negligible duration, infinite height at $t = 0$, but unit area. Find the amplitude spectrum of $\delta(t)$ as a limiting case by setting the area of the rectangular pulse in Prob. 12.2.2 equal to unity and then letting $\tau \to 0$. *Answer:* $F(\omega) = 1$.

12.2.4 Prove the similarity relation that if $F(\omega)$ is the Fourier transform of $f(t)$ then $(1/a)F(\omega/a)$ is the Fourier transform of $f(at)$, where a is a real, positive constant.

12.2.5 Show that the Fourier transform of $f(t) = e^{-(t/\tau)^2}$ is $e^{-(\omega\tau/2\pi)^2}$. Show that the duration of $f(t)$, defined by (12.2.4), is $\tau/2$ and that the width $\Delta\omega$ of its energy spectrum, defined by (12.2.5), is $1/\tau$ and hence that the product $\Delta t \, \Delta\omega$ has precisely the minimum value $\frac{1}{2}$ permitted by the uncertainty relation (12.2.7).

12.2.6 Prove that when $f(t)$ is real, $|F(\omega)|^2$ is necessarily an *even* function of ω. The energy spectrum of $f(t)$ is thus symmetrical about zero frequency, which justifies the form of (12.2.5).

12.2.7 Discuss the need for an amplifier to have a certain minimum bandwidth $\Delta\omega$ if it is to be used for amplifying pulses of duration Δt.

12.2.8 The voltage gain of a multistage low-pass amplifier is accurately approximated by the gaussian function

$$|\breve{G}(\omega)| = G_0 e^{-(\omega/2\pi\Delta\omega)^2}.$$

Show that $\Delta\omega$ is the rms bandwidth of the amplifier. Show also that the so-called *upper half-power frequency* ν_0 [the frequency at which $(|\breve{G}|/G_0)^2 = \frac{1}{2}$] is related to $\Delta\omega$ by

$$\nu_0 \approx 0.6\Delta\omega,$$

where ν_0 is an ordinary (cyclic) frequency. Show then that the uncertainty relation (12.2.7) implies that

$$\nu_0 \, \Delta t \approx \tfrac{1}{3}$$

when Δt, the duration of a narrow pulse whose frequency spectrum is limited by the amplifier, has its minimum permitted value. This form of the uncertainty relation expresses reasonably well the relation between the rise-time of an amplifier (see Prob. 11.4.3 for a definition of rise-time) and its bandwidth expressed by its upper half-power frequency.

12.2.9 The uncertainty relation (12.2.7) relates a waveform duration in the time domain to the bandwidth of its energy spectrum in the frequency domain. In the case of a traveling wave of finite duration, what can be said about the waveform extent Δx in the space domain and the spread $\Delta\kappa$ of the wave numbers in the wave-number domain? In quantum mechanics the energy E and momentum p of a particle are related to frequency and wave number by de Broglie's relations $E = \hbar\omega$ and $p = \hbar\kappa$, where $\hbar$ is Planck's constant divided by 2π. On the basis of these relations and the uncertainty relation, obtain Heisenberg's uncertainty principle $\Delta E\, \Delta t \geq \frac{1}{2}\hbar$ and $\Delta p\, \Delta x \geq \frac{1}{2}\hbar$. This principle is an inescapable part of a wave description of nature.

12.3 Amplitude-modulated Waves

Let us now investigate a sinusoidal wave whose amplitude varies, or is *modulated*, with time, and consequently with position. To be definite, suppose that the wave is a solution of a one-dimensional wave equation having a constant wave velocity independent of frequency. A positive-going amplitude-modulated (AM) sinusoidal wave of frequency ω_0 that satisfies such a wave equation can be written in the real form

$$\psi(x,t) = f(x - ct) = A(x - ct)\cos(\kappa_0 x - \omega_0 t), \qquad (12.3.1)$$

where $\kappa_0 = \omega_0/c$ is the wave number. For simplicity we suppose that the amplitude function $A(x - ct)$ is nonnegative. A typical wave of this sort is illustrated in Fig. 12.3.1, which also serves to illustrate its spatial variation at any instant in time. An AM radio wave is a familiar example of such a wave. Since the amplitude factor $A(x - ct)$ and the sinusoidal factor $\cos(\kappa_0 x - \omega_0 t)$ are both functions of $(x - ct)$, we can study the wave either as a function of x (at some time t_0) or as a function of t (at some position x_0). As in the previous two sections, we choose to do the latter since it appears easier to think about the wave in terms of frequency than in terms of wave number.

At the origin, for example, the wave amplitude varies according to

$$\psi(t) = A'(t)\cos\omega_0 t, \qquad (12.3.2)$$

where we have written $A'(t)$ for the amplitude function $A(-ct)$. Let us select for further study a wave modulated sinusoidally having the amplitude function

$$A'(t) = A_0(1 + m\sin\omega_m t), \qquad (12.3.3)$$

where the modulation frequency ω_m is ordinarily less than the frequency ω_0 of the unmodulated wave. For simplicity let the parameter m, the *degree of modulation*, be restricted to the range $0 \leq m \leq 1$. When m is zero, the wave is unmodulated and has the steady amplitude A_0. At the other extreme, when $m = 1$, the wave is said to be 100 per cent modulated; its amplitude then varies between 0 and $2A_0$ at the modulation frequency ω_m.

It is easy to show from trigonometric identities that the wave (12.3.2) with the amplitude function (12.3.3) can be expressed in the alternative form

$$\psi(t) = A_0[\cos\omega_0 t + \tfrac{1}{2}m\sin(\omega_0 + \omega_m)t - \tfrac{1}{2}m\sin(\omega_0 - \omega_m)t]. \qquad (12.3.4)$$

Accordingly we find that a sinusoidally modulated wave can be resolved into three steady sinusoidal waves, one at the original or *carrier* frequency ω_0,

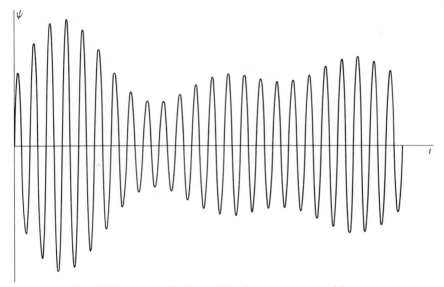

Fig. 12.3.1   An amplitude-modulated wave at some position x_0.

having the original (unmodulated) amplitude A_0; and two others at the *side-band* frequencies $(\omega_0 + \omega_m)$ and $(\omega_0 - \omega_m)$, each of these having the original wave amplitude reduced by the factor $\frac{1}{2}m$. The three sinusoidal waves constitute the spectrum of the AM wave, which can be expressed as either an amplitude or an energy spectrum. If we think of the modulation frequency as a variable parameter covering some frequency range, e.g., the audio-frequency range, the spectrum consists of the fixed carrier wave and a multitude of waves occupying two symmetrically disposed sidebands in which the individual sum and difference sideband frequencies lie, as illustrated in Fig. 12.3.2. For an AM wave to propa-

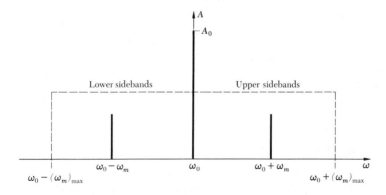

Fig. 12.3.2   Spectrum of an AM sinusoidal wave.

gate unchanged through a medium, the wave velocity must be independent of frequency over the frequency range $2(\omega_m)_{max}$ occupied by the two sidebands centered on the carrier frequency ω_0. In the case of AM radio waves, for example, the "medium" includes the electronic devices at the transmitter and receiver. Since 10 kHz is the authorized frequency spacing of adjacent AM radio channels in the standard broadcast band, the fidelity of music that can be broadcast by an AM radio station is limited to audio frequencies below 5 kHz.

When a modulation waveform is not sinusoidal, we have just seen that it can be regarded as the superposition of many sinusoidal waves using either a Fourier series, discussed in Secs. 1.7 and 12.1 for periodic waveforms, or a Fourier integral, discussed briefly in Sec. 12.2 for nonrecurrent waveforms. Hence the amplitude spectrum, giving the amplitude (and phase) of each frequency component that is present in an AM wave, can be found from the amplitude spectrum of its modulation waveform.

Problems

12.3.1   Verify that (12.3.4) is identical with (12.3.2) and (12.3.3).

12.3.2   Discuss the relative amounts of energy (or power) in the carrier and the two sideband waves.

12.3.3   How does a sinusoidally modulated wave differ from the wave illustrated in Fig. 4.7.1?

12.3.4   Show that the amplitude spectrum of a rectangular pulse of the sinusoidal oscillation $e^{-i\omega_0 t}$ lasting from $-\tau$ to $+\tau$ is given by

$$F(\omega) = 2a \, \frac{\sin(\omega - \omega_0)\tau}{(\omega - \omega_0)\tau}.$$

Relate this spectrum to that of the rectangular pulse discussed in Sec. 12.2. Generalize this result into a *modulation theorem*: if $F(\omega)$ is the transform of $f(t)$, then $F(\omega - \omega_0)$ is the transform of $f(t)e^{-i\omega_0 t}$. What form does the theorem take if $e^{-i\omega_0 t}$ is replaced by $\cos\omega_0 t$?

12.3.5   Light emitted by an atom consists of regular sinusoidal vibrations that last 10^{-9} sec or longer. Discuss the spread in frequencies on the basis of the modulation theorem of Prob. 12.3.4 and the uncertainty relation (12.2.7) and thus account for the *natural line width* of the radiation.

12.4   Phase-modulated Waves

We have just seen how an AM sinusoidal wave is mathematically (and physically!) equivalent to a number of steady sinusoidal waves that constitute its frequency spectrum. Let us next examine a sinusoidal wave whose amplitude remains constant but whose phase ϕ includes a function $\Phi(x - ct)$, in addition to the usual phase argument $\kappa_0 x - \omega_0 t$. Such a wave may be written

$$\psi(x,t) = A_0 \cos[\kappa_0 x - \omega_0 t + \Phi(x - ct)], \tag{12.4.1}$$

where κ_0 and $\omega_0 = \kappa_0 c$ pertain to the unmodulated wave. This *phase-modulated* wave is clearly a solution of the ordinary one-dimensional wave equation when the wave velocity c is a constant.

We can study the wave (12.4.1) either as a function of position or of time, and as before we choose to do the latter. At the origin, the wave displacement varies with time according to

$$\psi(t) = A_0 \cos[\omega_0 t + \Phi'(t)], \tag{12.4.2}$$

where we have written $\Phi'(t)$ for $-\Phi(-ct)$. We select for further study a wave for which $\Phi'(t)$ varies sinusoidally at some modulation frequency ω_m, putting

$$\Phi'(t) = m \sin\omega_m t, \tag{12.4.3}$$

where m is known as the *modulation index*. Let us now see how the function of time

$$\psi(t) = A_0 \cos(\omega_0 t + m \sin\omega_m t) \tag{12.4.4}$$

can be analyzed into a spectrum of steady sinusoidal vibrations with frequencies related to ω_0 and ω_m.

First expand (12.4.4) into the form

$$\psi(t) = A_0[\cos\omega_0 t \cos(m \sin\omega_m t) - \sin\omega_0 t \sin(m \sin\omega_m t)] \tag{12.4.5}$$

The periodic functions $\cos(m \sin\omega_m t)$ and $\sin(m \sin\omega_m t)$ have a period $2\pi/\omega_m$ and can be expanded in a Fourier series. Such an expansion, carried out by a straightforward application of the formulas for the Fourier coefficients (1.7.2), involves lengthy integrations. The following two Fourier series are usually derived by other means in a systematic treatment of Bessel functions:[†]

$$\cos(u \sin\theta) = J_0(u) + 2[J_2(u) \cos2\theta + J_4(u) \cos4\theta + \cdots] \tag{12.4.6}$$

$$\sin(u \sin\theta) = 2[J_1(u) \sin\theta + J_3(u) \sin3\theta + \cdots]. \tag{12.4.7}$$

The Fourier coefficients (except for the numerical factor 2) are seen to be Bessel functions, which are defined and briefly discussed in Sec. 2.3. Using trigonometric identities in conjunction with (12.4.6) and (12.4.7), we find that the function

---

[†] See A. Gray, G. B. Mathews, and T. M. MacRobert, "A Treatise on Bessel Functions and Their Applications to Physics," 2d ed., chap. 4, Dover Publications, Inc., New York, 1966.

(12.4.5) can be written in the form of the series

$$\psi(t) = A_0\{J_0(m) \cos\omega_0 t$$
$$+ J_1(m)[\cos(\omega_0 + \omega_m)t - \cos(\omega_0 - \omega_m)t]$$
$$+ J_2(m)[\cos(\omega_0 + 2\omega_m)t + \cos(\omega_0 - 2\omega_m)t]$$
$$+ J_3(m)[\cos(\omega_0 + 3\omega_m)t - \cos(\omega_0 - 3\omega_m)t] + \cdots\}. \quad (12.4.8)$$

The spectrum of a phase-modulated wave thus contains components at the sum and difference frequencies of the carrier frequency and the various *integral harmonics* of the phase-modulation frequency ω_m, as well as the carrier frequency ω_0. It is found from the properties of the Bessel functions that the amplitudes of the sideband frequency components fall off rapidly as soon as the harmonic order exceeds the modulation index m. When the phase-modulation function contains more than a single frequency, an extension of the present analysis reveals that the frequency spectrum contains not only harmonic components of each modulation frequency but also components obtained by combining (adding or subtracting) the modulation frequencies with various harmonic multipliers. Although the spectrum is then very complicated, we can still regard the phase-modulated wave as a linear superposition of many steady sinusoidal waves.

What is known in technical literature as a *frequency-modulated* (FM) radio wave is closely related to the phase-modulated wave that we have been examining. The distinction involves the way in which the modulation index m varies with modulation frequency. When a wave is frequency-modulated, the rate of change of phase, or *instantaneous frequency*, is varied in proportion to the amplitude of the modulating signal. For sinusoidal modulation, we can thus write that

$$\frac{d\phi}{dt} = \omega_0 + \Delta\omega \cos\omega_m t, \quad (12.4.9)$$

where $\Delta\omega$, the frequency deviation, is proportional to the amplitude of the modulating signal. We find by integration that the phase function is

$$\phi = \omega_0 t + \frac{\Delta\omega}{\omega_m} \sin\omega_m t. \quad (12.4.10)$$

Comparison with (12.4.4) shows that the modulation index m for a FM wave varies inversely with modulating frequency, while for a phase-modulated wave it is independent of frequency. From a practical standpoint this behavior is important. The decrease in index with increasing frequency tends to maintain an essentially constant width to the sidebands containing the important sideband frequency components. For a further discussion of frequency- and phase-modulated waves, the reader is referred to a technical account of the modulation of radio waves.†

Problems

★12.4.1 Verify the Fourier expansions (12.4.6) and (12.4.7).

12.4.2 Examine a numerical table of Bessel functions and verify that when the argument u of a Bessel function $J_n(u)$ exceeds its order n, the maximum excursions of $J_n(u)$ decrease rapidly. Use the table to construct charts, analogous to Fig. 12.3.2, showing quantitatively the frequency spectrum of a FM wave of constant frequency deviation $\Delta\omega$ at several values (for example, 4, 12, 24 rad) of the modulation index.

★12.4.3 Find the frequency spectrum of a wave phase-modulated simultaneously at two frequencies ω_1 and ω_2 with modulation indices m_1 and m_2.

12.5 The Motion of a Wave Packet in a Dispersive Medium

We have chosen so far to examine the spectrum analysis of waveforms in the time and frequency domains rather than in the space and wave-number domains. It is evident, however, that by replacing t by x and ω by κ, the various waveforms and their spectra equally well describe snapshots of particular plane waves progressing in the x direction. In terms of x and κ, the Fourier transform pair (12.2.1) and (12.2.2) takes the form

$$f(x) = \frac{1}{2\pi} \int_{-\infty}^{\infty} F(\kappa) e^{i\kappa x} \, d\kappa \tag{12.5.1}$$

$$F(\kappa) = \int_{-\infty}^{\infty} f(x) e^{-i\kappa x} \, dx. \tag{12.5.2}$$

In a dispersionless medium having the constant wave velocity c, we can modify (12.5.1) by replacing x by $x - ct$ to obtain an expression for a one-dimensional plane wave of arbitrary waveform traveling in the positive x direction. For such a wave

$$\begin{aligned}
f(x,t) &= \frac{1}{2\pi} \int_{-\infty}^{\infty} F(\kappa) e^{i\kappa(x-ct)} \, d\kappa \\
&= \frac{1}{2\pi} \int_{-\infty}^{\infty} F(\kappa) e^{i(\kappa x - \omega t)} \, d\kappa,
\end{aligned} \tag{12.5.3}$$

† See, for example, F. E. Terman, "Electronic and Radio Engineering," 4th ed., chap. 17, McGraw-Hill Book Company, New York, 1955.

where the substitution $\kappa c = \omega$ has been made. For a positive-going wave, ω must have the same sign as κ.

Equation (12.5.3) has the following interpretation. The wave $f(x,t)$ can be synthesized in all its details by the superposition of a continuum of sinusoidal plane waves $e^{i(\kappa x - \omega t)}$ of (complex) amplitude $(1/2\pi)F(\kappa)\,d\kappa$. The wave numbers required range continuously from $-\infty$ to $+\infty$. The Fourier transform $F(\kappa)$, which can be found from the waveshape at time $t = 0$ using (12.5.2), expresses the density of the (complex) amplitude spectrum as a function of wave number.

We know from our survey of wave-propagating media that wave equations usually can be considered as linear. Linearity, of course, is a necessary prerequisite for the validity of (12.5.3), which has its basis in the principle of superposition. We must relax the requirement of a constant wave (phase) velocity, however, to take dispersion into account. In Sec. 4.7 we have pointed out that dispersion causes a nonsinusoidal wave to change its shape as it advances. The velocity of travel of the beat pattern of two sinusoidal waves differing slightly in frequency and wave number occurs at the group velocity

$$c_g \equiv \frac{d\omega}{d\kappa}, \tag{12.5.4}$$

which can differ considerably from the phase velocity ω/κ in a dispersive medium. The derivative in (12.5.4) is to be evaluated at the mean wave number of the two waves. We can use (12.5.3) to investigate in greater detail the effect of dispersion on nonrecurrent wave trains by substituting for ω the *dispersion relation*

$$\omega = \kappa c_{\text{phase}}(\kappa), \tag{12.5.5}$$

where c_{phase}, the (positive) velocity of a sinusoidal wave, depends on wave number.

Let us suppose that the wave packet consists of a sinusoidal wave $e^{i(\kappa_0 x - \omega_0 t)}$ of constant frequency, amplitude-modulated by a pulse-shaped waveform, which initially restricts the wave packet to a finite spatial spread. For instance, we might choose a rectangular pulse for the envelope, giving rise to the wave packet examined in Prob. 12.3.4. A more interesting case, however, is the wave packet illustrated in Fig. 12.5.1; its envelope is the familiar bell-shaped gaussian function that avoids mathematical discontinuities, yet disappears rapidly at distances exceeding several standard deviations from its center. This waveform has the narrowest bandwidth (here a spectral width in the κ domain) permitted by the uncertainty relation (Prob. 12.2.5).

The gaussian wave packet at $t = 0$ may be expressed by the real part of

$$f(x) = A e^{-x^2/4(\Delta x_0)^2} e^{i\kappa_0 x}. \tag{12.5.6}$$

The modulation envelope is symmetrical about the origin, where it has the

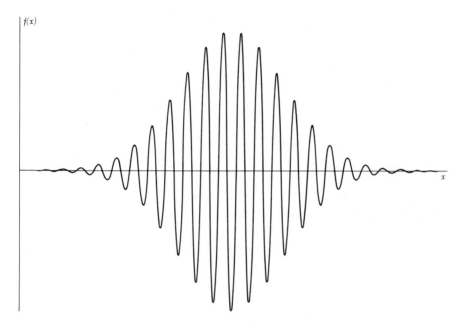

Fig. 12.5.1   A gaussian wave packet.

amplitude A; Δx_0 is the rms spatial spread of the energy distribution in the packet, which is proportional to $|f(x)|^2$ (Prob. 12.5.1). The Fourier transform (12.5.2) of the wave packet (12.5.6) is found to be (Prob. 12.5.2)

$$F(\kappa) = (4\pi)^{1/2} \Delta x_0 A e^{-(\Delta x_0)^2(\kappa-\kappa_0)^2}. \tag{12.5.7}$$

The wave at a later time is expressed by the inverse Fourier integral (12.5.3), using (12.5.7) for $F(\kappa)$. We must therefore evaluate

$$f(x,t) = \frac{\Delta x_0 A}{\pi^{1/2}} \int_{-\infty}^{\infty} e^{-(\Delta x_0)^2(\kappa-\kappa_0)^2} e^{i(\kappa x-\omega t)} \, d\kappa \tag{12.5.8}$$

after substituting for ω the appropriate dispersion relation connecting ω with κ.

Instead of choosing a particular form of dispersion relation, let us expand ω as a function of κ in a Taylor series about the wave number κ_0 of the un-modulated sine wave,

$$\omega = \omega_0 + \left(\frac{d\omega}{d\kappa}\right)_0 (\kappa - \kappa_0) + \frac{1}{2}\left(\frac{d^2\omega}{d\kappa^2}\right)_0 (\kappa - \kappa_0)^2 + \cdots, \tag{12.5.9}$$

where the subscript zero means that the derivatives are evaluated at the wave number κ_0, that is, at frequency ω_0. The coefficient of the linear term is the group velocity (12.5.4) evaluated at the frequency ω_0. Putting

$$\alpha \equiv \frac{1}{2}\left(\frac{d^2\omega}{d\kappa^2}\right)_0$$

(12.5.10)

for the coefficient of the quadratic term and discarding higher-order terms in the expansion, we establish that any well-behaved dispersion relation can be approximated by the general form

$$\omega = \omega_0 + c_g(\kappa - \kappa_0) + \alpha(\kappa - \kappa_0)^2$$

(12.5.11)

in the vicinity of a particular frequency and its related wave number. This expression is exact when ω is a linear or quadratic function of κ.

The Fourier integral (12.5.8) with the dispersion relation (12.5.11) is evaluated in Prob. 12.5.3, with the result that

$$f(x,t) = \frac{A\,\Delta x_0}{\Delta x}\,e^{-(x-c_g t)^2/4(\Delta x)^2}e^{-i(\kappa_0 x-\omega_0 t)},$$

(12.5.12)

where

$$(\Delta x)^2 \equiv (\Delta x_0)^2 + i\alpha t$$

(12.5.13)

defines a complex spatial spread of the wave packet. The physical wave is represented by the real part of (12.5.12); it has a fairly complicated mathematical form. In contrast, the square of the absolute magnitude of the wave, $|f(x,t)|^2$, which is proportional to its energy content (but does not exhibit phase relations in the wave), has the relatively simple form (Prob. 12.5.3)

$$|f(x,t)|^2 = \frac{A^2}{[1 + \alpha^2 t^2/(\Delta x_0)^4]^{1/2}}\,e^{-(x-c_g t)^2/2(\Delta x_0)^2[1+\alpha^2 t^2/(\Delta x_0)^4]}.$$

(12.5.14)

The effect of dispersion on the motion of the wave packet can be deduced from (12.5.14). For $t = 0$ the expression reduces, as it should, to the square of the absolute value of the initial wave packet (12.5.6). The maximum of the function, which continues to be gaussian in form, moves with the group velocity c_g. Although this result is not unexpected, it is of interest that the quadratic term in the Taylor series expansion of the dispersion relation does not contribute directly to the velocity of the wave packet. Hence group velocity is a much more useful concept than we might infer from the elementary derivation of $c_g = d\omega/d\kappa$ in Sec. 4.7.

The quadratic term makes itself felt in two related ways: (1) the spatial spread of the wave packet,

$$|\Delta x| = \Delta x_0\left[1 + \frac{\alpha^2 t^2}{(\Delta x_0)^4}\right]^{1/2},$$

(12.5.15)

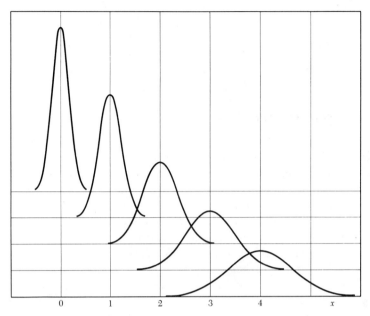

Fig. 12.5.2   The envelope of a gaussian wave packet as it progresses in a dispersive medium.

increases with time as the wave packet travels along, and (2) the amplitude of the wave packet simultaneously decreases with time, in such a manner that the area under $|f(x,t)|^2$ is constant. This behavior is illustrated in Fig. 12.5.2. Physically, the constancy of area means that the total energy content of the pulse is conserved. The spreading out of the pulse is independent of the sign of $\alpha = \frac{1}{2}(d^2\omega/d\kappa^2)_0$. In Sec. 4.7 we observed that the rate of energy transport by a sinusoidal wave traveling in a dispersive medium is given by the group velocity (*not* the phase velocity) times the average total energy density. Our finding here further illustrates this notion.

Problems

12.5.1   Define the spatial spread Δx of a waveform by an equation analogous to (12.2.4). Show that Δx_0 in (12.5.6) is the spatial spread of $f(x)$.

12.5.2   Show that the Fourier transform of (12.5.6) is stated correctly as (12.5.7). Define the spectral width $\Delta\kappa$ in the wave-number domain by an equation analogous to (12.2.5) and show that the spectral width of (12.5.7) is $1/2\Delta x_0$.

12.5.3   Obtain the wave (12.5.12) from (12.5.8) after substituting the dispersion relation (12.5.11). *Hint:* (1) Take the factor $e^{i(\kappa_0 x - \omega_0 t)}$ outside the integral; (2) change the integration variable to $\beta \equiv \kappa - \kappa_0$; (3) make the substitution (12.5.13); (4) complete the square of the quadratic function of β (by adding and subtracting a term independent of β); (5) change the variable of integration to obtain an integral of the form $\int_{-\infty}^{\infty} e^{-a^2 x^2}\, dx$, which has the value $\pi^{1/2}/a$.

12.6   The Fourier Transform Method

The analysis of a waveform into its spectral components by (12.2.1) and the converse synthesis of a waveform from its spectral components by (12.2.2) lead to a systematic method for treating many cases of wave propagation in which the waveform is not sinusoidal. The method is applicable whenever the differential equations describing a physical system are linear, so that the principle of superposition is valid. It is possible to understand the essential features of the Fourier transform method without discussing the many special results (theorems) that prove useful in particular situations.†

The Fourier transform method is basically a technique for constructing a solution (a time-varying solution as described here) of a differential equation (or a system of differential equations). When the differential equation involves spatial variables as well as time, the method of variable separation allows solutions to be constructed that satisfy the spatial boundary conditions and that also contain the separation constant of the time equation as a parameter. It is then possible using the Fourier transform method to construct a further solution that has a specified dependence on time at a particular position or spatial boundary. The transform method deals most easily with a single independent variable, such as the time t. Note, however, that in examining the motion of a wave packet in a dispersive medium, we used Fourier transforms for the spatial variable x.

Suppose that we wish to investigate some time-varying aspect of a linear physical system, which for simplicity we assume to depend on a single space variable, e.g., a vibrating string or rod or an electric transmission line or waveguide. The system is described by a number of related dependent variables (force, displacement, displacement velocity, a stress or strain component; or voltage, current, the peak electromagnetic field components in a waveguide, etc.) that are functions of a single position variable, together with the time variable. When the method is applied to a lumped-parameter electric (or acoustical) network, the "position variable" consists of a designation for the various elements or branches in the network.

Next suppose that the system under discussion is excited in some particular manner at a particular position or boundary (position 1) by an external source

---

† The reader is referred to Boas, *op. cit.*, Bracewell, *op. cit.*, and Sneddon, *op. cit.*, for systematic accounts of the use of Fourier transforms.

of power. In a (passive) electric network, the excitation could consist of a voltage varying in time, impressed at a pair of input terminals. In an elastic-wave problem, e.g., the longitudinally vibrating rod considered in Secs. 4.1 and 4.2, the excitation could consist of a time-varying force impressed at a specified position x_1 along the rod. Whatever its nature, we may term the excitation the *input*, the *cause*, or the *driving function* and denote it by $p(t)$.

As a result of the driving function $p(t)$, the system responds, sooner or later, throughout its extent. We are interested in the *output*, *effect*, or *response* at some other position or boundary in the system (position 2). (In certain cases position 1 and position 2 may coincide.) In the electric-network example, the response could be the voltage across a pair of output terminals or the current flowing in a particular branch of the system, etc. In the vibrating-rod example, the response could be the time-varying longitudinal stress produced at a particular position x_2 along the rod or the displacement velocity there, etc. We shall call the output, whatever its nature, the *response function* and denote it by $q(t)$. The term "output" is not meant to imply that power is necessarily removed from the system at position 2. We can summarize by stating that a driving function (or cause) $p(t)$ applied at position 1 of a linear physical system evokes the response function (or effect) $q(t)$ at position 2. Prior to the onset of the driving function we suppose that the system is quiescent.

The Fourier transform method includes, among its many attributes, a systematic method for calculating $q(t)$ for a specified $p(t)$. If $p(t)$ is a real function of time, then $q(t)$ obviously must also be a real function of time. If, on the other hand, we consider $p(t)$ to be complex using the real-part convention, then $q(t)$ will be complex, with the physical response given by the real part of $q(t)$. The use of complex functions of time for input waveforms is mostly restricted to AM sinusoidal waveforms, as in Prob. 12.3.4. It is important to note that the response $q(t)$ describes in a single expression both the *transient* and the *steady-state* response of the system.

The basic notions underlying this application of Fourier transforms are presented in Table 12.1.† The entries in the first row express two corollaries of linearity that make the method possible. The multiplication of a driving function $p_1(t)$ by a constant also multiplies the response function $q_1(t)$ by the same constant. The sum of any number of unrelated driving functions $p_i(t)$ gives a response function that is the sum of the respective response functions $q_i(t)$.

The entry in the second row indicates that if $p(t)$ is a sinusoidal input $\cos\omega t$ (or $\sin\omega t$), i.e., the real (or imaginary) part of $e^{j\omega t}$, then the response function is the real (or imaginary) part of a time function $\check{T}(\omega)e^{j\omega t}$ that depends on frequency. This input-output pair is a definition of the *complex transfer function*

---

† In the present section and in the one following, we revert to the time factor $e^{+j\omega t}$ used in Sec. 4.2, and in Chap. 8, since this sign convention is adopted in most accounts of the practical applications of Fourier transforms.

TABLE 12.1 Input-Output Pairs

	Driving function $p(t)$ at position 1	Response function $q(t)$ at position 2
Corollaries of linearity	$ap_1(t) \qquad a = \text{const}$ $p_1(t) + p_2(t) + \cdots$	$aq_1(t)$ $q_1(t) + q_2(t) + \cdots$
Steady-state response to sinusoidal input	$e^{j\omega t}$	$\breve{T}(\omega)e^{j\omega t}$
Periodic function with a discrete spectrum	$\displaystyle\sum_{-\infty}^{\infty} \breve{P}(\omega_n)e^{j\omega_n t} \qquad \omega_n = n\omega_1$	$\displaystyle\sum_{-\infty}^{\infty} \breve{P}(\omega_n)\breve{T}(\omega_n)e^{j\omega_n t}$
Nonperiodic function with a continuous spectrum	$\dfrac{1}{2\pi}\displaystyle\int_{-\infty}^{\infty} \breve{P}(\omega)e^{j\omega t}\,d\omega$	$\dfrac{1}{2\pi}\displaystyle\int_{-\infty}^{\infty} \breve{P}(\omega)\breve{T}(\omega)e^{j\omega t}\,d\omega$

$\breve{T}(\omega)$. The task of computing $\breve{T}(\omega)$ for a particular case is generally straightforward. Often $\breve{T}(\omega)$ is closely related to complex impedances that describe the system. In electric-network theory, the calculation of $\breve{T}(\omega)$ is a part of what is generally known as steady-state AC circuit analysis. A response function $\breve{T}(\omega) = |\breve{T}(\omega)|e^{j\phi(\omega)}$ can be measured experimentally by observing the magnitude of the response $|\breve{T}(\omega)|$ as a function of frequency and separately measuring the phase shift $\phi(\omega)$ as a function of frequency.

The entry in the third row indicates that when the transfer function is known for the discrete spectrum of frequencies $\omega_n = n\omega_1$, the response to a periodic driving function represented by the Fourier series

$$p(t) = \sum_{-\infty}^{\infty} \breve{P}(\omega_n)e^{j\omega_n t}, \tag{12.6.1}$$

where

$$\breve{P}(\omega_n) = \frac{1}{T_1}\int_{-T_1/2}^{T_1/2} p(t)e^{-j\omega_n t}\,dt \qquad T_1 \equiv \frac{2\pi}{\omega_1}, \tag{12.6.2}$$

is given by

$$q(t) = \sum_{-\infty}^{\infty} \breve{P}(\omega_n)\breve{T}(\omega_n)e^{j\omega_n t}. \tag{12.6.3}$$

Since we are using negative as well as positive frequencies, the real-part convention is not implied in any of these expressions. If, however, we choose to regard $p(t)$ as a complex periodic nonsinusoidal waveform, then we must use the real-part convention for both $p(t)$ and $q(t)$, a matter that is quite independent of the use of a complex Fourier series for finding $q(t)$ from $p(t)$.

The final entry in the table pertains when the driving function is nonperiodic. A driving function $p(t)$ then has a continuous frequency spectrum $\check{P}(\omega)$ such that

$$p(t) = \frac{1}{2\pi} \int_{-\infty}^{\infty} \check{P}(\omega) e^{j\omega t} \, d\omega, \tag{12.6.4}$$

where

$$\check{P}(\omega) = \int_{-\infty}^{\infty} p(t) e^{-j\omega t} \, dt. \tag{12.6.5}$$

In this case the response is given by

$$q(t) = \frac{1}{2\pi} \int_{-\infty}^{\infty} \check{P}(\omega) \check{T}(\omega) e^{-j\omega t} \, d\omega. \tag{12.6.6}$$

As in the discrete-frequency case for periodic functions of time, we are again synthesizing the response $q(t)$ by adding together (now by integration) a multitude of individual responses $\check{P}(\omega)\check{T}(\omega)e^{j\omega t} \, d\omega$ called forth by the complex frequency components $\check{P}(\omega)e^{j\omega t} \, d\omega$ of $p(t)$ in the frequency range ω to $\omega + d\omega$. We again remind the reader that if $p(t)$ is a real function of time, the integration in (12.6.6) must surely lead to a real response function. There is no need to invoke the real-part convention, because of the use of negative frequencies in synthesizing the response.

We have thus arrived at a particularly elegant and efficient way of calculating the response $q(t)$ of a physical system to an arbitrary but known excitation function $p(t)$ provided that we know the steady-state sinusoidal response (amplitude and phase) of the system as a function of frequency. The method assumes that the various definite integrals converge. Many Fourier integrals can be found evaluated in tables.† Although the Fourier transform method is very useful for treating problems that arise in wave motion, it finds its most important practical applications in discussing the transient response of electric networks, including electric transmission lines and waveguides, which of course involve wave propagation. As a result, many books presenting practical details of the Fourier transform method are found in the literature of electrical engineering.‡ The method is closely related to an operator method based on the Laplace transformation.

Although we have illustrated the Fourier transform method in terms of the response in time of a linear system, there exist many important applications of the method when space variables rather than a single time variable are involved.

---

† For example, Campbell and Foster, *op. cit.*

‡ See, for instance, F. F. Kuo, "Network Analysis and Synthesis," 2d ed., John Wiley & Sons, Inc., New York, 1966; E. A. Guillemin, "Theory of Linear Physical Systems," John Wiley & Sons, Inc., New York, 1963; L. A. Zadeh and C. A. Desoer, "Linear System Theory," McGraw-Hill Book Company, New York, 1963.

For example, the Fresnel-Kirchhoff diffraction formula in the Fraunhofer limit can be considered as a Fourier transform connecting an aperture transmission function (9.4.19) with the diffracted wave amplitude in a particular observation plane (see Prob. 12.6.3). In this example the inverse Fourier transform of the complex wave amplitude in the observation plane would be a reconstruction of the aperture transmission function. Thus if one could record the complex amplitude (magnitude and phase) of the diffraction pattern, the record would contain all the information needed to reconstruct an optical replica (image) of the aperture obstacles responsible for the diffraction pattern. Although one can record the intensity of the wave in the observation plane photographically, using the highly monochromatic light from a laser as a light source, such a record does not contain phase information. However, by the technique known as *holography*, the missing phase information is simultaneously recorded by photographing the interference pattern consisting of the diffracted wave pattern superposed on a uniform wavefield coming directly from the coherent light source. The cross-product terms in the combined wavefield contain the missing phase information. By suitably viewing a monochromatic light source through a *hologram* recorded in this way, one sees a *three-dimensional* virtual image of the objects originally scattering and diffracting the laser light.†

Problems

12.6.1 A sinusoidally varying force is applied at one end ($x = x_1 = 0$) of a long slender rod, terminated by its characteristic impedance at the other end ($x = l$) (see Sec. 4.2). Show that the transfer function for the force at any intermediate position x_2 is $T(\omega) = e^{-j\omega t_2}$, where $t_2 = x_2/c$, with c the wave velocity. If a force $f(t)$ is applied at $x_1 = 0$, show that the force function at x_2 is $f(t - t_2)$.

12.6.2 Find two or three other transfer functions for the rod of the previous problem, connecting dependent variables at $x_1 = 0$ with other dependent variables at position x_2. The "delay operator" $e^{-j\omega t_2}$ enters as a factor in each of these expressions.

12.6.3 Show that the Fraunhofer diffraction integral for a one-dimensional aperture, such as a single slit (Sec. 10.4) or a grating (Secs. 10.7 and 10.8) may be expressed in the form

$$\psi(s) = \check{C} \int_{-\infty}^{\infty} \check{T}(x) e^{isx} \, dx,$$

where $s \equiv \kappa(\sin\theta_s + \sin\theta_o)$ and $\check{T}(x)$ is the *aperture transmission function* (9.4.19), possibly

---

† For further details concerning holography and other applications of Fourier analysis to optical problems, see M. Born and E. Wolf, "Principles of Optics," 3 ded., sec. 8.10, Pergamon Press, New York, 1965; A. Papoulis, "Systems and Transforms with Applications in Optics," McGraw-Hill Book Company, New York, 1968; and J. W. Goodman, "Introduction to Fourier Optics," McGraw-Hill Book Company, New York, 1968.

complex, that expresses the wave-transmission properties of the aperture. Hence show that the diffracted-wave amplitude, except for a constant factor, is the Fourier transform of the aperture transmission function. It is thus no accident that the Fourier transform of a rectangular pulse (Prob. 12.2.2) is functionally the same as the Fraunhofer diffraction-amplitude function for a single slit, $\sin u/u$.

12.7   Properties of Transfer Functions

An interesting application of the Fourier-transform method consists in showing that various transfer functions $\check{T}(\omega)$, regardless of their origin, have a number of properties in common. Since the nature of the driving function $p(t)$ and of the response function $q(t)$ can be independently chosen, the physical dimensions of particular transfer functions differ widely. At first glance this variety would appear to preclude the existence of common properties. Such is not the case, as we can easily show. We must assume that the physical system is described by linear equations, so that superposition holds.

First recall that a driving function $p(t)$, having the Fourier transform (spectrum)

$$\check{P}(\omega) = \int_{-\infty}^{\infty} p(t)e^{-j\omega t} \, dt, \tag{12.7.1}$$

evokes the response $q(t)$ given by the inverse Fourier transform

$$q(t) = \frac{1}{2\pi} \int_{-\infty}^{\infty} \check{P}(\omega)\check{T}(\omega)e^{j\omega t} \, d\omega, \tag{12.7.2}$$

where $\check{T}(\omega)$ is the transfer function defined in Sec. 12.6. If $p(t)$ is a *real* function of time, the response $q(t)$ that it produces obviously must also be a *real* function of time. Let us see how this physical requirement puts a certain restriction on the mathematical form that $\check{T}(\omega)$ can have for any actual physical system.

For the function $p(t)$ let us choose the Dirac delta function (the impulse function) $\delta(t)$. Such a function is zero except at $t = 0$, where it becomes infinite in such a way that

$$\int_{-\infty}^{\infty} \delta(t) \, dt = 1. \tag{12.7.3}$$

In Prob. 12.2.3 we established that the Fourier transform (frequency spectrum) of $\delta(t)$ is unity. Hence the system response to a unit impulse at time $t = 0$ is

$$q_\delta(t) = \frac{1}{2\pi} \int_{-\infty}^{\infty} \check{T}(\omega)e^{j\omega t} \, d\omega, \tag{12.7.4}$$

that is, it is the inverse transform of the transfer function. The fact that $q_\delta(t)$ must be real, since $\delta(t)$ is real, puts a restriction on $\check{T}(\omega)$. If we divide $\check{T}(\omega)$ into its real and imaginary parts

$$\breve{T}(\omega) = T_r(\omega) + jT_i(\omega), \tag{12.7.5}$$

the Fourier integral (12.7.4) may be written in the form

$$q_\delta(t) = \frac{1}{2\pi} \int_{-\infty}^{\infty} [T_r(\omega) \cos\omega t - T_i(\omega) \sin\omega t] \, d\omega$$

$$+ \frac{j}{2\pi} \int_{-\infty}^{\infty} [T_r(\omega) \sin\omega t + T_i(\omega) \cos\omega t] \, d\omega. \tag{12.7.6}$$

Since the response $q_\delta(t)$ is to be real, the imaginary part of (12.7.6) must vanish identically for all values of t. The first integral in the imaginary term

$$\int_{-\infty}^{\infty} T_r(\omega) \sin\omega t \, d\omega$$

is an *odd* function of time, whereas the second integral

$$\int_{-\infty}^{\infty} T_i(\omega) \cos\omega t \, d\omega$$

is an *even* function of time. Therefore each of these integrals must *independently* equal zero for the imaginary part to vanish for all values of t. The first integral is identically zero if and only if the *real* part of $\breve{T}(\omega)$ is an *even* function of ω, and the second integral is also identically zero if and only if the *imaginary* part of $\breve{T}(\omega)$ is an *odd* function of ω. The evenness and oddness of the real and imaginary parts of $\breve{T}(\omega)$, respectively, is an important property possessed in common by all transfer functions.

We can go even further if we now introduce what may be called the *principle of causality*, which requires that the effect $q(t)$ must remain zero prior to the onset of the cause $p(t)$. Hence, when $p(t)$ is an impulse occurring at $t = 0$, its response $q_\delta(t)$ must be identically zero for negative values of time. The response is given by (12.7.6), which we may now write in the form

$$q_\delta(t) = \frac{1}{\pi} \int_0^{\infty} T_r(\omega) \cos\omega t \, d\omega - \frac{1}{\pi} \int_0^{\infty} T_i(\omega) \sin\omega t \, d\omega. \tag{12.7.7}$$

The evenness of $T_r(\omega)$ and the oddness of $T_i(\omega)$, which are required to make $q_\delta(t)$ real, also make each of the remaining integrands an even function of ω, permitting the change of the limits of integration. Moreover, the two terms of (12.7.7) are even and odd functions of time. Thus if $q_\delta(t)$ is to vanish for all negative values of t, (12.7.7) shows that

$$\frac{1}{\pi} \int_0^{\infty} T_r(\omega) \cos\omega t \, d\omega = -\frac{1}{\pi} \int_0^{\infty} T_i(\omega) \sin\omega t \, d\omega \tag{12.7.8}$$

for all *positive* values of t; that is, the two terms in (12.7.7) are each equal to $\frac{1}{2}q_\delta(t)$ when t is positive but precisely cancel when t is negative. Equation (12.7.8) constitutes another important restriction on the mathematical properties of the real and imaginary parts of a transfer function. Indeed if we know $T_r(\omega)$, (12.7.8) implies that we can compute $T_i(\omega)$, and conversely. This conclusion is independent of our use of the delta function since any physical driving function $p(t)$ can be synthesized by a superposition of delta functions.

To find an explicit relation between $T_r(\omega)$ and $T_i(\omega)$, let us first write down the Fourier transform inverse to (12.7.4), namely,

$$\breve{T}(\omega) = \int_{-\infty}^{\infty} q_\delta(t) e^{-j\omega t}\, dt = \int_0^\infty q_\delta(t) e^{-j\omega t}\, dt, \qquad (12.7.9)$$

where the lower limit has been advanced to zero since we know that $q_\delta(t) = 0$ for $t < 0$. Next write (12.7.9) as the sum of a real and imaginary part,

$$T_r(\omega) + jT_i(\omega) = \int_0^\infty q_\delta(t)\, \cos\omega t\, d\omega - j \int_0^\infty q_\delta(t)\, \sin\omega t\, dt, \qquad (12.7.10)$$

which shows that

$$T_r(\omega) = \int_0^\infty q_\delta(t)\, \cos\omega t\, dt \qquad (12.7.11)$$

$$T_i(\omega) = -\int_0^\infty q_\delta(t)\, \sin\omega t\, dt. \qquad (12.7.12)$$

These are known as Fourier cosine and sine transforms, respectively. If we now substitute

$$q_\delta(t) = \frac{2}{\pi} \int_0^\infty T_r(\omega)\, \cos\omega t\, d\omega \qquad (12.7.13)$$

from (12.7.8) into (12.7.12), we find that

$$T_i(\omega) = -\frac{2}{\pi} \int_0^\infty \int_0^\infty T_r(\omega')\, \cos\omega' t\, \sin\omega t\, d\omega'\, dt, \qquad (12.7.14)$$

which gives $T_i(\omega)$ explicitly as a function of $T_r(\omega)$. If we carry out the time integration (Prob. 12.7.2), we find that

$$T_i(\omega) = \frac{2\omega}{\pi} \int_0^\infty \frac{T_r(\omega')\, d\omega'}{\omega'^2 - \omega^2}. \qquad (12.7.15)$$

By a similar calculation, based on (12.7.8) and (12.7.11),

$$T_r(\omega) = \frac{2}{\pi} \int_0^\infty \frac{\omega' T_i(\omega')\, d\omega'}{\omega'^2 - \omega^2}. \qquad (12.7.16)$$

If the integrand of (12.7.15) or (12.7.16) possesses a singularity, it is necessary to take (Cauchy) *principal values* of the integral.† We thus have found explicit formulas relating the two parts of a transfer function. If the real part of the

† See Boas, *op. cit.*, chap. 11, sec. 7.

transfer function has an additive constant term (independent of frequency), it does not contribute to $T_i(\omega)$ according to (12.7.15), since the principal value of the integral $\int_0^\infty [1/(\omega'^2 - \omega^2)] \, d\omega'$ is zero (Prob. 12.7.4). Hence (12.7.16) gives only the frequency-dependent part of $T_r(\omega)$. Any constant part must be supplied separately. The two formulas, (12.7.15) and (12.7.16), are often called the *Kramers-Kronig relations*. They show that the real and imaginary parts of all response functions of linear systems are related in a particular way, insofar as their dependence on frequency is concerned. These relations have numerous applications in physics and are of considerable importance in the theory and design of electric networks. They apply of course to the transfer functions arising in the theory of wave motion. A discussion of the conditions necessary for the Kramers-Kronig relations to be valid has been given by Sharnoff.†

Problems

12.7.1   Show that a necessary and sufficient condition for $p(t)$ to be a real function of time is that its Fourier transform $\breve{P}(\omega)$ have the symmetry property $\breve{P}(-\omega) = \breve{P}^*(\omega)$, which shows that the real part of $\breve{P}(\omega)$ is an even function of ω but that its imaginary part is an odd function of ω. Then show that a necessary and sufficient condition for $q(t)$, as given by (12.7.2) also to be real is for $\breve{P}(\omega)\breve{T}(\omega)$, and hence $\breve{T}(\omega)$, to obey the same symmetry property, so that $\breve{T}(-\omega) = \breve{T}^*(\omega)$. Note that this argument establishes the evenness and oddness of the real and imaginary parts of a transfer function without using the delta function.

12.7.2   Establish the integral

$$\int_0^\infty \sin at \, \cos bt \, dt = \frac{a^2}{a^2 - b^2}$$

by first multiplying the integrand by e^{-st} and allowing $s \to 0$ after integration. This, or a similar artifice, is often necessary in evaluating Fourier transform integrals. A Fourier transform so evaluated is then described as a *limiting form*.

12.7.3   Apply the Kramers-Kronig relations (12.7.15) and (12.7.16) to the two parts of the (delay) transfer function of Prob. 12.6.1, $\breve{T}(\omega) = \cos\omega t_2 - j \sin\omega t_2$, to verify that one part implies the other, and vice versa.

12.7.4   Verify that the principal value of the integral $\int_0^\infty [1/(\omega'^2 - \omega^2)] \, d\omega$ is zero.

12.7.5   What limitations must be put on the mathematical behavior of a transfer function

† M. Sharnoff, *Am. J. Phys.*, **32**: 40 (1964). See also C. H. Holbrow and W. C. Davidon, *Am. J. Phys.*, **32**: 762 (1964).

at an infinite frequency if the Kramers-Kronig integrals are to converge? *Answer: $T_r(\infty)$* remains finite, $T_i(\infty) = 0$.

12.7.6 A sinusoidal voltage $V = V_0 \cos\omega t$, whose frequency can be varied, is applied to the terminals of a network, and the component of the current *in phase with* V is found to be $I_r = [V_0/R(1 + \omega^2\tau^2)] \cos\omega t$. Use (12.7.15) to find the out-of-phase component of current. What is the input impedance of the network?

12.8 Partial Coherence in a Wavefield

The theory of the diffraction of waves at obstacles and apertures discussed in Chaps. 9 to 11 is based on an idealized point (or line) source of monochromatic waves. The wavefield from such a source is *coherent;* i.e., the instantaneous wave amplitude can be expressed by a unique sinusoidal function of time and position. Stated another way, the ratio of the complex wave amplitude at two points in the wavefield, no matter how far apart, does not change with time. A coherent wavefield is produced, for example, by a radio antenna broadcasting a sinusoidal electromagnetic wave of steady amplitude and frequency. A laser generates light waves that have a high degree of coherence. At the other extreme, a completely *incoherent* wavefield would exhibit no correlation whatever between the wave amplitude at two neighboring points in space, no matter how close. Thermal radiation (blackbody radiation) in an isothermal enclosure is an example of a wavefield that has a high degree of incoherence.

Most optical wavefields formed by light from a pinhole or narrow-slit aperture illuminated by a nominally monochromatic light source are only approximately like the ideal coherent wavefield. Such a wavefield is *partially coherent.* The lack of coherency comes from the range of frequencies present in a single spectral line of the light from a gas-discharge lamp and from the finite area of the source or its image on the source aperture. Such a source is populated by atoms that emit wave trains having no phase connection with each other (except in a laser).

Suppose that a particular spectral line is isolated by means of an optical filter, e.g., the green line (546 nm) of mercury. A high-resolution spectroscope shows that such a spectral line is not perfectly monochromatic but has an intensity spectrum of the sort illustrated in Fig. 12.8.1. A number of effects contribute to the broadening of the line:

(1) The frequency of the light radiated by an atom depends on the velocity of the atom relative to the observer (*Doppler* or *thermal broadening*).

(2) Neighboring atoms or molecules may perturb an atom while it is radiating (*pressure broadening*).

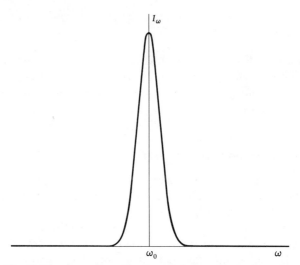

Fig. 12.8.1 Spectral profile of a quasi-monochromatic source.

(**3**) The finite duration of the wave train associated with a photon causes a *natural line width* (see Prob. 12.3.5). The duration of the wave train is related to the lifetime of the two atomic energy states connected with the emission of the photon.

As a result of these and perhaps other effects (such as the existence of isotopes, hyperfine level splitting, Stark and Zeeman effects, etc.), a spectral line has a finite *spectral bandwidth* $\Delta\omega$. Hence light waves passing a point in a wavefield from such a *quasi-monochromatic* source retain a reasonably definite phase and amplitude only for a time somewhat less than a minimum time interval Δt related to $\Delta\omega$ by the uncertainty relation (12.2.7), namely,

$$\Delta t\,\Delta\omega \geq \tfrac{1}{2}. \tag{12.8.1}$$

Associated with the minimum value of Δt defined by (12.8.1), which is termed the *coherence time*, is a *coherence length*

$$\Delta l = c\,\Delta t \tag{12.8.2}$$

in the direction of wave travel. Experimentally Δl sets the order of magnitude of the maximum path difference in an interferometer for which clear interference fringes can be observed. A typical coherence length for quasi-monochromatic light from a gas-discharge tube is somewhat less than 1 m, whereas it can be hundreds of kilometers for light from a gas laser.

Temporal coherence, through the related coherence length Δl of (12.8.2), establishes the *longitudinal* extent of a region surrounding a point in the wavefield in which the wave pattern at any instant is similar to the ideally coherent wavefield of a strictly monochromatic source. The *transverse* extent of the region of coherency for a gas-discharge spectral source depends on the lateral dimensions of the source. The analysis for transverse coherence proceeds on the assumption that each element of area of the source radiates Huygens wavelets having a phase unrelated to that of wavelets from neighboring elements and a frequency that is statistically consistent with the spectrum profile, such as that illustrated in Fig. 12.8.1. The average intensity at any point in the wavefield is found by adding together the amplitude contributions from the many independently radiated Huygens wavelets, to obtain a local instantaneous amplitude, which is then squared and averaged both over time and over the independently radiating elements constituting the source.

Let $\psi_i(t)$ be the (real) amplitude at an observation point P_o of the Huygens wavelet coming from an element of area ΔS_i of an extended quasi-monochromatic source. The amplitude of the wave disturbance at P_o from the entire source is then

$$\psi(t) = \sum_{i=1}^{N} \psi_i(t),$$

and the intensity is proportional to

$$I(P_o) \propto \overline{[\psi(t)]^2} = \sum_i \overline{[\psi_i(t)]^2} + 2 \sum_{i>j} \overline{\psi_i(t)\psi_j(t)}. \tag{12.8.3}$$

Since each source element is assumed to radiate quasi-monochromatic waves independently of the other source elements, it is evident that when $i \neq j$,

$$\overline{\psi_i(t)\psi_j(t)} = 0. \tag{12.8.4}$$

Hence, in this case, the intensity at P_o is the *sum of the intensities* coming from the individual source elements, that is,

$$I(P_o) = \sum_i I_i(P_o). \tag{12.8.5}$$

For example, if the N source elements each contribute the same intensity I_1 to the total intensity at P_o, then

$$I(P_o) = NI_1. \tag{12.8.6}$$

If, however, the waves from the many source elements are radiated in phase with each other, rather than with random phases, the interference term

$$2\Sigma\overline{\psi_i(t)\psi_j(t)}$$

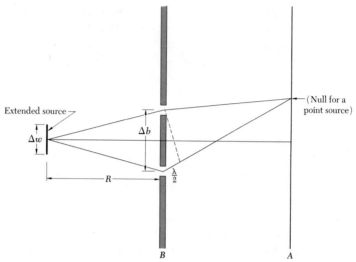

Fig. 12.8.2   Geometry for discussing transverse coherence.

is nonzero. The intensity at P_o can then be as great as

$$I(P_o) = N^2 I_1 \tag{12.8.7}$$

or as small as zero.†

We may use the notion just established that wave intensities, rather than wave amplitudes, are to be added in combining contributions from various area elements of an extended quasi-monochromatic source to find the transverse extent of the region of coherency. Suppose that the source of width Δw is situated a distance $R \gg \lambda$ from a plane B in which the transverse extent Δb of the region of coherency is to be estimated. An experimental test for coherency of the wavefields at the termini of Δb consists in placing an opaque screen at B with small pinholes at the termini of Δb and observing the interference pattern on a screen A, as indicated in Fig. 12.8.2. This experimental arrangement was used by Thomas Young in 1802 to demonstrate the wave nature of light. If the source width Δw is very small, clear interference fringes are observed. When Δw is increased in size, the fringes become indistinct and soon disappear. The reason for their disappearance is that the wave disturbances at the termini of Δb no

---

† As an example, consider the Fraunhofer single-slit diffraction problem of Fig. 10.2.3, regarding the aperture as an extended coherent source. When the observation point P_o is located at the center of the diffraction pattern, the lenses equalize all optical paths from the monochromatic point source P_s to the point P_o. At other observation points P_o, however, the diffracted intensity may be zero (Fig. 10.4.3), i.e., the second (interference) term of (12.8.3) just cancels the first (incoherent) term.

longer have an approximately constant ratio of complex wave amplitude; i.e., the wavefield has reached its limit of transverse coherency.

The relationship connecting Δb with Δw, R, and λ, where λ is the average wavelength of the quasi-monochromatic light from the source, is easily found by considering that each element of area of the extended source gives rise to its own pattern of interference fringes on the screen A. According to (12.8.5), the total intensity pattern is the sum of the individual intensity patterns of the many source elements. It is easy to establish (Prob. 12.8.1) that the fringe pattern becomes indistinct when

$$\Delta b \approx \frac{R}{\Delta w} \, \lambda. \tag{12.8.8}$$

In a practical optical setup for studying diffraction phenomena, the source slit is normally many wavelengths wide, and the distance R is rarely greater than 1 m. The transverse extent of the region of coherency is then considerably less than the longitudinal extent of the region. From the present point of view, the Michelson stellar interferometer, discussed briefly in Sec. 10.6, serves to measure the angular width of the source $\Delta w/R$ by finding the transverse extent of the region of wavefield coherency.

The reader is referred to Born and Wolf and to other accounts of the theory of partial coherence and of interferometry for further discussion, especially of the statistical aspects of the subject.†

Problem

12.8.1 Establish (12.8.8) by using a simple argument to find the angle subtended by one period of the interference pattern at A, assuming the paraxial approximation. Then find how wide the source can be made so that the superposition of the array of intensity patterns gives uniform illumination (see Prob. 10.4.2).

---

† M. Born and E. Wolf, *op. cit.*, chap. 10; see also W. H. Steel, "Interferometry," chap. 5, Cambridge University Press, New York, 1967; M. J. Beron and G. B. Parrent, Jr., "Theory of Partial Coherence," Prentice-Hall, Inc., Englewood Cliffs, N.J., 1964; and M. Françon, "Optical Interferometry," chap. 1, Academic Press Inc., New York, 1966.

appendix a

Vector Calculus†

In physics the term *field* denotes a physical quantity that is a continuous function of position, within a certain region of space. We distinguish *scalar fields*, such as temperature or electrostatic potential, from *vector fields*, such as the velocity of a moving fluid or the electric field **E**. A vector field assigns a direction, as well as a magnitude, to each point in space. Since a vector can be represented by three components, a vector field is specified analytically by a set

---

† More careful and thorough discussions of vector calculus may be found in many textbooks, including M. L. Boas, "Mathematical Methods in the Physical Sciences," pp. 222–268, John Wiley & Sons, Inc., New York, 1966; E. A. Kraut, "Fundamentals of Mathematical Physics," chap. 3, McGraw-Hill Book Company, New York, 1967; J. B. Marion, "Principles of Vector Analysis," Academic Press Inc., New York, 1965; and B. Green, "Vector Calculus: A Programmed Course," Appleton-Century-Crofts, Inc., New York, 1967.

of three functions of position, subject to the transformation properties discussed in Sec. 7.1. The fields that arise in physics are almost always such that they and their derivatives are finite, continuous, and single-valued, thereby avoiding mathematical subtleties associated with less well-behaved functions.

Scalar fields can be described pictorially by *surfaces of constant value.* For instance, the contour lines on a topographical map locate constant values of the "elevation field" in the two-dimensional space of the earth's surface. Equipotential surfaces are familiar in electrostatics. For single-valued fields, the contour loops and equipotential surfaces do not intersect; hence they are *nested.* Vector fields, on the other hand, are pictured by *streamlines,* or *lines of force.* These lines show the local direction of the field; also the density of lines can often be arranged to indicate the magnitude; i.e., the lines of force are close together where the field is strong, farther apart where it is weak. Both these visualizations, of scalar and vector fields, are incomplete in the sense that they introduce a sampling size or granularity; i.e., only a few representative equipotentials or force lines can be drawn even though we understand that in principle an equipotential or force line passes through every point in the space.†

Since scalar and vector fields are functions of position, they must have spatial derivatives and integrals. However, the multidimensionality of space puts these analytical operations on a more subtle level than, for instance, the derivative or integral of a field with respect to time. Consider first a scalar field $\phi(x,y,z)$. The spatial derivative, i.e., the rate of change in ϕ for small step away from the point (x,y,z), clearly depends upon the direction of the step: the derivative operation has vectorlike properties. Specifically, for an infinitesimal step of magnitude ds in the direction specified by the unit vector $\hat{\mathbf{n}}$,

$$\hat{\mathbf{n}} \, ds = \mathbf{i} \, dx + \mathbf{j} \, dy + \mathbf{k} \, dz, \tag{A.1}$$

the change in ϕ is

$$d\phi = \frac{\partial \phi}{\partial x} dx + \frac{\partial \phi}{\partial y} dy + \frac{\partial \phi}{\partial z} dz$$

$$= \left(\mathbf{i} \frac{\partial \phi}{\partial x} + \mathbf{j} \frac{\partial \phi}{\partial y} + \mathbf{k} \frac{\partial \phi}{\partial z} \right) \cdot \hat{\mathbf{n}} \, ds = (\boldsymbol{\nabla} \phi) \cdot \hat{\mathbf{n}} \, ds, \tag{A.2}$$

where we introduce the *vector derivative operator del*

$$\boldsymbol{\nabla} \equiv \mathbf{i} \frac{\partial}{\partial x} + \mathbf{j} \frac{\partial}{\partial y} + \mathbf{k} \frac{\partial}{\partial z}. \tag{A.3}$$

---

† Many physical fields involve a granularity on a more fundamental level. For instance, while temperature is a continuous scalar field in the macroscopic world, it loses its statistically defined meaning as the microscopic, molecular scale is approached. Similarly, electric field and potential are statistical-average quantities inside a dielectric medium.

Thus the derivative of ϕ in the direction $\hat{n}$ is

$$\frac{d\phi}{ds} = \hat{n} \cdot \boldsymbol{\nabla}\phi. \tag{A.4}$$

The vector field $\boldsymbol{\nabla}\phi$ is known as the *gradient* of ϕ

$$\mathbf{grad}\,\phi \equiv \boldsymbol{\nabla}\phi = \mathbf{i}\frac{\partial\phi}{\partial x} + \mathbf{j}\frac{\partial\phi}{\partial y} + \mathbf{k}\frac{\partial\phi}{\partial z}. \tag{A.5}$$

The gradient is thus a vector derivative, having the magnitude and direction of the greatest space rate of change of the scalar field ϕ. By (A.4), the derivative in an arbitrary direction is simply the component of the gradient vector in that direction. It is easy to see that the gradient is directed perpendicular to the surface $\phi = $ const. If the field is expressed in a noncartesian coordinate system (cylindrical, spherical, etc.; see Table A.1), the form of the derivatives in (A.2), (A.3), and (A.5) is altered but the meaning of the gradient is unchanged.

The spatial derivatives of a vector field, such as

$$\mathbf{A}(x,y,z) \equiv \mathbf{i}A_x(x,y,z) + \mathbf{j}A_y(x,y,z) + \mathbf{k}A_z(x,y,z), \tag{A.6}$$

are considerably more complicated since both the derivative and the field, independently, have properties that depend on direction. We suspect that various combinations of the vector derivative operator (A.3) with the field $\mathbf{A}$ have interesting properties. Thus, following the rules of vector algebra, we identify the scalar field $\boldsymbol{\nabla} \cdot \mathbf{A}$ and the vector field $\boldsymbol{\nabla} \times \mathbf{A}$ (and the dyadic field $\boldsymbol{\nabla}\mathbf{A}$) as possibilities. Rather than exploring the physical or geometric significance of these particular derivatives directly, however, we can provide a more intuitive understanding by first shifting our attention to spatial *integrals* of the field $\mathbf{A}$.

The vector field $\mathbf{A}$ can be integrated over a volume Δv, a surface ΔS, or a line Δl. Within the rules of ordinary vector algebra, excluding dyadic quantities, we may thus list five forms of spatial integrals:

Volume integral:

$$\int_{\Delta v} \mathbf{A}\,dv \tag{A.7}$$

Surface integrals:

$$\int_{\Delta S} \mathbf{A} \cdot d\mathbf{S} \tag{A.8}$$

$$\int_{\Delta S} \mathbf{A} \times d\mathbf{S} \tag{A.9}$$

Line integrals:

$$\int_{\Delta l} \mathbf{A} \cdot d\mathbf{l} \tag{A.10}$$

$$\int_{\Delta l} \mathbf{A} \times d\mathbf{l} \tag{A.11}$$

The volume integral (A.7) is straightforward; it arises, for instance, in finding the average value of **A** over a finite volume Δv. The remaining forms involve vector differentials and are more difficult to visualize. The scalar surface integral (A.8) is known as the *flux* of **A** through the surface ΔS; a familiar example is the magnetic flux $\Phi = \int \mathbf{B} \cdot d\mathbf{S}$. The scalar line integral (A.10) also occurs frequently in physical problems, in the definitions of work and of electrostatic potential, for example. The vector-product integrals (A.9) and (A.11) do not arise naturally in physical problems and are of less interest. Important special cases are the flux integral $\oint_S \mathbf{A} \cdot d\mathbf{S}$ over a *closed* simple surface S and the line integral $\oint_L \mathbf{A} \cdot d\mathbf{l}$ around a *closed* loop L. The latter form is often called the *circulation* of **A**; it vanishes if **A** is a *conservative* vector field. The fact that the integration is carried out over a closed domain is signified by a circle superimposed on the integral sign.

With this perspective, we introduce an operator for vector fields related to the flux integral (A.8): the *divergence* of a vector field **A** is a scalar field defined by

$$\mathrm{div}\mathbf{A} \equiv \lim_{\Delta v \to 0} \frac{1}{\Delta v} \oint_S \mathbf{A} \cdot d\mathbf{S}, \tag{A.12}$$

where the closed surface S encloses the volume Δv.† The divergence may be visualized as the number of lines of force originating inside the (infinitesimal) volume Δv; thus the divergence is nonzero only when there are local sources or sinks of lines of force. An important theorem, known as the *divergence* or *Gauss'* *theorem*, follows directly from the definition (A.12) by integration over volume,

$$\int_{\Delta v} \mathrm{div}\mathbf{A}\, dv = \oint_S \mathbf{A} \cdot d\mathbf{S} \tag{A.13}$$

where the finite closed surface S surrounds the finite volume Δv. It remains for us to see how to express $\mathrm{div}\mathbf{A}$ explicitly in terms of derivatives. Consider the elementary parallelepiped of Fig. A.1. The outward flux through side 1, for

† The vector direction ascribed to the surface element $d\mathbf{S}$ is perpendicular to the tangent plane, with sense taken to be *outward* with respect to the closed surface.

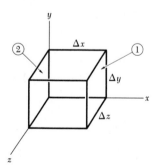

Fig. a.1 Parallelepiped used to evaluate the divergence.

example, is

$$\mathbf{A} \cdot \Delta\mathbf{S}_1 = \left(A_x + \frac{\partial A_x}{\partial x} \Delta x + \cdots\right)(\Delta y\, \Delta z),$$

while that through the opposite side 2 is

$$\mathbf{A} \cdot \Delta\mathbf{S}_2 = (-A_x)(\Delta y\, \Delta z).$$

The net outward flux through this pair of sides is

$$\frac{\partial A_x}{\partial x} \Delta x\, \Delta y\, \Delta z.$$

Thus the divergence (A.12), i.e., the flux generation per unit volume, may be written

$$\text{div}\mathbf{A} = \frac{\partial A_x}{\partial x} + \frac{\partial A_y}{\partial y} + \frac{\partial A_z}{\partial z} = \nabla \cdot \mathbf{A}. \tag{A.14}$$

A similar argument in cylindrical or spherical coordinates establishes that the divergence operator may always be written in the form $\nabla\cdot$, where ∇ is the same vector differential operator used for the gradient (see Table A.1).

  If the flux-density vector $\mathbf{A}$ represents the flow of mass or charge, the closed flux integral $\oint \mathbf{A} \cdot d\mathbf{S}$ represents the loss of a conserved quantity from the volume, which may be expressed as a negative time derivative. For instance, for a fluid of mass density $\rho(x,y,z,t)$ moving with a velocity field $\mathbf{v}(x,y,z,t)$, we can write

$$\oint_S (\rho\mathbf{v}) \cdot d\mathbf{S} = -\int_{\Delta v} \frac{\partial \rho}{\partial t}\, dv \tag{A.15}$$

where the closed surface S encloses the volume Δv. Using Gauss' theorem (A.13) and the fact that the region Δv can be chosen arbitrarily, we obtain the *equation of continuity*

$$\nabla \cdot (\rho\mathbf{v}) = -\frac{\partial \rho}{\partial t}, \tag{A.16}$$

which is a statement of conservation of mass.

  We also introduce a second operator for vector fields, related to the line integral (A.10): the *curl* or *rotation* of a vector field $\mathbf{A}$ is another vector field whose component in the $\hat{\mathbf{n}}$ direction is defined by

$$(\mathbf{curl}\ \mathbf{A})_n \equiv (\mathbf{rot}\ \mathbf{A})_n \equiv \lim_{\Delta S \to 0} \frac{1}{\Delta S} \oint_{L_n} \mathbf{A} \cdot d\mathbf{l}, \tag{A.17}$$

TABLE A.1

Cartesian (rectangular) coordinates:

Gradient $\qquad \boldsymbol{\nabla}\Phi = \mathbf{i}\dfrac{\partial\Phi}{\partial x} + \mathbf{j}\dfrac{\partial\Phi}{\partial y} + \mathbf{k}\dfrac{\partial\Phi}{\partial z}$

Divergence $\qquad \boldsymbol{\nabla}\cdot\mathbf{A} = \dfrac{\partial A_x}{\partial x} + \dfrac{\partial A_y}{\partial y} + \dfrac{\partial A_z}{\partial z}$

Curl $\qquad \boldsymbol{\nabla}\times\mathbf{A} = \begin{vmatrix} \mathbf{i} & \mathbf{j} & \mathbf{k} \\ \dfrac{\partial}{\partial x} & \dfrac{\partial}{\partial y} & \dfrac{\partial}{\partial z} \\ A_x & A_y & A_z \end{vmatrix} = \mathbf{i}\left(\dfrac{\partial A_z}{\partial y} - \dfrac{\partial A_y}{\partial z}\right) + \cdots$

Laplacian $\qquad \nabla^2\Phi = \dfrac{\partial^2\Phi}{\partial x^2} + \dfrac{\partial^2\Phi}{\partial y^2} + \dfrac{\partial^2\Phi}{\partial z^2}$

Cylindrical coordinates:

Gradient $\qquad \boldsymbol{\nabla}\Phi = \hat{\mathbf{r}}\dfrac{\partial\Phi}{\partial r} + \hat{\boldsymbol{\theta}}\dfrac{1}{r}\dfrac{\partial\Phi}{\partial\theta} + \mathbf{k}\dfrac{\partial\Phi}{\partial z}$

Divergence $\qquad \boldsymbol{\nabla}\cdot\mathbf{A} = \dfrac{1}{r}\dfrac{\partial}{\partial r}(rA_r) + \dfrac{1}{r}\dfrac{\partial A_\theta}{\partial\theta} + \dfrac{\partial A_z}{\partial z}$

Curl $\qquad \boldsymbol{\nabla}\times\mathbf{A} = \begin{vmatrix} \dfrac{\hat{\mathbf{r}}}{r} & \hat{\boldsymbol{\theta}} & \dfrac{\mathbf{k}}{r} \\ \dfrac{\partial}{\partial r} & \dfrac{\partial}{\partial\theta} & \dfrac{\partial}{\partial z} \\ A_r & rA_\theta & A_z \end{vmatrix}$

Laplacian $\qquad \nabla^2\Phi = \dfrac{1}{r}\dfrac{\partial}{\partial r}\left(r\dfrac{\partial\Phi}{\partial r}\right) + \dfrac{1}{r^2}\dfrac{\partial^2\Phi}{\partial\theta^2} + \dfrac{\partial^2\Phi}{\partial z^2}$

Spherical coordinates:

Gradient $\qquad \boldsymbol{\nabla}\Phi = \hat{\mathbf{r}}\dfrac{\partial\Phi}{\partial r} + \hat{\boldsymbol{\theta}}\dfrac{1}{r}\dfrac{\partial\Phi}{\partial\theta} + \hat{\boldsymbol{\phi}}\dfrac{1}{r\sin\theta}\dfrac{\partial\Phi}{\partial\phi}$

Divergence $\qquad \boldsymbol{\nabla}\cdot\mathbf{A} = \dfrac{1}{r^2}\dfrac{\partial}{\partial r}(r^2 A_r) + \dfrac{1}{r\sin\theta}\dfrac{\partial}{\partial\theta}(\sin\theta\, A_\theta) + \dfrac{1}{r\sin\theta}\dfrac{\partial A_\phi}{\partial\phi}$

Curl $\qquad \boldsymbol{\nabla}\times\mathbf{A} = \begin{vmatrix} \dfrac{\hat{\mathbf{r}}}{r^2\sin\theta} & \dfrac{\hat{\boldsymbol{\theta}}}{r\sin\theta} & \dfrac{\hat{\boldsymbol{\phi}}}{r} \\ \dfrac{\partial}{\partial r} & \dfrac{\partial}{\partial\theta} & \dfrac{\partial}{\partial\phi} \\ A_r & rA_\theta & r\sin\theta\, A_\phi \end{vmatrix}$

Laplacian $\qquad \nabla^2\Phi = \dfrac{1}{r^2}\dfrac{\partial}{\partial r}\left(r^2\dfrac{\partial\Phi}{\partial r}\right) + \dfrac{1}{r^2\sin\theta}\dfrac{\partial}{\partial\theta}\left(\sin\theta\dfrac{\partial\Phi}{\partial\theta}\right) + \dfrac{1}{r^2\sin^2\theta}\dfrac{\partial^2\Phi}{\partial\phi^2}$

where the closed loop L_n lies in a plane perpendicular to $\hat{\mathbf{n}}$ and encloses the area ΔS.† Three mutually perpendicular plane loops (or three projections of a non-plane loop) suffice to define the full vector field **curl A.** The geometric interpretation of the curl is more subtle than for the divergence. As an example, if **A** represents the velocity field of a flowing liquid, then nonzero curl at a point implies that the liquid would cause a small paddlewheel to rotate about an axis parallel to the curl. By integration of (A.17) over a surface, *Stokes' theorem* follows directly,

$$\oint_{\Delta S} \text{curl } \mathbf{A} \cdot d\mathbf{S} = \oint_{L} \mathbf{A} \cdot d\mathbf{l} \qquad (A.18)$$

where the closed loop L bounds the finite open surface ΔS (L and ΔS need not lie in a plane). Finally we wish to express **curl A** explicitly in terms of derivatives. From Fig. A.2, the line integral along side 1 is

$$\mathbf{A} \cdot \Delta \mathbf{l}_1 = \left(A_y + \frac{\partial A_y}{\partial x} \Delta x + \cdots \right) \Delta y,$$

while that along the opposite side 2 is

$$\mathbf{A} \cdot \Delta \mathbf{l}_2 = (-A_y)\, \Delta y.$$

Accordingly the z component of the curl (A.17), i.e., the circulation per unit area about the z axis, may be written

$$(\text{curl } \mathbf{A})_z = \frac{\partial A_y}{\partial x} - \frac{\partial A_x}{\partial y} = (\nabla \times \mathbf{A})_z, \qquad (A.19)$$

and by extension

$$\text{curl } \mathbf{A} = \nabla \times \mathbf{A} = \begin{vmatrix} \mathbf{i} & \mathbf{j} & \mathbf{k} \\ \dfrac{\partial}{\partial x} & \dfrac{\partial}{\partial y} & \dfrac{\partial}{\partial z} \\ A_x & A_y & A_z \end{vmatrix}. \qquad (A.20)$$

† The positive sense of traversing the loop L_n is related to the sense of $\hat{\mathbf{n}}$ by the right-hand rule.

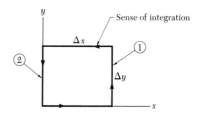

Fig. a.2 Rectangle used to evaluate the z component of the curl.

By a more general argument for orthogonal curvilinear coordinate systems, it can be shown that the curl operator may be written in the form $\nabla \times$, where again ∇ is the same vector differential operator used for the gradient (see Table A.1).

We thus establish that the formal del operations $\nabla \cdot \mathbf{A}$ and $\nabla \times \mathbf{A}$ are equivalent to the integral definitions of the divergence (A.12) and the curl (A.17), respectively.† The explicit del formulations are more useful in a manipulative sense, while the definitions in terms of flux generation per unit volume and circulation per unit area afford more insight into the meaning of these operations.

It is useful to consider briefly *second-order* spatial derivatives. The forms of double del operations permitted by the rules of vector algebra (excluding dyadic and higher-order tensor quantities) are

$$\nabla \cdot \nabla \phi \equiv \nabla^2 \phi \equiv \text{laplacian of } \phi \tag{A.21}$$
$$\nabla \times \nabla \phi \equiv 0 \tag{A.22}$$
$$\nabla (\nabla \cdot \mathbf{A}) \tag{A.23}$$
$$\nabla \cdot \nabla \times \mathbf{A} \equiv 0 \tag{A.24}$$
$$\nabla \times \nabla \times \mathbf{A} \equiv \nabla (\nabla \cdot \mathbf{A}) - \nabla^2 \mathbf{A} \tag{A.25}$$
$$\nabla \cdot \nabla \mathbf{A} \equiv \nabla^2 \mathbf{A} \equiv \text{laplacian of } \mathbf{A}. \tag{A.26}$$

Of this list, the curl of a gradient, (A.22), and the divergence of a curl, (A.24), can be shown to vanish identically. The remaining nontrivial second-order derivative of a scalar field, the laplacian (A.21), is itself a scalar field. The three nontrivial second-order derivatives of a vector field are themselves vector fields and are interrelated by the *BAC-CAB* rule applied to the double curl (A.25), as shown. Various species of third- and higher-order derivatives may, of course, be generated by repeated use of the del operator. Derivatives of the product of two fields can be expanded by satisfying both the chain rule for differentiation of a product and the rules of vector algebra; thus

$$\nabla(\phi\psi) = \phi \nabla\psi + \psi \nabla\phi \tag{A.27}$$
$$\nabla \cdot (\phi\mathbf{A}) = \phi \nabla \cdot \mathbf{A} + \mathbf{A} \cdot \nabla\phi \tag{A.28}$$
$$\nabla \times (\phi\mathbf{A}) = \phi \nabla \times \mathbf{A} - \mathbf{A} \times \nabla\phi \tag{A.29}$$
$$\nabla \cdot (\mathbf{A} \times \mathbf{B}) = \mathbf{B} \cdot \nabla \times \mathbf{A} - \mathbf{A} \cdot \nabla \times \mathbf{B} \tag{A.30}$$
$$\nabla \times (\mathbf{A} \times \mathbf{B}) = \mathbf{A} \nabla \cdot \mathbf{B} - \mathbf{B} \nabla \cdot \mathbf{A} + \mathbf{B} \cdot \nabla\mathbf{A} - \mathbf{A} \cdot \nabla\mathbf{B}. \tag{A.31}$$

† We note that an equivalent, but not particularly useful, definition of the curl is

$$\mathbf{curl\ A} = -\lim_{\Delta v \to 0} \frac{1}{\Delta v} \oint_S \mathbf{A} \times d\mathbf{S},$$

which involves a surface integral of the form (A.9). Similarly, the gradient (A.5) may be defined by

$$\mathbf{grad}\phi = \lim_{\Delta v \to 0} \frac{1}{\Delta v} \oint_S \phi \, d\mathbf{S}.$$

We state without proof a group of three fundamental theorems, often collectively called *Helmholtz' theorem*, which apply to well-behaved vector fields that vanish at infinity.

(**1**)   Any vector field can be written as the sum of a *solenoidal* field, whose divergence is zero, and an *irrotational* field, whose curl is zero.

(**2**)   A vector field is uniquely determined if both its divergence and curl are specified.

(**3**)   An irrotational (curl-free) vector field can be expressed as the gradient of a scalar field, known as its *scalar potential*. A solenoidal (divergence-free) vector field can be expressed as the curl of another vector field, known as its *vector potential*.

The vector fields that occur in physical problems often turn out to be either purely irrotational, e.g., the electrostatic field, or purely solenoidal, e.g., the magnetic field.

appendix b

The Smith Calculator†

The Smith transmission-line calculator is a slide rule for making numerical calculations involving the transmission-line equations of Secs. 4.2 and 8.1, namely,

$$\check{Z}_g = Z_0 \frac{\check{Z}_l + jZ_0 \tan\kappa l}{Z_0 + j\check{Z}_l \tan\kappa l} \qquad \begin{matrix} (4.2.16) \\ (8.1.15) \end{matrix} \Bigg\} \quad (B.1)$$

$$\check{R} \equiv |\check{R}|e^{j\phi} = \frac{\check{Z}_l - Z_0}{\check{Z}_l + Z_0} \qquad \begin{matrix} (4.2.15)\ddagger \\ (8.1.16) \end{matrix} \Bigg\} \quad (B.2)$$

$$|\check{R}| = \frac{\text{VSWR} - 1}{\text{VSWR} + 1} \qquad \qquad (8.1.24) \quad (B.3)$$

$$\phi = 2\kappa \, \Delta z_{\max} \mp n(2\pi) \qquad \qquad (8.1.26) \quad (B.4)$$

† P. H. Smith, *Electronics*, **17**(1):130, 318 (January, 1944).
‡ See footnote, page 102.

We examine here the theory of the calculator and show how it applies to wave propagation in a one-dimensional medium to which (B.1) to (B.4) apply.

The transmission-line equations are first put in dimensionless form by introducing *normalized-impedances* $\check{Z}/Z_0 \equiv \check{z} \equiv r + jx$, that is, by dividing each physical impedance by the (real) characteristic impedance Z_0 of the line. If $\tan \kappa l$ is then expressed in terms of $e^{\pm j \kappa l}$, (B.1) becomes

$$\check{z}_g = \frac{1 + \check{R}e^{-2j\kappa l}}{1 - \check{R}e^{-2j\kappa l}}, \tag{B.5}$$

where now

$$\check{R} = \frac{\check{z}_l - 1}{\check{z}_l + 1} \tag{B.6}$$

is the complex reflection coefficient for force or voltage. If we introduce a new variable

$$\check{W} \equiv U + jV \equiv \check{R}e^{-2j\kappa l}, \tag{B.7}$$

which is basically a function of $\check{z}_l$ and l, the (normalized) input impedance seen by a hypothetical generator at $z = 0$ is expressed by

$$\check{z}_g = \frac{1 + \check{W}}{1 - \check{W}}, \tag{B.8}$$

which can be inverted to read

$$\check{W} = \frac{\check{z}_g - 1}{\check{z}_g + 1}. \tag{B.9}$$

Comparison of (B.9) with (B.6) shows that $\check{W}$ is the complex (voltage) reflection coefficient of the impedance $\check{Z}_g$ considered as a load terminating a line of characteristic impedance Z_0.

A relation between two complex variables such as (B.9) is known as a *conformal transformation*. Its maps each point in the complex $\check{z}_g$ plane into a corresponding point in the complex $\check{W}$ plane, such that the angle between two intersecting lines in one plane is preserved in the other. The Smith calculator exploits the properties of the particular conformal transformation (B.9).

The nature of the transformation (B.9) can be discovered by equating separately its real and its imaginary parts,

$$(1 + r)U - xV = 1 + r \tag{B.10}$$
$$xU + (1 + r)V = x. \tag{B.11}$$

If we now eliminate x, then r, in turn, between (B.10) and (B.11), we find that the equations

$$\left(U - \frac{r}{1+r}\right)^2 + V^2 = \left(\frac{1}{1+r}\right)^2 \tag{B.12}$$

$$(U - 1)^2 + \left(V - \frac{1}{x}\right)^2 = \left(\frac{1}{x}\right)^2 \tag{B.13}$$

give the loci of points of constant r and of constant x, respectively, in the $\check{W}$ plane, as shown in Fig. B.1. Equation (B.12) represents a family of circles with r as a parameter. The centers are at $[r/(1+r), 0]$, and the radii are $1/(1+r)$. Similarly (B.13) represents a second family of circles, now with x as a parameter. The centers are at $(1, 1/x)$, and the radii are $1/|x|$. All circles of both families pass through the point $(1,0)$. The two families of circles are *orthogonal*, reflecting the fact that the transformation is conformal (angle-preserving) and that the straight lines of constant x and of constant r in the $\check{z}_g$ plane are at right angles. All points in the entire $\check{z}_g$ half-plane (with $r > 0$) map within the circle of unit radius in the $\check{W}$ plane, with its center at the origin. Commerically printed Smith calculators show a large enough number of circles to enable an arbitrary impedance $r + jx$ to be located with good accuracy ($\sim \frac{1}{2}$ percent).

The locus of a point of constant $|\check{W}|$ and therefore of constant $|\check{R}|$ is a circle centered on the origin in the $\check{W}$ plane, which is at the center of the calculator. Such a circle passes through all the complex values of $\check{z}_l$ that can possibly give rise to this particular magnitude of reflection coefficient, and therefore of the related VSWR, as given by (B.3). If the transmission line has a *known* normalized load impedance $\check{z}_l$, its value can be set on the calculator and the corresponding VSWR read on a graduated radial arm on the calculator. To find the input or generator impedance of a line of length l/λ fractional wavelengths, the arm, with a fiducial marker set at the appropriate VSWR, is rotated clockwise (designated "toward generator" on the calculator) through the angle $2\kappa l$ rad as required by (B.5) and (B.7). For this purpose there is a circular scale graduated in fractional wavelengths outside the unit circle corresponding to $r = 0$. After rotation, the fiducial point on the arm lies on the value of $\check{z}_g$ that is the normalized impedance looking into a line of fractional length l/λ, terminated by the normalized impedance $\check{z}_l$. One complete revolution of the arm corresponds to half a wavelength, which of course is the period of $\check{Z}_g$ as a function of l, as given by (B.1).

If the terminating impedance is unknown, measured values of the VSWR and the distance (in fractional wavelengths) of a maximum (or minimum) from the end of the line can be used to find $\check{z}_l$ (see Prob. 1.4.3 and 8.1.8). First set the measured VSWR on the radial arm by moving its slider. Next set the radial fiducial line of the arm on the $x = 0$ line on the side with $r > 1$ for a voltage maximum (or with $r < 1$ for a minimum). Now turn the arm counterclockwise

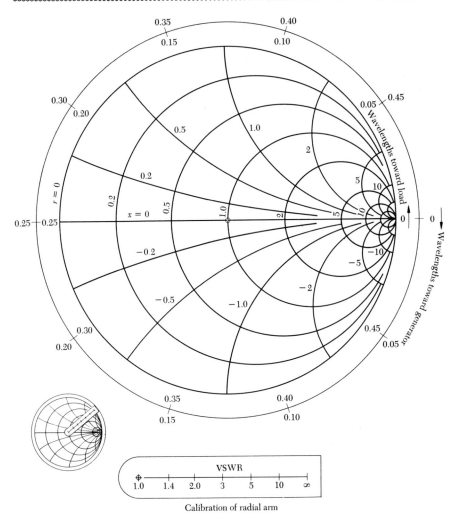

Fig. b.1 The Smith calculator (abbreviated).

("toward load") an amount corresponding to the distance in fractional wave-lengths from the maximum (or minimum) to the end of the line. The unknown normalized load impedance then lies under the fiducial point on the radial arm. Note that we have used the fact that the maxima (minima) of a standing-wave pattern occur at positions on the line where $\check{z}_g$ is real and greater (or less) than $r = 1$.

The form of the transformation (B.9) is such that a given impedance $\check{z}_g$ and its reciprocal plot in the $\check{W}$ plane at the same radius but 180° apart in angle. Thus the Smith chart may be used as a slide rule for computing reciprocals of

arbitrary complex numbers by simply reading off the coordinates of the dia-
metrically opposite point. Furthermore, as it is often convenient to make imped-
ance calculations in terms of the reciprocal impedance, or *admittance*, $\breve{Y} \equiv \breve{Z}^{-1}$,
the Smith chart can equally well be thought of as being plotted in admittance
coordinates, i.e., with the loci of constant r and x now representing constant
values of the real and imaginary parts of the normalized admittance $\breve{y}_g$.

appendix c

Proof of the Uncertainty Relation

The uncertainty relation (12.2.7),

$$\Delta t \, \Delta \omega \geq \tfrac{1}{2}, \tag{C.1}$$

connects the rms duration Δt of a wave packet with its rms bandwidth $\Delta \omega$. The quantities Δt and $\Delta \omega$ are defined formally by (12.2.4) and (12.2.5). The proof of (C.1) rests on two mathematical theorems which we now establish.

(a) Parseval's Theorem

Given the Fourier transform pair

$$G(\omega) = \int_{-\infty}^{\infty} g(t) e^{i\omega t} \, dt \tag{C.2}$$

$$g(t) = \frac{1}{2\pi} \int_{-\infty}^{\infty} G(\omega) e^{-i\omega t} \, d\omega \tag{C.3}$$

and the complex conjugate

$$g^*(t) = \frac{1}{2\pi} \int_{-\infty}^{\infty} G^*(\omega)e^{i\omega t}\, d\omega, \tag{C.4}$$

consider the integral

$$\int_{-\infty}^{\infty} |g(t)|^2\, dt = \int_{-\infty}^{\infty} gg^*\, dt$$

$$= \frac{1}{(2\pi)^2} \int_{-\infty}^{\infty} \int_{-\infty}^{\infty} G(\alpha)G^*(\beta)\, d\alpha\, d\beta \int_{-\infty}^{\infty} e^{-i(\alpha-\beta)t}\, dt, \tag{C.5}$$

where the dummy variables α and β distinguish the two frequency integrals. Problem 12.2.3 establishes the transform pair for the Dirac delta function,

$$G_\delta(\omega) = \int_{-\infty}^{\infty} \delta(t)e^{i\omega t}\, dt = 1 \tag{C.6}$$

$$\delta(t) = \frac{1}{2\pi} \int_{-\infty}^{\infty} e^{-i\omega t}\, d\omega. \tag{C.7}$$

Since a transform pair is symmetrical (except for the factor 2π), we can use (C.7) to evaluate the time integral in (C.5), obtaining

$$\int_{-\infty}^{\infty} e^{-i(\alpha-\beta)t}\, dt = 2\pi\, \delta(\alpha - \beta); \tag{C.8}$$

that is, the integral vanishes unless $\alpha = \beta$. The integral over α (or over β) in (C.5) can now be carried out easily by invoking the fundamental property of the Dirac delta function that for any function $f(x)$

$$\int_{-\infty}^{\infty} f(x)\, \delta(x - a)\, dx = f(a). \tag{C.9}$$

Accordingly, (C.5) becomes *Parseval's theorem*

$$\int_{-\infty}^{\infty} |g(t)|^2\, dt = \frac{1}{2\pi} \int_{-\infty}^{\infty} |G(\omega)|^2\, d\omega. \tag{C.10}$$

A corollary to Parseval's theorem is obtained by noting from (C.3) that the Fourier transform of dg/dt is given by $-i\omega G(\omega)$. It then follows that

$$\int_{-\infty}^{\infty} \left|\frac{dg(t)}{dt}\right|^2\, dt = \frac{1}{2\pi} \int_{-\infty}^{\infty} \omega^2 |G(\omega)|^2\, d\omega. \tag{C.11}$$

(b) Schwarz' Inequality

Consider the obvious inequality

$$0 \le \int_{-\infty}^{\infty} |g(t) + \epsilon h(t)|^2\, dt$$

$$= \int_{-\infty}^{\infty} [g(t) + \epsilon h(t)][g^*(t) + \epsilon h^*(t)]\, dt$$

$$= \epsilon^2 \int_{-\infty}^{\infty} hh^*\, dt + \epsilon \int_{-\infty}^{\infty} (gh^* + g^*h)\, dt + \int_{-\infty}^{\infty} gg^*\, dt, \tag{C.12}$$

where ϵ is an arbitrary real number and g and h are arbitrary functions. The quantity in (C.12) is of the form

$$\phi(\epsilon) \equiv a\epsilon^2 + b\epsilon + c, \qquad (C.13)$$

where a, b, c stand for real integrals. Since $a \equiv \int |h|^2 \, dt$ is positive, the parabola represented by $\phi(\epsilon)$ is open upward. Moreover since $\phi(\epsilon)$ is nonnegative, the roots of $\phi(\epsilon) = 0$ must be complex (or a single degenerate real root). Thus, necessarily,

$$b^2 - 4ac \leq 0, \qquad (C.14)$$

which constitutes *Schwarz' inequality*

$$\int_{-\infty}^{\infty} |g(t)|^2 \, dt \cdot \int_{-\infty}^{\infty} |h(t)|^2 \, dt \geq \tfrac{1}{4} \left\{ \int_{-\infty}^{\infty} [g(t)h^*(t) + g^*(t)h(t)] \, dt \right\}^2. \qquad (C.15)$$

We are now prepared to establish the uncertainty relation. For simplicity, choose a time origin at $t = \bar{t}$. We then have, from the definitions (12.2.4) and (12.2.5),

$$(\Delta t)^2 (\Delta \omega)^2 = \frac{\int_{-\infty}^{\infty} t^2 f(t) f^*(t) \, dt \cdot \int_{-\infty}^{\infty} \omega^2 F(\omega) F^*(\omega) \, d\omega}{\int_{-\infty}^{\infty} f(t) f^*(t) \, dt \cdot \int_{-\infty}^{\infty} F(\omega) F^*(\omega) \, d\omega}. \qquad (C.16)$$

First eliminate the two frequency integrals using Parseval's theorem (C.10) and (C.11), so that (C.16) becomes

$$(\Delta t)^2 (\Delta \omega)^2 = \frac{\int_{-\infty}^{\infty} t^2 ff^* \, dt \cdot 2\pi \int_{-\infty}^{\infty} \dot{f}\dot{f}^* \, dt}{2\pi \left(\int_{-\infty}^{\infty} ff^* \, dt \right)^2} \qquad (C.17)$$

where $\dot{f} \equiv df/dt$. Next use the Schwarz inequality (C.15), letting $tf(t)$ replace $g(t)$ and $\dot{f}(t)$ replace $h(t)$, obtaining

$$(\Delta t)^2 (\Delta \omega)^2 \geq \frac{\tfrac{1}{4} \left[\int_{-\infty}^{\infty} (tf\dot{f}^* + tf^*\dot{f}) \, dt \right]^2}{\left(\int_{-\infty}^{\infty} ff^* \, dt \right)^2} = \frac{\tfrac{1}{4} \left[\int_{-\infty}^{\infty} t \frac{d}{dt} (ff^*) \, dt \right]^2}{\left(\int_{-\infty}^{\infty} ff^* \, dt \right)^2}. \qquad (C.18)$$

Now integrate the integral in the numerator by parts,

$$\int_{-\infty}^{\infty} t \frac{d}{dt} (ff^*) \, dt = \left[t(ff^*) \right]_{-\infty}^{\infty} - \int_{-\infty}^{\infty} ff^* \, dt. \qquad (C.19)$$

For the pulse waveforms of interest to us here, we may assume that $|f(t)|^2$ vanishes faster than $1/t$ as t goes to $\pm \infty$ and drop the integrated term. Thus the numerator and denominator of (C.18) become identical, except for the factor $\tfrac{1}{4}$, and we have established the uncertainty relation (C.1).

Index

This book was set in Bruce Old Style, printed on permanent paper by The Maple Press Company, and bound by The Maple Press Company. The designer was Marsha Cohen; the drawings were done by Felix Cooper. The editors were Bradford Bayne and Eva Marie Strock. Peter D. Guilmette supervised the production.